Systemic Vasculitides: Current Status and Perspectives

Franco Dammacco • Domenico Ribatti
Angelo Vacca
Editors

Systemic Vasculitides: Current Status and Perspectives

Springer

Editors
Franco Dammacco
University of Bari Medical School
Bari, Italy

Domenico Ribatti
University of Bari Medical School
Bari, Italy

Angelo Vacca
University of Bari Medical School
Bari, Italy

ISBN 978-3-319-40134-8 ISBN 978-3-319-40136-2 (eBook)
DOI 10.1007/978-3-319-40136-2

Library of Congress Control Number: 2016954956

Printed on acid-free paper

This Springer imprint is published by Springer Nature
The registered company is Springer International Publishing AG Switzerland

To be fully aware of one's own ignorance is an irrepressible incitation to the pursuit of knowledge.

Preface

This volume seeks to provide a comprehensive overview of the systemic vasculitides, an extremely heterogeneous group of diseases characterized by inflammation and necrosis of different-sized blood vessels. With a few exceptions (e.g., HCV-related cryoglobulinemic vasculitis and HBV-positive polyarteritis nodosa), the etiology of these clinical conditions remains unknown. In spite of their relatively low prevalence, the systemic vasculitides have been the object of recent, intensive, basic, and clinical studies. For this reason, it can be safely stated that this group of diseases is one of the most rapidly progressing areas of clinical medicine, as evidenced by the dramatic achievements in terms of clinical remission and overall prognostic improvement.

The pathophysiology of the vasculitides is multifactorial and thus in most cases poorly defined. Among the many potential influences on disease expression, sex, ethnicity, and genetic as well as environmental factors are likely to play a role. In addition, the vascular damage characteristic of the systemic vasculitides may be the result of autoimmune responses, such as antineutrophil cytoplasmic autoantibodies, anti-endothelial cell autoantibodies, immune complex deposition, an immune response to foreign antigens or infectious agents, and T-lymphocyte responses with granuloma formation.

The general aims of this book are:

(i) To provide an in-depth update of the major pathogenetic, genetic, and clinical advances in the field encompassing the vasculitides, including, for each condition, a summary of the most cogent information scattered in the medical literature but not always readily retrievable
(ii) To describe not only conventional treatments, but also the more recently developed and tested drugs as well as efforts at patient-tailored therapies, especially for patients with refractory and relapsing disease
(iii) To point out future directions of research that, while challenging, are likely to be profitable in terms of improved diagnosis and therapy

If this book serves as a stimulating resource for basic and clinical researchers and specialists in related disciplines, as well as practicing physicians and advanced medical students interested in this fascinating branch of medicine, our efforts as editors will have been fully rewarded. The lion's share of the merit should, however, be given to the international contributors who accepted our invitation to collaborate in this project and who, in doing so, were able to impart the knowledge gained during their multiyear experience in this field.

Bari, Italy

Franco Dammacco
Domenico Ribatti
Angelo Vacca

Contents

Part I
Biology of Blood Vessels, Experimental Models and Nomenclature of Vasculitides

Chapter 1
Morphofunctional Aspects of Endothelium

Domenico Ribatti

Abstract Blood vessels represent an essential component of all organs. The vascular tree develops early during embryogenesis and progresses into a highly branched system of vascular channels lined by endothelial cells (ECs) and surrounded by mural cells. A highly hierarchical vascular architecture is established which comprises distinct arterial, capillary and venous segments as well as organ- and tissue-specific vascular beds. EC heterogeneity plays a role in directing disease processes to distinct vascular territories and characterization of the molecules involved in creating vascular heterogeneity might eventually allow refinement of diagnostic and therapeutic strategies aimed at targeting distinct segments of the vascular tree.

Keywords Endothelial cells • Vascular heterogeneity • Vascular diseases

1.1 Endothelial Cell Heterogeneity and Organ Specificity

Endothelial cells (ECs) form a continuous monolayer between the blood and the interstitial fluid. The EC surface in an adult human is composed of approximately 1.6×10^{13} cells and covers a surface area of approximately 7 m^2 [1]. Quiescent ECs generate an active antithrombotic surface through the expression of tissue factor pathway inhibitors, heparan sulphate proteoglycans that can interfere with thrombin-controlled coagulation, and thrombomodulin that facilitate transit of plasma and cellular constituents throughout the vasculature. Perturbations and dysfunctions (Table 1.1) induce ECs to create a prothrombotic and antifibrinolytic microenvironment.

An increase of blood flow into a capillary induces local recruitment of smooth muscle cells and leads to a differentiation into an artery or vein, while cessation of blood flow causes vessel regression [2].

D. Ribatti (✉)
Department of Basic Medical Sciences, Neurosciences and Sensory Organs, University of Bari Medical School, Policlinico – Piazza G. Cesare, 11, 70124 Bari, Italy

National Cancer Institute "Giovanni Paolo II", 70124 Bari, Italy
e-mail: domenico.ribatti@uniba.it

F. Dammacco et al. (eds.), *Systemic Vasculitides: Current Status and Perspectives*, DOI 10.1007/978-3-319-40136-2_1

Table 1.1 Consequences of endothelial dysfunctions

Inflammation
Fibrosis
Cardiovascular diseases
Pulmonary hypertension
Atherosclerosis
Hyperlipidemia
Thrombosis
Immune reactions
Peripheral vascular diseases
Angiogenesis

There are differences between ECs derived from various microvascular beds/organs, ascribed to genetic and microenvironmental influences [3], including extracellular matrix components, locally produced pro- and anti-angiogenic molecules, interactions with neighboring cells, and mechanical forces. Interactions may occur through the release of cytokines and the synthesis and organization of matrix proteins on which the endothelium adheres and grows. Moreover, ECs release and express on the cell surface many signaling molecules that can affect the density of developing neighboring tissue cells [4].

ECs lining the capillaries of different organs are morphologically distinct. The vasculature of liver, spleen and bone marrow sinusoids is highly permeable because vessels are lined by discontinuous ECs, capillaries in the brain and retinal capillaries, dermis, bone tissue, skeletal muscle, myocardium, testes and ovaries are continuous, and ECs in endocrine glands and kidney are fenestrated. EC heterogeneity is also appreciable in individual organs. For example, the kidney contains fenestrated ECs in its peritubular capillaries, discontinuous ECs in its glomerular capillaries, and continuous ECs in other regions. The phenotype of ECs is unstable and likely to change when they are removed from their microenvironment [5]. Endothelial heterogeneity is also responsible for different responses across different vascular beds to pathological stimuli and disease states [6].

Antigens are differentially expressed on ECs of certain organs and tissues [7]. For example, the von Willebrand factor (vWF) marker is expressed at higher levels on the venous rather than on the arterial side of the capillary circulation, while it is largely absent from sinusoidal ECs. Moreover, vWF may play a role in tumor cell dissemination, as significantly higher levels have been reported in metastatic cancers [8].

1.2 Arterial and Venous Endothelial Cell Distinctions

The discovery that members of the ephrin family are differentially expressed in arteries (Fig. 1.1) and veins from very early stages of development is an indication that artery-vein identity is intrinsically programmed. Ephrin-B2 is expressed in arterial ECs, large arteries within the embryo, and in the endocardium of the developing

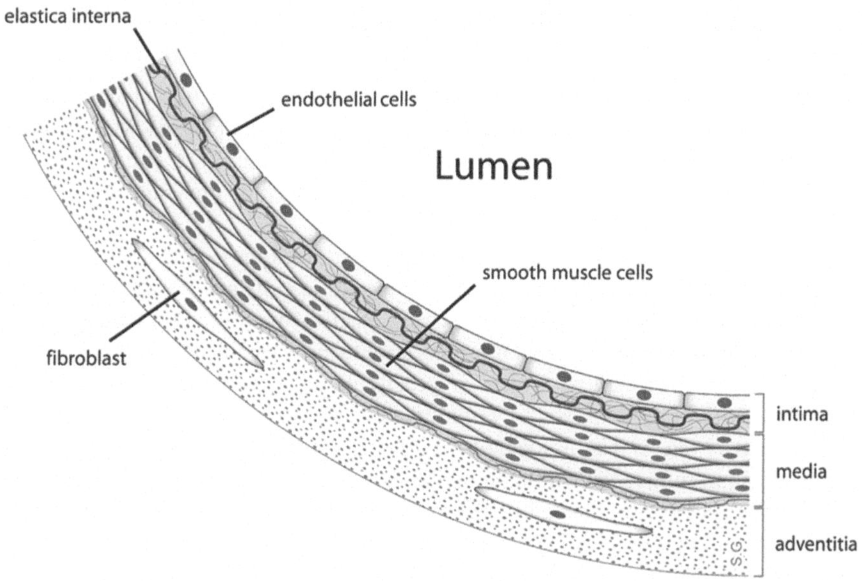

Fig. 1.1 Schematic drawing showing the general organization of the wall of an arterial vessel (Reproduced from "Endotelijalna ćelija" by D. Rosenbach at English Wikipedia)

heart, while the receptor for Ephrin-B2, Eph-B4, displays a reciprocal expression pattern in embryonic veins, large veins and also in the endocardium. Remodelling of the primary vascular plexus into arteries and veins was arrested in both Ephrin-B2 and Eph-B4 mutants, suggesting important roles for Ephrin-B2/Eph-B4 interactions on arterial and venous ECs differentiation, respectively [2].

Other specific markers for the arterial system include neuropilin-1 (NRP-1) and members of the Notch family, Notch-3, DDL4 and GRIDLOCK (Grl), while venous markers include NRP-2 Notch signalling is necessary for remodelling the primary plexus into mature vascular beds and maintaining arterial fate, and is essential for the homeostatic functions of fully differentiated arteries. During vascular development, defects in signalling through the Notch pathway, including ligands such as Jagged-1, Jagged-2, and Delta-like-4 and receptors, such as Notch-1, Notch-2, and Notch-4, disrupt normal differentiation into arteries or veins, resulting in loss of artery specific markers [9].

Le Noble et al. [10] studying arterial-venous differentiation in the developing yolk sac of the chick embryo, observed that prior to the onset of flow, EC expressing arterial and venous specific markers are localized in a posterior-arterial and anterior-venous pole. Ligation of one artery by means of a metal clip, lifting the artery, and arresting arterial flow distal to the ligation site could morphologically transform the artery into a vein. When the arterial flow was restored by removal of the metal clip, arterial maker was re-expressed, suggesting that the genetic fate of arterial EC is plastic and controlled by hemodynamic forces.

1.3 Vascular Diversity in Pathological Conditions

Endothelium plays a major role in the pathophysiology of different conditions, including inflammation and cancer. EC activation by inflammatory cytokines results in the increased expression of adhesion molecules, including E-selectin, intercellular adhesion molecule-1 (ICAM-1), and vascular cell adhesion molecule-1 (VCAM-1) [11].

It has long been recognized that systemic vasculitides impact distinct segments and branches of the vascular tree, and smooth muscle cells and dendritic cells are involved in the pathogenesis of these diseases. Necrotizing sarcoid granulomatosis, Takayasu's arteritis, and giant cell arteritis cause macrovascular compromise, while cryoglobulinemic vasculitis affects microcirculation. Some diseases such as Behçet's syndrome (a small vessel vasculitis that can affect venules) and Wegener's granulomatosis (an ANCA-associated vasculitis generally affecting small arteries and veins) compromise the whole pulmonary vasculature. Churg-Strauss syndrome (a necrotizing ANCA-associated vasculitis) targets medium-sized arteries and veins of the macrocirculation, whereas microscopic polyangiitis impacts arterioles, capillaries and venules of the microcirculation [12].

A dual role of angiogenesis in vasculitides has been proposed. On the one hand, angiogenesis may be a compensatory response to ischemia and to the increased metabolic activity in acute phase of the disease. On the other hand, ECs of newly-formed vessels express adhesion molecules and produce colony-stimulating factors and chemokines for leukocytes [12].

Tumor vessels exhibit chaotic blood flow, have focal regions that lack ECs or basement membrane. Qualitative differences exist in the tumor vasculature at different stages [13]. Distinct tumor vessels may need specific vascular growth factors and cytokines at defined tumor stages. There are vascular tumors that derive from ECs and express unique autonomous properties. In infantile hemangioma, molecular profiling has provided evidence for a placental derivation of ECs [14]. Kaposi's sarcoma, an AIDS-defining vascular tumor, involves a phenotypically unique spindle cell that appears to derive from lymphatic ECs [15]. The vasculature of tumors tends to acquire characteristics similar to those of the host environment. The microvasculature of murine mammary carcinoma, rhabdomyosarcoma, and human glioblastoma implanted s.c. in nude mice become extensively fenestrated [16]. On the contrary, the same tumors implanted in the brain acquire a microvasculature resembling more closely the brain microvasculature phenotype.

Differential pattern of expression of angiogenic genes accompany and are probably responsible for the host-environment-induced differences in vascularisation. St. Croix et al. [17] have identified markers specifically induced in ECs from human colorectal carcinoma through a comparison of gene expression profiles of ECs isolated from human colorectal carcinoma and normal human colorectal tissue. Ria et al. [18] have identified genes differentially expressed in multiple myeloma ECs compared to ECs of monoclonal gammopathy of undetermined significance. Deregulated genes are involved in extracellular matrix formation and bone remodeling, cell adhesion, chemotaxis, angiogenesis, resistance to apoptosis, and cell-cycle regulation.

Coronary artery disease is one example of a disease that targets the arterial ECs. In response to hypercholesterolemia, myocardial ECs increase the expression of adhesion molecules, which leads to intimal thickening and plaque formation. ECs lack preferential cell alignment and often show a polygonal morphology in zones of disturbed vascular flow, such as regions susceptible of atherogenesis including the aortic arch or heart valves. Up-regulation of genes associated with endoplasmic reticulum processing of proteins, endoplasmic reticulum stress and unfolded protein response, contribute to enhanced endothelial permeability via focally increased EC proliferation in these regions [19, 20].

Selective EC activation may be responsible for the development of some brain pathologies, including blood–brain barrier dysfunction linked to Alzheimer's disease [21].

1.4 Concluding Remarks

EC diversity has crucial implications for the development of vascular diseases. Systemic vasculitides target distinct segments and branches of the vascular tree as well as selective vascular beds. Even thrombotic or hemorrhagic conditions recognize specific vascular beds as the sites of disease occurrence. Potential implications for the pathogenesis of vascular metabolic diseases like atherogenesis are also strong. EC differences exist in the tumor vasculature at different stages, a situation which may profoundly affect the efficacy of tumor treatment.

Understanding how early, basic ECs can differentiate into a specialized assortment of organ- and tissue-associated ECs is essential for appreciating the complexity of vascular disorders and for establishing critically designed strategies of vascular diseases' treatment, through the use of glucocorticoids, immunosuppressive, immunomodulator, and anti-inflammatory drugs. Indeed, identification of vascular-bed specific molecular profiles should facilitate the development of molecular imaging for diagnosis and surveillance as well as the improvement of "intelligent" molecules targeting selected vascular districts.

Acknowledgements The research leading to these results has received funding from the European Union Seventh Framework Programme (FP7/2007- 2013) under Grant agreement no. 278570.

References

1. Sumpio BE, Riley JT, Dardik A (2002) Cells in focus: endothelial cell. Int J Biochem Cell Biol 34:1508–1512
2. Ribatti D (2006) Genetic and epigenetic mechanisms in the early development of the vascular system. J Anat 208:139–152
3. Cleaver O, Melton DA (2003) Endothelial signaling during development. Nat Med 9:661–668

4. Lammert E, Cleaver O, Melton D (2003) Role of endothelial cells in early pancreas and liver development. Mech Dev 120:59–64
5. Crivellato E, Nico B, Ribatti D (2007) Contribution of endothelial cells to organogenesis: a modern reappraisal of an old Aristotelian concept. J Anat 211:415–427
6. Molema G (2010) Heterogeneity in endothelial responsiveness to cytokines, molecular causes, and pharmacological consequences. Semin Thromb Hemost 36:246–264
7. Auerbach R, Alby L, Morisey L et al (1985) Expression of organ-specific antigens on capillary endothelial cells. Microvasc Res 29:401–411
8. Damin DC, Rosito MA, Gus P et al (2002) Von Willebrand factor in colorectal cancer. Int J Colon Dis 17:42–45
9. Lawson ND, Scheer N, Pham VN et al (2001) Notch signaling is required for arterial-venous differentiation during embryonic vascular development. Development 128:3675–3683
10. Le Noble F, Moyon D, Pardanaud L et al (2004) Flow regulates arterial-venous differentiation in the chick embryo yolk sac. Development 131:361–375
11. Banks RE, Gearing AJ, Hemingway IK et al (1993) Circulating intercellular adhesion molecule-1 (ICAM-1), E-selectin and vascular cell adhesion molecule-1 (VCAM-1) in human malignancies. Br J Cancer 68:122–124
12. Maruotti N, Cantatore FP, Nico B et al (2008) Angiogenesis in vasculitides. Clin Exp Rheumatol 26:476–483
13. Langenkamp E, Molema G (2009) Microvascular endothelial cell heterogeneity: general concepts and pharmacological consequences for anti-angiogenic therapy of cancer. Cell Tissue Res 335:205–222
14. Barnes CM, Huang S, Kaipainen A et al (2005) Evidence by molecular profiling for a placental origin of infantile hemangioma. Proc Natl Acad Sci U S A 102:19097–19102
15. Wang HW, Trotter MW, Lagos D et al (2004) Kaposi sarcoma herpes virus-induced cellular reprogramming contributes to the lymphatic endothelial gene expression in Kaposi sarcoma. Nat Genet 36:687–693
16. Roberts WG, Delaat J, Nagane M et al (1998) Host microvasculature influence on tumor vascular morphology and endothelial gene expression. Am J Pathol 153:1239–1248
17. St Croix B, Rago C, Velculescu V et al (2000) Gene expressed in human tumor and endothelium. Science 289:1197–1202
18. Ria R, Todoerti K, Berardi S et al (2009) Gene expression profiling of bone marrow endothelial cells in patients with multiple myeloma. Clin Cancer Res 15:5369–5378
19. Schwartz SM, Benditt EP (1976) Clustering of replicating cells in the aortic endothelium. Proc Natl Acad Sci U S A 73:651–653
20. Chen YL, Jan KM, Lin HS et al (1995) Ultrastructural studies on macromolecular permeability in relation to endothelial cell turnover. Atherosclerosis 118:89–104
21. Girolamo F, Coppola C, Ribatti D et al (2014) Angiogenesis in multiple sclerosis and experimental autoimmune encephalomyelitis. Acta Neuropathol Commun 2:84

Chapter 2
Animal Models of ANCA-Associated Vasculitides

Domenico Ribatti and Franco Dammacco

Abstract Antibodies against neutrophil proteins myeloperoxidase (MPO) and proteinase-3 (PR3) are responsible for the development of anti-neutrophil cytoplasmic autoantibodies (ANCA)-associated vasculitides (AAV). Although the knowledge of these conditions is remarkably improved in the last few years, their etiology and pathogenetic mechanism(s) are still poorly understood. The establishment of experimental models has been repeatedly attempted with the aim of achieving a deeper understanding of their human counterpart. Here, we discuss the principal animal models currently used to investigate the mechanisms underlying the onset of AAV.

Keywords Animal models • Anti-neutrophil cytoplasmic autoantibodies • Vasculitis

2.1 Introduction

A number of *in vitro* and *in vivo* studies, focusing on different aspects of the neutrophil biology and function, have clearly demonstrated the potential role that neutrophils can exert in the modulation of innate and adaptive immune responses [1].

Anti-neutrophil cytoplasmic autoantibodies (ANCA) were first recognized by van der Woude et al. [2], who described circulating autoantibodies that reacted with cytoplasmic antigens of neutrophils and monocytes in patients with granulomatosis with polyangiitis (GPA). ANCA-associated vasculitides (AAV) are systemic autoimmune disorders characterized by inflammatory necrosis of small blood vessels affecting joints, lungs, kidneys, skin and other tissues [3]. Neutrophils are cardinal

D. Ribatti (✉)
Department of Basic Medical Sciences, Neurosciences and Sensory Organs, University of Bari Medical School, Policlinico – Piazza G. Cesare, 11, 70124 Bari, Italy

National Cancer Institute "Giovanni Paolo II", 70124 Bari, Italy
e-mail: domenico.ribatti@uniba.it

F. Dammacco
Department of Biomedical Sciences and Human Oncology, Section of Internal Medicine, University of Bari Medical School, 70124 Bari, Italy

F. Dammacco et al. (eds.), *Systemic Vasculitides: Current Status and Perspectives*, DOI 10.1007/978-3-319-40136-2_2

cells in the pathophysiological process underlying AAV since they are both effector cells responsible for endothelial damage and targets of autoimmunity. It should, however, be emphasized that some patients showing similar disease manifestations as those who are ANCA-positive are nonetheless ANCA-negative [4].

Four diseases are characterized by the presence of ANCA, namely GPA (formerly called Wegener's granulomatosis), microscopic polyangiitis (MPA), eosinophilic granulomatosis with polyangiitis (EGPA, formerly termed Churg-Strauss syndrome), and the necrotizing crescenting glomerulonephritis (NCGN). The etiological factors responsible for the production of vessel-damaging ANCA are unknown. Although infectious agents have been repeatedly suspected and *Staphylococcus aureus* has long been known to be associated with GPA, their precise immunologic link with AAV has not been proven.

In the 1980s, autoantibodies to cytoplasmic components of myeloid cells were detected in patients with pauci-immune necrotizing small vessel vasculitis. In AAV, the autoimmune response is directed against neutrophil and monocyte lysosomal enzymes, including myeloperoxidase (MPO) and proteinase 3 (PR3) [5]. MPO is abundantly expressed and exclusively found in azurophilic granules, and is a key component of the phagocyte oxygen-dependent intracellular microbicidal system [6]. On the other hand, PR3, also called myeloblastin, belongs to the neutrophil serine protease family and is classically localized in azurophilic granules. Following phagocytosis of pathogens, PR3 is secreted in the phagolysosome to play its crucial microbicidal function [7].

Clinical and experimental studies have provided extensive evidence for the involvement of autoantibodies to MPO and PR3 in the pathogenesis of AAV [8], thus leading to treatment strategies aimed at ANCA removal. Plasma exchange, for example, has been shown to remove plasma constituents as well as ANCA and to increase the chances of renal recovery in severe renal vasculitis [9].

The crucial factors required for animal models of vasculitis are the similarities to the clinical and pathologic phenotypes of human diseases, with the obvious assumption that their study may contribute to the pathogenetic elucidation of human vasculitis. Here, we will briefly discuss the principal animal models currently used to investigate the mechanism(s) of vascular injury in AAV.

2.2 Animal Models Involving Anti-MPO Immune Response

In spite of the large body of *in vitro* studies, unequivocal evidence that ANCA are pathogenic *in vivo* was obtained only recently [10]. The pathogenicity of ANCA has been investigated in mouse models by exploring both passive transfer and active immunization strategies in order to reproduce systemic vasculitis.

The first animal model resembling the human disease was introduced by Xian et al. [11]. They reported that injection of splenocytes, derived from MPO-deficient mice immunized with mouse MPO, into recipient mice lacking mature T and B cells (RAG2-deficient mice) caused severe necrotizing glomerulonephritis. In a second

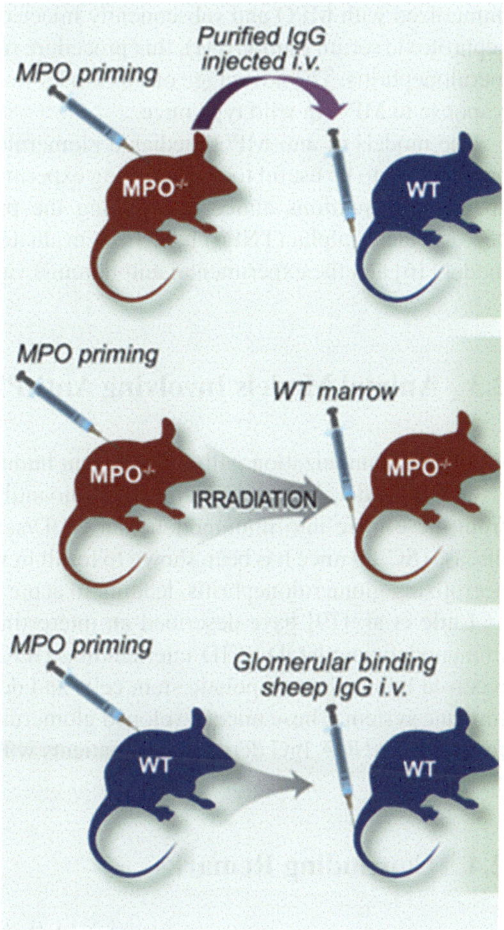

Fig. 2.1 Summary of three mouse models that have been used to study pathogenetic mechanisms in ANCA-associated vasculitis (Modified from Coughlan et al. Clin Exp Immunol. 2012; 169: 229–37)

approach, IgG were isolated from MPO-deficient mice immunized with MPO and passively transferred into wild type and RAG2−/− mice, resulting in a pauci-immune glomerulonephritis mimicking the human disease (Fig. 2.1) [11], thus confirming that neutrophil is the primary effector cell in anti-MPO-induced glomerulonephritis [12].

In an additional model, MPO-deficient mice were immunized with murine MPO; after production of anti-MPO IgG, the animals were lethally irradiated and transplanted with bone marrow from MPO-positive wild type mice (Fig. 2.1) [13]. By 8 weeks after bone marrow transplantation, the mice developed a pauci-immune glomerulonephritis with urine abnormalities. The transfer of anti-MPO lymphocytes into immune-deficient mice has also resulted in necrotizing glomerulonephritis with glomerular immune deposits [14].

A third mouse model is based on the induction of both humoral and cellular autoimmune responses to MPO (Fig. 2.1) [15]. Wild type mice were in fact immunized with MPO and subsequently injected with a sub-nephritogenic dose of nephrotoxic serum (anti-GBM), this procedure resulting in the development of glomerulonephritis. The advantage of this model was the generation of an autoimmune response to MPO in wild type mice.

The models of anti-MPO-mediated glomerulonephritis shortly described above have proven to be useful tools for testing experimental therapies. For example, therapeutic interventions aimed at blocking the pro-inflammatory effects of tumor necrosis factor-alpha (TNFα) have been evaluated in both the MPO-ANCA mouse model [16] and the experimental autoimmune vasculitis rat model [17].

2.3 Animal Models Involving Anti-PR3 Immune Response

Following immunization with recombinant human mouse PR3, non-obese diabetic (NOD) mice develop specific anti-PR3 autoantibodies. The transfer of splenocytes from these mice into immunodeficient NOD/severe combined immunodeficiency disease (SCID) mice has been shown to result in vasculitis and severe segmental and necrotizing glomerulonephritis, leading to acute kidney failure and death [18].

Little et al. [19] have described an interesting model consisting of humanized immunodeficient NOD/SCID interleukin-2 (IL-2)- receptor knockout mice, which received human hematopoietic stem cells and developed a human-mouse chimeric immune system. These mice developed glomerulonephritis following passive transfer of PR3-ANCA IgG derived from patients with severe systemic vasculitis [19].

2.4 Concluding Remarks

Interesting animal models of anti-MPO-related vasculitis, closely resembling clinical and pathological features in humans, have been established. By inducing an abnormal immune response to MPO, these models mimic the clinical aspects of the human MPO-AAV and are contributing to elucidate how ANCA cause vasculitis.

Animal models of anti-PR3 associated disease are much less advanced, and generation of an experimental model that implies both anti-PR3-associated vasculitis and granuloma formation is a major challenge in this field. Further investigations are needed to identify the molecular mechanisms that control the complex neutrophil/endothelium interactions and to establish whether they are dysregulated in AAV [20].

References

1. Mantovani A, Cassatella MA, Costantini C et al (2011) Neutrophils in the activation and regulation of innate and adaptive immunity. Nat Rev Immunol 11:519–531
2. van der Woude FJ, Rasmussen N, Lobatto S et al (1985) The TH. Autoantibodies against neutrophils and monocytes: tool for diagnosis and marker of disease activity in Wegener's granulomatosis. Lancet 1:425–429
3. Jennette JC, Falk RJ (1997) Small-vessel vasculitis. N Engl J Med 337:1512–1523
4. Eisenberger U, Fakhouri F, Vanhille P et al (2005) ANCA-negative pauci-immune renal vasculitis: histology and outcome. Nephrol Dial Transplant 20:1392–1399
5. Lepse N, Abudalahad WH, Kallenberg CG et al (2011) Immune regulatory mechanisms in ANCA-associated vasculitides. Autoimmun Rev 11:77–83
6. Nauseef WM (2007) How human neutrophils kill and degrade microbes: an integrated view. Immunol Rev 219:88–102
7. Hajjar E, Broemstrup T, Kantari C et al (2010) Structures of human proteinase 3 and neutrophil elastase so similar yet so different. FEBS J 277:2238–2254
8. Kallenberg CG (2011) Anti-neutrophil cytoplasmic antibody (ANCA)-associated vasculitis: where to go? Clin Exp Immunol 164(Suppl. 1):1–3
9. Gregersen JW, Kristensen T, Krag SR et al (2012) Early plasma exchange improves outcome in PR3-ANCA-positive renal vasculitis. Clin Exp Rheumatol 30:S39–S47
10. Gan PY, Ooi JD, Kitching AR et al (2015) Mouse models of anti-neutrophil cytoplasmic antibody-associated vasculitis. Curr Pharm Des 21:2380–2390
11. Xiao H, Heeringa P, Liu Z et al (2002) Antineutrophil cytoplasmic autoantibodies specific for myeloperoxidase cause glomerulonephritis and vasculitis in mice. J Clin Invest 110:955–963
12. Xiao H, Heeringa P, Liu Z et al (2005) The role of neutrophils in the induction of glomerulonephritis by anti-myeloperoxidase antibodies. Am J Pathol 167:39–45
13. Schreiber A, Xiao H, Falk RJ et al (2006) Bone marrow-derived cells are sufficient and necessary targets to mediate glomerulonephritis and vasculitis induced by anti-myeloperoxidase antibodies. J Am Soc Nephrol 17:3355–3364
14. Jennette JC, Xiao H, Falk R et al (2011) Experimental models of vasculitis and glomerulonephritis induced by antineutrophil cytoplasmic autoantibodies. Contrib Nephrol 169:211–220
15. Ruth AJ, Kitching AR, Kwan RYQ et al (2006) Anti-neutrophil cytoplasmic antibodies and effector CD4+ cells play nonreduntant roles in anti-myeloperoxidase crescentic glomerulonephritis. J Am Soc Nephrol 17:1940–1949
16. Huugen D, Xiao H, van Esch A et al (2005) Aggravation of anti-myeloperoxidase antibody-induced glomerulonephritis by bacterial lipopolysaccharide: role of tumor necrosis factor-alpha. Am J Pathol 167:47–58
17. Little MA, Bhangal G, Smyth CL et al (2006) Therapeutic effect of anti-TNF-alpha antibodies in an experimental model of anti-neutrophil cytoplasm antibody-associated systemic vasculitis. J Am Soc Nephrol 17:160–169
18. Primo VC, Marusic S, Farnklin CC et al (2010) Anti-PR3 immune responses induce segmental and necrotizing glomerolonephritis. Clin Exp Immunol 159:327–337
19. Little MA, Al-Ani B, Ren S et al (2012) Anti-proteinase 3 anti-neutrophil cytoplasm autoantibodies recapitulate systemic vasculitis in mice with a humanized immune system. PLoS One 7, e28626
20. Woodfin A, Voisin MB, Nourshargh S (2010) Recent developments and complexities in neutrophil transmigration. Curr Opin Hematol 17:9–17

Chapter 3
Nomenclature of Vasculitides: 2012 Revised International Chapel Hill Consensus Conference

Nomenclature of Vasculitides and Beyond

J. Charles Jennette, Ronald J. Falk, and Marco A. Alba

Abstract A nomenclature system provides names and definitions for diseases, and provides the framework for establishing classification criteria for groups of patients and diagnostic criteria for individual patients. The International Chapel Hill Consensus Conference Nomenclature of Vasculitides (CHCC) provides standardized names and definitions for different classes of vasculitis, but does not provide validated criteria for classifying cohorts of patients into these classes, or for diagnosing (classifying) an individual patient. The CHCC nomenclature and definitions are useful for communication among health care providers, understanding the medical literature, guiding development of classification and diagnostic criteria, and facilitating research on cohorts of patients with vasculitis. Names and definitions evolve more slowly than classification and diagnostic criteria because the latter must change as new diagnostic technologies and clinical laboratory testing are available. For example, the discovery of anti-neutrophil cytoplasmic autoantibodies (ANCA) added a new criterion for classifying vasculitis. The most robust ongoing effort to develop classification and diagnostic criteria for vasculitis is by the Diagnostic and Classification Criteria for Vasculitis (DCVAS) study group. Once data are collected from large vasculitis patient cohorts, identifying the most clinically and biologically relevant classes, and the most accurate and precise diagnostic criteria, may require the application of supervised and unsupervised machine learning algorithms. It will be interesting to see how machine generated vasculitis classes agree (or not) with the CHCC classes that were devised by mere mortals.

Keywords Algorithms • Classification criteria • Diagnostic criteria • Nomenclature system • Systemic vasculitides

---

J.C. Jennette (✉) • R.J. Falk • M.A. Alba
Department of Pathology and Laboratory Medicine/Department of Medicine, School of Medicine, University of North Carolina at Chapel Hill,
308 Brinkhous-Bullitt Building, CB#7525, Chapel Hill, NC 27599-7525, USA
e-mail: jcj@med.unc.edu

F. Dammacco et al. (eds.), *Systemic Vasculitides: Current Status and Perspectives*, DOI 10.1007/978-3-319-40136-2_3

3.1 Introduction

A nomenclature system provides names and definitions for diseases. A classification system classifies cohorts of patients into distinct classes, and provides the framework for establishing classification criteria for groups of patients and diagnostic criteria for individual patients. The International Chapel Hill Consensus Conference Nomenclature of Vasculitides (CHCC) provides standardized names and definitions for different classes of vasculitis [1, 2] (Table 3.1). The CHCC does not provide validated criteria for classifying cohorts of patients into these classes, or for diagnosing (classifying) an individual patient. The CHCC nomenclature and classification is of value for:

- Communicating among health care providers involved in the care of patients with vasculitis
- Writing and understanding medical literature pertinent to vasculitis
- Guiding the development of classification and diagnostic criteria to be validated in cohorts of vasculitis patients
- Facilitating clinical and basic research on cohorts (classes) of patients with distinct forms of vasculitis

The name and definition of a disease are specified in an accepted nomenclature system, for example the CHCC system. Effective nomenclature/classification/diagnostic systems are based on up to date clinical and pathobiological data, especially etiology and pathogenesis when known. Classification criteria are the data that are used to place groups of patients into standardized classes. Diagnostic criteria are data that demonstrate or confidently predict the presence of the defining features of a disease in a specific patient. Classification criteria and diagnostic criteria must be tested and validated by studying actual cohorts of patients, and comparing them to disease controls and healthy individuals. Classification and diagnostic criteria evolve most quickly, driven by advances in diagnostic technologies and clinical laboratory testing. For example, new biomarkers that are validated as useful clinical laboratory tests are added to existing classification and diagnostic criteria, as was the case when anti-neutrophil cytoplasmic autoantibodies (ANCA) were discovered, and the presence or absence of ANCA were added as criteria for the classification and diagnosis of small vessel vasculitis [1–3]. Effective names and definitions may persist indefinitely, such as myocardial infarction defined as focal ischemic necrosis of myocardium; whereas the diagnostic and classification criteria for myocardial infarction change over time as new imaging, electrophysiological and laboratory tests are developed.

The 1994 International Chapel Hill Consensus Conference on the Nomenclature of Systemic Vasculitides (CHCC 1994) proposed names and definitions for some of the most common variants of vasculitis [1], and was widely adopted throughout the world. As expected, following publication of the CHCC 1994 article, there were many advances in the understanding of vasculitis. The CHCC 1994 purposefully was confined to proposing names and definitions for a limited number of vasculitides,

Table 3.1 Names for vasculitides adopted by the 2012 International Chapel Hill consensus conference on the nomenclature of vasculitides

Large vessel vasculitis (LVV)
Takayasu arteritis (TAK)
Giant cell arteritis (GCA)
Medium vessel vasculitis (MVV)
Polyarteritis nodosa (PAN)
Kawasaki disease (KD)
Small vessel vasculitis (SVV)
Antineutrophil cytoplasmic antibody (ANCA)–associated vasculitis (AAV)
Microscopic polyangiitis (MPA)
Granulomatosis with polyangiitis (Wegener's) (GPA)
Eosinophilic granulomatosis with polyangiitis (Churg-Strauss) (EGPA)
Immune complex SVV
Anti-glomerular basement membrane (anti-GBM) disease
Cryoglobulinemic vasculitis (CV)
IgA vasculitis (Henoch-Schönlein) (IgAV)
Hypocomplementemic urticarial vasculitis (HUV) (anti-C1q vasculitis)
Variable vessel vasculitis (VVV)
Behcet's disease (BD)
Cogan's syndrome (CS)
Single-organ vasculitis (SOV)
Cutaneous leukocytoclastic angiitis
Cutaneous arteritis
Primary central nervous system vasculitis Isolated aortitis
Others
Vasculitis associated with systemic disease
Lupus vasculitis
Rheumatoid vasculitis
Sarcoid vasculitis
Others
Vasculitis associated with probable etiology
Hepatitis C virus–associated cryoglobulinemic vasculitis
Hepatitis B virus–associated vasculitis
Syphilis-associated aortitis
Drug-associated immune complex vasculitis
Drug-associated ANCA-associated vasculitis
Cancer-associated vasculitis
Others

Modified from Jennette et al. [2]
The items highlighted in red are changes or additions compared to the 1994 International Chapel Hill Consensus Conference on the Nomenclature of Vasculitides

and thus did not address many other forms of vasculitis, including vasculitis associated with and presumably secondary to a number of systemic diseases. Over the past 10 years there also has been an impetus to reduce the use of eponyms in medical terminology. The 2012 International Chapel Hill Consensus Conference (CHCC 2012) had the goals of changing vasculitis names and definitions as appropriate, and adding important categories of vasculitis that were not included in CHCC 1994 [2] (Table 3.1, Fig. 3.1). Table 3.1 shows the new or modified categories of vasculitis highlighted on red. These changes and additions will be discussed in the following section.

As a basic simplified approach to classifying vasculitis into broad general categories, the first level of classification is into large vessel vasculitis (LVV), medium vessel vasculitis (MVV), small vessel vasculitis (SVV) and variable vessel vasculitis (VVV). Unfortunately, these broad classes are interpreted too literally and simplistically by some. This terminology refers to classes of vasculitis that unquestionably cannot be identified merely by measuring the caliber of involved

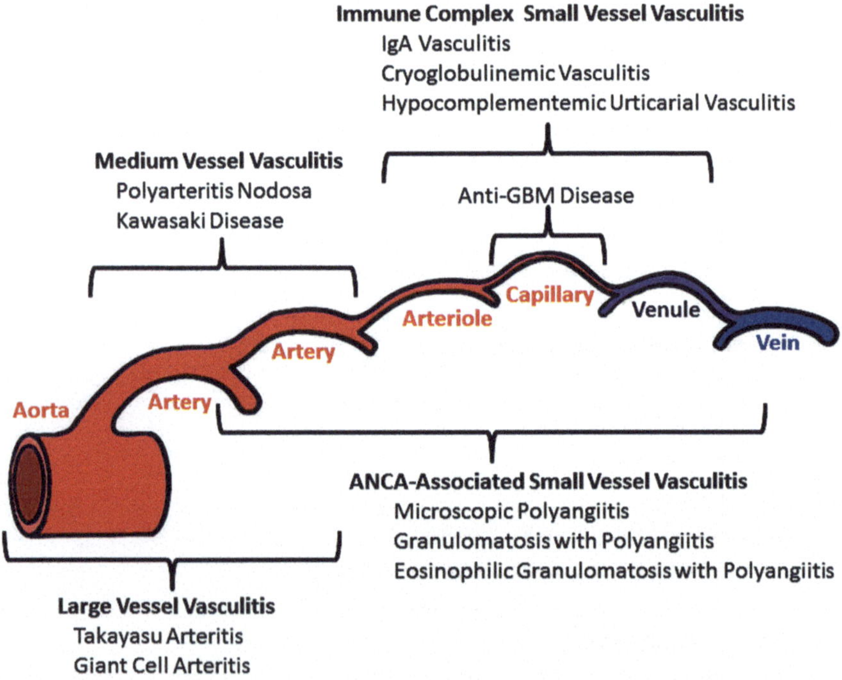

Fig. 3.1 Diagram depicting the predominant distribution of vessel involvement by different CHCC categories of vasculitis. Note that large vessel vasculitis, medium vessel vasculitis and small vessel vasculitis all can affect arteries, however, only small vessel vasculitis affects capillaries and venules (Reproduced from Ref. [2] with permission)

vessels, but rather are the names for classes of vasculitis that are defined by the integration of multiple different parameters.

3.2 Large Vessel Vasculitis (LVV)

The two major variants of LVV are Takayasu arteritis (TAK) and giant cell arteritis (GCA). CHCC 2012 defines LVV as vasculitis affecting the aorta and its major branches more often than other vasculitides although, any size artery may be affected [2]. Importantly, this definition does not state that LVV affects predominantly large vessels because in many patients the actual number of large arteries affected is less than the number of medium and small arteries affected. For example, in a patient with GCA, only a few branches of the carotid arteries may be affected when many more small branches extending into the head and neck are affected. However, LVV definitely affect the aorta and its major branches more often than MVV or SVV (Fig. 3.2).

Both TAK and GCA are arteritis, often granulomatous, that most often affects the aorta and its major branches [2]. The most reliable distinction between the two is the age of the patient at the time of onset (not the time of diagnosis). TAK has onset in patients younger than 50 years old, whereas GCA has onset in patients greater than 50 years old. There are different trends in the distribution and pathology of vasculitis in GCA compared to TAK, but these do not provide clear-cut criteria for classification or diagnosis. GCA has a predilection for branches of the carotid and temporal arteries, and polymyalgia rheumatica frequently accompanies GCA. The relationship of isolated aortitis with pathologic features indistinguishable from GCA and TAK is controversial, although at least some appear to be TAK or GC limited to the aorta.

3.3 Medium Vessel Vasculitis (MVV)

Two major classes of MVV are polyarteritis nodosa (PAN) and Kawasaki disease (KD) [2]. MVV is vasculitis predominantly affecting medium arteries, which are the main visceral arteries and their initial branches in the parenchyma. Importantly, any size artery may be affected, including the smallest arteries [2]. The histopathology of the acute necrotizing lesions overlaps, although, PAN tends to have more conspicuous fibrinoid necrosis and neutrophils, whereas KD tends to have more monocytes and less fibrinoid necrosis [4].

PAN is necrotizing arteritis primarily affecting medium or small arteries without glomerulonephritis or vasculitis in arterioles, capillaries, or venules; and not associated with ANCA [2]. The absence or presence of ANCA, respectively, is a distinguishing feature between PAN and microscopic polyangiitis [3]. KD is arteritis occurring in infants and young children that is associated with mucocutaneous

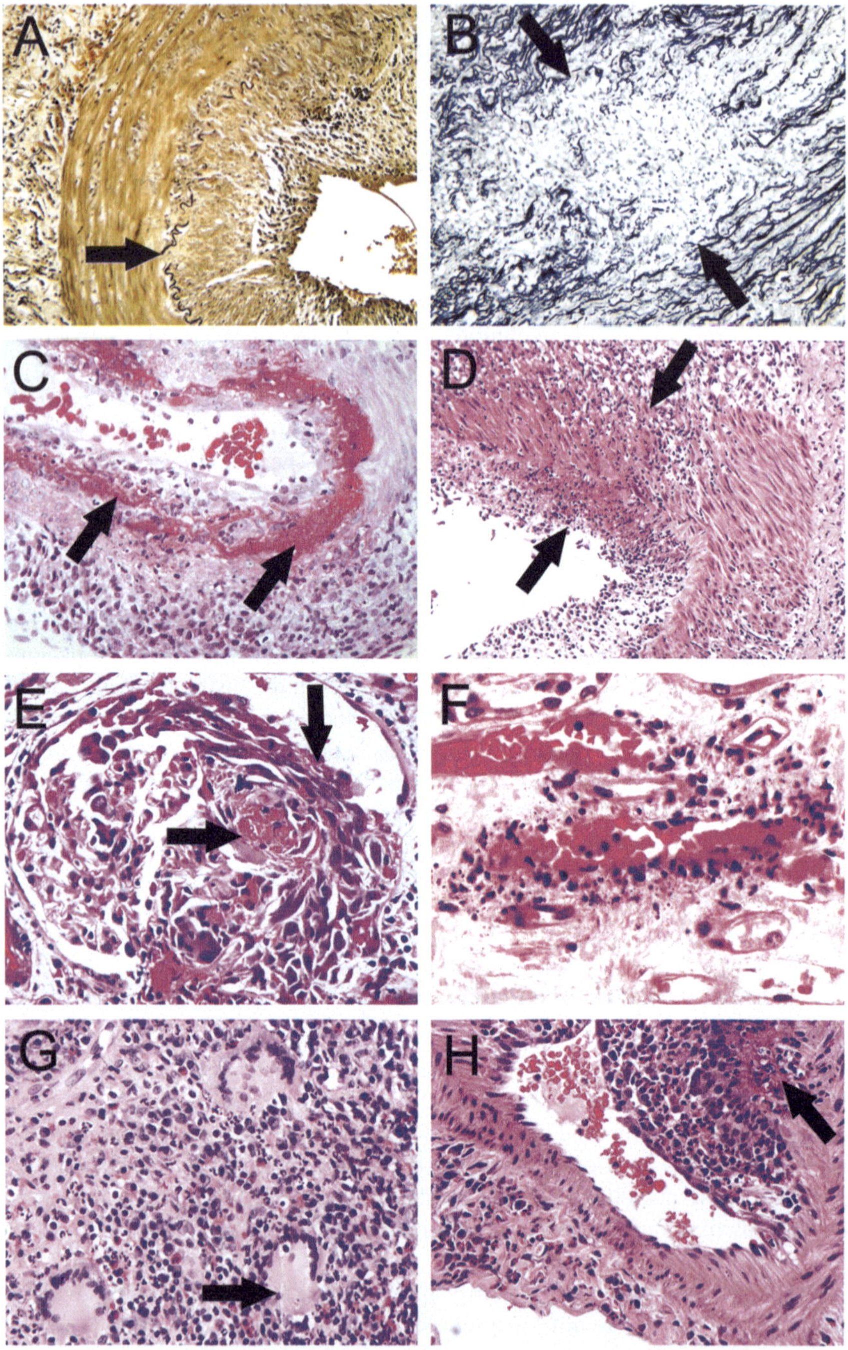

Fig. 3.2 Photomicrographs showing (**a**) giant cell arteritis with fragmentation of the internal elastica (*arrow*) and transmural inflammation (temporal artery, elastic tissue stain); (**b**) Takayasu arteritis of the aorta causing focal fragmentation of the elastic media (medial laminar necrosis) (elastic tissue stain); (**c**) arteritis in a skeletal muscle biopsy with prominent fibrinoid necrosis that is

lymph node syndrome. Coronary arteries are a frequent target, with a predilection for causing aneurysms (pseudoaneurysms) at the origin of carotid arteries from the aorta.

PAN in essence is a diagnosis of exclusion. When necrotizing arteritis or its chronic sequelae are identified in a patient, a diagnosis of PAN is appropriate only after KD, arteritis as a component of a SVV or VVV, and arteritis secondary to a system disease are ruled out.

3.4 Small Vessel Vasculitis (SVV)

New in CHCC 2012 compared to CHCC 1994 is the classification of SVV as either anti-neutrophil cytoplasmic antibody (ANCA)-associated vasculitis (AAV) or immune complex SVV [2]. AAV is characterized by circulating ANCA and a paucity of immunoglobulin and complement in vessel walls, whereas immune complex SVV is characterized by the absence of ANCA in the circulation and substantial immunoglobulin and complement in vessel walls. Of note is the observation that some patients have concurrence of ANCA and immune complex disease, for example ANCA disease plus anti-glomerular basement membrane (anti-GBM) disease.

AAV is necrotizing vasculitis, with few or no immune deposits, predominantly affecting small vessels, and usually, but not always, accompanied by positive serologic assays for ANCA specific for myeloperoxidase (MPO-ANCA) or proteinase 3 (PR3-ANCA) [2]. AAV is subdivided into microscopic polyangiitis (MPA), granulomatosis with polyangiitis (Wegener's) (GPA), and eosinophilic granulomatosis with polyangiitis (Churg-Strauss) (EGPA) [2]. The eponyms Wegener's granulomatosis and Churg-Strauss syndrome were replaced in CHCC 2012 with the more descriptive terms GPA and EGPA respectively. The change to GPA was influenced by a decision that had already been made by several medical societies [5]. An AAV diagnosis should include both ANCA serology status (MPO-ANCA, PR3-ANCA or ANCA-negative) as well as the clinicopathologic phenotype if it is identifiable. Examples of appropriate diagnostic terms are MPO-ANCA MPA and ANCA-negative GPA. Designating the ANCA antigen specificity is important because the ANCA specificity, along with other clinical and laboratory findings, helps predict patient course and outcome [6, 7], and ANCA antigen specificity correlates with other biomarkers of disease including genetic markers [8]. The relevance to ANCA

Fig. 3.2 (continued) consistent with multiple types of vasculitis such as PAN, MPA or GPA (H&E stain); (**d**) Kawasaki disease arteritis with transmural inflammation (*arrows*) with minimal fibrinoid material (renal lobar artery, H&E stain); (**e**) ANCA-associated glomerulonephritis with fibrinoid necrosis (*horizontal arrow*) and crescent (*vertical arrow*) (H&E stain); (**f**) ANCA-associated leukocytoclastic angiitis in the renal medulla (H&E stain); (**g**) Wegener's granulomatosis acute inflammation with multinucleated giant cells (*arrow*) and numerous neutrophils (nasal mucosa, H&E stain); (**h**) Churg-Strauss syndrome arteritis with necrosis (*arrow*) and transmural inflammation with numerous eosinophils (lung, H&E stain) (Reproduced from Ref. [4] with permission)

in the classification of SVV also is supported by the strong evidence that these autoantibodies are pathogenic [9].

MPA is necrotizing vasculitis, with few or no immune deposits, predominantly affecting small vessels [2]. Necrotizing arteritis involving small and medium arteries may be present. Necrotizing and crescentic glomerulonephritis is very common. ANCA-positive pauci-immune necrotizing and crescentic glomerulonephritis also occurs as a renal limited disease. In some clinical trials, renal-limed ANCA-associated glomerulonephritis is grouped with MPA, in essence as a form of renal-limited MPA.

GPA is necrotizing granulomatous inflammation, usually involving the upper and lower respiratory tract, with necrotizing vasculitis affecting predominantly small to medium vessels [2]. ANCA associated with GPA usually is PR3-ANCA, but some patients with classic GPA are MPO-ANCA positive.

EGPA is eosinophil-rich and necrotizing granulomatous inflammation often involving the respiratory tract, with necrotizing vasculitis predominantly affecting small to medium vessels, and associated with asthma and eosinophilia [2]. Over half of patients who would fit into the broad spectrum of Churg-Strauss syndrome are ANCA-negative at diagnosis. However, 75 % or more of EGPA patients with glomerulonephritis are ANCA-positive [10]. This suggests that the eosinophil-rich inflammatory component of EGPA may have a pathogenetically distinct prodromal phase that is separate from the ANCA-associated glomerulonephritis and vasculitis.

Immune complex SVV is vasculitis predominantly affecting small vessels with moderate to marked vessel wall deposits of immunoglobulin and complement [2]. Glomerulonephritis is frequent. Involvement of arteries is much more frequent in ANCA SVV than in immune complex SVV. In CHCC 2012, immune complex SVV includes anti-glomerular basement membrane (anti-GBM) disease, cryoglobulinemic vasculitis (CV), IgA vasculitis (IgAV), and hypocomplementemic urticarial vasculitis (HUV) (anti-C1q vasculitis). Anti-GBM disease is vasculitis affecting glomerular capillaries, pulmonary capillaries, or both, with capillary basement membrane deposition of anti-basement membrane autoantibodies that appears linear by immunofluorescence microscopy, in contrast to other forms of immune complex disease that have granular staining of vessel walls [3, 4]. In anti-GBM disease, immune complexes form in situ in capillary walls by the complexing of anti-GBM antibodies with GBM antigens. Cryoglobulinemic vasculitis is vasculitis with immune deposits of cryoglobulins affecting small vessels and associated with cryoglobulins in serum [2]. IgAV is vasculitis with IgA1-dominant immune deposits affecting small vessels. [2]. IgA-dominant glomerulonephritis that is indistinguishable from IgA nephropathy is often a component of IgAV. There is strong evidence that IgAV and IgA nephropathy are caused by abnormally glycosylated IgGA1, possibly complexed with anti-IgA1 autoantibodies [11]. In accord with the movement to drop eponyms when more specific identifiers or specific etiologic factors are known, CHCC 2012 eliminated the Henoch-Schönlein eponym [2]. Hypocomplementemic urticarial vasculitis (HUV) was the least common SVV immune complex vasculitis that was included in CHCC 2012. HUV is vasculitis accompanied by urticaria and hypocomplementemia affecting small vessels, very often if not always accompanied by circulating anti-C1q autoantibodies [2, 12]. Glomerulonephritis, arthritis, obstructive pulmonary disease, and ocular inflammation are common. Overlap with systemic lupus erythematosus may occur [12].

3.5 Variable Vessel Vasculitis (VVV)

VVV vasculitis is a new CHCC 2012 category that has no predominant type of vessel involved but rather can affect vessels of any size (small, medium, and large) and type (aorta, arteries, veins, and capillaries) [2]. Behcet's disease (BD) and Cogan's syndrome (CS) are distinct forms of VVV.

Behcet's disease (BD) is simply defined as vasculitis occurring in patients with BD that can affect arteries or veins [2]. BD is characterized clinically by recurrent oral and/or genital aphthous ulcers accompanied by cutaneous, ocular, articular, gastrointestinal, and/or central nervous system inflammatory lesions. The spectrum of vasculitis observed in BD includes small vessel leukocytoclastic vasculitis, thromboangiitis, thrombosis, arteritis, and arterial aneurysms.

VVV occurs in patients with Cogan's syndrome, which is characterized by ocular inflammatory lesions, including interstitial keratitis, uveitis, and episcleritis, and inner ear disease, including sensorineural hearing loss and vestibular dysfunction [2]. The spectrum of vasculitis observed in CS includes aortitis, aortic aneurysms, aortic and mitral valvulitis, and arteritis affecting small, medium, or large arteries [2]. The ocular and inner ear disease also may be manifestations of small vessels in the tissues of the eyes and ears.

3.6 Single-Organ Vasculitis (SOV)

SOV also is a new CHCC 2012 category and is vasculitis in a single organ. A very important caveat is that SOV does not have features that indicate that it is a limited expression of one of the other CHCC 2012 specific categories of systemic vasculitis [2]. For example, AAV limited to the kidneys or to the lungs should not be called SOV but rather should be considered renal-limited or pulmonary-limited AAV, respectively.

The SOV diagnosis should specify the organ and/or vessel type that is targeted (e.g. testicular arteritis, cutaneous small vessel vasculitis, central nervous system vasculitis). Some patients with a diagnosis of SOV may later develop additional disease manifestations that indicate one of the systemic vasculitides. For example, a patient may initially appear to have SOV cutaneous arteritis but subsequently be found to have arteritis in the kidney and gut indicative of systemic PAN.

3.7 Vasculitis Associated with Systemic Disease or Likely Etiology

This category of vasculitis is characterized by an accompanying systemic disease or other factor that is the likely cause for the vasculitis. The diagnosis for this vasculitis should have a prefix that identifies the associated systemic disease or etiology. Examples are rheumatoid vasculitis, lupus vasculitis, hydralazine-associated MPO-ANCA MPA, and hepatitis C virus-associated cryoglobulinemic vasculitis. These latter two examples show that this category of vasculitis is not mutually exclusive of the other CHCC categories, but rather allows the recognition of a likely cause for a particular variant of vasculitis.

3.8 Beyond CHCC 2012: Modifications and Criteria for Classification and Diagnosis

As noted already, although CHCC 2012 provides definitions for distinct classes of vasculitis, and thus establishes the structure for a classification system, the CHCC does not provide validated classification or diagnostic criteria. Thus, beyond CHCC 2012 is the need to develop and validate classification or diagnostic criteria. The most robust ongoing effort to do this is by the Diagnostic and Classification Criteria for Vasculitis (DCVAS) study, which is a multinational study designed to develop and validate diagnostic and classification criteria for primary systemic vasculitis [13]. DCVAS aims to formulate these criteria on the basis of data from >2000 patients with six different forms of vasculitis (GPA, MPA, EGPA, GCA, TAK and PAN) and 1500 comparators that will be recruited across medical centers in Europe, North America and Asia. Criteria will be based not only on pathologic features but also on detailed clinical, laboratory and imaging data, disease activity and damage index scores [13]. The data from this large and well-designed study also is likely to inform future adjustments to the CHCC nomenclature and classification system.

A notable earlier attempt to develop a workable classification system resulted in the European Medicines Agency (EMA) algorithm. The EMA algorithm was originally developed by a consensus group of experts on the epidemiology and clinical manifestations of AAV and PAN [14]. Their main objective was to provide a harmonized system for the classification of EGPA, GPA, MPA and PAN by combining 1994 CHCC nomenclature and definitions [1, 2], 1990 American College of Rheumatology classification criteria [15], and 1984 Lanham criteria for Churg-Strauss syndrome [16]. The algorithm was initially developed in a cohort of 99 Caucasians patients in a stepwise hierarchichal approach. The algorithm was reasonably successful at classifying patients into a single class [14].

Subsequently, The EMA algorithm was validated further in other populations [17, 18]. The original cohort also has been reassessed using the CHCC 2012

Table 3.2 EULAR/PRINTO/PRES endorsed names for childhood vasculitides

Predominantly large vessel vasculitis
Takayasu arteritis
Predominantly medium sized vessel vasculitis
Childhood polyarteritis nodosa
Cutaneous polyarteritis
Kawasaki disease
Predominantly small vessels vasculitis
Granulomatous
Wegener's granulomatosis
Churg-Strauss syndrome
Non-Granulomatous
Microscopic polyangiitis
Henoch-Schönlein purpura
Isolated cutaneous leucocytoclastic vasculitis
Hypocomplementic urticarial vasculitis
Other vasculitides
Behcet disease
Vasculitis secondary to infection, malignancies, and drugs
Vasculitis associated with connective tissue diseases
Isolated vasculitis of the central nervous system
Cogan syndrome
Unclassified

Modified from Ozen et al. [21]

definitions showing that the EMA algorithm remains valid using the CHCC 2012 definitions for classifying patients as GPA, MPA or EGPA [19].

A Registry for Childhood Vasculitis e-entry (ARChiVe) cohort study, which is a Childhood Arthritis and Rheumatology Research Alliance initiative, evaluated the classification of GPA in children using multiple adult-derived classification systems (including the EMA algorithm) compared to GPA classification using a validated pediatric system [20]. Multiple problems were identified with classification sensitivity, specificity and inter-system comparability. This study demonstrated that current approaches to classification are not ideal.

The classification and diagnosis of vasculitis in children has been addressed more globally by the European League against Rheumatism, the Paediatric Rheumatology International Trial Organization, and the Paediatric Rheumatology European Society (EULAR/PRINTO/PRES) [21]. The proposed nomenclature and classification categories are similar to the CHCC categories (Table 3.2). The EULAR/PRINTO/PRES group also established and endorsed consensus criteria for the classification of childhood IgAV (Henoch-Schonlein purpura), KD, PAN, WG, or TAK [21]. These classification criteria subsequently were modified and further validated using a retrospective/prospective web-data collection method derived from evaluation of 1398 children. Once again, the system demonstrated good specificity and sensitivity for childhood IgAV (Henoch-Schonlein purpura), KD, PAN, WG, or TAK [22].

Valuable parameters to use as classification and diagnostic criteria are continually being developed, which is why the most effective classification and diagnostic criteria will always evolve over time, even if the diagnostic categories (classes) remain more stable. For example, diagnostic and classification criteria for LVV need to adjust to rapidly developing imaging techniques that can assess inflammation in large vessels. Recent studies have demonstrated that imaging studies can detect otherwise unsuspected inflammation of the aorta and its major branches in GCA patients even when they lack classic cranial disease or biopsy evidence for GCA [23]. The practical use of imaging criteria for assessing GCA patients is illustrated by a clinical trial designed to test the efficacy of tocilizumab to sustain remission in GCA that uses angiography or cross-sectional imaging studies such as magnetic resonance angiography, computed tomography angiography, or positron emission tomography as classification criteria for inclusion in the study [24].

Another approach that has been proposed for improving classification and diagnosis is the use of an artificial neural network (ANN). For example, 23 clinical parameters were evaluated in a cohort of 240 patients with WG and 78 patients with MPA to generate classification criteria using ANN, compared to traditional approaches based on a classification tree and logistic regression [25]. Validation was performed by applying the same approaches to an independent cohort of 46 patients with WG and 21 patients with MPA. On the basis of 4 clinical variables (pulmonary nodules, and involvement of nose, sinuses and ears), ANN was able to distinguish between GPA and MPA with an accuracy of >90 %. This was superior to using CHCC 1994 definitions, ACR classification criteria or Sørensen diagnostic criteria [25]. Given this demonstration of the potential of current rudimentary ANN technology, this approach has great promise for converting large sets of data, such as the DCVAS dataset [13], into effective classification and diagnostic criteria for categories (classes) of vasculitis that have biological and clinical relevance. Machine learning algorithms, including ANN technology, can be either unsupervised or supervised. For example, a supervised learning algorithm could utilize training data derived from patients with a set of gold standard findings that fulfill the CHCC definitions for different classes of vasculitis. It will be interesting to compare the classes of vasculitis identified by unsupervised compared to supervised machine learning algorithms. It also will be interesting to see how machine generated vasculitis classes correspond to the CHCC classes devised by mere mortals.

References

1. Jennette JC, Falk RJ, Andrassy K, Bacon PA, Churg J, Gross WL, Hagen EC, Hoffman GS, Hunder GG, Kallenberg CGM, McCluskey RT, Sinico RA, Rees AJ, van Es LA, Waldherr R, Wiik A (1994) Nomenclature of systemic vasculitides: the proposal of an international consensus conference. Arthritis Rheum 37:187–192

2. Jennette JC, Falk RJ, Bacon PA, Basu N, Cid MC, Ferrario F, Flores-Suarez LF, Gross WL, Guillevin L, Hagen EC, Hoffman GS, Jayne DR, Kallenberg CGM, Lamprecht P, Langford CA, Luqmani RA, Mahr AD, Matteson EL, Merkel PA, Ozen S, Pusey CD, Rasmussen N, Rees AJ, Scott DGI, Specks U, Stone JH, Takahashi K, Watts RA (2013) 2012 revised international Chapel Hill consensus conference nomenclature of Vasculitides. Arthritis Rheum 65:1–11
3. Jennette JC, Falk RJ (1997) Small vessel vasculitis. N Engl J Med 337:1512–1523
4. Jennette JC, Falk RJ (2007) Pathologic classification of vasculitis. Pathol Case Rev 12:179–185
5. Falk RJ, Gross WL, Guillevin L, Hoffman G, Jayne DRW, Jennette JC et al (2011) "Granulomatosis with polyangiitis (Wegener's)": an alternative name for "Wegener's granulomatosis". A joint proposal of the American College of Rheumatology, the American Society of Nephrology, and the European League Against Rheumatism. Ann Rheum Dis 70:704. J Am Soc Nephrol 22:587–588. Arthritis Rheum 63:863–864
6. Lionaki S, Blyth ER, Hogan SL, Hu Y, Senior JBA, Jennette CE, Nachman PH, Jennette JC, Falk RJ (2012) Classification of antineutrophil cytoplasmic autoantibody vasculitides: the role of antineutrophil cytoplasmic autoantibody specificity for myeloperoxidase or proteinase 3 in disease recognition and prognosis. Arthritis Rheum 64:3452–3462
7. Mahr A, Katsahian S, Varet H, Guillevin L, Hagen EC, Höglund P, Merkel PA, Pagnoux C, Rasmussen N, Westman K, Jayne DR, French Vasculitis Study Group (FVSG) and the European Vasculitis Society (EUVAS) (2013) Revisiting the classification of clinical phenotypes of anti-neutrophil cytoplasmic antibody-associated vasculitis: a cluster analysis. Ann Rheum Dis 72:1003–1010
8. Lyons PA, Rayner TF, Trivedi S, Holle JU, Watts RA, Jayne DR, Baslund B, Brenchley P, Bruchfeld A, Chaudhry AN, Cohen Tervaert JW, Deloukas P, Feighery C, Gross WL, Guillevin L, Gunnarsson I, Harper L, Hrušková Z, Little MA, Martorana D, Neumann T, Ohlsson S, Padmanabhan S, Pusey CD, Salama AD, Sanders JS, Savage CO, Segelmark M, Stegeman CA, Tesař V, Vaglio A, Wieczorek S, Wilde B, Zwerina J, Rees AJ, Clayton DG, Smith KG (2012) Genetically distinct subsets within ANCA-associated vasculitis. N Engl J Med 367:214–223
9. Jennette JC, Falk RJ (2014) Pathogenesis of antineutrophil cytoplasmic autoantibody-mediated disease. Nat Rev Rheumatol 10:463–473
10. Sinico RA, Di Toma L, Maggiore U, Tosoni C, Bottero P, Sabadini E et al (2006) Renal involvement in Churg-Strauss syndrome. Am J Kidney Dis 47:770–779
11. Suzuki H, Kiryluk K, Novak J, Moldoveanu Z, Herr AB, Renfrow MB, Wyatt RJ, Scolari F, Mestecky J, Gharavi AG, Julian BA (2011) The pathophysiology of IgA nephropathy. J Am Soc Nephrol 22:1795–1803
12. Kallenberg CG (2008) Anti-C1q autoantibodies. Autoimmun Rev 7:612–615
13. Craven A, Robson J, Ponte C, Grayson PC, Suppiah R, Judge A, Watts R, Merkel PA, Luqmani RA (2013) ACR/EULAR-endorsed study to develop diagnostic and classification criteria for vasculitis (DCVAS). Clin Exp Nephrol 17:619–621
14. Watts R, Lane S, Hanslik T, Hauser T, Hellmich B, Koldingsnes W, Mahr A, Segelmark M, Cohen-Tervaert JW, Scott D (2007) Development and validation of a consensus methodology for the classification of the ANCA-associated vasculitides and polyarteritis nodosa for epidemiological studies. Ann Rheum Dis 66:222–227
15. Fries JF, Hunder GG, Bloch DA, Michel BA, Arend WP, Colabrese LH (1990) The American College of Rheumatology 1990 criteria for the classification of vasculitis: summary. Arthritis Rheum 33:1135–1136
16. Lanham JG, Elkon KB, Pusey CD, Hughes GR (1984) Systemic vasculitis in asthma and eosinophilia: a clinical approach to the Churg Strauss syndrome. Medicine (Baltimore) 63:65–81
17. Kamali S, Artim-Esen B, Erer B, Ozdener L, Gul A, Ocal L, Aral O, Inanc M (2012) Re-evaluation of 129 patients with systemic necrotizing vasculitides by using classification algorithm according to consensus methodology. Clin Rheumatol 31:325–328

18. Liu LJ, Chen M, Yu F, Zhao MH, Wang HY (2008) Evaluation of a new algorithm in classification of systemic vasculitis. Rheumatology (Oxford) 47:708–712
19. Abdulkader R, Lane SE, Scott DG, Watts RA (2013) Classification of vasculitis: EMA classification using CHCC 2012 definitions. Ann Rheum Dis 72:1888
20. Uribe AG, Huber AM, Kim S, O'Neil KM, Wahezi DM, Abramson L, Baszis K, Benseler SM, Bowyer SL, Campillo S, Chira P, Hersh AO, Higgins GC, Eberhard A, Ede K, Imundo LF, Jung L, Kingsbury DJ, Klein-Gitelman M, Lawson EF, Li SC, Lovell DJ, Mason T, McCurdy D, Muscal E, Nassi L, Rabinovich E, Reiff A, Rosenkranz M, Schikler KN, Singer NG, Spalding S, Stevens AM, Cabral DA, A Registry for Children with Vasculitis e-entry (ARChiVe) Network (2012) Increased sensitivity of the European medicines agency algorithm for classification of childhood granulomatosis with polyangiitis. J Rheumatol 39:1687–1697
21. Ozen S, Ruperto N, Dillon MJ, Bagga A, Barron K, Davin JC, Kawasaki T, Lindsley C, Petty RE, Prieur AM, Ravelli A, Woo P (2006) EULAR/PReS endorsed consensus criteria for the classification of childhood vasculitides. Ann Rheum Dis 65:936–941
22. Ozen S, Pistorio A, Iusan SM, Bakkaloglu A, Herlin T, Brik R, Buoncompagni A, Lazar C, Bilge I, Uziel Y, Rigante D, Cantarini L, Hilario MO, Silva CA, Alegria M, Norambuena X, Belot A, Berkun Y, Estrella AI, Olivieri AN, Alpigiani MG, Rumba I, Sztajnbok F, Tambic-Bukovac L, Breda L, Al-Mayouf S, Mihaylova D, Chasnyk V, Sengler C, Klein-Gitelman M, Djeddi D, Nuno L, Pruunsild C, Brunner J, Kondi A, Pagava K, Pederzoli S, Martini A, Ruperto N, Paediatric Rheumatology International Trials Organisation (PRINTO) (2010) EULAR/PRINTO/PRES criteria for Henoch-Schönlein purpura, childhood polyarteritis nodosa, childhood Wegener granulomatosis and childhood Takayasu arteritis: Ankara 2008. Part II: Final classification criteria. Ann Rheum Dis 69:798–806
23. Ponte C, Grayson PC, Suppiah R, Robson J, Anthea C, Judge A, Merkel PA, Watts RA, Luqmani RA (2014) Development of the classification criteria for giant cell arteritis in the diagnostic and classification criteria for vasulitis study: a pilot study using a panel review methodology. Ann Rheum Dis 73(Suppl 2):555 (abstract)
24. Unizony SH, Dasgupta B, Fisheleva E, Rowell L, Schett G, Spiera R, Zwerina J, Harari O, Stone JH (2013) Design of the tocilizumab in giant cell arteritis trial. Int J Rheumatol 2013, 912562
25. Linder R, Orth I, Hagen EC, van der Woude FJ, Schmitt WH (2011) Differentiation between Wegener's granulomatosis and microscopic polyangiitis by an artificial neural network and by traditional methods. J Rheumatol 38:1039–1047

Chapter 4
Search for Autoantibodies in Systemic Vasculitis: Is It Useful?

Joice M.F.M. Belem, Bruna Savioli, and Alexandre Wagner Silva de Souza

Abstract The detection of autoantibodies is a useful tool for the diagnosis of some small-vessel systemic vasculitides and may be an alternative when tissue biopsy is not conclusive or not available. Antineutrophil cytoplasmic antibodies (ANCA) are biomarkers of ANCA-associated vasculitis and are detected by indirect immunofluorescence (IIF) with three main patterns described as follows: cytoplasmic (C-ANCA), perinuclear (P-ANCA) and atypical (A-ANCA). ANCA specificity may be determined by enzyme-linked immunosorbent assay (ELISA) for anti-proteinase 3 (anti-PR3) and anti-myeloperoxidase (anti-MPO) antibodies. ANCA is not only important for the diagnosis of ANCA-associated vasculitis, their specificity has also been associated with disease phenotype and with relapse risk. ANCA are also detected in inflammatory bowel diseases, primary sclerosing cholangitis, autoimmune hepatitis and in drug-induced ANCA-associated vasculitis. Other autoantibodies that may be detected in sera from patients with small vessel vasculitis are anti-glomerular basement membrane (anti-GBM) antibodies and anti-C1q antibodies which may be useful in establishing the diagnosis of anti-GBM antibody disease and hypocomplementemic urticarial vasculitis, respectively. The investigation of cryoglobulinemic vasculitis includes the detection of cryoglobulins and rheumatoid factor. Although, some autoantibodies have been described in patients with medium and large vessel vasculitides (e.g. antiendothelial cell antibodies, anti-aorta and anti-ferritin antibodies), the search for these autoantibodies is not useful for the investigation of systemic vasculitis.

Keywords Systemic vasculitis • Autoantibodies • Antineutrophil cytoplasmic antibodies

J.M.F.M. Belem • B. Savioli • A.W.S. de Souza (✉)
Rheumatology Division, Universidade Federal de São Paulo – Escola Paulista de Medicina (Unifesp-EPM), 3rd Floor, Rua Botucatu, 04023-900 São Paulo-SP, Brazil
e-mail: alexandre_wagner@uol.com.br

F. Dammacco et al. (eds.), *Systemic Vasculitides: Current Status and Perspectives*, DOI 10.1007/978-3-319-40136-2_4

4.1 Introduction

Systemic vasculitides are a group of heterogeneous, multisystem diseases, characterized by inflammation and necrosis in vessel walls. There is a broad spectrum of signs and symptoms of systemic vasculitis, and diagnosis rely on recognizing a clinical pattern of the disease and is supported by appropriate investigations [1, 2]. Prompt diagnosis of a systemic vasculitis is of the utmost importance, since delays in commencing therapy may lead to damage accrual and impact prognosis, resulting in increased morbidity and even mortality [3]. Depending on vessel size predominantly affected by the vasculitic process, the precise diagnosis of a systemic vasculitis may be confirmed by imaging studies (e.g. conventional angiography or magnetic resonance angiography) for patients with large and medium vessel vasculitis or by tissue biopsy in patients presenting manifestations suggestive of medium and small vessel vasculitis. The search for autoantibodies is important to ascertain the diagnosis of specific sub-groups of small vessel vasculitis, such as antineutrophil cytoplasmic antibodies-associated vasculitis, anti-glomerular basement membrane (anti-GBM) disease and hypocomplementemic urticarial vasculitis (HUV) [1–3].

Cryoglobulins are immunoglobulins that precipitate *in vitro* at temperatures <37 °C and solubilize after re-warming, they may be monoclonal immunoglobulins, immune complex containing antigen and mono/polyclonal antibodies or only polyclonal antibodies. Cryoglobulinemic vasculitis is frequently associated with hepatitis C-virus infection [4]. The investigation of cryoglobulins will be reviewed elsewhere in this book.

4.2 Antineutrophil Cytoplasmic Antibodies (ANCA)

ANCA are antibodies against constituents of primary (azurophilic) granules of neutrophils and lysosomes of monocytes [5]. The first description of ANCA dates back to 1982, when Davies et al. described eight patients with segmental necrotizing glomerulonephritis who presented in their sera a factor that stained the cytoplasm of neutrophils at indirect immunofluorescence (IIF). The cause of this glomerulonephritis was attributed to the infection by Ross River virus and not much attention was given to ANCA until 1985, when van der Woude et al. described the association between ANCA and granulomatosis with polyangiitis (GPA) (formerly Wegener's) [6, 7]. Two patterns of ANCA, cytoplasmic and perinuclear (i.e. C-ANCA and P-ANCA, respectively) were then recognized (Fig. 4.1a and b) and antigen specificity for P-ANCA with anti-myeloperoxidase (anti-MPO) antibodies was discovered in 1988, whereas the antigen specificity for C-ANCA with anti-proteinase 3 (anti-PR3) antibodies was found in 1990 [8, 9]. Moreover, atypical patterns of ANCA (A-ANCA) (Fig. 4.1c), were recognized to be associated with antibodies to other

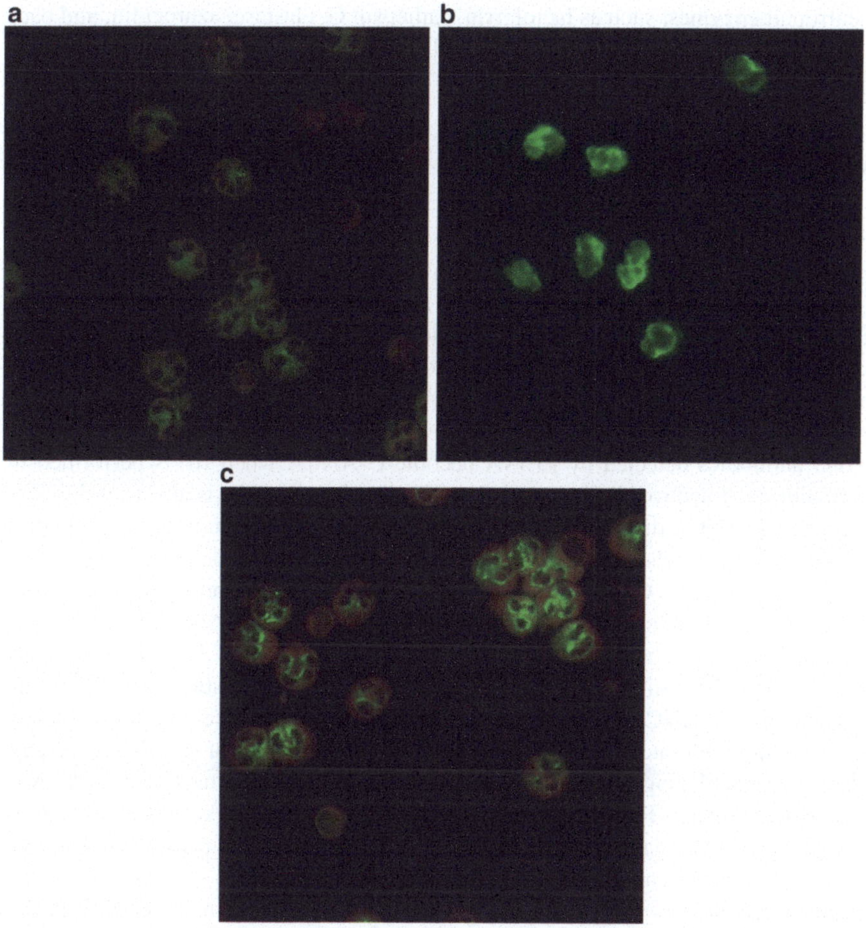

Fig. 4.1 Antineutrophil cytoplasmic antibodies detected by indirect immunofluorescence in ethanol-fixed human neutrophils

This figure illustrates the C-ANCA pattern displaying a coarse cytoplasmic granular staining with inter-lobular enhancement (**a**), in patients with a positive ANCA test and a negative ANA test, no fluorescence is observed in lymphocytes present in the slide. P-ANCA pattern is a perinuclear staining on neutrophils with nuclear extension (**b**) while in atypical perinuclear A-ANCA no nuclear extension is observed (**c**). *ANA* antinuclear antibodies, *ANCA* antineutrophil cytoplasmic antibodies

neutrophil enzymes, such as lactoferrin, cathepsin G, elastase, azurocidin, and bacterial permeability increasing protein (BPI) [10].

4.2.1 Methods

The main assays for detecting ANCA in sera are IIF on ethanol-fixed human neutrophils and on formalin-fixed neutrophils, as well as enzyme-linked immunosorbent assay (ELISA) for anti-MPO and anti-PR3 antibodies [11]. Granule's permeability is increased in neutrophils treated with ethanol and this process leads to redistribution of the cationic MPO from the cytoplasm to the perinuclear area while PR3 remains scattered in the cytoplasm. Since P-ANCA is an artifact due to ethanol fixation of neutrophils, patients presenting P-ANCA by IIF in ethanol-fixed slides and anti-MPO antibodies detected by ELISA become C-ANCA when IIF is performed in formalin-fixed neutrophils. IIF with formalin-fixed neutrophils is also useful to differentiate P-ANCA due to anti-MPO antibodies from positive antinuclear antibodies (ANA) in Hep-2 cells when the ELISA technique is not available [12].

The International Consensus Statement on Testing and Reporting ANCA recommends the combination of IIF and ELISA for the detection of ANCA when investigating ANCA-associated vasculitides. Up to 10 % of ANCA positive patients present only IFI positive results. As stated above, the main patterns in IIF are the cytoplasmic (C-ANCA) pattern (Fig. 4.1a) that shows a diffuse granular cytoplasmic staining with interlobular accentuation, and the perinuclear (P-ANCA) pattern (Fig. 4.1b) displaying perinuclear fluorescence with nuclear extension. C-ANCA is associated with anti-PR3 antibodies and P-ANCA is associated with anti-MPO antibodies [11, 13]. Regarding the A-ANCA pattern, it appears as perinuclear staining without nuclear extension (Fig. 4.1c) or diffuse flat cytoplasmic staining or the combination of both cytoplasmic and nuclear/perinuclear staining on neutrophils [11].

In clinical practice, the investigation of ANCA specificity by ELISA includes the search for anti-PR3 and anti-MPO antibodies [14]. Currently, there are three generations of ELISA tests for detecting ANCA, the first generation ELISA applies absorption coating methods with target antigens directly immobilized to the surface of the ELISA plate. However, this may induce pitfalls that decrease sensitivity by masking and deformation of epitopes of PR3 and MPO [11, 14]. Second and third generation ELISA tests were developed to improve sensitivity lost by the first generation ELISA. The second generation ELISA or capture ELISA uses capture molecules mainly monoclonal antibodies to bind the ANCA antigen to the ELISA plate and the third generation ELISA uses anchor molecules to bind the antigen to the plate [11, 13–15].

More recently, novel techniques have been developed to detect ANCA, including automated fluorescent techniques with the acquisition of high-resolution digital images that are analyzed by software programs which determine positivity and ANCA pattern. To detect antigen specificity for ANCA, assays other than ELISA have also been developed such as chemoiluminescent immunoassay and bead-based flow cytometry assays. All of them have been tested to detect anti-PR3 and anti-MPO antibodies but are not available in most laboratories [11, 15].

4.2.2 ANCA-Associated Vasculitides

ANCA are the serological biomarkers of a group of small-vessel necrotizing vasculitides with few or no immune deposits in vessel walls (i.e. pauci-immune vasculitis) termed as ANCA-associated vasculitides (AAV). GPA, microscopic polyangiitis (MPA), eosinophilic granulomatosis with polyangiitis (EGPA) and renal limited vasculitis (RLV) are the main AAV [16].

The detection of ANCA is useful for the diagnosis of a patient with suspect AAV. ANCA is positive in approximately 90 % of GPA patients with generalized disease and in up to 40 % of GPA patients with localized disease. In GPA patients with a positive ANCA test, 80–95 % present C-ANCA by IIF and anti-PR3 antibodies by ELISA (i.e. PR3-ANCA), whereas 5–20 % present P-ANCA and anti-MPO antibodies (i.e. MPO-ANCA) [17]. In other AAV, most ANCA positive patients present MPO-ANCA, ANCA positivity is approximately 70 % in MPA, and 38 % in EGPA, whereas in RVL it ranges from 70 to 90 % [18–20]. In the appropriate clinical context with a high pre-test probability (i.e. patient with suspect AAV presenting manifestations of small vessel vasculitis) the presence of PR3-ANCA yields a specificity of 98 % for the diagnosis of GPA while the positivity for MPO-ANCA has a specificity of 99.4 % for AAV. However, a negative result of ANCA test does not rule out AAV [21, 22].

Although the role of ANCA is well established for the diagnosis of AAV, positivity of ANCA is not adequate for monitoring disease activity and following its titers is controversial regarding the prediction of disease relapses. Instead, disease activity in AAV is best evaluated by the Birmingham Vasculitis Activity Score (BVAS) rather than by serum ANCA titers [12, 23]. A recent meta- analysis demonstrated that a rise in ANCA titers or their persistence is only modestly associated with predicting disease relapse [24]. Indeed, the presence of PR3-ANCA is independently associated with an increased risk of relapses while specific AAV diagnosis (e.g. GPA, MPA or RLV) defined by the Chapel Hill Consensus Conference (CHCC) and by European Medicines Agency (EMA) system were not predictive of disease relapses. Furthermore, ANCA specificity is also associated with disease phenotypes in AAV, patients with RLV or with vasculitic manifestations without evidence of granulomatous inflammation are more likely to present MPO-ANCA, whereas patients with necrotizing granulomatous inflammation are more likely to present PR3-ANCA [25]. Genetic background in AAV is more linked with ANCA specificity rather than with disease phenotypes. A genome-wide association study (GWAS) showed that PR3-ANCA is associated with HLA-DP and with genes encoding α1-anti-trypsin (SERPINA1) and PR3 (PRTN3). MPO-ANCA is, in turn, associated with HLA-DQ [26].

In EGPA, ANCA positivity is also associated with disease phenotypes. ANCA-positive patients present a higher frequency of vasculitic manifestations such as glomerulonephritis, purpura and mononeuritis multiplex, whereas disease manifestations in ANCA-negative patients are more linked to tissue infiltration of eosinophils that often leads to cardiopulmonary involvement [27].

4.2.3 ANCA Positivity in Other Diseases

At presentation, up to 40 % of patients with anti-glomerular basement membrane (GBM) antibody disease are ANCA positive, mainly with anti-MPO antibodies [28]. The meaning of this association is not completely understood, one study showed that patients with anti-GBM antibody disease and MPO-ANCA present unusual features for anti-GBM disease such as purpura and joint pain, while in another study, MPO-ANCA were associated with worse renal prognosis [29–31].

ANCA are also frequently found in some autoimmune gastrointestinal diseases such as inflammatory bowel diseases, especially ulcerative colitis, primary sclerosing cholangitis and autoimmune hepatitis. In those diseases, A-ANCA is usually observed with specificity mostly against antigens other than MPO or PR3 [5, 15]. ANCA are also positive in drug-induced ANCA-associated vasculitis and in the vasculopathy induced by the use of levamisole-contaminated cocaine. Drugs usually associated with the development of ANCA are propylthiouracil, hydralazine, minocycline, penicillamine, procainamide and allopurinol, to name but a few. In drug-induced ANCA-associated vasculitis, anti-MPO, anti-elastase and anti-lactoferrin antibodies are most commonly found [32, 33].

4.2.4 Anti-Lysosome Associated Membrane Protein-2 Antibodies

Anti-lysosome associated membrane protein-2 (anti-LAMP-2) antibodies are considered a subtype of ANCA due to the close relation between its antigen with PR3 and MPO in intracellular vesicles of neutrophils. However, LAMP-2 is also expressed in lysosomes of monocytes, neutrophils and endothelial cells. There is a strong molecular mimicry between the epitope P_{41-49} of LAMP-2 and an epitope in FimH, a bacterial adhesin from fimbriated bacteria such as *Escherichia coli* [14]. Anti-LAMP-2 antibodies were first described in the context of AAV by a small study that evaluated 15 RVL patients and amongst them, 13 patients (87 %) were positive for anti-LAMP-2 antibodies by a western blot technique [34]. Subsequently, two large multicenter studies performed in six European cohorts confirmed the high prevalence of anti-LAMP-2 antibodies in active patients with different AAV subsets (i.e. GPA, MPA and RLV) ranging from 80 to 93 %. Conversely, anti-LAMP-2 antibodies were present in only 7 % of AAV patients in remission [35, 36]. Nonetheless, a study performed in AAV patients from the US that included either patients with active disease and in remission found a prevalence of anti-LAMP-2 antibodies of only 21 %. The reasons for the differences in these studies may be the use of different techniques to detect anti-LAMP-2 antibodies and the inclusion of patients in different disease phases (i.e. active and remission) [14, 37]. More recently, a study performed in ANCA negative AAV patients found anti-LAMP-2 antibodies in 73 % of them and IgG from their sera were shown to bind to normal human kidney sections and to human endothelial cells in culture. This finding suggests a potential role of anti-LAMP-2 antibodies in the pathogenesis of AAV in ANCA-negative patients [38]. Despite the amount of evidence regarding the potential role of anti-LAMP-2 antibodies in AAV, this issue is still a matter of controversy and the search for these antibodies is not available in routine clinical practice [14].

4.3 Anti-Glomerular Basement Membrane Antibodies

Anti-GBM antibodies are biomarkers for the anti-GBM antibody disease (formerly Goodpasture's disease) which has been recently included as a small-vessel immune complex vasculitis by the 2012 CHCC. The conventional term Goodpasture's disease is often reserved for those patients with renal and pulmonary manifestations in the presence of anti-GBM antibodies [16]. The typical presentation of pulmonary-renal syndrome is found in 40–60 % of the patients with anti-GBM antibody disease. Renal involvement is due to crescentic glomerulonephritis and is manifested as rapidly progressive glomerulonephritis with nephritic urinary sediment. Pulmonary involvement in anti-GBM antibody disease ranges from dyspnea and cough to overt pulmonary hemorrhage [39].

The detection of anti GBM antibodies is mandatory for the diagnosis of anti-GBM antibody disease in a patient with pulmonary-renal syndrome or with rapidly progressive glomerulonephritis. These antibodies may be detected in the kidney tissue or in serum, the latter is only reserved for patients with contra-indications to perform a renal biopsy. To detect the presence of anti-GBM antibodies in renal tissue, the direct immunofluorescence needs to be performed and a linear deposition of IgG is observed on glomerular capillaries [40]. Alternatively, the ELISA test is used to detect circulating anti-GBM antibodies and its sensitivity ranges from 65 to 100 %. ELISA assays that use purified or recombinant alpha-3 chain of collagen IV present the best sensitivity. In a review of 77 patients, the titers of antibodies directed to the N-terminal domain of NC1 were correlated with renal survival. Antigen specificity may also be confirmed by western blot but IIF rarely needs to be performed [41].

4.4 Anti-C1q Antibodies

Anti-C1q antibodies are autoantibodies against the classic pathway complement component "C1q". These antibodies have been considered biomarkers for HUV [16]. However, they are not specific for HUV, since they can also be detected in other autoimmune rheumatic diseases, such as lupus nephritis, and in infectious diseases, including hepatitis C and HIV infection [42, 43].

HUV is an entity characterized by an inflammatory injury of vessels from the skin and other organs and systems, associated with complement consumption. Recently, HUV has been classified as an immune complex small-vessel vasculitis by the 2012 CHCC [16]. Besides cutaneous lesions (e.g. wheals that persist for more than 24 h), common features of HUV include glomerulonephritis, arthralgia or arthritis, lung involvement, gastrointestinal vasculitis, and ocular inflammation. Skin biopsy of urticarial lesions display cutaneous leukocytoclastic vasculitis [42].

Anti-C1q antibodies are usually detected by the ELISA technique. Although previous small studies had shown up to 100 % positivity of anti C1q in HUV [42–44], a recent French nationwide retrospective study was performed in 57 HUV patients and found a prevalence of 55 % of anti-C1q antibodies. On the other hand, this study indicated that low serum C1q levels in the setting of HUV is a more sensitive marker than anti-C1q antibodies. Furthermore, anti-C1q antibodies in HUV patients are associated with specific clinical features, including a higher frequency of angioedema, livedo reticularis, ocular involvement, musculoskeletal and renal involvement with less frequent pulmonary and gastrointestinal manifestations [45].

4.5 Other Autoantibodies in Systemic Vasculitis

Several autoantibodies have been described in patients with systemic vasculitis (Table 4.1). However, to date the clinical relevance of many of these antibodies needs to be determined and, thus, they are not available for routine clinical practice [14]. For more than one decade, the presence of anti-endothelial cell antibodies (AECA) have been evaluated in different systemic vasculitides, including Takayasu arteritis, giant cell arteritis, polyarteritis nodosa, GPA, MPA, EGPA, Kawasaki disease, IgA vasculitis and Behçet's disease [46]. AECA comprise a heterogeneous family of antibodies against diverse target antigens present in endothelial cell membrane [47]. Several techniques were developed with different antigenic substrates to detect AECA, including ELISA, IIF, western blot and radioimmunoassay. The prevalence of AECA in different vasculitides is highly variable, with those antibodies being more prevalent in Takayasu arteritis and GPA, and least found in EGPA and polyarteritis nodosa [46]. Anti-ferritin antibodies were described in giant cell arteritis and in Takayasu arteritis, whereas in the latter anti-aorta, anti-monocytes and anti-annexin V antibodies were also described [48–52]. However, their clinical relevance is still unknown.

Table 4.1 Autoantibodies in primary systemic vasculitides

Vasculitis	Autoantibodies
Takayasu arteritis	AECA
	Anti-aorta antibodies
	Anti-ferritin antibodies
	Anti-annexin V antibodies
	Anti-monocyte antibodies
Giant cell arteritis	AECA
	Anti-ferritin antibodies
Polyarteritis nodosa	AECA
Kawasaki disease	AECA
ANCA-associated vasculitides	AECA
	PR3-ANCA
	MPO-ANCA
	Anti-LAMP-2 antibodies
Immune complex vasculitides	Anti-C1q antibodies
	Anti-GBM antibodies
	Cryoglobulins
	IgA-ANCA
	IgA-Anticardiolipin
	Rheumatoid factor

Adapted from Silva de Souza [14]
AECA – Antiendothelial cell antibodies; ANCA – Antineutrophil cytoplasmic antibodies; GBM – glomerular basement membrane; LAMP-2 – Lysosome associated membrane protein-2; MPO – Myeloperoxidase; PR3 – Proteinase 3

IgA vasculitis is a small-vessel vasculitis caused predominantly by IgA1 immune complex deposition in vessel walls [16]. High levels of IgA are observed in 50–70 % of patients with active disease and may be associated with renal involvement. Immune deposits in vessel walls of patients with IgA vasculitis comprise mainly IgA and C_3 [53, 54]. Abnormal glycosylation of IgA1 (e.g. galactose deficient IgA1) is implicated in the pathogenesis of IgA vasculitis with an increased synthesis and a decreased clearance of IgA, as well as the production of circulating IgA1-IgA and IgA1-IgG immune complexes, IgA rheumatoid factor (RF), IgA ANCA and even IgA anticardiolipin antibodies. However, the diagnosis of IgA vasculitis is established by tissue biopsy showing neutrophil infiltration and direct immunofluorescence with IgA deposit in small vessels in a patient presenting features suggestive of IgA vasculitis [54]. To date, there is no evidence that the detection of autoantibodies would be useful for the diagnosis of this small-vessel vasculitis.

RF is classically an antibody against the Fc fraction of IgG; in clinical practice RF is mostly detected as IgM antibodies against IgG [55]. However, as stated previously, IgA RF is associated with IgA vasculitis and IgG RF has been described in rheumatoid vasculitis [54, 55]. In systemic vasculitis, monoclonal IgM RF is frequently detected at high levels in patients with type II and type III mixed cryoglobulinemia [4].

4.6 Conclusion

The search for autoantibodies is a valuable diagnostic tool for the diagnosis of small vessel vasculitides such as AAV, anti-GBM antibody disease and HUV. To date, the most relevant autoantibodies for clinical practice are ANCA, anti-MPO and anti-PR3 antibodies, anti-GBM antibodies, anti-C1q antibodies, cyoglobulins and rheumatoid factor. The usefulness of other autoantibodies for the diagnosis of systemic vasculitis has yet to be determined.

Acknowledgements Authors thank prof. Luis Eduardo Coelho Andrade, Silvia Helena Rodrigues and Marcele Grecco for their valuable contribution to this book chapter.

References

1. Nataraja A, Mukhtyar C, Hellmich B, Langford C, Luqmani R (2007) Outpatient assessment of systemic vasculitis. Best Pract Res Clin Rheumatol 21:713–732. doi:10.1016/j.berh.2007.01.004
2. Jayne D (2009) The diagnosis of vasculitis. Best Pract Res Clin Rheumatol 23:445–453. doi:10.1016/j.berh.2009.03.001
3. Miller A, Chan M, Wiik A, Misbah SA, Luqmani RA (2010) An approach to the diagnosis and management of systemic vasculitis. Clin Exp Immunol 160:143–160. doi:10.1111/j.1365-2249.2009.04078.x
4. Ramos-Casals M, Stone JH, Cid MC, Bosch X (2012) The cryoglobulinaemias. Lancet 379:348–360. doi:10.1016/S0140-6736(11)60242-0
5. Bosch X, Guilabert A, Font J (2006) Antineutrophil cytoplasmic antibodies. Lancet 368:404–418
6. Davies DJ, Moran JE, Niall JF, Ryan GB (1982) Segmental necrotising glomerulonephritis with antineutrophil antibody: possible arbovirus aetiology? Br Med J (Clin Res Ed) 285:606
7. van der Woude FJ, Rasmussen N, Lobatto S, Wiik A, Permin H, van Es LA et al (1985) Autoantibodies against neutrophils and monocytes: tool for diagnosis and marker of disease activity in Wegener's granulomatosis. Lancet 1:425–429
8. Falk RJ, Jennette JC (1988) Anti-neutrophil cytoplasmic autoantibodies with specific city for myelopreoxidase in patients with systemic vasculitis and idiopathic necrotizing and crescentic glomerulonephritis. N Engl J Med 318:1651–1657
9. Jennette JC, Hoidal JR, Falk RJ (1990) Specificity of anti-neutrophil cytoplasmic autoantibodies for proteinase 3. Blood 75:2263–2264
10. Wiik AS (2001) Clinical use of serological tests for antineutrophil cytoplasmic antibodies. What do the studies say? Rheum Dis Clin North Am 27:799–813
11. Csernok E, Moosig F (2014) Current and emerging techniques for ANCA detection in vasculitis. Nat Rev Rheumatol 10:494–501. doi:10.1038/nrrheum.2014.78
12. Sinico RA, Radice A (2014) Antineutrophil cytoplasmic antibodies (ANCA) testing: detection methods and clinical application. Clin Exp Rheumatol 32(3 Suppl 82):S112–S117

13. Savige J, Gillis D, Benson E, Davies D, Esnault V, Falk RJ et al (1999) International consensus statement on testing and reporting of Antineutrophil Cytoplasmic Antibodies (ANCA). Am J Clin Pathol 111:507–513
14. Silva de Souza AW (2015) Autoantibodies in systemic vasculitis. Front Immunol 6:184. doi:10.3389/fimmu.2015.00184
15. Schulte-Pelkum J, Radice A, Norman GL, López Hoyos M, Lakos G, Buchner C et al (2014) Novel clinical and diagnostic aspects of antineutrophil cytoplasmic antibodies. J Immunol Res 2014, 185416. doi:10.1155/2014/185416
16. Jennette JC, Falk RJ, Bacon PA, Basu N, Cid MC, Ferrario F et al (2013) 2012 revised International Chapel Hill consensus conference nomenclature of Vasculitides. Arthritis Rheum 65:1–11. doi:10.1002/art.37715
17. Hoffman GS, Specks U (1998) Antineutrophil cytoplasmic antibodies. Arthritis Rheum 41:1521–1537
18. Guillevin L, Durand-Gasselin B, Cevallos R, Gayraud M, Lhote F, Callard P et al (1999) Microscopic polyangiitis: clinical and laboratory findings in eighty-five patients. Arthritis Rheum 42:421–430
19. Sablé-Fourtassou R, Cohen P, Mahr A, Pagnoux C, Mouthon L, Jayne D et al (2005) French Vasculitis Study Group. Antineutrophil cytoplasmic antibodies and the Churg-Strauss syndrome. Ann Int Med 143:632–638
20. Rutgers A, Sanders JS, Stegeman CA, Kallenberg CG (2010) Pauci-immune necrotizing glomerulonephritis. Rheum Dis Clin North Am 36:559–572. doi:10.1016/j.rdc.2010.05.002
21. Rao JK, Weinberger M, Oddone EZ, Allen NB, Landsman P, Feussner JR (1995) The role of antineutrophil cytoplasmic antibody (c-ANCA) testing in the diagnosis of Wegener granulomatosis. A literature review and meta-analysis. Ann Int Med 123:925–932
22. Choi HK, Liu S, Merkel PA, Colditz GA, Niles JL (2001) Diagnostic performance of antineutrophil cytoplasmic antibody tests for idiopathic vasculitides: meta-analysis with a focus on antimyeloperoxidase antibodies. J Rheumatol 28:1584–1590
23. Mukhtyar C, Lee R, Brown D, Carruthers D, Dasgupta B, Dubey S et al (2009) Modification and validation of the Birmingham vasculitis activity score (version 3). Ann Rheum Dis 68:1827–1832. doi:10.1136/ard.2008.101279
24. Tomasson G, Grayson PC, Mahr AD, Lavalley M, Merkel PA (2012) Value of ANCA measurements during remission to predict a relapse of ANCA-associated vasculitis-a meta-analysis. Rheumatology (Oxford) 51:100–109. doi:10.1093/rheumatology/ker280
25. Lionaki S, Blyth ER, Hogan SL, Hu Y, Senior BA, Jennette CE et al (2012) Classification of antineutrophil cytoplasmic autoantibody vasculitides: the role of antineutrophil cytoplasmic autoantibody specificity for myeloperoxidase or proteinase 3 in disease recognition and prognosis. Arthritis Rheum 64:3452–3462. doi:10.1002/art.34562
26. Lyons PA, Rayner TF, Trivedi S, Holle JU, Watts RA, Jayne DR et al (2012) Genetically distinct subsets within ANCA-associated vasculitis. N Engl J Med 367:214–223. doi:10.1056/NEJMoa1108735
27. Sinico RA, Bottero P (2009) Churg-Strauss angiitis. Best Pract Res Clin Rheumatol 23:355–366. doi:10.1016/j.berh.2009.02.004
28. Hellmark T, Niles JL, Collins AB, McCluskey RT, Brunmark C (1997) Comparison of anti-GBM antibodies in sera with or without ANCA. J Am Soc Nephrol 8:376–385
29. Jayne DR, Marshall PD, Jones SJ, Lockwood CM (1990) Autoantibodies to GBM and neutrophil cytoplasm in rapidly progressive glomerulonephritis. Kidney Int 37:965–970
30. Yang G, Tang Z, Chen Y, Zeng C, Chen H, Liu Z et al (2005) Antineutrophil cytoplasmic antibodies (ANCA) in Chinese patients with anti-GBM crescentic glomerulonephritis. Clin Nephrol 63:423–428
31. Rutgers A, Slot M, van Paassen P, van Breda VP, Heeringa P, Tervaert JW (2005) Coexistence of anti-glomerular basement membrane antibodies and myeloperoxidase-ANCAs in crescentic glomerulonephritis. Am J Kidney Dis 46:253–262

32. Wiik A (2008) Drug-induced vasculitis. Curr Opin Rheumatol 20:35–39. doi:10.1097/BOR.0b013e3282f1331f
33. Pendergraft WF 3rd, Niles JL (2014) Trojan horses: drug culprits associated with antineutrophil cytoplasmic autoantibody (ANCA) vasculitis. Curr Opin Rheumatol 26:42–49. doi:10.1097/BOR.0000000000000014
34. Kain R, Matsui K, Exner M, Binder S, Schaffner G, Sommer EM et al (1995) A novel class of autoantigens of anti-neutrophil cytoplasmic antibodies in necrotizing and crescentic glomerulonephritis: the lysosomal membrane glycoprotein h-lamp-2 in neutrophil granulocytes and a related membrane protein in glomerular endothelial cells. J Exp Med 181:585–597
35. Kain R, Tadema H, McKinney EF, Benharkou A, Brandes R, Peschel A et al (2012) High prevalence of autoantibodies to hLAMP-2 in anti-neutrophil cytoplasmic antibody-associated vasculitis. J Am Soc Nephrol 23:556–566. doi:10.1681/ASN.2011090920
36. Kain R, Exner M, Brandes R, Ziebermayr R, Cunningham D, Alderson CA et al (2008) Molecular mimicry in pauci-immune focal necrotizing glomerulonephritis. Nat Med 14:1088–1096. doi:10.1038/nm.1874
37. Roth AJ, Brown MC, Smith RN, Badhwar AK, Parente O, Chung H et al (2012) Anti-LAMP-2 antibodies are not prevalent in patients with antineutrophil cytoplasmic autoantibody glomerulonephritis. J Am Soc Nephrol 23:545–555. doi:10.1681/ASN.2011030273
38. Peschel A, Basu N, Benharkou A, Brandes R, Brown M, Dieckmann R et al (2014) Autoantibodies to hLAMP-2 in ANCA-negative pauci-immune focal necrotizing GN. J Am Soc Nephrol 25:455–463. doi:10.1681/ASN.2013030320
39. Lazor R, Bigay-Gamé L, Cottin V, Cadranel J, Decaux O, Fellrath JM et al (2007) Groupe d'Etudes et de Recherche sur les Maladies Orphelines Pulmonaires (GERMOP); Swiss Group for Interstitial and Orphan Lung Diseases (SIOLD). Alveolar hemorrhage in anti-basement membrane antibody disease: a series of 28 cases. Medicine (Baltimore) 86:181–193
40. Boyce NW, Holdsworth SR (1986) Pulmonary manifestations of the clinical syndrome of acute glomerulonephritis and lung hemorrhage. Am J Kidney Dis 8:31–36
41. Greco A, Rizzo MI, De Virgilio A, Gallo A, Fusconi M, Pagliuca G et al (2015) Goodpasture's syndrome: a clinical update. Autoimmun Rev 14:246–253. doi:10.1016/j.autrev.2014.11.006
42. Kallenberg CG (2008) Anti-C1q autoantibodies. Autoimmun Rev 7:612–615. doi:10.1016/j.autrev.2008.06.006
43. Jara LJ, Navarro C, Medina G, Vera-Lastra O, Saavedra MA (2009) Hypocomplementemic urticarial vasculitis syndrome. Curr Rheumatol Rep 11:410–415
44. Grotz W, Baba HA, Becker JU, Baumgärtel MW (2009) Hypocomplementemic urticarial vasculitis syndrome: an interdisciplinary challenge. Dtsch Arztebl Int 106:756–763. doi:10.3238/arztebl.2009.0756
45. Jachiet M, Flageul B, Deroux A, Le Quellec A, Maurier F, Cordoliani F et al (2015) French Vasculitis Study Group. The clinical spectrum and therapeutic management of hypocomplementemic urticarial vasculitis: data from a French nationwide study of fifty-seven patients. Arthritis Rheumatol 67:527–534. doi:10.1002/art.38956
46. Belizna C, Duijvestijn A, Hamidou M, Tervaert JW (2006) Antiendothelial cell antibodies in vasculitis and connective tissue disease. Ann Rheum Dis 65:1545–1550
47. Domiciano DS, Carvalho JF, Shoenfeld Y (2009) Pathogenic role of anti-endothelial cell antibodies in autoimmune rheumatic diseases. Lupus 18:1233–1238. doi:10.1177/0961203309346654
48. Régent A, Ly KH, Blet A, Agard C, Puéchal X, Tamas N et al (2013) Contribution of antiferritin antibodies to diagnosis of giant cell arteritis. Ann Rheum Dis 72:1269–1270. doi:10.1136/annrheumdis-2012-202963
49. Große K, Witte T, Moosig F, Hoyer BF, Lansche C, Schmidt RE, Baerlecken NT (2014) Association of ferritin antibodies with Takayasu arteritis. Clin Rheumatol 33:1523–1526. doi:10.1007/s10067-014-2764-2
50. Tripathy NK, Sinha N, Nityanand S (2003) Anti-annexin V antibodies in Takayasu's arteritis: prevalence and relationship with disease activity. Clin Exp Immunol 134:360–364

51. Tripathy NK, Sinha N, Nityanand S (2003) Antimonocyte antibodies in Takayasu's arteritis: prevalence of and relation to disease activity. J Rheumatol 30:2023–2026
52. Chauhan SK, Tripathy NK, Nityanand S (2006) Antigenic targets and pathogenicity of anti-aortic endothelial cell antibodies in Takayasu arteritis. Arthritis Rheum 54:2326–2333
53. Davin JC, Ten Berge IJ, Weening JJ (2001) What is the difference between IgA nephropathy and Henoch-Schönlein purpura nephritis? Kidney Int 59:823–834
54. Yang YH, Chuang YH, Wang LC, Huang HY, Gershwin ME, Chiang BL (2008) The immunobiology of Henoch-Schönlein purpura. Autoimmun Rev 7:179–184. doi:10.1016/j.autrev.2007.11.012
55. Mannik M, Nardella FA (1985) IgG rheumatoid factors and self-association of these antibodies. Clin Rheum Dis 11:551–572

Chapter 5
Pathogenic Role of ANCA in Small Vessel Inflammation and Neutrophil Function

Giuseppe A. Ramirez and Angelo A. Manfredi

Abstract Once thought as the mere epiphenomenon of a largely unknown inflammatory process, Anti-Neutrophil Cytoplasmic Antibodies (ANCA) have gained growing attention in recent years as key players in the pathogenesis of small vessel inflammation. Genetic studies as well as animal and *in vitro* models suggest that multiple factors induce the generation of ANCA, which in turn have the ability to promote neutrophil activation and progression towards neutrophil extracellular trap (NET) development. A vicious circle emerges as NETs endowed with ANCA targets promote ANCA generation. Besides consolidated pathogenic data, less is yet known about the precise role of ANCA in the clinical practice.

Keywords Small vessel vasculitis • ANCA • Neutrophils • NET • Autophagy

5.1 Introduction

Anti-Neutrophil Cytoplasmic Antibodies (ANCA) constitute a set of autoantibodies with the ability to recognize constituents of the oxidative and digestive machinery of neutrophils, such as myeloperoxidase (MPO), proteinase 3 (PR3) or lysosomal-associated membrane protein 2 (LAMP2). ANCA have been linked to the so-called pauci-immune small-vessel vasculitides including granulomatosis with polyangiitis (GPA, formerly Wegener's granulomatosis), eosinophilic granulomatosis with polyangiitis (EGPA, formerly Churg-Strauss' syndrome) and microscopic polyangiitis (MPA). However, ANCA can be detected in other inflammatory conditions such as systemic lupus erythematosus (SLE), inflammatory bowel diseases and other small vessel vasculitides. In the setting of GPA, EGPA and MPA, the detection

G.A. Ramirez (✉)
IRCCS Ospedale San Raffaele and Università Vita-Salute San Raffaele, 20132 Milan, Italy
e-mail: ramirez.giuseppealvise@hsr.it

A.A. Manfredi
Università Vita-Salute San Raffaele, Via Olgettina 60, 20132 Milan, Italy

San Raffaele Scientific Institute, Via Olgettina 60, 20132 Milan, Italy

F. Dammacco et al. (eds.), *Systemic Vasculitides: Current Status and Perspectives*, DOI 10.1007/978-3-319-40136-2_5

of ANCA has been historically considered as an epiphenomenon of a deeper pathogenic process. However, this paradigm is being progressively overturned in recent years. In fact, increasing evidence suggests a direct involvement of ANCA in the development of vessel inflammation. In addition, genetic surveys indicate that inherited risk factors for the development of vasculitis could better correlate with the serological profile rather than with the clinical shape, maybe suggesting the opportunity to classify patients as having anti-PR3/MPO-associated or ANCA-negative vasculitis [1].

5.2 Genetics

Small candidate-gene and larger genome-wide association studies (GWAS) are progressively revealing the presence of specific genetic backgrounds underlying the development of ANCA-associated vasculitides (AAV), besides the known contribution of environmental triggers such as bacteria (e.g. *Staphylococcus aureus* or *Klebsiella pneumoniae*), silica dust or drugs (e.g. levamisole or propylthiouracil) [2]. Two large GWAS were performed in European and American Caucasian patients with GPA and MPA (while a first EGPA-GWAS is ongoing). These studies revealed a strong association between AAV and six single nucleotide polymorphisms (SNPs), located at chromosome 5, 6,14 and 19, on the loci encoding for semaphorin 6A, some HLA variant (HLA-DPB1, HLA-DPA1), PR3 and its inhibitor α1-antitrypsin [1, 3]. In the European GWAS, PR3 and α1-antitrypsin gene variants were more strongly associated with GPA and in particular with anti-PR3-positive vasculitis, whereas a subset-analysis of patients with anti-MPO-positive vasculitis unmasked an unexpected association with the HLA-DQ locus [1]. These data are of particular interest, since they are consistent with the abnormal constitutive expression of PR3 on neutrophils in GPA patients [4] and suggest a pathogenic model in which altered PR3 expression and impaired PR3 functional suppression by α1-antitrypsin prompt the generation of anti-PR3 antibodies through permissive HLA presentation and eventually lead to AAV development. An association with HLA variants was reported in both studies. Candidate-gene approaches have also revealed significant associations with the protein tyrosine phosphatase N22 (PTPN22), the cytotoxic T lymphocyte antigen 4 (CTLA4) and with the IgG receptor family (FcγR) [5]. PTPN22 encodes for a crucial protein for the activation of T and B cells and can drive the immune response towards increased antibody production, if dysfunctional. CTLA4 is a known T cell-inhibitor with a pivotal function in immune-regulation. FcγRs mediate the recognition of antibodies by circulating leukocytes and, if mutated, could exacerbate the activating effects of ANCA towards neutrophils and monocytes (see below). Taken together, these data seem to indicate that strictly autoimmune mechanisms, possibly driven by abnormalities of specific autoantigens and leading to the generation of pathogenic autoantibodies, predominate in the development of ANCA-associated vasculitides [5].

5.3 Murine and *In Vitro* Models

Most studies performed on animal models of AAV had the aim of replicating the development of the disease by immunizing mice with neutrophil cytoplasmic antigens (NCA). In contrast to the recent data provided by genetic studies, the earliest and strongest evidence for a role of NCA and ANCA in the pathogenesis of AAV came from studies involving MPO. In particular, MPO-deficient mice immunized with murine MPO and later transplanted with the bone marrow of MPO-positive wild-type mice developed a pauci-immune necrotizing crescentic glomerulonephritis resembling that observed in severe cases of human AAV [6]. A recent study involving the use of the proteasome inhibitor bortezomib to deplete MPO-specific plasma cells, further emphasized the pathogenic impact of anti-MPO ANCA in the development of experimental pauci-immune glomerulonephritis [7]. Attempts to reproduce the results obtained with MPO by immunizing mice with PR3 led to the generation of anti-PR3 antibodies. However, unexpectedly, no clinical evidence of disease was observed, except for a single report describing glomerulonephritis development in NOD-SCID but not RAG-1-\- immunodeficient mice after transfer of splenocytes from PR3-immunized mice [8]. The reason for this failure could be found in the lower homology between murine and human PR3, when compared to MPO, and to the fact that murine PR3 is not expressed on the surface of unstimulated neutrophils in contrast to human PR3. Indeed, transfer of ANCA (with known capacity to activate neutrophils *in vitro*) from human patients with severe nephropulmonary disease into humanized mice replicated AAV phenotype [9]. Another research line focused on the highly glycosylated protein LAMP2 and on its bacterial analogue FimH (a component of the fimbriae of Gram negative bacteria). Rats immunized with recombinant FimH developed antibodies that cross-reacted with rat and human LAMP2 and generated positive ANCA immunofluorescence. In addition, 9/10 rats developed focal crescentic glomerulonephritis and 2/10 hemorrhagic pulmonary vasculitis. Histological analyses provided evidence of a direct role of anti-FimH/LAMP2 antibodies in endothelial and glomerular injury [10]. Furthermore, this study provided a formal support to the idea that molecular mimicry towards components of common pathogens has a major role in the development of ANCA. Additional studies investigated more deeply the ways by which ANCA coordinate and sustain the immune responses in small vessel inflammatory diseases.

5.3.1 ANCA Activate Neutrophils and Generate NETs

Neutrophils primed by Toll-like receptor stimulation, complement anaphylotoxins or inflammatory cytokines such as tumor necrosis factor (TNF), express NCA on the cell surface and can interact with ANCA to become fully activated and undergo reactive oxygen species (ROS) production and degranulation (Fig. 5.1) [11–13].

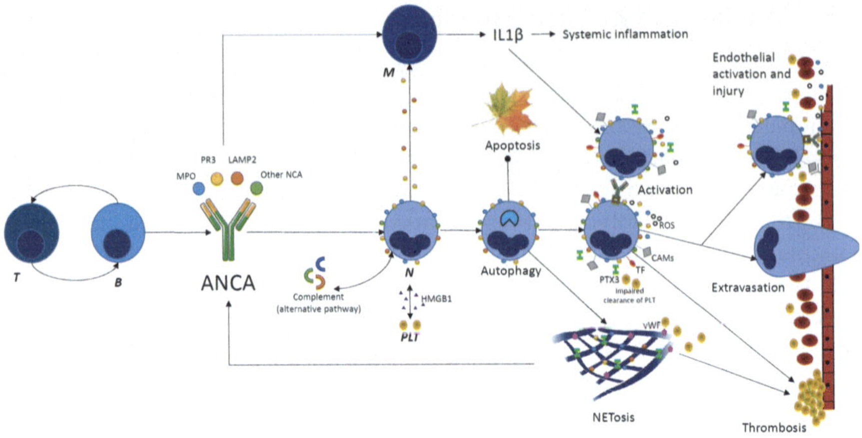

Fig. 5.1 ANCA in the pathogenesis of small vessel vasculitides
Genetic and environmental triggers prime neutrophils to become activated and prompt the reciprocal activation of T and B lymphocytes to sustain systemic inflammation. B cells in particular are responsible for the generation of ANCA, which can recognize a set of neutrophil cytoplasmic antigens (NCA) such as MPO, PR3 and LAMP2. Primed neutrophils, expressing NCA on the cell surface interact with ANCA and undergo further activation. The alternative pathway of complement activation plays a critical role in priming neutrophils to ANCA stimulation; on the other hand, neutrophils promote the activation of complement in a feed-forward loop. Neutrophils stimulated by ANCA become resistant to apoptosis through the adoption of autophagic programs and eventually undergo NETosis or express their effector function at the level of small blood vessels. Extensive NETosis favors itself the generation of ANCA since NCA are widely captured within chromatin threads and can be productively taken up by myeloid dendritic cells. Under inflammatory conditions, neutrophils interact also productively with activated platelets: in this setting the high mobility box 1 protein (HMGB1), released by platelets, plays a pivotal role in prompting the progression of neutrophils towards autophagy and NETosis. Activated neutrophils of patients with systemic vasculitides showed also limited capacity to phagocytose activated platelets, maybe as the consequence of extensive release of PTX3 from neutrophil granules. This phenomenon, together with the ability of NETs to recruit pro-thrombotic factors such as von Willebrand factor (vWF) and to segregate some circulating anticoagulants, prompts the development of the so-called immuno-thrombosis. Neutrophils receive additional activating signals through the recognition of NCA-ANCA complexes on the surface of other neutrophils by the Fc-γ-receptor (FcγR). Extensive neutrophil degranulation causes direct injury to the endothelial walls and facilitate endothelial activation. Activated endothelial cells, decorated with neutrophil-derived NCA, prompt further the activation of neutrophils, facilitate cell adhesion and eventually extravasation and further contribute to the development of thrombosis. ANCA also interact with monocytes facilitating the generation of IL1β and systemic inflammation; in this setting neutrophil-derived proteases are also required to ensure the activation of IL1β. Abbreviations: *B, T* B, T lymphocyte, *CAMs* cell adhesion molecules, *HMGB1* high mobility group protein 1, *LAMP2* lysosomal-associated membrane protein 2 *M* monocyte, *MPO* myeloperoxidase, *N* neutrophil, *NCA* neutrophil cytoplasmic antigens, *NET* neutrophil extracellular traps, *PLT* platelets, *PR3* proteinase 3, *PTX3* pentraxin 3, *ROS* reactive oxygen species, *TF* tissue factor, *vWF* von Willebrand Factor

Reciprocal recognition of NCA-ANCA complexes on the neutrophil cellular membrane involves the expression of FcγR. Downstream FcγR, the signalling cascade is dependent on the gamma isoform of phosphoinositol-3 kinase, at least in experimental models [14]. Intact function of the alternative pathway of complement activation seems also important for neutrophil priming and for the full establishment of ANCA-induced glomerulonephritis in mice [13]. ANCA-induced respiratory burst and degranulation are directly responsible for endothelial and tissue injury at sites of inflammation. In addition, they cause endothelial activation, which promotes thrombosis. Recent studies highlighted the ability of ANCA to induce neutrophils to form neutrophil extracellular traps (NETs) and elucidated the role of ANCA-induced NETosis in the pathogenesis of AAV. NETs are threads of decondensed neutrophil-derived chromatin, enriched with microbicidal and pro-thrombotic moieties including NCA. Under physiological conditions, NETs enhance the host response by promoting pathogens capture and immuno-thrombosis. Primed neutrophils challenged with ANCA from patients with AAV have been shown to undergo extensive NETosis *in vitro* [15, 16]. Furthermore, signs of NETosis were detected at sites of inflammation in kidney biopsies [15]. Notably, NETosis-prone subsets of neutrophils have been associated with vasculitic manifestations in other inflammatory diseases, such as SLE [17]. Subsequent studies revealed that extended neutrophil survival and progression of neutrophils towards NETosis are granted by the implementation of an autophagic program, which can in turn be induced by ANCA stimulation [18]. Evidence from our group suggests that this process could also be attributable to HMGB1 signalling from activated platelets [19], which are increased in number and defectively cleared out of the bloodstream by leukocytes in AAV patients [20]. Activated platelets probably also cooperate with ANCA-induced NETs in causing thrombosis, a known complication of AAV [21].

5.3.2 Neutrophil Activation and NETosis Induce the Generation of ANCA

The generation of NETs after ANCA stimulation increases the burden of autoantigens and natural adjuvants exposed to the immune system. In particular, MPO and PR3 could be taken up from NETs by myeloid dendritic cells and promote the development of ANCA through stimulation of specific T and B cells (Fig. 5.1) [22]. In addition, the serum of patients with anti-MPO-positive MPA impairs NET degradation, possibly through antibodies interfering with DNAse activity, as described in SLE [23].

5.3.3 Activation of Monocytes and Macrophages in Association with ANCA

Human and murine anti-MPO and anti-PR3 antibodies can also interact with monocytes and induce interleukin 1β (IL1β) production and secretion. This process is apparently dependent on the presence of neutrophil-derived serine proteases. The release of IL1β is probably responsible for the systemic inflammatory symptoms associated to AAV and is required for the development of glomerular inflammation in mouse models [24]. Additional sources of IL1β in AAV could be constituted by tissue-resident macrophages challenged with NETs generated at sites of inflammation or by activated neutrophils [25]. Extensive release of IL1β has also a feed-forward effect on neutrophil priming/activation (Fig. 5.1).

5.4 Clinical Significance

Besides the growing evidence about the role of ANCA in the pathogenesis of small vessel inflammation, the place of ANCA testing in the clinical practice is still controversial. ANCA detection by immunofluorescence and quantitative methods (such as enzyme-linked or chemiluminescence assays) for the specific recognition of anti-MPO, anti-PR3 and anti-LAMP2 antibodies have an established role in the diagnosis of AAV. On the other hand, the usefulness of ANCA to monitor disease activity, predict relapse and measure response to therapy is less clear [26]. Anti-LAMP2 antibodies, in particular, have been shown to be very sensitive to the initiation of immunosuppressive drugs [27]. Recent studies suggest that the employment of novel technologies with full native antigens embedded in the testing plates could better correlate with disease activity, especially for anti-PR3 antibodies and for renal disease monitoring [28, 29]. The ongoing MAINRITSAN 2 study (NCT01731561) has been designed to elucidate the opportunity to modulate rituximab treatment on the basis of ANCA and CD19 lymphocytes count versus systematic infusions.

References

1. Lyons PA, Rayner TF, Trivedi S, Holle JU, Watts RA, Jayne DR et al (2012) Genetically distinct subsets within ANCA-associated vasculitis. N Engl J Med 367(3):214–223. doi:10.1056/NEJMoa1108735
2. Flint SM, McKinney EF, Smith KG (2015) Emerging concepts in the pathogenesis of antineutrophil cytoplasmic antibody-associated vasculitis. Curr Opin Rheumatol 27(2):197–203. doi:10.1097/BOR.0000000000000145
3. Xie G, Roshandel D, Sherva R, Monach PA, Lu EY, Kung T et al (2013) Association of granulomatosis with polyangiitis (Wegener's) with HLA-DPB1*04 and SEMA6A gene variants: evidence from genome-wide analysis. Arthritis Rheum 65(9):2457–2468. doi:10.1002/art.38036

4. Witko-Sarsat V, Lesavre P, Lopez S, Bessou G, Hieblot C, Prum B et al (1999) A large subset of neutrophils expressing membrane proteinase 3 is a risk factor for vasculitis and rheumatoid arthritis. J Am Soc Nephrol 10(6):1224–1233
5. Alberici F, Martorana D, Bonatti F, Gioffredi A, Lyons PA, Vaglio A (2014) Genetics of ANCA-associated vasculitides: HLA and beyond. Clin Exp Rheumatol 32(3 Suppl 82):S90–S97
6. Schreiber A, Xiao H, Falk RJ, Jennette JC (2006) Bone marrow-derived cells are sufficient and necessary targets to mediate glomerulonephritis and vasculitis induced by anti-myeloperoxidase antibodies. J Am Soc Nephrol 17(12):3355–3364. doi:10.1681/ASN.2006070718
7. Bontscho J, Schreiber A, Manz RA, Schneider W, Luft FC, Kettritz R (2011) Myeloperoxidase-specific plasma cell depletion by bortezomib protects from anti-neutrophil cytoplasmic autoantibodies-induced glomerulonephritis. J Am Soc Nephrol 22(2):336–348. doi:10.1681/ASN.2010010034
8. Primo VC, Marusic S, Franklin CC, Goldmann WH, Achaval CG, Smith RN et al (2010) Anti-PR3 immune responses induce segmental and necrotizing glomerulonephritis. Clin Exp Immunol 159(3):327–337. doi:10.1111/j.1365-2249.2009.04072.x
9. Little MA, Al-Ani B, Ren S, Al-Nuaimi H, Leite M Jr, Alpers CE et al (2012) Anti-proteinase 3 anti-neutrophil cytoplasm autoantibodies recapitulate systemic vasculitis in mice with a humanized immune system. PLoS One 7(1), e28626. doi:10.1371/journal.pone.0028626
10. Kain R, Exner M, Brandes R, Ziebermayr R, Cunningham D, Alderson CA et al (2008) Molecular mimicry in pauci-immune focal necrotizing glomerulonephritis. Nat Med 14(10):1088–1096. doi:10.1038/nm.1874
11. Falk RJ, Terrell RS, Charles LA, Jennette JC (1990) Anti-neutrophil cytoplasmic autoantibodies induce neutrophils to degranulate and produce oxygen radicals in vitro. Proc Natl Acad Sci U S A 87(11):4115–4119
12. Holle JU, Windmoller M, Lange C, Gross WL, Herlyn K, Csernok E (2013) Toll-like receptor TLR2 and TLR9 ligation triggers neutrophil activation in granulomatosis with polyangiitis. Rheumatology 52(7):1183–1189. doi:10.1093/rheumatology/kes415
13. Schreiber A, Xiao H, Jennette JC, Schneider W, Luft FC, Kettritz R (2009) C5a receptor mediates neutrophil activation and ANCA-induced glomerulonephritis. J Am Soc Nephrol: JASN 20(2):289–298, ASN.2008050497 [pii]10.1681/ASN.2008050497
14. Schreiber A, Rolle S, Peripelittchenko L, Rademann J, Schneider W, Luft FC et al (2010) Phosphoinositol 3-kinase-gamma mediates antineutrophil cytoplasmic autoantibody-induced glomerulonephritis. Kidney Int 77(2):118–128. doi:10.1038/ki.2009.420
15. Kessenbrock K, Krumbholz M, Schonermarck U, Back W, Gross WL, Werb Z et al (2009) Netting neutrophils in autoimmune small-vessel vasculitis. Nat Med 15(6):623–625, nm.1959 [pii]10.1038/nm.1959
16. Zhang Y, Shi W, Tang S, Li J, Yin S, Gao X et al (2013) The influence of cathelicidin LL37 in human anti-neutrophils cytoplasmic antibody (ANCA)-associated vasculitis. Arthritis Res Ther 15(5):R161. doi:10.1186/ar4344
17. Villanueva E, Yalavarthi S, Berthier CC, Hodgin JB, Khandpur R, Lin AM et al (2011) Netting neutrophils induce endothelial damage, infiltrate tissues, and expose immunostimulatory molecules in systemic lupus erythematosus. J Immunol 187(1):538–552. doi:10.4049/jimmunol.1100450
18. Tang S, Zhang Y, Yin S, Gao X, Shi W, Wang Y et al (2015) Neutrophil extracellular trap formation is associated with autophagy-related signaling in ANCA-associated vasculitis. Clin Exp Immunol 180(3):408–18. doi:10.1111/cei.12589
19. Maugeri N, Campana L, Gavina M, Covino C, De Metrio M, Panciroli C et al (2014) Activated platelets present high mobility group box 1 to neutrophils, inducing autophagy and promoting the extrusion of neutrophil extracellular traps. J Thromb Haemost 12(12):2074–2088
20. Maugeri N, Rovere-Querini P, Evangelista V, Godino C, Demetrio M, Baldini M et al (2012) An intense and short-lasting burst of neutrophil activation differentiates early acute myocardial infarction from systemic inflammatory syndromes. PLoS One 7(6), e39484. doi:10.1371/journal.pone.0039484PONE-D-12-03569[pii]

21. Allenbach Y, Seror R, Pagnoux C, Teixeira L, Guilpain P, Guillevin L et al (2009) High frequency of venous thromboembolic events in Churg-Strauss syndrome, Wegener's granulomatosis and microscopic polyangiitis but not polyarteritis nodosa: a systematic retrospective study on 1130 patients. Ann Rheum Dis 68(4):564–567. doi:10.1136/ard.2008.099051
22. Sangaletti S, Tripodo C, Chiodoni C, Guarnotta C, Cappetti B, Casalini P et al (2012) Neutrophil extracellular traps mediate transfer of cytoplasmic neutrophil antigens to myeloid dendritic cells toward ANCA induction and associated autoimmunity. Blood 120(15):3007–3018. doi:10.1182/blood-2012-03-416156
23. Nakazawa D, Shida H, Tomaru U, Yoshida M, Nishio S, Atsumi T et al (2014) Enhanced formation and disordered regulation of NETs in myeloperoxidase-ANCA-associated microscopic polyangiitis. J Am Soc Nephrol: JASN 25(5):990–997. doi:10.1681/ASN.2013060606
24. Schreiber A, Pham CT, Hu Y, Schneider W, Luft FC, Kettritz R (2012) Neutrophil serine proteases promote IL-1beta generation and injury in necrotizing crescentic glomerulonephritis. J Am Soc Nephrol: JASN 23(3):470–482. doi:10.1681/ASN.2010080892
25. Kahlenberg JM, Carmona-Rivera C, Smith CK, Kaplan MJ (2013) Neutrophil extracellular trap-associated protein activation of the NLRP3 inflammasome is enhanced in lupus macrophages. J Immunol 190(3):1217–1226. doi:10.4049/jimmunol.1202388
26. Chen M, Kallenberg CG (2010) ANCA-associated vasculitides – advances in pathogenesis and treatment. Nat Rev Rheumatol 6(11):653–664, nrrheum.2010.158 [pii] 10.1038/nrrheum.2010.158
27. Kain R (2013) L29. Relevance of anti-LAMP-2 in vasculitis: why the controversy. Presse Med 42(4 Pt 2):584–588. doi:10.1016/j.lpm.2013.01.029
28. Schulte-Pelkum J, Radice A, Norman GL, Lomicronpez Hoyos M, Lakos G, Buchner C et al (2014) Novel clinical and diagnostic aspects of antineutrophil cytoplasmic antibodies. J Immunol Res 2014, 185416. doi:10.1155/2014/185416
29. Kemna MJ, Damoiseaux J, Austen J, Winkens B, Peters J, van Paassen P et al (2014) ANCA as a predictor of relapse: useful in patients with renal involvement but not in patients with nonrenal disease. J Am Soc Nephrol: JASN 26(3):537–42. doi:10.1681/ASN.2013111233

Part II
Primary Systemic Vasculitides

Chapter 6
Takayasu Arteritis: When Rarity Maintains the Mystery

Enrico Tombetti, Elena Baldissera, Angelo A. Manfredi, and Maria Grazia Sabbadini

Abstract Takayasu arteritis (TA) is a rare vasculitis involving the large arteries and often recognized with significant delay. The etiology of TA is unknown. Young women are more frequently affected, with significant mortality and morbidity. The clinical picture is heterogeneous and mainly determined by arterial involvement, in terms of arterial steno-occlusions or dilatations. Unfortunately, TA is still an orphan disease, and many issues remain open. Multiple reasons limit the studies in this field: the rarity of TA, the difficulty in obtaining tissue specimens, the lack of informative animal models, the need of accurate outcome measures and of adequate techniques to evaluate TA proteiform activity. Here we review principles of TA management and discuss some major open issues, underlining the need to overcome the simplistic view of "TA as a purely inflammatory condition", to recognise the important role of arterial remodelling as a consequence of arterial inflammation and injury.

Keywords Large vessel vasculitis • Takayasu arteritis • Pathogenesis of Takayasu arteritis • Disease activity in Takayasu arteritis • Therapy of Takayasu arteritis • Arterial remodelling

6.1 Introduction to Large-Vessel Vasculitides

Diseases of the arterial tree represent the first cause of mortality and morbidity in the Western countries. Within this group of diseases, atherosclerosis is by far the most frequent, followed by arterial thromboembolism, aneurismatic disease and congenital malformations. However, these conditions do not account for the whole entirety of arterial diseases, as primarily inflammatory diseases represent a smaller

E. Tombetti (✉) • A.A. Manfredi • M.G. Sabbadini
Università Vita-Salute San Raffaele, Via Olgettina 60, 20132 Milan, Italy

San Raffaele Scientific Institute, Via Olgettina 60, 20132 Milan, Italy
e-mail: Tombetti.enrico@hsr.it

E. Baldissera
San Raffaele Scientific Institute, Via Olgettina 60, 20132 Milan, Italy

F. Dammacco et al. (eds.), *Systemic Vasculitides: Current Status and Perspectives*, DOI 10.1007/978-3-319-40136-2_6

but clearly distinct group. In these conditions, inflammation can either affect the arterial wall, as in arteritis, or be limited to the *tunica adventitia* and the surrounding perivascular connective tissues, as in peri-arteritis. Arteritis and peri-arteritis can be further distinguished in different entities, rare or very rare diseases on themselves but not as a group.

6.2 Large-Vessel Vasculitides

Vasculitides are distinguished according to the type and size of the vessels they mainly involve [1]. *Large-vessel vasculitides* (LVVs) are rare conditions characterized by idiopathic inflammation within the wall of large-sized arteries (lumen >5 mm) [2]. The essential features of LVVs are discrete arteritic lesions that tend to involve all the three *tunicas*, resulting in wall thickening. Lesions can cause steno-occlusions or ectasias/aneurysms, and have a patchy nature with a highly variable spatial distribution. Takayasu arteritis (TA) and giant cells arteritis (GCA) are by far the two most frequent large-vessel vasculitides of the adult. *Isolated aortitis* is another condition that may frequently represent a localized variant of GCA and TA and it is still unclear whether it constitutes a separate disease [3–5]. TA typically affects young women (<40 years-old), but it has been described from childhood [6] to the seventh decade [7]. Inflammation primarily localizes in large conductance arteries, such as the aorta and its main branches [8, 9], usually with a focal and symmetrical pattern [10].

On the other hand, GCA affects patients older than 50 years and is associated to *polymyalgia rheumatica* (PMR) in about half of subjects. Moreover, GCA can involve also medium-sized arteries, resulting in a "cephalic" phenotype, characterized by steno-occlusion of the extra-cranial branches of the external carotid arteries. A rarer "systemic" GCA phenotype, sometimes indistinguishable from elderly-onset TA, has been only recently recognised [11]. Cranial and systemic phenotypes sometimes coexist and can both combine with ascending aortitis. Although there is some overlap between TA and GCA, the prototypical clinical and epidemiologic features differ between the two [12]. This chapter will describe TA, highlighting some of the many unresolved issues that maintain the mystery around this rare but clinically significant condition.

6.3 Takayasu Arteritis

6.3.1 Arterial Involvement

As in other LVVs, arterial inflammation in TA has a patchy distribution. The arteries involved by vasculitis are highly variable from patient to patient, and geographical differences in the pattern of arterial involvement have been described [13]. As a general rule, the subclavian arteries are the most frequently involved sites (about 80 % of patients), followed by the carotids, thoraco-abdominal aorta and celiac trunk [8, 10, 13]. Involvement of the pulmonary [14] or coronary [15] arteries is not rare.

Also, the extent of arterial disease is highly heterogeneous, ranging from a single involved arterial district to a widespread disease affecting most of the large arteries. Classifications of the pattern of arterial involvement, not reflecting the actual disease burden, have been proposed [16].

Arterial lesions result in thickening of the vessel wall, which is usually (but not always) associated with alterations in the luminal calibre (Figs. 6.1 and 6.2): steno-occlusions is the most frequent evolution, while ectasias or aneurysms occur in about 10–25 % of patients [8, 13].

6.3.2 Diagnosis

One of the main issues about LVVs and peri-arteritis is their frequent misdiagnosis for their more prevalent cognate non-inflammatory conditions. As they require a specific management, misdiagnosis has frequently serious consequences [17]. Table 6.1 highlights red flags that suggest the need of a more exhaustive differential diagnosis when facing arterial diseases. Identification of typical arterial wall thickening has frequently high importance for diagnosing LVVs.

The diagnosis of TA is further hindered by its rarity and the frequent absence of systemic acute-phase response. As for many other autoimmune idiopathic conditions, there is no diagnostic test for TA. Classification criteria for TA have been proposed [18], but they have a low sensitivity: clinical diagnosis of TA should not be limited to their fulfilment, and should be determined on clinical judgement.

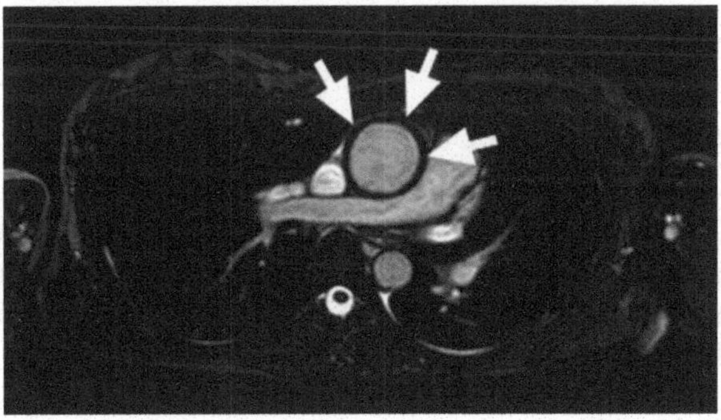

Fig. 6.1 MRI axial T2 weighted image of a young lady with TA. An involvement of the ascending aorta with mild ectasia and typical wall thickening (*white arrows*) is here demonstrated

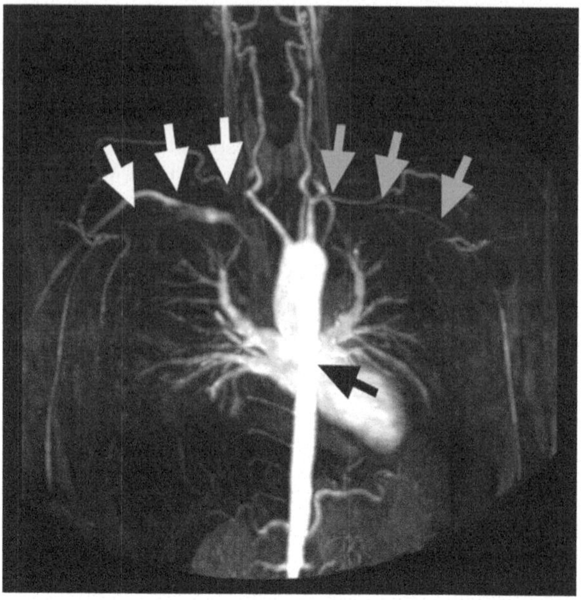

Fig. 6.2 Coronal Maximum Intensity Projection (MIP) of the thoracic and cervical arteries, obtained with Contrast Enhanced Magnetic Resonance Angiography. Note the occlusion of the right subclavian-axillary axis (*white arrows*), the occlusion of the left subclavian artery with long stenosis of the left axillary artery (*grey arrows*), and the mild stenosis of the descending thoracic aorta (*black arrow*)

Table 6.1 Red flags for infective or inflammatory arteritis or peri-arteritis

Systemic inflammatory responses
Presence of other chronic inflammatory conditions/infections (ie, TB, syphilis, etc.)
Circumferential thickening of the arterial wall/peri-arterial structures
Co-occurrence of steno-occlusive disease and dilative disease
Distribution within the arterial tree atypical for atherosclerosis

6.3.3 Disease Course and Clinical Picture

TA natural history comprises an active phase with highly variable duration and course that eventually subsides with residual arterial scarring. TA can range from a monophasic illness in 20 % of patients to a chronic-relapsing condition that lasts up to many years [8, 13, 19].

Morbidity and mortality are significant in TA [13, 20], and the clinical picture (Table 6.2) is usually dominated by (i) constitutional symptoms, sometimes configuring a classical systemic inflammatory syndrome. As said, the absence of inflammatory features is not rare, possibly hampering TA diagnosis; (ii) inflammatory arterial signs (tenderness over actively inflamed arteries); (iii) arterial insufficiency (anisosphygmia, end-organ ischemias, systemic hypertension, precapillary pulmo-

Table 6.2 Clinical manifestations of TA

Constitutional	Arterial inflammation	Arterial insufficiency	Arterial dilation	Complications of the previous
Fatigue	Carotidodynia	Arterial bruits	Aneurysms	Accelerated atherosclerosis
Fever	Subclaviodynia	Anisosphygmia	Aortic valve insufficiency	Cardiomyopathy (hypertensive or valvular)
Night sweats		End-organ ischemia (limbs, encephalon, eyes, kidney, bowel, heart)	Arterial rupture	Post-capillary pulmonary hypertension
Arthralgias		Pulmonary precapillary hypertension	Arterial dissection	Chronic renal failure (ischemic or hypertensive)
Arthritis		Systemic hypertension due to baroreflex dysfunction or renal artery stenosis	Thrombotic emboli	
Myalgias		Systolic systemic hypertension due to reduced total arterial compliance		
Weight loss (not otherwise explained)				

nary hypertension); (iv) dilative arterial disease and (v) complication of the above (cardiomyopathy, accelerated atherosclerosis, chronic renal failure). Extra-arterial inflammation, involving the joints or the heart, can at times occur during active phases. Other inflammatory conditions, most notably inflammatory bowel diseases, sarcoidosis, or relapsing polychondritis [21], can associate to TA, often with an independent onset and course.

6.3.4 *Histology and Pathogenesis*

TA is a pan-arteritis: histology of active lesions typically shows adventitial thickening, degeneration with focal leukocytic infiltration of the *media* with frequent granulomatous organisation and intimal hyperplasia [22, 23]. Laminar necrosis of the *media* is frequent. Inflammatory infiltrate is most intense in proximity to *vasa vasorum* and comprises macrophages, $CD4^+$ and $CD8^+$ T cells, γδ T cells, NK cells, dendritic cells (DCs) and rare B-cells. Granulocytes rarely infiltrate arterial lesions. Disruption of the internal elastic lamina is typical. Multinucleated giant cells can be present at the *media-intima* border.

Healed, burn-out lesions are characterized by scarring with fibrosis and an attempt of repair in the *media*, with muddled regeneration of vascular smooth muscle cells (VSMCs) and loss of elastic fibres.

TA pathogenesis is still highly mysterious (reviewed in [24]). TA is an idiopathic condition in which a heterogeneous genetic predisposition combines with unknown environmental factors to cause the vasculitic phenotype. Current models are based on histologic studies, the observed response to immunological therapies and studies on GCA. There are no disease-specific animal models and studies are further complicated by difficulties in obtaining tissues from living patients. Weyand and colleagues have focused on DCs, T-cells and macrophages in GCA and have developed a chimeric mouse model of GCA by temporal artery implants into immunodeficient mice [25–34]. Some results obtained in this model might have implications for TA as well.

Steno-occlusions are believed to derive from intimal hyperplasia and adventitial fibrosis. According to the currently accepted model [24], steno-occlusive arterial remodelling is caused by the local production of growth factors, including platelet-derived growth-factor (PDGF), vascular endothelial growth factor (VEGF) and transforming growth factor- (TGF)-β [9, 12, 35]. On the contrary, dilatory involvement results from medial degeneration and the action of proteases including matrix metalloproteinases. The macrophage-derived pivotal inflammatory cytokines, such as Tumor Necrosis Factor-α (TNFα) and interleukin-6 (IL6), are believed to be responsible for systemic inflammatory symptoms, but their role within the arterial wall lesions is not clear.

Arteries have intrinsic mechanisms to protect the *media* and the *intima* from potentially dangerous immune responses. These protective mechanisms apparently fail in LVVs, and leukocytes get into the arterial wall via *vasa vasorum* [22]. Vascular DCs, γδ- and αβ T-lymphocytes are supposed to be responsible for the vasculitic process. Recently, roles also for B cells [36] and, within $CD4^+$ T cells population, for Th1 and Th17 lymphocytes have been proposed [37]. Under lymphocyte drive, macrophages and the stromal components of the arterial wall would cause the injuries and the remodelling that are observed histologically and that are responsible for the clinical manifestations [24].

6.3.5 Conundrums in TA Assessment

Therapeutic decisions regarding TA patients are complicated by the difficulties in assessing disease activity [35], as a consequence of poor knowledge of TA pathogenesis. There is no wide agreement on clinical measures to evaluate activity or damage in TA.

In clinical practice, prevention of the progression of the arterial involvement represents an important therapeutic goal [35, 38], although there is no evidence-based definition of clinically-relevant arterial progression. However, it is currently unclear how to identify patients undergoing progression before it has significantly occurred:

acute-phase reactants and multi-item activity criteria do not accurately correlate with arterial progression, probably because both of them detect mainly the systemic inflammation associated with TA. Indeed, arterial progression appears to be the result of different processes, of which only some associated with systemic inflammatory responses. The importance to overcome a "systemic inflammation only" view of TA is remarked by the observation of arterial progression in up to 60 % of patients believed to be in remission [8] and in those on biologic agents blocking IL-6 [39]. Local inflammatory pathways and non-inflammatory remodelling of the arterial wall likely contribute to progression [38, 40]. In the absence of accurate markers, we evaluate TA activity integrating clinical, laboratory and radiological data of already-occurred progression. Serial radiological imaging is thus fundamental [40] and usually requires integration of different techniques, most notably magnetic resonance and Doppler ultrasonography [41]. Novel biomarkers, such as the long pentraxin PTX3, might be useful to identify vessel-associated events such as smouldering inflammation and remodelling [42].

6.3.6 Conundrums in TA Treatment

TA is still an orphan disease, as its rarity and the absence of adequate outcome measurements have hindered randomized clinical trials. Accordingly, there is poor evidence to guide clinical management [43] and no definitive advice can be given as to dosing, duration and choice of therapeutic agents.

Differently from atherosclerosis, the fulcrum of the management of active TA is medical therapy, rather than interventional procedures. Steroids are the mainstay of medical therapy of TA, usually with an initial prednisone daily dose of 0.5–1 mg/kg. Numerous tapering schemes have been empirically proposed, but the majority of patients relapse as steroids are tapered [8, 13], needing steroid-sparing agents such as methotrexate, azathioprine, mycophenolate mofetil and leflunomide [9, 35]. Nonetheless, clinical relapses and progression of the vascular involvement remain frequent despite these combined regimens. Cyclophosphamide, used for life-threatening disease in the pre-biologic era, is now less commonly considered because of its toxicity.

Patients refractory to combined regimens can benefit from biologic therapies. TNFα-inhibitors are the most widely studied biologic agents in TA, with more than 100 patients reported in the literature. Most of these patients were on infliximab, but there is no clear evidence of differences in efficacy among various TNFα-inhibitors. Response, relapse and progression rate on TNFα-inhibitors are about 85–90 %, 35–60 % and 15–35 % respectively [44–48]. IL-6 blockade with tocilizumab is another possible choice for refractory TA [39, 49], although it has to be considered that tocilizumab-induced normalisation of systemic inflammatory responses might not always parallel prevention of arterial progression. Although modulation over time has been reported with long-term therapy with tocilizumab [50], relapses are common after discontinuation of TNFα and IL-6 blockage, suggesting that these agents control TA without curing it. Other proposed targeted therapies include rituximab [36] and abatacept [51], which tackle B and T cells respectively.

Non-immunologic therapies cannot be overlooked in TA [35]: aspirin and other anti-platelet agents might be used on the basis of retrospective data showing reduced frequency of ischemic events in TA [52]. Accelerated atherosclerosis is another feature of TA [53], and atherosclerotic risk factors should be controlled accordingly.

Interventional procedures on arteries have worse long-term patency rate in TA than in atherosclerosis. Revascularisations failures in TA are mainly caused by restenosis, which reflects myointimal hyperplasia similar to non-vasculitis-associated restenosis [35]. However, the inflammatory milieu appears to enhance this remodelling hyperplastic process [35], as long-term outcomes improve if procedures are performed during quiescent phases and if they are followed by immunosuppressive therapies. Reasons are unclear, since histology reveals scarce leukocytic infiltrate in restenotic lesions. Compared to surgery, endovascular revascularisation for TA occlusive arterial disease is less invasive but has worse long-terms results [35], and stents have still not been shown to improve the outcome of balloon angioplasty [54]. The best timing of revascularisation and the choice between endovascular (less invasive but less durable) and surgical techniques are still open issues. We feel that the decision must be taken on a individualized basis in a qualified Center with experience in TA management.

6.3.7 Concluding Remarks

In conclusion, we believe that specialized Centers should be involved in the management of TA, while all Clinicians dealing with arterial diseases should be aware of this conditions and its major unresolved aspects. Our hope is that networking between specialized Centers will enable to overcome its rarity and pave the way to enlighten its mysteries and improve management of these patients.

References

1. Jennette JC et al (2013) 2012 revised international Chapel Hill consensus conference nomenclature of vasculitides. Arthritis Rheum 65:1–11
2. Weyand CM et al (2013) Immune mechanisms in medium and large-vessel vasculitis. Nat Rev Rheumatol 9:731–740
3. Stone JR et al (2015) Consensus statement on surgical pathology of the aorta from the society for cardiovascular pathology and the association for European cardiovascular pathology: I. Inflammatory diseases. Cardiovasc Pathol 24:267–278
4. Burke AP et al (2008) Aortitis and ascending aortic aneurysm: description of 52 cases and proposal of a histologic classification. Hum Pathol 39:514–526
5. Ryan C et al (2015) Non-infectious aortitis of the ascending aorta: a histological and clinical correlation of 71 cases including overlap with medial degeneration and atheroma-a challenge for the pathologist. J Clin Pathol 68:898

6. Szugye HS et al (2014) Takayasu Arteritis in the pediatric population: a contemporary United States-based single center cohort. Pediatr Rheumatol Online J 12:21
7. Sato T et al (2014) Diagnosis of late-onset Takayasu arteritis for elderly adults using fluorine-18 fluorodeoxyglucose positron emission tomography/computed tomography. J Am Geriatr Soc 62:2463–2464
8. Kerr GS et al (1994) Takayasu arteritis. Ann Intern Med 120:919–929
9. Mason JC (2010) Takayasu arteritis–advances in diagnosis and management. Nat Rev Rheumatol 6:406–415
10. Grayson PC et al (2012) Distribution of arterial lesions in Takayasu's arteritis and giant cell arteritis. Ann Rheum Dis 71:1329–1334
11. Prieto-Gonzalez S et al (2012) Large vessel involvement in biopsy-proven giant cell arteritis: prospective study in 40 newly diagnosed patients using CT angiography. Ann Rheum Dis 71:1170–1176
12. Weyand CM et al (2014) Giant-cell arteritis and polymyalgia rheumatica. N Engl J Med 371:1653
13. Maksimowicz-McKinnon K et al (2007) Limitations of therapy and a guarded prognosis in an American cohort of Takayasu arteritis patients. Arthritis Rheum 56:1000–1009
14. Toledano K et al (2011) Pulmonary arteries involvement in Takayasu's arteritis: two cases and literature review. Semin Arthritis Rheum 41:461–470
15. Soto ME et al (2011) Coronary CT angiography in Takayasu arteritis. JACC Cardiovasc Imaging 4:958–966
16. Hata A et al (1996) Angiographic findings of Takayasu arteritis: new classification. Int J Cardiol 54 Suppl:S155–S163
17. Nazareth R et al (2011) Takayasu arteritis: severe consequences of delayed diagnosis. QJM 104:797–800
18. de Souza AW et al (2014) Diagnostic and classification criteria of Takayasu arteritis. J Autoimmun 48–49:79–83
19. Freitas DS et al (2012) Takayasu arteritis: assessment of response to medical therapy based on clinical activity criteria and imaging techniques. Rheumatol Int 32:703–709
20. Ishikawa K et al (1994) Long-term outcome for 120 Japanese patients with Takayasu's disease. Clinical and statistical analyses of related prognostic factors. Circulation 90:1855–1860
21. Di Chio MC, Sartorelli S, Tombetti E, Aiello P, Manfredi AA, Sabbadini MG, Baldissera E (2014) High prevalence of inflammatory comorbidities in a 70 cases-cohort of Takayasu arteritis: not only by chance? EULAR 2014, Paris
22. Hotchi M (1992) Pathological studies on Takayasu arteritis. Heart Vessels Suppl 7:11–17
23. Inder SJ et al (2000) Accumulation of lymphocytes, dendritic cells, and granulocytes in the aortic wall affected by Takayasu's disease. Angiology 51:565–579
24. Tombetti E, Mason JC (in press) Pathophysiology of vasculitides. ESC textbook of vascular biology
25. Brack A et al (1997) Giant cell vasculitis is a T cell-dependent disease. Mol Med 3:530–543
26. Ma-Krupa W et al (2004) Activation of arterial wall dendritic cells and breakdown of self-tolerance in giant cell arteritis. J Exp Med 199:173–183
27. Pryshchep O et al (2008) Vessel-specific Toll-like receptor profiles in human medium and large arteries. Circulation 118:1276–1284
28. Deng J et al (2010) Th17 and Th1 T-cell responses in giant cell arteritis. Circulation 121:906–915
29. Wagner AD et al (1994) Functional profile of tissue-infiltrating and circulating CD68+ cells in giant cell arteritis. Evidence for two components of the disease. J Clin Invest 94:1134–1140
30. Weyand CM et al (1996) Correlation of the topographical arrangement and the functional pattern of tissue-infiltrating macrophages in giant cell arteritis. J Clin Invest 98:1642–1649
31. Nikkari ST et al (1996) Macrophages contain 92-kd gelatinase (MMP-9) at the site of degenerated internal elastic lamina in temporal arteritis. Am J Pathol 149:1427–1433
32. Rittner HL et al (1999) Tissue-destructive macrophages in giant cell arteritis. Circ Res 84:1050–1058

33. Kaiser M et al (1998) Platelet-derived growth factor, intimal hyperplasia, and ischemic complications in giant cell arteritis. Arthritis Rheum 41:623–633
34. Kaiser M et al (1999) Formation of new vasa vasorum in vasculitis. Production of angiogenic cytokines by multinucleated giant cells. Am J Pathol 155:765–774
35. Tombetti E et al (2013) Management options for Takayasu arteritis. Expert Opin Orphan Drugs 1:685–693
36. Hoyer BF et al (2012) Takayasu arteritis is characterised by disturbances of B cell homeostasis and responds to B cell depletion therapy with rituximab. Ann Rheum Dis 71:75–79
37. Saadoun D et al (2015) Th1 and Th17 cytokines drive Takayasu Arteritis inflammation. Arthritis Rheumatol 67:1353–1360
38. Tombetti E, Di Chio MC, Sartorelli S, Bozzolo E, Sabbadini MG, Manfredi AA, Baldissera E (2014) Anti-cytokine treatment for Takayasu arteritis: state of the art. Intractable Rare Dis Res 3:29–33
39. Tombetti E et al (2013) Treatment of refractory Takayasu arteritis with tocilizumab: 7 Italian patients from a single referral center. J Rheumatol 40:2047–2051
40. Youngstein T et al (2013) Interleukin 6 targeting in refractory Takayasu arteritis: serial noninvasive imaging is mandatory to monitor efficacy. J Rheumatol 40:1941–1944
41. Mavrogeni S et al (2013) The role of multimodality imaging in the evaluation of Takayasu arteritis. Semin Arthritis Rheum 42:401–412
42. Tombetti E et al (2014) Systemic pentraxin-3 levels reflect vascular enhancement and progression in Takayasu arteritis. Arthritis Res Ther 16:479
43. Mukhtyar C et al (2009) EULAR recommendations for the management of large vessel vasculitis. Ann Rheum Dis 68:318–323
44. Hoffman GS et al (2004) Anti-tumor necrosis factor therapy in patients with difficult to treat Takayasu arteritis. Arthritis Rheum 50:2296–2304
45. Molloy ES et al (2008) Anti-tumour necrosis factor therapy in patients with refractory Takayasu arteritis: long-term follow-up. Ann Rheum Dis 67:1567–1569
46. Schmidt J et al (2012) Tumor necrosis factor inhibitors in patients with Takayasu arteritis: experience from a referral center with long-term followup. Arthritis Care Res (Hoboken) 64:1079–1083
47. Mekinian A et al (2012) Efficacy and tolerance of infliximab in refractory takayasu arteritis: French multicentre study. Rheumatology (Oxford) 51:882–886
48. Quartuccio L et al (2012) Long-term efficacy and improvement of health-related quality of life in patients with Takayasu's arteritis treated with infliximab. Clin Exp Rheumatol 30:922–928
49. Abisror N et al (2013) Tocilizumab in refractory Takayasu arteritis: a case series and updated literature review. Autoimmun Rev 12:1143–1149
50. Nakaoka Y et al (2013) Tocilizumab for the treatment of patients with refractory Takayasu arteritis. Int Heart J 54:405–411
51. Abatacept for treating adults with giant cell arteritis and Takayasu's arteritis.
52. de Souza AW et al (2010) Antiplatelet therapy for the prevention of arterial ischemic events in Takayasu arteritis. Circ J 74:1236–1241
53. Seyahi E et al (2013) Aortic and coronary calcifications in Takayasu arteritis. Semin Arthritis Rheum 43:96–104
54. Park HS et al (2013) Long term results of endovascular treatment in renal arterial stenosis from Takayasu arteritis: angioplasty versus stent placement. Eur J Radiol 82:1913–1918

Chapter 7
Takayasu Arteritis and Ulcerative Colitis: A Frequent Association?

Chikashi Terao

Abstract Previous studies have revealed that Takayasu arteritis (TAK) and ulcerative colitis (UC) sometimes develop in the same patients. However, the low prevalence of TAK has made it difficult to accumulate the number of cases to establish the ratio of co-occurrence. Recently, our Japanese group analyzed the data of 470 consecutive patients with TAK and revealed that 6.4 % of patients with TAK suffered from UC. This ratio is much higher than the prevalence of UC in Japan. HLA-B*52:01, the strongest susceptibility allele to both disease, showed strong enrichment in patients with these two diseases. *IL12B*, a common non-HLA susceptibility locus to both diseases, also showed a trend of enrichment. We also showed that other genetic determinants were shared between the two diseases. The high ratio of co-occurrence and genetic overlap suggest common molecular pathways centering on HLA-B*52:01 and IL12p40 underlying the two diseases.

Keywords Takayasu arteritis • Ulcerative colitis • Crohn's disease • Genetic overlap • Common complication • Common molecular pathway

C. Terao (✉)
Center for Genomic Medicine, Kyoto University Graduate School of Medicine, Kyoto, Japan

Center for the Promotion of Interdisciplinary Education and Research, Kyoto, Japan

Division of Rheumatology, Immunology, and Allergy, Harvard Medical School, Boston, MA, USA

Division of Genetics, Brigham and Women's Hospital, Harvard Medical School, Boston, MA, USA

Program in Medical and Population Genetics, Broad Institute, Cambridge, MA, USA
e-mail: a0001101@kuhp.kyoto-u.ac.jp

F. Dammacco et al. (eds.), *Systemic Vasculitides: Current Status and Perspectives*, DOI 10.1007/978-3-319-40136-2_7

7.1 Takayasu Arteritis and Ulcerative Colitis

Takayasu arteritis (TAK) is a relatively rare vasculitis firstly reported from Japan by Dr. Mikito Takayasu in 1908 in spite of the controversy of the first actual literal case report [1, 2]. Its prevalence is estimated to be about 0.005–0.01 % in Japan which is one of the countries with a higher number of patients with TAK. Ulcerative colitis (UC) is one of inflammatory bowel diseases (IBD) which affects mainly lower gastrointestinal tract. The occurrence of IBD has been increased due to spread of high fat diet in Japan [3]. The prevalence of UC is 0.1 % in Japan and is increasing [4]. The co-occurrence of TAK and UC was firstly reported by Kawasaki from Japan in 1964 [5]. Since the first report, more than 50 reports of 60 patients were published mainly from Japan (Table 7.1).

Table 7.1 represents the summary of the case reports or case series for co-occurrence of TAK and UC. As shown, most of the reports are single case reports, suggesting difficulty of collection of samples in a single institution. In addition, there are many other case reports including reports in academic conference or Japanese journals not listed in Table 7.1.

7.2 Origin of the Case Reports of Co-occurrence

We should note that the reports have also been made from outside of Japan, including Korea, India, Turkey, Spain and US (Table 7.1). Thus, the co-occurrence can be observed all over the world. About 70–80 % of the reports for co-occurrence of the two diseases were made from Japan. Is it due to advanced studies dealing with TAK in Japan? Or is it the co-occurrence observed more frequently in Japan? To address this question, it is appropriate to draw a two dimensional plot of prevalence of TAK and co-occurrence of TAK and UC in the world to assess deviation of co-occurrence in Japan from distribution in other countries. Since the prevalence of TAK in each country has not been reported in detail, we cannot assert enrichment of co-occurrence in Japan. Although we should restrain from a final conclusion, considering the genetic architecture underlying the co-occurrence discussed below, it seems reasonable that the co-occurrence can be generalized beyond ethnicity and frequently observed in Japan due to the prevalence of HLA-B*52:01.

7.3 Ratio of Co-occurrence

To address the question whether the co-occurrence of these two diseases are coincidental or beyond coincidence, accumulating a large number of data of patients is essential. However, it is quite a challenge to recruit consecutive patients with TAK and assess overlapping of UC. Horai et al. estimated the co-occurrence ratio as

Table 7.1 Case reports of co-occurrence

Reporter	Year	Age	Sex	Precede	HLA-B52	Country	Ref number
Kawasaki	1964	19	F	TAK	NA	Japan	[29]
Soloway	1970	33	F	UC	NA	USA	[30]
Sugishita	1973	20	F	TAK	NA	Japan	[31]
Tsuchiya	1976	20	F	UC	NA	Japan	[32]
Yassinger	1976	15	F	UC	NA	USA	[33]
Chapman	1978	17	F	UC	NA	Pakistan	[34]
Miwa	1979	23	F	TAK	B*5	Japan	[35]
Beau	1980	17	F	UC	NA	France	[36]
Tamanaha	1981	15	M	UC	NA	Japan	[37]
Moroe	1983	39	F	TAK	NA	Japan	[38]
Nishimura	1984	38	F	Si	1	Japan	[39]
Ookata	1985	26	M	UC	NA	Japan	[40]
Achar	1986	35	F	UC	B*5	Srilanka	[41]
Yamaguchi	1988	19	M	UC	NA	Japan	[42]
Ichikawa	1988	12	F	UC	1	Japan	[43]
Ikenaga	1989	8	F	UC	NA	Japan	[44]
Sakhuja	1990	31	F	UC	NA	India	[45]
Sakhuja	1990	13	M	UC	NA	India	[45]
Goto	1991	10	F	Si	1	Japan	[46]
Yazawa	1992	19	M	UC	NA	Japan	[47]
Yoshida	1992	21	M	TAK	1	Japan	[48]
Ishikawa	1993	27	F	UC	1	Japan	[49]
Oyanagi	1994	25	F	UC	1	Japan	[50]
Sato	1994	13	F	UC	1	Japan	[51]
So	1995	21	F	UC	1	Japan	[52]
Morita	1996	19	F	UC	1	Japan	[53]
Aoyagi	1998	14	M	UC	0	Japan	[17]
Kanaya	1998	26	M	UC	0	Japan	[18]
Ariizumi	1999	28	F	UC	1	Japan	[54]
Kawashima	1999	32	F	UC	1	Japan	[55]
Ito	2001	15	F	UC	1	Japan	[56]
Suzuki	2001	14	M	Si	1	Japan	[57]
Shibata	2002	36	F	UC	1	Japan	[58]
Fukunaga	2002	13	F	UC	1	Japan	[59]
Masuda	2002	10	F	TAK	1	Japan	[60]
Masuda	2002	13	F	TAK	1	Japan	[60]
Segawa	2002	22	F	UC	1	Japan	[61]
Bansal	2003	15	F	TAK	1	India	[62]
Hokama	2003	27	F	UC	0	Japan	[63]
Gearry	2003	42	F	TAK	0	New Zealand	[64]

(continued)

Table 7.1 (continued)

Reporter	Year	Age	Sex	Precede	HLA-B52	Country	Ref number
Ohta	2003	23	F	UC	NA	Japan	[65]
Nakano	2004	39	M	TAK	1	Japan	[66]
Katsinelos	2005	33	F	TAK	1	non-asian	[67]
Sood	2006	18	F	Si	NA	Inida	[68]
Callejas-Rubio	2006	17	M	UC	0	Spain	[69]
Balamtekin	2009	14	F	UC	0	Turkey	[70]
Ooka	2009	17	F	si	NA	Japan	[11]
Asano	2010	34	F	UC	NA	Japan	[71]
Kudo	2010	14	F	UC	1	Japan	[72]
Kashima	2010	15	F	UC	1	Japan	[73]
Inagaki	2010	36	F	TAK	NA	Japan	[74]
Takahashi	2011	20	M	UC	1	Japan	[75]
Gecse	2011	20	F	UC	1	Hungary	[76]
Horai	2011	46	F	UC	1	Japan	[6]
Horai	2012	22	F	UC	1	Japan	[77]
Azak	2012	24	F	UC	NA	Turkey	[78]
Kim	2012	9	F	UC	NA	Korea	[79]
Chae	2013	25	M	UC	1	Korea	[80]
Pyo	2013	38	F	si	NA	Korea	[81]
Watanabe	2014	23	F	TAK	NA	Japan	[7]
Watanabe	2014	35	F	UC	NA	Japan	[7]
Watanabe	2014	28	F	UC	NA	Japan	[7]
Watanabe	2014	18	F	UC	NA	Japan	[7]
Watanabe	2014	22	F	TAK	NA	Japan	[7]
Khoshnama	2014	35	F	UC	NA	Iran	[82]
Sum		22.9 ± 9.2	F:52 M:13	UC:45 TAK:14	1:28 0:6 B5:2	Japan:46 Others:19	

0.783 % based on the number of case reports and registry of the two diseases in Japan [6]. However, the ratio is apparently underestimated since their calculation was based on the assumption that all of the subjects developing the two diseases were reported in manuscripts. We should pay attention to selection bias as well as to estimate a collection of reported cases.

There are two studies recruiting more than 50 patients with TAK to address the co-occurrence of the two diseases. Both are from Japan. One is reported from Tohoku University [7] and the other comes from our group in Kyoto University [4]. The former is a single-center study and the latter is a multi-center study. The samples in the two studies did not overlap each other. The conclusion and estimates of co-occurrence ratio are quite similar between the two studies.

The study from Tohoku University reports that 5 out of 82 cases with TAK had UC (6.1 % (95 % confidence interval:0.92 %–11.3 %)) [7]. The study from Kyoto

University reports that 30 out of 470 patients with TAK had UC (6.4 % (95 % CI: 4.3 %–9.0 %)) without evidence of deviation of co-occurrence among the multiple institutions [4]. In spite of the estimation from cross-sectional study, these studies, especially the study from Kyoto University, offer the estimated ratio with relatively narrow confidence interval. When we summed both studies, the co-occurrence is 6.3 % (95 % CI:4.3 %–8.4 %).

It should be noted that the estimated co-occurrence ratio is nearly 100 times of prevalence of UC [4]. Thus, the ratio cannot be coincidental co-occurrence. While patients with UC or TAK who visit hospitals or clinics may have higher chance to be detected other diseases in comparison with general population based on the chances to undergo work-up, it cannot explain this big difference in ratio.

What about the co-occurrence of TAK and UC among patients with UC? There is a small number of reports addressing this issue. An old Japanese study, in which researchers aggressively searched complication of UC, reports that 3 out of 1,433 patients with UC had TAK [8]. It might be interesting that the co-occurrence ratio (0.21 %, 95 % CI 0 %–0.045 %) is also about twenty to forty times of general prevalence of TAK, suggesting enrichment of co-occurrence even in UC patients in spite of wide confidence interval. Since this study did not aim specifically to establish the overlapping of the two diseases, this ratio of overlapping in patients with UC should be carefully evaluated.

7.4 Characteristics of the Co-occurrent Cases

What are the clinical characteristics of patients with both TAK and UC? The study from our group in Kyoto University reported that the patients of co-occurrence develop TAK on average 6 years younger than patients without UC. We did not observe clear differences between patients with and without UC in the ratio of aortic regurgitation (AR) development, severity of AR, renal stenosis, sex distribution, ischemic heart disease, stroke and usage of biologic agents. We did not find difference of total colectomy in patients with UC and TAK, indicating no strong evidence of severe UC in the co-occurrent cases. The previous studies detected significant associations between HLA-B*52:01 or *IL12B* and clinical manifestations including development and severity of AR [9]. As discussed below, the concurrent cases show enrichment of these genetic variants. Thus, there is a possibility that lack of the significant associations between co-occurrence and TAK manifestations is due to underpower of this study.

What about long outcome and mortality in the patients with both diseases? While these data are not available in the previous studies, the prospective observation of the concurrent cases would reveal important information regarding these questions.

Which occurs earlier, TAK or UC, in the patients with both diseases? Generally, TAK and UC develop at age younger than 30 while UC can develop at any age [10]. Forty-five out of 65 cases in Table 7.1 reported preceding onset of UC to TAK. Six out of the 65 cases were diagnosed in the same year and the remaining 14 cases developed TAK earlier than UC. Although these case reports are not recruiting consecutive patients and may contain some bias, binomial test provided a significant difference in a precedent disease (45 over 59 cases, 76.3 %, p = 0.000065). Ooka et al. [11] mentioned that 68 % of 74 cases in the world reported between 1964 and 2005 developed UC first.

However, the study from our group in Kyoto University in 2015 did not find deviation of preceding UC to TAK (Table 7.2) and even observed a trend of developing TAK earlier. It is still inconclusive that UC precedes TAK in patients developing the two diseases. The previous studies may have time lag to diagnose TAK considering its low prevalence outside of Japan.

Table 7.2 30 cases reported from Kyoto University in 2015

Reporter	Year	Age	Sex	Precede	HLA-B52	Country	Ref number
Terao	2015	29	F	TAK	1	Japan	[4]
Terao	2015	16	M	NA	1	Japan	[4]
Terao	2015	43	F	NA	1	Japan	[4]
Terao	2015	28	F	NA	1	Japan	[4]
Terao	2015	22	F	NA	1	Japan	[4]
Terao	2015	37	F	NA	1	Japan	[4]
Terao	2015	32	F	NA	1	Japan	[4]
Terao	2015	24	F	NA	1	Japan	[4]
Terao	2015	30	F	NA	1	Japan	[4]
Terao	2015	10	F	NA	1	Japan	[4]
Terao	2015	8	F	NA	1	Japan	[4]
Terao	2015	NA	F	NA	NA	Japan	[4]
Terao	2015	37	F	NA	NA	Japan	[4]
Terao	2015	50	F	UC	0	Japan	[4]
Terao	2015	30	F	NA	1	Japan	[4]
Terao	2015	21	M	NA	1	Japan	[4]
Terao	2015	20	F	NA	1	Japan	[4]
Terao	2015	26	F	NA	1	Japan	[4]
Terao	2015	20	F	si	1	Japan	[4]
Terao	2015	17	M	UC	1	Japan	[4]
Terao	2015	36	F	si	0	Japan	[4]
Terao	2015	24	F	TAK	1	Japan	[4]
Terao	2015	21	F	TAK	1	Japan	[4]
Terao	2015	35	F	TAK	1	Japan	[4]
Terao	2015	11	F	TAK	1	Japan	[4]
Terao	2015	19	M	UC	1	Japan	[4]
Terao	2015	23	M	TAK	1	Japan	[4]
Terao	2015	38	F	UC	1	Japan	[4]
Terao	2015	20	F	TAK	NA	Japan	[4]
Terao	2015	15	F	TAK	1	Japan	[4]
Sum		25.6 ± 10.1	F:25 M:5	UC:4 TAK:8	1:25 0:2		

B*52:01 was reported to be the strongest susceptibility allele to UC in Japanese patients [12]. B*52:01 was reported to be associated with early onset of UC [12]. Considering the strong enrichment of B*52:01 in the co-occurrence discussed below, patients with the two diseases may develop earlier than UC patients without co-occurrence. Since the patients with the two diseases also develop TAK at an earlier stage of life than patients without UC [4], further accumulation of data would provide the clear evidence for a preceding disease.

7.5 Genetic Architectures of the Co-occurrence

The HLA locus is associated with various autoimmune diseases. TAK and UC are also associated with the HLA locus [13]. HLA-B*52:01 is the strongest susceptibility allele to TAK [14]. HLA-B*52:01 is also most strongly associated with UC in the Japanese population [12, 15]. Parts of the case reports or case series reported HLA-B alleles in patients with UC and TAK. A total of 28 out of 36 patients (77.8 %) whose HLA-B alleles were reported from 35 manuscripts were positive for HLA-B*52:01 or B52 (Table 7.1). Furthermore, 2 out of the 8 patients who were not confirmed to have B*52:01, were reported to have B5, containing B52. In fact, our Japanese study recently published showed that 25 out of 27 patients (92.6 %) of the patients with UC and TAK were positive for HLA-B*52:01 or B*52 with OR of 12.14 (95 % CI: 2.96–107.23) over patients without UC. Based on the strong effect size of B52:01 on the co-occurrence among patients with TAK, HLA-B*52:01 should play a central role in the co-occurrence of TAK and UC. Evidence of HLA genotypes in co-occurrent cases outside of Japan is not enough. Future studies should focus on this point especially in the context of shared HLA alleles or amino acid residues between TAK and UC.

It should be notable that a dose-dependent effect of HLA-B*52:01 on the co-occurrence was not reported. Case reports giving information of the two alleles of HLA-B did not report homozygous patients for HLA-B*52:01 developing both UC and TAK. Even a trend to increase the co-occurrence was not observed in the Japanese study, suggesting that HLA-B*52:01 effect on co-occurrence should be in a dominant-manner.

Then, what about other HLA alleles susceptible to TAK or UC? HLA-B*67:01 is another susceptibility allele to TAK recently reported from two different cohorts [13, 16]. The association between B*67:01 and UC has not been reported. To date, case reports by Kanaya et al. and Aoyagi et al. were the ones among studies addressing the co-occurrence of TAK and UC to report cases carrying B*67 [17, 18]. The enrichment of B*67 in co-occurrent cases is inconclusive.

The enrichment of HLA-B*52:01 in co-occurrent cases raises the possibility of enrichment of other genetic components. What about non-HLA susceptibility loci? Recent technological development has enabled us to perform genome-wide association study using DNA microarrays for single nuclear polymorphisms (SNPs) to identify susceptibility genes in an unbiased fashion. The genetic studies to date have identified a total of more than 100 susceptibility loci in non-HLA loci to IBD [19]. UC and Crohn's disease (CD) seem to have large parts of these loci in common [19]. A total of 6 susceptibility non-HLA loci to TAK have been identified to date by genetic studies from different populations [9, 20, 21]. *IL12B* is an established susceptibility locus beyond populations. *IL12B* is also associated with IBD [19]. In addition, both of them displayed the strongest association in the same SNP, rs6871626. Our Japanese study reported that a trend of enrichment of rs6871626 in *IL12B* was observed in the co-occurrent cases (p = 0.16, OR 2.47[95 % CI 0.72–13.11]) while the enrichment did not reach a significant level due to the limited number of subjects [4]. Further, our Japanese study showed that the UC-susceptibility alleles showed the common direction of susceptibility effects and the effect sizes are significantly correlated between UC and TAK [4]. The study also showed that the higher the effect sizes of UC-susceptibility alleles carried, the higher the rate of common directions of SNPs observed. Our group also showed the clear departure of p-values of UC-susceptibility alleles for the association with TAK from expected distribution of p-values.

These results strongly indicate that TAK and UC share genetic architecture not only in the HLA-locus but also in the non-HLA loci.

Recent genetic studies showed that rare variants are associated with UC and CD and might confer substantial proportion of heritability [22]. Whether rare variant burden is also observed in TAK susceptibility and the co-occurrent cases is an interesting topic to address in the future with large number of samples.

7.6 Co-Occurrence of TAK and Crohn's Disease

Then, what about co-occurrence between TAK and CD, the other IBD? Since the genetic architecture of CD is very similar to that of UC, it should share the genetic characteristics with TAK as well. On this point, the Japanese reports of co-occurrence of TAK and CD are much lesser than the reports of UC and TAK [23]. Most of the reports of co-occurrence of CD and TAK were made from outside of Japan [23].

This may be explained by (1) the difference in distribution of CD and UC between Japan and Western countries, (2) the difference in sex predominance in CD in Japan, and (3) genetic difference between UC and CD observed in the Japanese.

US has comparable prevalence of CD and UC as 250 each in adults [24]. On the contrary, the prevalence of CD is about one third to one fourth of that of UC in Japan [3]. The less frequent prevalence of CD may partly explain the small number of case reports about the co-occurrence of CD and TAK.

While UC has similar sex distribution, CD dominantly affects men in Japan. Thus, TAK patients, 90 % of them are females, may have lower chance to develop CD than UC in the Japanese population.

In spite of the genetic overlap between CD and UC, HLA haplotype of B*52-DRB1*15:02 was reported to have an opposing effect between susceptibilities to

CD and UC in the Japanese population [15]. The protective effect of the haplotype against CD may explain the fewer case reports of co-occurrence of TAK and CD in the Japanese population. Since B*52:01 seems critical to co-occurrence of TAK and UC, the occurrence of CD and TAK beyond coincidence in the Japanese population may require confirmation from future large-scale studies.

Since about 70 % of the case reports of TAK and UC were made from Japan, the co-occurrence of CD and TAK seemed more frequently observed than that of UC and TAK outside Japan. Further accumulation of data will hopefully reveal whether patients with TAK have a higher chance to develop CD than UC outside Japan.

7.7 Animal Models of the Diseases

Do animal experiments support the co-occurrence of the two diseases? To date, no previous studies have addressed this co-occurrence. Since there are many animal models of IBD including Il10 KO mice, Rag KO mice to which CD4+ CD45RBhi naïve T cells were transferred, an acute trinitrobenzenesulfonic acid (TNBS) model of colitis, it would be interesting to analyze involvement of aorta in various conditions.

Interestingly, no transgenic mice of HLA-B52 was successfully produced and observed in detail. Development of HLA-B52-transgenic mice would support common molecular pathways between TAK and UC in the animal model.

7.8 Clinical Applications Based on the Co-occurrence

The co-occurrence of two diseases beyond coincidence and shared genetic characteristics strongly suggest that common molecular networks underlie the two diseases. This raises the possibility that treatment for one disease may be applicable to the other and vice versa. *IL12B* is the common genetic determinant between the two diseases and *IL12B* plays important roles on progression of TAK, including development of AR and severity of AR in addition to susceptibility to TAK. *IL12B* encodes IL12p40, a common subunit of IL12 and IL23. Ustekinumab is a biological agent targeting IL12p40. Ustekinumab has been used for patients with psoriasis or CD [25–27]. While usage of ustekinumb in patients with UC is not prevailed, previous studies have revealed that CD and UC share large parts of genetic architectures. Thus, ustekinumab is a promising candidate of the new treatment options for patients with TAK. To address this point, our team has recently published a manuscript reporting a pilot study of usage of ustekinumab for patients with TAK [28]. In the manuscript, a total of three patients with TAK who were refractory to conventional treatment received ustekinumab. All three patients responded to ustekinumab without any severe side effects. However, the effectiveness of ustekinumab in patients with TAK is still inconclusive. Furthermore, it is ambiguous whether ustekinumab is effective for long-term outcome of vascular inflammation. It should be noted that none of the biological agents were proven to be effective to prevent inflammatory vascular stenosis in patients with TAK. Large clinical trials with long observation periods are necessary to prove efficacy of biological agents, especially ustekinumab,in TAK patients.

Acknowledgement I'd like to thank to Dr. Yohei Kirino at the Department of Internal Medicine and Clinical Immunology, Yokohama City University Graduate School of Medicine, Yokohama for his kind provision of table of previous case series or reports of co-occurrence.

References

1. Terao C, Yoshifuji H, Mimori T (2014) Recent advances in Takayasu arteritis. Int J Rheum Dis 17(3):238–247. doi:10.1111/1756-185X.12309
2. Terao C (2014) History of Takayasu arteritis and Dr. Mikito Takayasu. Int J Rheum Dis 17(8):931–935. doi:10.1111/1756-185X.12576
3. Thia KT, Loftus EV Jr, Sandborn WJ, Yang SK (2008) An update on the epidemiology of inflammatory bowel disease in Asia. Am J Gastroenterol 103(12):3167–3182. doi:10.1111/j.1572-0241.2008.02158.x
4. Terao C, Matsumura T, Yoshifuji H, Kirino Y, Maejima Y, Nakaoka Y, Takahashi M, Amiya E, Tamura N, Nakajima T, Origuchi T, Horita T, Matsukura M, Kochi Y, Ogimoto A, Yamamoto M, Takahashi H, Nakayamada S, Saito K, Wada Y, Narita I, Kawaguchi Y, Yamanaka H, Ohmura K, Atsumi T, Tanemoto K, Miyata T, Kuwana M, Komuro I, Tabara Y, Ueda A, Isobe M, Mimori T, Matsuda F (2015) Takayasu arteritis and ulcerative colitis-high concurrence ratio and genetic overlap. Arthritis Rheumatol. doi:10.1002/art.39157
5. Kawasaki K (1964) A case of ulcerative colitis associated with symptoms of multiple arterial stenosis. J Jpn Soc Intern Med 53:588
6. Horai Y, Miyamura T, Shimada K, Takahama S, Minami R, Yamamoto M, Suematsu E (2011) A case of Takayasu's arteritis associated with human leukocyte antigen A24 and B52 following resolution of ulcerative colitis and subacute thyroiditis. Intern Med 50(2):151–154
7. Watanabe R, Ishii T, Nakamura K, Shirai T, Fujii H, Saito S, Harigae H (2014) Ulcerative colitis is not a rare complication of Takayasu arteritis. Mod Rheumatol 24(2):372–373. doi:10.3109/14397595.2013.854045
8. Oshitani N, Watanabe K, Nakamura S, Higuchi K, Arakawa T (2005) Extraintestinal complications in patients with ulcerative colitis. Nihon Rinsho 63(5):874–878
9. Terao C, Yoshifuji H, Kimura A, Matsumura T, Ohmura K, Takahashi M, Shimizu M, Kawaguchi T, Chen Z, Naruse TK, Sato-Otsubo A, Ebana Y, Maejima Y, Kinoshita H, Murakami K, Kawabata D, Wada Y, Narita I, Tazaki J, Kawaguchi Y, Yamanaka H, Yurugi K, Miura Y, Maekawa T, Ogawa S, Komuro I, Nagai R, Yamada R, Tabara Y, Isobe M, Mimori T, Matsuda F (2013) Two susceptibility loci to Takayasu arteritis reveal a synergistic role of the IL12B and HLA-B regions in a Japanese population. Am J Hum Genet 93(2):289–297. doi:10.1016/j.ajhg.2013.05.024
10. da Silva BC, Lyra AC, Rocha R, Santana GO (2014) Epidemiology, demographic characteristics and prognostic predictors of ulcerative colitis. World J Gastroenterol 20(28):9458–9467. doi:10.3748/wjg.v20.i28.9458
11. Ooka SWM (2009) IBD Res 3(3):234–241
12. Aizawa H, Kinouchi Y, Negoro K, Nomura E, Imai G, Takahashi S, Takagi S, Kakuta Y, Tosa M, Mochida A, Matsumura Y, Endo K, Shimosegawa T (2009) HLA-B is the best candidate of susceptibility genes in HLA for Japanese ulcerative colitis. Tissue Antigens 73(6):569–574. doi:10.1111/j.1399-0039.2009.01241.x

13. Terao C, Yoshifuji H, Ohmura K, Murakami K, Kawabata D, Yurugi K, Tazaki J, Kinoshita H, Kimura A, Akizuki M, Kawaguchi Y, Yamanaka H, Miura Y, Maekawa T, Saji H, Mimori T, Matsuda F (2013) Association of Takayasu arteritis with HLA-B 67:01 and two amino acids in HLA-B protein. Rheumatology (Oxford) 52(10):1769–1774. doi:10.1093/rheumatology/ket241
14. Terao C (2015) Revisited HLA and non-HLA genetics of Takayasu arteritis-where are we? J Hum Genet 61(1):27–32. doi:10.1038/jhg.2015.87
15. Okada Y, Yamazaki K, Umeno J, Takahashi A, Kumasaka N, Ashikawa K, Aoi T, Takazoe M, Matsui T, Hirano A, Matsumoto T, Kamatani N, Nakamura Y, Yamamoto K, Kubo M (2011) HLA-Cw*1202-B*5201-DRB1*1502 haplotype increases risk for ulcerative colitis but reduces risk for Crohn's disease. Gastroenterology 141(3):864–871 e861–865. doi:10.1053/j.gastro.2011.05.048
16. Takamura C, Ohhigashi H, Ebana Y, Isobe M (2012) New human leukocyte antigen risk allele in Japanese patients with Takayasu arteritis. Circ J 76(7):1697–1702. doi:DN/JST.JSTAGE/circj/CJ-12-0089 [pii]
17. Aoyagi S, Akashi H, Kawara T, Ishihara K, Tanaka A, Kanaya S, Koga Y, Ishikawa R (1998) Aortic root replacement for Takayasu arteritis associated with ulcerative colitis and ankylosing spondylitis – report of a case. Jpn Circ J 62(1):64–68
18. kanaya S, Tokuda K, Tanaka S, Hirano K, Koga Y, Imaizumi T, Suzuki S, Aoyagi S, Kato S, Koga A, Ajisaka H (1998) Rare case of ulcerative colitis and ankylosing spondylitis associated with aortitis and severe aortic valve insufficiency. Nihon Naika Gakkai Zasshi 87(11):2316–2318
19. Jostins L, Ripke S, Weersma RK, Duerr RH, McGovern DP, Hui KY, Lee JC, Schumm LP, Sharma Y, Anderson CA, Essers J, Mitrovic M, Ning K, Cleynen I, Theatre E, Spain SL, Raychaudhuri S, Goyette P, Wei Z, Abraham C, Achkar JP, Ahmad T, Amininejad L, Ananthakrishnan AN, Andersen V, Andrews JM, Baidoo L, Balschun T, Bampton PA, Bitton A, Boucher G, Brand S, Buning C, Cohain A, Cichon S, D'Amato M, De Jong D, Devaney KL, Dubinsky M, Edwards C, Ellinghaus D, Ferguson LR, Franchimont D, Fransen K, Gearry R, Georges M, Gieger C, Glas J, Haritunians T, Hart A, Hawkey C, Hedl M, Hu X, Karlsen TH, Kupcinskas L, Kugathasan S, Latiano A, Laukens D, Lawrance IC, Lees CW, Louis E, Mahy G, Mansfield J, Morgan AR, Mowat C, Newman W, Palmieri O, Ponsioen CY, Potocnik U, Prescott NJ, Regueiro M, Rotter JI, Russell RK, Sanderson JD, Sans M, Satsangi J, Schreiber S, Simms LA, Sventoraityte J, Targan SR, Taylor KD, Tremelling M, Verspaget HW, De Vos M, Wijmenga C, Wilson DC, Winkelmann J, Xavier RJ, Zeissig S, Zhang B, Zhang CK, Zhao H, International IBDGC, Silverberg MS, Annese V, Hakonarson H, Brant SR, Radford-Smith G, Mathew CG, Rioux JD, Schadt EE, Daly MJ, Franke A, Parkes M, Vermeire S, Barrett JC, Cho JH (2012) Host-microbe interactions have shaped the genetic architecture of inflammatory bowel disease. Nature 491(7422):119–124. doi:10.1038/nature11582
20. Saruhan-Direskeneli G, Hughes T, Aksu K, Keser G, Coit P, Aydin SZ, Alibaz-Oner F, Kamali S, Inanc M, Carette S, Hoffman GS, Akar S, Onen F, Akkoc N, Khalidi NA, Koening C, Karadag O, Kiraz S, Langford CA, McAlear CA, Ozbalkan Z, Ates A, Karaaslan Y, Maksimowicz-McKinnon K, Monach PA, Ozer HT, Seyahi E, Fresko I, Cefle A, Seo P, Warrington KJ, Ozturk MA, Ytterberg SR, Cobankara V, Onat AM, Guthridge JM, James JA, Tunc E, Duzgun N, Bicakcigil M, Yentur SP, Merkel PA, Direskeneli H, Sawalha AH (2013) Identification of multiple genetic susceptibility loci in Takayasu arteritis. Am J Hum Genet 93(2):298–305. doi:10.1016/j.ajhg.2013.05.026
21. Renauer P, Saruhan-Direskeneli G, Coit P, Adler A, Aksu K, Keser G, Alibaz-Oner F, Aydin SZ, Kamali S, Inanc M, Carette S, Cuthbertson D, Hoffman GS, Akar S, Onen F, Akkoc N, Khalidi NA, Koening C, Karadag O, Kiraz S, Langford CA, Maksimowicz-McKinnon K, McAlear CA, Ozbalkan Z, Ates A, Karaaslan Y, Duzgun N, Monach PA, Ozer HT, Erken E, Ozturk MA, Yazici A, Cefle A, Onat AM, Kisacik B, Pagnoux C, Kasifoglu T, Seyahi E, Fresko I, Seo P, Sreih A, Warrington KJ, Ytterberg SR, Cobankara V, Cunninghame-Graham DS, Vyse TJ, Pamuk ON, Tunc E, Dalkilic E, Bicakcigil M, Yentur SP, Wren JD, Merkel PA,

Direskeneli H, Sawalha AH (2015) Genome-wide association study identifies susceptibility loci in IL6, RPS9/LILRB3, and an intergenic locus on chromosome 21q22 in Takayasu's arteritis. Arthritis Rheumatol 67(5):1361–1368. doi:10.1002/art.39035
22. Momozawa Y, Mni M, Nakamura K, Coppieters W, Almer S, Amininejad L, Cleynen I, Colombel JF, de Rijk P, Dewit O, Finkel Y, Gassull MA, Goossens D, Laukens D, Lemann M, Libioulle C, O'Morain C, Reenaers C, Rutgeerts P, Tysk C, Zelenika D, Lathrop M, Del-Favero J, Hugot JP, de Vos M, Franchimont D, Vermeire S, Louis E, Georges M (2011) Resequencing of positional candidates identifies low frequency IL23R coding variants protecting against inflammatory bowel disease. Nat Genet 43(1):43–47. doi:10.1038/ng.733
23. Sy A, Khalidi N, Dehghan N, Barra L, Carette S, Cuthbertson D, Hoffman GS, Koening CL, Langford CA, McAlear C, Moreland L, Monach PA, Seo P, Specks U, Sreih A, Ytterberg SR, Assche GV, Merkel PA, Pagnoux C, Vasculitis Clinical Research C, the Canadian Vasculitis N (2015) Vasculitis in patients with inflammatory bowel diseases: a study of 32 patients and systematic review of the literature. Semin Arthritis Rheum 45(4):475–482. doi:10.1016/j.semarthrit.2015.07.006
24. Kappelman MD, Moore KR, Allen JK, Cook SF (2013) Recent trends in the prevalence of Crohn's disease and ulcerative colitis in a commercially insured US population. Dig Dis Sci 58(2):519–525. doi:10.1007/s10620-012-2371-5
25. Reddy M, Torres G, McCormick T, Marano C, Cooper K, Yeilding N, Wang Y, Pendley C, Prabhakar U, Wong J, Davis C, Xu S, Brodmerkel C (2010) Positive treatment effects of ustekinumab in psoriasis: analysis of lesional and systemic parameters. J Dermatol 37(5):413–425. doi:10.1111/j.1346-8138.2010.00802.x
26. Sandborn WJ, Gasink C, Gao LL, Blank MA, Johanns J, Guzzo C, Sands BE, Hanauer SB, Targan S, Rutgeerts P, Ghosh S, de Villiers WJ, Panaccione R, Greenberg G, Schreiber S, Lichtiger S, Feagan BG, Group CS (2012) Ustekinumab induction and maintenance therapy in refractory Crohn's disease. N Engl J Med 367(16):1519–1528. doi:10.1056/NEJMoa1203572
27. Yeilding N, Szapary P, Brodmerkel C, Benson J, Plotnick M, Zhou H, Goyal K, Schenkel B, Giles-Komar J, Mascelli MA, Guzzo C (2012) Development of the IL-12/23 antagonist ustekinumab in psoriasis: past, present, and future perspectives – an update. Ann N Y Acad Sci 1263:1–12. doi:10.1111/j.1749-6632.2012.06670.x
28. Terao C, Yoshifuji H, Nakajima T, Yukawa N, Matsuda F, Mimori T (2015) Ustekinumab as a therapeutic option for Takayasu arteritis: from genetic findings to clinical application. Scand J Rheumatol 45:1–3. doi:10.3109/03009742.2015.1060521
29. Kawasaki K (1964) A case of ulcerative colitis associated with symptoms of multiple arterial stenosis. J Jpn Soc Internal Med 53:588
30. Soloway M, Moir TW, Linton DS Jr (1970) Takayasu's arteritis. Report of a case with unusual findings. Am J Cardiol 25(2):258–263
31. Sugishita T, Kasai K, Kikuchi T (1973) A case of aortitis syndrome associated with ulcerative colitis. Naika. 31:517–521
32. Tsuchiya MOI, Mizuno O, Asakura H, Hiramatsu K (1976) Aortitis syndrome associated with ulcerative colitis. Vasc Surg 10:176–181
33. Yassinger S, Adelman R, Cantor D, Halsted CH, Bolt RJ (1976) Association of inflammatory bowel disease and large vascular lesions. Gastroenterology 71(5):844–846
34. Chapman R, Dawe C, Whorwell PJ, Wright R (1978) Ulcerative colitis in association with Takayasu's disease. Am J Dig Dis 23(7):660–662
35. Miwa Y, Nagasako K, Sasaki H, Ichikawa K, Yaguchi T, Hasegawa K, Kubokura T (1979) Aortitis syndrome associated with ulcerative colitis: report of a case. Gastroenterol Jpn 14(5):492–495
36. Beau B, Colasse W, Le Bihan G, Bourreille J (1980) Association of Takayasu's arteritis and inflammatory bowel disease (author's transl). Sem Hop 56(43–44):1841–1845
37. E. T (1981) Shoni Shinryo 44(9):1439–1442
38. K. M (1983) Nihon Naika Gakkai Zasshi 72(1):102

39. Nishimura JYA, Araki H, Tomoike H, Takeshita A, Nakamura M (1984) A case of Takayasu's disease associated with ulcerative colitis: an important role of HLA-Bw52in its pathogenesis. J Jpn Soc Int Med 75:1341
40. H. O (1985) Nihon Naika Gakkai Zasshi 74(1):82
41. Achar KN, Al-Nakib B (1986) Takayasu's arteritis and ulcerative colitis. Am J Gastroenterol 81(12):1215–1217
42. Yamaguchi A, Kitamura N, Otaki M, Minoji T, Tamura H, Kurata A (1988) A case of Bentall's operation for annulo aortic ectasia complicated with ulcerative colitis and aortitis syndrome. Nippon Kyobu Geka Gakkai Zasshi 36:578–582
43. Ichikawa M, Onizuka T, Ichikawa K, Ohama A, Takahara O, Hoshino H (1988) A case of ulcerative colitis associated with Takayasu's arteritis. Nihon Shokakibyo Gakkai Zasshi 85(9):1690–1693
44. Ikenaga H, Ogihara T, Iyori S, Kou S, Yoshikawa H, Okura MM (1989) Does a common pathophysiological basis exist in the association of ulcerative colitis and Takayasu's aortitis? Report of a case. Postgrad Med J 65(768):761–764
45. Sakhuja V, Gupta KL, Bhasin DK, Malik N, Chugh KS (1990) Takayasu's arteritis associated with idiopathic ulcerative colitis. Gut 31(7):831–833
46. Goto MNI, Sakai K (1991) A case of ulcerative colitis with aortitis syndrome, scroilitis, and pancreatic cancer. Kyushu Ryumachi 10:144
47. Yazawa MTK, Takahashi M, Sato Y (1992). [Zasshi] [Journal] Nihon Kyobu Geka Gakkai 40:459–465
48. Yoshida H, Tomokuni T, Tamura K, Watase K, Matsuwaka R, Matsuda H (1992) A surgical case of thoracic aortic aneurysm due to Takayasu's aortitis associated with ulcerative colitis. Nihon Kyobu Geka Gakkai 40(7):1135–1139
49. Ishikawa H, Kondo Y, Yusa Y, Ejiri Y, Sato Y, Miyata M, Obara K, Nishimaki T, Kasukawa R, Saito A et al (1993) An autopsy case of ulcerative colitis associated with Takayasu's disease with a review of 13 Japanese cases. Gastroenterol Jpn 28(1):110–117
50. Oyanagi H, Ishihata R, Ishikawa H, Suzuki S, Kondo Y, Miyata M, Obara K, Nishimaki T, Kasukawa R, Tanno Y et al (1994) Ulcerative colitis associated with Takayasu's disease. Intern Med 33(2):127–129
51. Sato R, Sato Y, Ishikawa H, Oshima Y, Suzuki T, Watanabe S, Oyanagi H, Sekine K, Kondo Y, Miyata M et al (1994) Takayasu's disease associated with ulcerative colitis. Intern Med 33(12):759–763
52. So S, Okabe N, Maeda K, Okada M, Tanaka M, Okumura M, Fujita K, Tanaka K (1995) A case of ulcerative colitis associated with aortitis syndrome. Nihon Shokakibyo Gakkai zasshi 92(2):175–179
53. Morita Y, Yamamura M, Suwaki K, Mima A, Ishizu T, Hirohata M, Kashihara N, Makino H, Ota Z (1996) Takayasu's arteritis associated with ulcerative colitis; genetic factors in this association. Intern Med 35(7):574–578
54. Ariizumi KSA, Hashimoto R, Tada Y (1999) Carotid reconstruction for cerebral ischemia in aortitis syndrome combined with ulcerative colitis. Jpn J Cardiovasc Surg 28:339–342
55. Kawashima M, Koike R, Nishio J, Nanki T, Kohsaka H, Kubota T, Miyasaka N (1999) A case of Takayasu's arteritis with ulcerative colitis diagnosed by carotodynia and MRI findings. Nihon Rinsho Meneki Gakkai Kaishi 22(5):317–323
56. Ito Y, Kume H, Goto K, Ichikawa M, Nishiyama O, Kato K, Ito Y, Suzuki R, Yamaki K (2001) Aortitis syndrome associated with ulcerative colitis, preceded by pulmonary infarction involvement. Nihon Kokyuki Gakkai Zasshi 39(4):287–292
57. Suzuki TKH, Tabata M, Tomomasa T, Morikawa A (2001) Association of aortitis syndrome and ulcerative colitis: case report and review of literatures. Jpn J Pediatr Gastroenterol Hepatol Nutr 15(2):88–93
58. Shibata C, Funayama Y, Naito H, Matsuno S, Sasaki I (2002) Takayasu's arteritis after total proctocolectomy for ulcerative colitis: report of a case. Dis Colon Rectum 45(3):422–424

59. Fukunaga K, Sawada K, Fukuda Y, Matoba Y, Onishi K, Fukui S, Yamamura M, Satomi M, Shimoyama T (2002) A case report: first case of filtration leukocytapheresis for a patient of aortitis syndrome associated with ulcerative colitis. Ther Apher 6(1):93–98
60. Masuda H, Ishii U, Aoki N, Nakayama H, Sato F, Karube H, Suzuki S, Kondo T (2002) Ulcerative colitis associated with Takayasu's disease in two patients who received proctocolectomy. J Gastroenterol 37(4):297–302
61. Segawa TUK, Taniguchi Y, Tachibana H, Satou H, Kawazoe K, Hiramori K (2002) A case of Bentall's operation for aortitis syndrome associated with ulcerative colitis. Respir Circ 50(6):631–636
62. Bansal R, Aggarwal P, Handa R, Biswas A, Bandhu S, Wali JP (2003) Ulcerative colitis associated with Takayasu arteritis. Int J Cardiol 88(1):91–93
63. Hokama A, Kinjo F, Arakaki T, Matayoshi R, Yonamine Y, Tomiyama R, Sunagawa T, Makishi T, Kawane M, Koja K, Saito A (2003) Pulseless hematochezia: Takayasu's arteritis associated with ulcerative colitis. Intern Med 42(9):897–898
64. Gearry R, Carne P, Frizelle F (2003) Takayasu's arteritis and ulcerative colitis. N Z Med J 116(1170):U359
65. Ohta Y, Ohya Y, Fujii K, Tsuchihashi T, Sato K, Abe I, Iida M (2003) Inflammatory diseases associated with Takayasu's arteritis. Angiology 54(3):339–344
66. Nakano H, Ikeuchi H, Uchino M, Nakamura M, Noda M, Yanagi H, Yamamura T (2004) Takayasu's arteritis associated with ulcerative colitis. J Jpn Soc Coloproctol 57(7):396–401
67. Katsinelos P, Tsolkas P, Beltsis A, Chatzimavroudis G, Papaziogas B, Kapelidis P, Katsinelos T, Atmatzidis K, Paroutoglou G (2005) Takayasu's arteritis and ulcerative colitis in a young non-Asian woman: a rare association. Vasa 34(4):272–274. doi:10.1024/0301-1526.34.4.272
68. Sood A, Midha V, Sood N, Bansal M (2006) Coexistence of Takayasu's arteritis with ulcerative colitis. J Assoc Physicians India 54:151–152
69. Callejas-Rubio JL, Lopez-Nevot MA, Martin-Ibanez J, Ortego-Centeno N (2006) Takayasu arteritis associated with ulcerative colitis. Med Clin (Barc) 126(13):518
70. Balamtekin N, Gurakan F, Ozen S, Oguz B, Talim B (2009) Ulcerative colitis associated with Takayasu's arteritis in a child. Acta Paediatr 98(8):1368–1371. doi:10.1111/j.1651-2227.2009.01330.x
71. Asano Y, Morita S, Iguchi K, Kasugai H, Inamori M, Uchiyama T, Iida H, Endo H, Hosono K, Sakamoto Y, Fujita K, Yoneda M, Takahashi H, Koide T, Tokoro C, Goto A, Abe Y, Kobayashi N, Kubota K, Nakajima A (2010) Ulcerative colitis with Takayasu disease. Digestion 82(4):261. doi:10.1159/000297160
72. Kudo T, Aoyagi Y, Fujii T, Ohtsuka Y, Nagata S, Shimizu T (2010) Ulcerative colitis and aortitis syndrome. Pediatr Int 52(1):e43–e45. doi:10.1111/j.1442-200X.2009.02985.x
73. Kashima K, Kawasaki D, Yotsumoto G, Hatake S, Yamashita E, Nagayoshi S, Yoshishige Y, Tanoue K, Nagano S, Tanaka H, Nuruki N, Sonoda M (2010) Rapid progression of aortic regurgitation with thoracic aortic aneurysm due to Takayasu arteritis associated with ulcerative colitis. Intern Med 49(11):1007–1011
74. Inagaki TKT, Naganuma H, Hashimoto K, Morikawa T, Hano H (2010) Takayasu arteritis associated with nonspecific interstitial pneumonia and ulcerative colitis. Jikeikai Med J 57:33–38
75. Takahashi N, Tanabe K, Sugamori T, Sato M, Kitamura J, Sato H, Yoshitomi H, Ishibashi Y, Shimada T (2011) Association between Takayasu arteritis and ulcerative colitis – case report and review of serological HLA analysis. Med Sci Monit 17(7):CS81–CS84
76. Gecse K, Ruzsa Z, Nagy F, Wittmann T, Molnar T (2011) Successful infliximab treatment in a patient with Takayasu arteritis associated with ulcerative colitis or migration does not override genetics. Inflamm Bowel Dis 17(7):E69–E70. doi:10.1002/ibd.21731
77. Horai Y, Satoru O, Lapalme-Remis S, Sumiyoshi R, Nakashima Y, Suzuki T, Okada A, Kawashiri SY, Ichinose K, Tamai M, Yamasaki S, Nakamura H, Takeshima F, Origuchi T, Kawakami A (2013) Takayasu arteritis developing during treatment of ulcerative colitis with infliximab. Mod Rheumatol 23(3):572–576. doi:10.1007/s10165-012-0684-z

78. Azak A, Huddam B, Kocak G, Kilic F, Kocak E, Duranay M (2012) Takayasu arteritis and ulcerative colitis; coexistence or misdiagnosis? Sarcoidosis Vasc Diffuse Lung Dis 29(1):53–54
79. Kim GB, Kwon BS, Bae EJ, Noh CI (2012) Takayasu arteritis presenting as dilated cardiomyopathy with left ventricular thrombus in association with ulcerative colitis. J Am Coll Cardiol 60(14):e25. doi:10.1016/j.jacc.2011.11.080
80. Chae MJ, Yu CW, Lee SY, Jang DH, Hyun JY, Jeong SJ, Kim MH (2013) Takayasu's disease in a patient with ulcerative colitis. Korean Circ J 43(2):135–138. doi:10.4070/kcj.2013.43.2.135
81. Pyo JY, Park JS, Song CH, Lee SW, Park YB, Lee SK (2013) Takayasu arteritis associated with ulcerative colitis and optic neuritis: first case in Korea. Korean J Intern Med 28(4):491–496. doi:10.3904/kjim.2013.28.4.491
82. Khoshnama E, Taghipour M, Sharifian R, Saravi OE (2014) Coexistence of Takayasu's arteritis, ulcerative colitis, and stroke: a letter to the editor. J Res Med Sci 19(3):284–285

Chapter 8
Giant Cell Arteritis

Silvia Laura Bosello, Elisa Gremese, Angela Carbonella, Federico Parisi, Francesco Cianci, and Gianfranco Ferraccioli

Abstract Giant cell arteritis (GCA) is vasculitis characterized by a granulomatosus infiltrate, that typically occurs in medium and large arteries. This systemic autoimmune disease can cause sudden and potentially bilateral sequential vision loss in the elderly. GCA frequently occurs together with polymyalgia rheumatica. Both are syndromes of unknown cause, but genetic and environmental factors might have a role in their pathogenesis. The clinical findings in GCA are broad, but commonly include visual loss, headache, scalp tenderness, jaw claudication, cerebrovascular accidents, aortic arch syndrome, thoracic aorta aneurysm, and dissection. Glucocorticosteroids are the cornerstone of treatment, but some patients have a chronic course and might need glucocorticosteroids for several years. Adverse events of glucocorticosteroids affect more than 50 % of patients. Trials of steroid-sparing drugs have yielded conflicting results. The understanding of the molecular mechanisms involved in the pathogenesis of GCA recently has indicated new targets for therapy.

8.1 Definition and Position in Nomenclature of Vasculitis

Giant cell arteritis (GCA) is vasculitis characterized by a granulomatosus infiltrate, that typically occurs in medium and large arteries with well developed wall layers and advential vasa vasorum [1, 2]. According to the new proposed Chapel Hill classification, GCA is one of the two main subtypes of large vessel vasculitis (LVV), together with Takayasu arteritis [3]. The term "temporal arteritis" is not anymore considered a suitable alternative for GCA because not all patients have temporal artery involvement, and other vasculitis can affect the temporal arteries [3]. The two LVV affect different age groups of patients and have different disease burden. Aortic aneurysms are more common in GCA, while stenotic changes of the aorta are more common in Takayasu. Although many of the clinical features and pathological findings in GCA and Takayasu can overlap, they are distinct entities with different

S.L. Bosello • E. Gremese • A. Carbonella • F. Parisi • F. Cianci • G. Ferraccioli (✉)
Fondazione Policlinico Universitario Agostino Gemelli,
Università Cattolica del Sacro Cuore, Rome, Italy
e-mail: gianfranco.ferraccioli@unicatt.it

F. Dammacco et al. (eds.), *Systemic Vasculitides: Current Status and Perspectives*, DOI 10.1007/978-3-319-40136-2_8

pathophysiologic mechanisms; in fact when a direct comparison between Takayasu and GCA with upper extremities involvement has been performed, many differences in clinical manifestations and in imaging findings emerged [4].

In GCA the aorta and its extracranial branches are primarily affected. The vascular beds, that are usually involved, include the external carotid branches (e.g., temporal and occipital arteries), the ophthalmic, vertebral, distal subclavian, and axillary arteries, the aorta (especially thoracic segment) and, rarely, lower limb arteries [1, 2, 5, 6]. Intracranial arteries are rarely associated with the inflammatory process. Even though LVV affect large arteries much more often than does vasculitis of any other category, large artery injury is not the cause of the most significant morbidity in GCA, because for every large artery that is affected there may be many smaller branches affected (especially medium arteries), like when blindness is due to injury to smaller branches of the ophthalmic arteries [3].

Vasculitis leads to luminal occlusion and therefore cranial ischemic complications such as ischemic optic neuropathy are the most feared aspects of GCA, that causes vision loss in 10–15 % of patients [1, 2, 5, 6]. Aortitis can be complicated by dissection and aneurysm formation. GCA frequently occurs with polymyalgia rheumatica (PMR). These two conditions are closely related disorders that affect people of middle age and older, and their clinical and epidemiological connections have suggested that they are different manifestations of the same disease process [1, 2, 7].

8.2 Epidemiology

The incidence rates of GCA and PMR increase progressively after 50 years of age, with a mean age of onset around 72 years. The prevalence ranges from 1.47 to 20 per 100,000 people over 50 years. The reported rates for GCA are the highest in northern European countries and in North America, lower in Mediterranean countries and the lowest in Arabian and Asian countries (Japan). Women are affected 2–3 times more commonly than men [1, 2, 5–7]. The incidence of GCA has increased over the past 20–40 years, possibly because of raised awareness and as the population throughout the world continues to age, an increased prevalence of the disease should be expected. The cyclic pattern of yearly incidence rates and the seasonal variations reported by some studies might suggest an environmental-infectious etiology. The decreasing incidence with a north–south gradient – the highest incidence being in Scandinavian and in US communities with a Scandinavian ethnic background – and the occasional familial cases support a possible role for both environmental and genetic factors. An association with HLA-DRB*04 alleles was found and there is a potential relevance of gene polymorphisms that influence the immune and inflammatory responses [8].

Forty-fifty percent of GCA patients have polymyalgic symptoms and 5–15 % of PMR patients will be diagnosed of GCA [1, 2, 7].

Aorta involvement occurs in 15 % of patients with GCA, already at diagnosis. Although patients with GCA have an increased risk of developing aortic aneurysm and dissection and cerebrovascular accidents, most studies of long-term survival have shown no excess mortality. GCA is not a self-limiting condition and vasculitis is persistent without treatment, but it is not associated with life-threatening complications [1, 2, 5–7].

8.3 Pathogenesis

GCA etiology remains unknown but could be associated with exposure to bacterial or viral antigens following infection or vaccination, in genetically predisposed patients. Both the innate and adaptive immune systems contribute to GCA pathogenesis. Granulomatous inflammatory infiltrates composed of CD4 T cells, activated macrophages, and multinucleated giant cells induce intimal hyperplasia and luminal destruction. Dendritic cells located at the adventitia-media border of the artery have a crucial role in initiation of vasculitis. The activated dendritic cells become chemokine-producing effector cells, which recruit macrophages and CD4 T cells into the vascular wall through the vasa vasorum. T cells and macrophages are critical players in the vasculitic process sustaining the granulomatous inflammation in the arteries [1, 9]. Two major immune-response networks have been identified: the interleukin-12–type-1 helper T-cell (Th1)–interferon-γ axis and the interleukin-6–type 17 helper T-cell (Th17)–interleukin-17 or interleukin-21 axis; the latter (but not the former) is effectively suppressed with glucocorticoid treatment. Th17 immunity seems to be more important for the acute manifestations, both systemically and in the blood vessels; while Th1 immunity is associated with chronically persistent vascular lesions [10]. The T-regulator cells Foxp3+, which normally limit immune response, are reduced in GCA. Vascular lesions in inflamed temporal arteries contain an array of cytokines and inflammatory mediators. High levels of IL-6, a pleiotropic cytokine, produced by T cell, B cell, macrophages, and fibroblasts, have been associated with disease activity and therapeutic response. The IL-6 pathway represents the intersection of the innate and acquired immune systems and, when unregulated, maintains inflammation in GCA. Besides IL-6, circulating IL-1 and IL-12 were elevated in untreated patients. Effector cytokines released into the arterial wall activate inflammatory cells and target endothelial cells, vascular smooth-muscle cells, and fibroblasts, leading to lumen-obstructive intimal hyperplasia. The endothelial cells participate in the activation of T cells, the terminal differentiation of B cells, the survival of plasmocytes, the differentiation of Th17 lymphocytes, and the inhibition of Treg-cell differentiation and function [1, 9].

In the adventitia, macrophages produce the inflammatory cytokines IL-1 and IL-6, whereas in the media they release metalloproteinases and reactive oxygen intermediates. These inflammatory mediators lead to the fragmentation of the internal elastic lamina and trigger repair mechanisms such as intimal hyperplasia and neoangiogenesis, which are regulated by platelet-derived growth factor and vascular endothelial growth factor. Platelet-derived growth factor seems to be crucial in induction of intimal hyperplasia and vessel occlusion [1, 9].

Autoantibodies were not consistently found in GCA, although plasma cells were found in the adventitia in 7–24 % of temporal artery biopsies from patients with GCA. The exception was in few studies the antiphospholipid antibodies, which were found in 30–80 % of GCA cases, however, most studies did not find significant correlations between the presence of antiphospholipid antibodies and ischemic complications in GCA patients [7–9]. Recently it has been reported that most patients with GCA have autoantibodies to human and bacterial ferritin peptides, but the etiological, pathogenetic and clinical significance of these findings need to be studied further [11].

8.4 Histopatological Findings

In GCA inflammation mainly affects the large-sized and medium-sized muscular arteries, especially the proximal aorta and its branches. These arteries have a prominent internal elastic membrane and vasa vasorum. The histological analysis is performed at the temporal artery, in some atypical manifestation the occipital or cervical arteries can be biopsied. The temporal artery biopsy should be performed, as soon as the GCA has been suspected, to start as soon as possible therapy and thus avoid visual complications. Multiple cross sections of the temporal artery will need to be examined in GCA due to the occurrence of skip lesions [1, 2, 5–7].

Histopathologically, two main patterns are considered diagnostic of GCA: those with inflammation of the vessel wall and those with post-inflammatory alterations.

GCA is characterized by:

- Granulomatous inflammation and multinucleated giant cells at the junction of media and intima,
- A mononuclear transmural infiltrate without giant cells.
- Vasculitis involving small vessels close to a non-inflamed temporal artery.
- False channel, inflammatory infiltrate, media modifications, and abnormalities in smooth muscle cells, associated with phagocytic activity.

The inflammatory infiltrate in the affected vessels is predominantly composed of lymphocytes and macrophages with or without multinucleated giant cells. Sparse neutrophils and eosinophils can be present. The majority of lymphocytes are CD4 positive cells, but also CD8 positive cells are involved [1, 2, 5, 7]. The presence of B cells in temporal artery biopsy specimens is scarce, but abnormalities in circulating B cells have been recently reported in GCA and can contribute to the enhanced IL-6 response in the disease [12]. Inflammation tends to affect the arteries in a segmental fashion and the inflammatory process is usually most severe in the inner portion of the media adjacent to the internal elastic lamina. In some cases the inflammation is restricted to vasa vasorum to periadventitial small vessels or both. In a chronic damage, arteritis lacks the active inflammation but shows fibrosis of the vessel wall with disruption of the internal elastic lamina in the media. The media necrosis is characterized by areas with loss of smooth cells and collapse of the

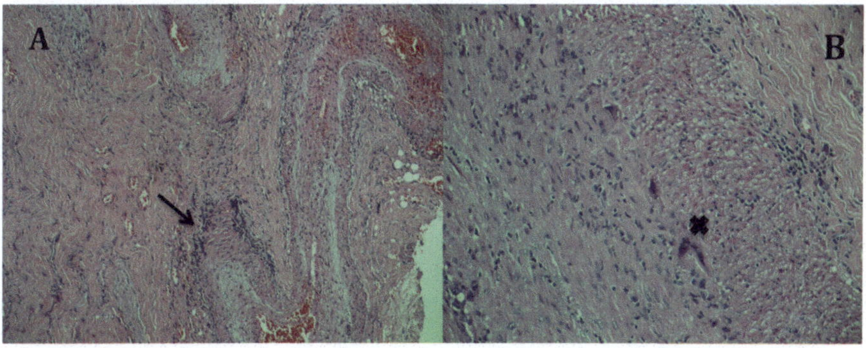

Fig. 8.1 Histopathological findings at temporal artery. Legend: Granulomatous inflammation and multinucleated giant cells at the junction of media and intima. (**A**) section of vasa vasorum of temporal artery from a patient with GCA. Mononuclear cell infiltrate (*arrow*) and (**B**) giant cells (*asterisk*)

elastic lamellae often bordered by proliferating vasa vasorum with perivascular inflammation. Elastin stains can be used to highlight the fragmentation and loss of internal elastic lamina. The intima shows reactive hyperplasia that can become fibrotic with or without subsequent atherosclerosis over time. A variable thickening and inflammation of the adventia can be noted, but milder that the one present in Takayasu disease [1, 2, 5, 7]. In few reported cases the temporal artery biopsy excluded the diagnosis of GCA, showing diffuse amyloid infiltration of the artery (Fig. 8.1).

8.5 Clinical Manifestations

GCA is characterized by a wide range of clinical manifestations, and can present with atypical features, resulting in challenging diagnosis.

Severe headache is probably the most common onset symptom and is present in about two-thirds of the patients; scalp tenderness is usually limited to the temporal arteries, but it may also involve larger areas. Pain is usually continuous throughout the day, and often interferes with sleep, and responds incompletely to analgesic drugs. The temporal arteries when involved appear thickened, nodulous, painful, and erythematous with impaired or absent pulse. Almost half of the patients present "jaw claudication" (due to ischemia of the muscles of mastication), tongue pain, dysphagia and impaired swallowing; seldom a severe vascular thickening can lead to scalp or tongue infarction. Jaw claudication is a high predictor of GCA, but is not pathognomonic. Scalp tenderness arises in around half of patients; it is usually worsened by brushing or combing the hair. These symptoms are most often in patients with headache. Permanent visual loss, partial or total, occurs in up to 20 % of the patients and is often the first manifestation of the disease. Once established, the visual deficiency is usually permanent. Amaurosis fugax precedes permanent loss in 44 % of the patients. Transient diplopia is present in around 6 % of the patients [6].

Polymyalgia is the most frequent muscoloskeletal manifestation in GCA and polymyalgic symptoms can occur before, during or after the vasculitis diagnosis. Since some patients with PMR have subclinical vasculitis or a condition that progresses to vasculitic complications, follow-up evaluation is needed. Polymyalgia is from 3 to 10 times more frequent in patients with GCA. Distal symptoms, such a peripheral arthritis and distal swelling with pitting edema and Raynaud's phenomenon can arise in 25 % of the patients [1, 6, 9].

One or more systemic manifestations, including fever, malaise, anorexia, and weight loss, are present in most patients. Fever is usually of low grade, but it reaches 39–40 °C in about 15 % of patients and might be the presenting manifestation or the only feature of giant-cell arteritis. Neurological manifestations occur in about 30 % of patients, in around 14 % of all patients they consist of neuropathies and peripheral mononeuropathies and peripheral polyneuropathies of the upper and lower extremities. Respiratory tract symptoms including cough, sore throat and hoarseness occur in about 10 % of the patients. GCA patients may also suffer from ear pain, and dizziness, due to vestibular or cochlear dysfunction. Cranial nerve palsies or ischemic myopathy may rarely occur [1, 9].

Large vessel vasculitis occurs in 25 % of patients with GCA and aortic arch syndrome in about 10–15 % of patients presenting with claudication of the arms; bruits over the carotid, subclavian, axillar and brachial arteries; and absent or decreased pulses in the neck or arms. Thoracic aortic aneurysm and dissection of the aorta are important late complications of GCA, but, unfortunately, there are no clinical predictors across studies that allow clinicians to identify this risk of large vessel involvement. Disease recurrences and an incomplete response to standard doses of steroids would raise the possibility of large vessel involvement [1, 5].

8.6 Ocular Manifestations

Sudden, severe, and sequential vision loss is the hallmark of GCA. The vision loss is usually discovered upon awakening in the morning. Visual acuity is usually less than 20/200 in more than 60 % of patients with vision loss. The fellow eye usually gets involved within days to weeks of the initial eye. In addition to causing a sudden permanent vision loss, GCA can present weeks earlier with amaurosis fugax or a temporary vision loss, which is due to partial occlusion of the short posterior ciliary arteries or central retinal artery causing transient ischemia. GCA may initially also present with diplopia or eye pain [1, 6].

A significant decline in the incidence of visual manifestations has been reported over the past 5 decades. Chances of recovery were poor in patients with anterior ischemic optic neuropathy or complete vision loss, but patients diagnosed in the later decades were more likely to recover from visual symptoms because of an earlier treatment. Thus, any elderly patient presenting to the doctor with visual

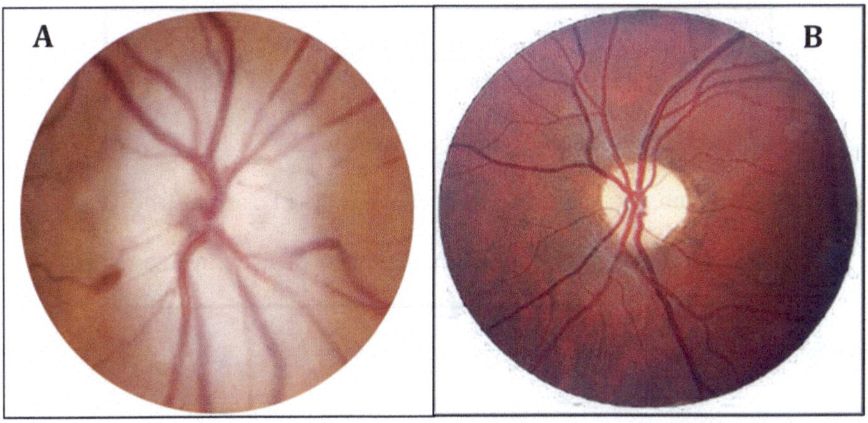

Fig. 8.2 Optic disc in patients with GCA. Legend: (**a**) Optic disc edema in initial phases. (**b**) Optic disc atrophy and flamed shape hemorrhages

symptoms or eye pain should be considered to be a GCA until proven otherwise in order to minimize permanent vision loss in GCA patients [13].

Sudden vision loss in GCA occurs most often due to an inflammatory thrombosis of the short posterior ciliary arteries, which form a fine vascular network that supplies the optic disk. When these vessels become thrombosed with inflammation, a stroke to the optic disk occurs. This is called anterior ischemic optic neuropathy or AION, that is characterized by a swollen optic disk accompanied by hemorrhages and sometimes exudates. The swollen optic disk may have a chalky white appearance in GCA. This pallid swelling is due to the extreme ischemia of GCA, in the early phases. Jaw claudication, diplopia and temporal abnormalities are predictive features of anterior ischemic optic complication [6].

Rarely, the ischemia to the optic nerve occurs posteriorly, and therefore there is no disk swelling. In this instance, it is called posterior ischemic optic neuropathy or PION. GCA may also sometimes cause a central retinal artery occlusion (CRAO). About 5 % of patients over age 50 with CRAO have GCA. In this case a classic cherry red spot in the macula occurs, but no cholesterol or calcific embolus will be seen since again this is due to an inflammatory thrombosis. GCA may also rarely cause a cilio-retinal artery occlusion or ocular ischemic syndrome, that is characterized by eye pain, iritis, and hypotonia. Cotton wool spots may also be seen in the retina and indicate concurrent retinal ischemia. Fluorescein angiography (FA) may be helpful to identify choroidal hypoperfusion and aid in the timely diagnosis of GCA (Figs. 8.2 and 8.3) [6].

AION: inflammatory thrombosis of the short ciliary arteries, that normally supply the optic disc

CRAO: central retinal artery occlusion

PION: posterior ischemic optic nerve, without disk swelling

Cilio retinal artery occlusion: eye pain, iritis and hypotony

Fig. 8.3 Ocular manifestations in GCA

Table 8.1 Clinical subtypes in GCA disease

	Polymyalgia Rheumatica	Temporal arteritis	Aortitis	Inflammatory arteritis syndrome
Clinical symptoms	Inflammatory muscle pain, morning stiffness	Headache, amaurosis fugax, jaw or tongue claudication, scalp tenderness and/or hyperesthesia	Claudication, pulse differences	ESR-FUO Weight loss Night sweating Fatigue, malaise
Involved Arteries		Branches of A. carotid A. vertebralis	Subclavian arteries, Carotid, Vertebrobasilar Iliac arteries Aorta	
Complications	Excessive pain medication	Loss of vision Ischemic tongue	Aneurysm, Arterial dilatation, Aortic valve insufficiency	Loss of vision

8.7 Subtypes of GCA Disease

The clinical characteristics of GCA spectrum can be summarized in four main clinical subtypes that are associated with peculiar complications and are summarized in Table 8.1 [5].

- **PMR:** is characterized by predominantly proximal muscle pain and stiffness, significant morning stiffness, and much less commonly a peripheral inflammatory arthritis. The large vessels involvement is often asymptomatic.
- **Cranial (temporal) arteritis:** This is characterized by temporal headache, amaurosis fugax, jaw and/or tongue claudication, scalp tenderness and/or

hyperesthesia. The affected temporal artery is often thickened, tortuous, tender, nodular, or have diminished pulsation. The main concern of cranial GCA is "arteritic" AION of the eye. If left untreated, the other eye is also likely to be affected within weeks. Rarely, patients can present with scalp necrosis.

- **Extracranial disease manifesting as LVV :** It has become evident in the last few decades that large artery involvement is a common but under-recognized manifestation of GCA, and tends to occur in up to a third of patients with this disease. This could be in the form of (a) arterial stenoses, occlusions or ectasias affecting subclavian and axillary arteries, carotid and vertebro-basilar arteries, and also sometimes the iliac arteries and their distal branches, and (b) aortitis, which is often asymptomatic, manifesting as ectasias and aneurysm formation in the thoracic and less commonly the abdominal aorta. This can sometimes be complicated by aortic rupture or dissection. As it is mostly asymptomatic, vascular imaging studies can be useful to detect the large vessels involvement in patients with atypical symptoms and raised inflammatory biomarkers.
- **Inflammatory arteritis syndrome with constitutional symptoms:** are common in both GCA and PMR or can be present as only symptoms: fever, malaise, fatigue, weight loss, anorexia, night sweats.

8.8 Diagnosis

GCA is considered on the basis of the medical history, clinical evaluation, and laboratory and imaging tests, and it is confirmed on the basis of histologic findings.

The diagnosis of GCA is particularly challenging and should be considered in a patient over the age of 50 who develops [14]:

- new headache,
- jaw claudication,
- acute onset of visual disturbances,
- symptoms of polymyalgia rheumatica,
- unexplained fever or anemia, or cough or cachexia,
- elevated acute phase reactants,
- temporal artery abnormality on examination.

A high percentage (40 %) of patients presents nonspecific symptoms. Except for histopathology of the artery wall, neither laboratory data nor specific signs or symptoms exist for the diagnosis of temporal arteritis. Only 4 % of patients with confirmed GCA had both a normal ESR and a normal CRP level at the time of diagnosis. The classification criteria proposed by the American College of Rheumatology are currently used (Table 8.2). For each of these criteria, the presence of all the characteristics is needed in order to formulate the diagnosis of GCA. The presence of three or more criteria leads to a 93.5 % sensibility and to a 91.2 % specificity [9, 15].

Table 8.2 Classification criteria for the diagnosis of GCA (ACR 1990) [15]

Criteria	Definition
Age	Onset of the symptoms at age 50 or older
Recent onset of severe headache	Onset of a new type of pain located in the head
Anomalies of temporal artery	Pain in the palpation of temporal artery or reduced pulsatility, not connected to the atherosclerosis of the cervical arteries
Increase of ESR	ESR > 40 mmHg
Typical abnormalities of the biopsy of the temporal artery	Vasculitis typically characterized by an infiltrate of mononuclear cells or inflammation of granuloma with multinucleated cells.

The superficial temporal artery biopsy represents the gold standard for the diagnosis of GCA, while a negative result does not exclude it; 10–15 % of the biopsies are falsely negative.

EULAR recommendations for the management LVV recommended that "*A temporal biopsy should be performed whenever a diagnosis of GCA is suspected, but this should not delay the treatment. A contralateral biopsy is not routinely indicated*" [16]. Thus, temporal artery biopsy should be performed soon when cranial GCA is suspected, ideally prior to starting glucocorticoids, on the most symptomatic side. However, due to the segmental inflammatory involvement of the temporal artery, a contralateral biopsy may be considered to confirm a pathologic diagnosis of this vasculitis in patients in whom clinical suspicion of GCA is high. This simple procedure has no absolute contraindication and complications are uncommon. Thus, if not done earlier, it is recommended to get the biopsy even up to 4 weeks after initiation of therapy. Due to segmental involvement, sufficient specimen length (>1 cm) is also important. If the temporal arteries are not abnormal on examination but the facial or occipital arteries are, then these arteries should be biopsied instead. This may also be deemed necessary if bilateral temporal artery biopsies are negative and the diagnosis of cranial-GCA is still being considered [5].

The ultrasound of the temporal artery is considered to be helpful in the diagnosis. In the ultrasound color-doppler, a "halo" effect around the temporal artery is a specific aspect in patients with clinically active arteritis and confirms the diagnosis; however its sensitivity is low (40 %); therefore the absence of this sign does not exclude it [9]. If an extracranial involvement of GCA is suspected, an arteriography, a computed tomography (CT) scan, and a magnetic resonance (MRI) are required as diagnostic tests [5, 7, 9].

8.9 Laboratory and Instrumental Diagnosis

Laboratory findings in GCA are non-specific but indicate the inflammatory nature of this syndrome [1]. Increased ESR and C–reactive protein (CRP) can be associated with thrombocytosis. The combination of ESR and CRP increase is 97 %

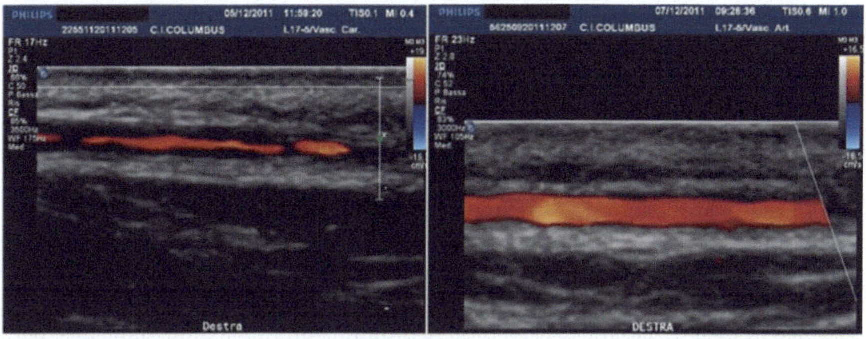

Fig. 8.4 Temporal artery US Halo sign. Legend: Halo sign in US of the temporal artery: a dark hypoechoic circumferential wall thickening around the artery lumen predominantly due to an acute inflammatory process of the arterial wall. In this figure the halo sign tends to disappear after starting steroid treatment

specific for the diagnosis of GCA. Tests for rheumatoid factor and for anti-cyclic citrullinated peptide antibodies are usually negative. A moderate anemia of chronic disease (ie, normocytic anemia) is present in most patients, and concentrations of liver enzymes—particularly alkaline phosphatase—can be mildly raised [1]. Serum interleukin-6 concentrations tend to correlate with disease activity and may be a better predictor of relapse than ESR [5], but no recommended biomarkers are available for diagnosis and monitoring.

The role of imaging studies in diagnosis and follow-up has been insufficiently defined and the validation of imaging techniques for diagnosis and monitoring is in the research agenda for GCA. Ultrasounds (US), computed tomography (CT), magnetic resonance imaging (MR), contrast angiography and positron emission tomography (PET) can be useful for corroborating the clinical diagnosis, to rule out asymptomatic large vessel involvement or asymptomatic progression and to evaluate a worsening due to inflammation and/or mechanical factors. The sensitivity and specificity of this high-cost imaging method for diagnosing and monitoring GCA have not been established. Cost and accessibility must be weighed against the concerns of radiation and contrast toxicity [5].

US can be useful to identify the vessel wall edema (halo sign) and irregularity in the artery (Fig. 8.4). The main specific US finding is the halo sign, a dark hypoechoic circumferential wall thickening around the artery lumen predominantly due to an acute inflammatory process of the arterial wall, that tends to disappear after starting steroid treatment [17].

MR and CT of the aortic arch and its major branches are useful in patients in whom GCA has been confirmed on biopsy, in order to assess the extent of arterial involvement (including the presence of stenosis, dissection, and aneurysms) and to monitor vascular lesions for any signs of progression. MR or CT may also be used to identify large-vessel involvement in patients with suspected GCA that has not been confirmed on biopsy and in whom there is clinical evidence of peripheral ischemic disease. CT can identify structural abnormalities, thickening and enhancement of the vessel wall. Intramural leaky microvessels give raised delayed enhancement of the arterial wall, which is consistent with, but not specific for, inflammatory activity. Wall thickening and increased intra-wall blood pooling may not be reversible with treatment and should not be used to assess the inflammatory burden or disease activity. Given the effective use of MR and CT, traditional angiography is now reserved for planning revascularization procedures, when required [2, 5, 7].

Conventional imaging tools (US, CT, MR, contrast angiography) provide anatomic and morphological information, while PET/CT provide metabolic assessment of GCA. ^{18}F-Fluorodeoxyglucose (^{18}F-FDG) PET is an effective tool for the diagnosis, grading, and follow-up of patients affected by GCA involving the aorta and its proximal branches, and it has been proposed for quantifying the inflammatory burden, but the lack of a standardized method for the assessment of vascular inflammation remains a critical issue, potentially leading to misclassification. Although there is no perfect method for the diagnosis of GCA at present, FDG-PET may play a role in the management of GCA when the ultrasonography is negative. Nowadays, ^{18}F-FDG with PET/CT cannot be relied on to distinguish vasculitis from non-vasculitic inflammatory lesions (e.g., atherosclerotic changes in vessel walls), and its sensitivity for smoldering and treated vasculitis is limited; thus, routine use is not recommended (Fig. 8.5) [18].

8.10 Open Questions: The Cardiovascular Risk and Large Vessel Involvement Evaluation

An increased cardiovascular and thromboembolic risk in vasculitis is reported and seem to be related to the period of higher activity of the disease. The majority of cerebrovascular ischemic events are due to the involvement of internal carotid or vertebro-basilar arteries, while arteritis of the intracranial district is rare and limited to a subset of GCA patients with a poor prognosis and a worse outcome with corticosteroids therapy. The role of traditional risk factors for cerebrovascular ischemic event in this disease is still debated, but considering the age of onset of GCA the investigation of cardiovascular risk factor can have important implications for clinical care both immediately after the diagnosis and in long-term treatment [9]. A recent large cohort study on 3,408 GCA patients with a median follow-up of 4 years evaluated the cardiovascular risk in GCA patients as compared to a sample of controls and found an increased incidence of cardiovascular events (myocardial infarction, cerebrovascular accident, peripheral vascular disease) with an incidence rate ratio of 1.68 (95% CI:1.49–1.89) [19]. The association of GCA with cardiovascular disease was higher 1 month after the diagnosis than during follow-up. These findings are consistent with those studies of GCA reporting an increased risk for blindness – also an ischemic vascular event- occurring early in the disease. These data support treatment with a low-dose of aspirin in clinical practice to prevent ischemic events [9] and the importance of treatment of all classical cardiovascular risk factors (arterial hypertension, diabetes, smoke) during corticosteroids therapy in GCA.

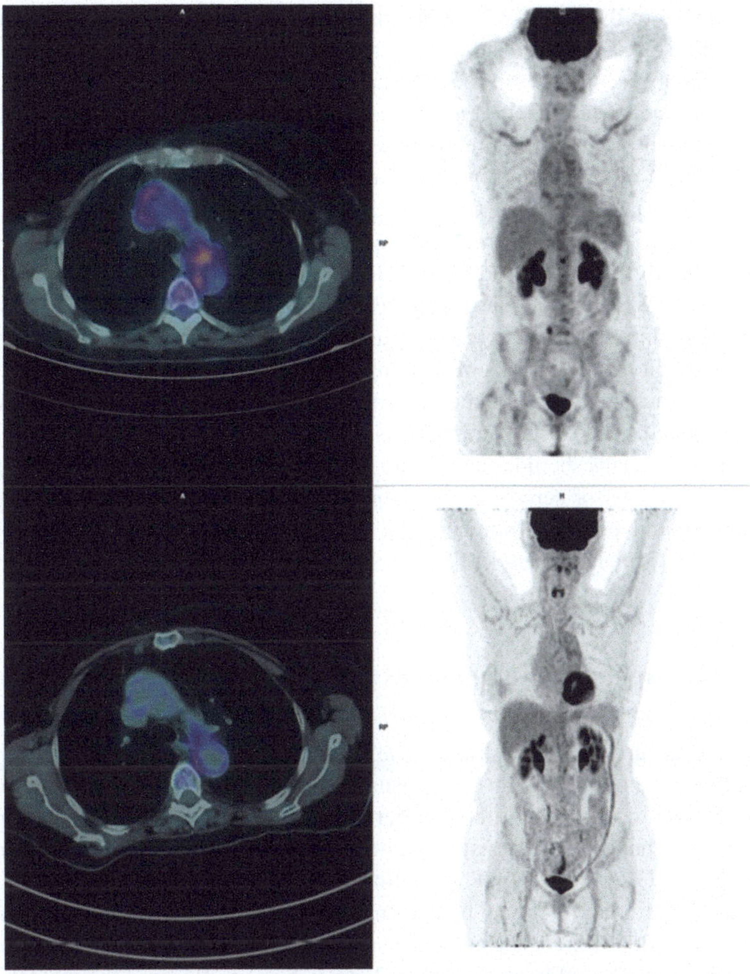

Fig. 8.5 PET/CT images of a patient with GCA and aortitis. Legend: PET/CT images of a patient with GCA and aortitis. *Upper panel*: Before treatment, wall thickening was demonstrated in the ascending aorta and aortic arch with increased FDG uptake, consistent with active inflammation. *Lower panel*: After glucocorticoid treatment, significant reduction in degree of FDG uptake associated with the wall of aortic arch, when compared to previous study. This finding is consistent with a positive response of the hypermetabolic inflammation of the aortic wall to therapy

Patients with GCA appear to have an elevated incidence of aortic aneurysms, particularly thoracic aortic aneurysm (TAA), compared with the general population; the aneurysm may only be discovered some years after GCA diagnosis, in the event of (often fatal) dissection or rupture. This late aneurysm development in GCA might be a consequence of cumulative inflammatory damage to the smooth muscle and/or elastic laminae of the aortic wall, although it has also been proposed that the aortic inflammation in GCA is secondary to atrophy of the aortic smooth muscle. In a recent metanalysis, the literature data support an association between GCA and TAA or thoracic aorta dilatation (TAD) compared with age-matched controls, but the true relative risk and the time course of that risk remain unclear because no controlled trials of aneurysm screening in GCA are available [7, 20]. Current UK guidelines for management of GCA suggest a chest radiograph every 2 years to screen for TAA, but the chest radiography is not a very sensitive test. Conversely, America Guidelines recommend CT or MRI of the thoracic aorta in the initial evaluation of GCA. Whether routine imaging can be generally recommended must also take into account resource use (costs, scanning time, patient acceptability of each imaging modality, and the risks of ionising radiation or intravenous contrast), considering that metanalysis states that in order to detect a previously unknown aneurysm, we would need to screen as few as five to ten GCA patients. So in clinical practice the clinicians should retain a high index of suspicion for aortic pathology in patients with GCA for screening the patients for TAA. Before ordering imaging, clinicians should consider whether, and how, detecting aortic pathology would affect a patient's management [20].

8.11 Therapeutic Management

Corticosteroids are the cornerstone of the therapy in GCA. Their use has dramatically reduced the frequency of severe visual ischemic complications in this disease. Current therapy with glucocorticoids offers prompt suppression of some inflammatory pathways, but probably resistant pathogenic pathways sustain chronic vascular remodelling in GCA. No absolute guidelines exist as to the length of treatment with corticosteroids for GCA. It may be reasonable to maintain the patient on treatment for 2 years to reduce the chances of relapses. Even then, relapses have been reported. Some patients may need treatment for as long as 5 years. Because the incidence of new visual damage appears to decrease with disease duration, repeat a temporal artery biopsy could be considered before restarting corticosteroids in patients who relapse after 18–24 months.

It is not clear whether more aggressive, longer term immunosuppression could improve outcomes. The coexistence of several vasculogenic immune abnormalities has complicated the development of new, glucocorticoid-sparing therapies.

8.11.1 Glucocorticoid Therapy

Glucocorticoids have been the gold standard treatment for GCA since their first use for the condition in the 1950s. EULAR recommends early initiation of high-dose glucocorticoid therapy for induction of remission in large vessel vasculitis [16] and the use of low-dose aspirin that can modulate the transcriptional activation of INF-ϒ gene in experimental model. Clinical trials on tromboprophylaxis for primary prevention of vascular outcomes in GCA are still lacking [16].

Given the risk of irreversible ischemic complications, new-onset clinical manifestations of disease indicating an unstable supply of blood to the eyes or the central nervous system (e.g., arteritic optic neuropathy, evolving visual loss, amurosis fugax) are typically managed with intravenous pulse therapy (e.g., 1000 mg of methylprednisolone per day for 3 consecutive days) to optimize immunosuppression and suppress tissue edema. Once tissue necrosis occurs (e.g., optic-nerve ischemia with blindness for several hours), it is irreversible [14].

To summarize the therapeutic approach:

- High-dose pulsed intravenous corticosteroids in patients with visual impairment
- Prednisolone 1 mg/kg/day (maximum 60 mg/day) as initial dose maintained for 1 month (in uncomplicated GCA without jaw claudication and visual disturbance)
- Established vision loss: 60 mg prednisolone daily to protect the contralateral eye.
- Tapering gradually corticosteroids: 10 mg every 2 weeks to 20 mg/die, then 2.5 mg every 2–4 weeks to 10 mg/die, then 1 mg every month
- Not alternate day therapy (more likely relapse)
- Therapy should last 1–2 years
- Low dose aspirin (75–150 mg per day), in the absence of contraindications.

The symptoms of GCA should respond rapidly to high-dose glucocorticosteroid treatment, followed by resolution of the inflammatory response. Failure to do so should raise the question of an alternative diagnosis. Monitoring of therapy for large vessel vasculitis should be clinical and supported by the measurements of inflammatory markers. Glucocorticosteroid reduction should be considered only in the absence of clinical symptoms, signs and laboratory abnormalities suggestive of active disease.

Patients should be monitored for evidence of relapse, disease-related complications and glucocorticosteroid related complications. Proton pump inhibitors for gastrointestinal protection should be considered [15]. Bone density measurement is recommended when therapy is initiated. Calcium, vitamin D supplements, and bisphosphonates are also necessary. Patients should be warned about other common side effects of long-term glucocorticoid use such as weight gain, glucose intolerance, hypertension, and opportunistic infections.

Even with gradual reduction of doses of glucocorticosteroids, clinical flares have been reported to occur in more than 50 % of patients, particularly during the first 12–16 months, when the prednisone dose is reduced to about 5–10 mg per day [1]. Disease relapse should be suspected in patients with a return of symptoms of GCA, ischemic complications, unexplained fever or polymyalgic symptoms and in particular, the following features should be sought: jaw and tongue claudication, visual symptoms, vascular claudication of limbs, bruits and asymmetrical pulses, polymyalgic symptoms, osteoporotic risk factors and fractures. A rise in ESR/CRP is usually seen with relapse, but relapse can be seen with normal inflammatory markers.

8.11.2 Combination Therapy

During a 10-year follow-up of a population-based cohort of patients with GCA, more than 80 % had at least one complication related to glucocorticoid treatment. Although mortality is not increased in cohorts of patients with GCA, corticosteroids usage is associated with a seven-fold increase in the risk of severe opportunistic infections compared to the rest of the population. A combination therapy approach has been proposed in GCA to reduce the relapse, reach lower dose of prednisolone, sparing oral glucocorticoids, reducing adverse event. Hydroxychloroquine, Methotrexate, Cyclophosphamide, Leflunomide, Azathioprine have been reported for the treatment of GCA, but their efficacy, though often used in this setting, has not been established.

EULAR recommendations and British guidelines recommend that an immunosuppressive agent should be considered for use in LVV as adjunctive therapy, especially in recurrent relapses [14, 16], but the results from a recent meta-analysis on combination therapy show no clear benefit from using adjunct therapy with corticosteroids for the treatment [21].

This metanalysis considering three placebo-controlled randomized trials involving patients with newly diagnosed GCA showed that a regimen of glucocorticoid therapy plus methotrexate as compared with glucocorticoids alone conferred a significant but modest benefit in lowering the relapse rate and in reducing the cumulative dose of glucocorticoids, without increased risk for infection but without reducing also the side effects of the glucocorticoids [21].

8.11.3 Biologic Therapy

The use of anti-TNF agents is limited. The infliximab trial was stopped early due to an increased risk of infection in those individuals receiving infliximab. The other anti-TNF trial involving adalimumab revealed a significantly increased number of patients with overall infections in the intervention arm as compared to control, and this reached statistical significance.

Therapy targeted at disrupting the function of IL-6 is currently undergoing clinical testing. In a small number of patients with GCA, treatment with the IL-6 receptor antagonist tocilizumab at a dose of 8 mg per kilogram per month resulted in rapid suppression of systemic inflammation [22]. However, it is not certain whether IL-6 blockade is effective for the treatment of vascular inflammation and it is important to keep in mind the risk of infection when using this drug in patients with GCA.

New drugs blocking IL17 might achieve interesting results in GCA. Adequately powered randomized controlled trials to assess the role of immunosuppressive therapy (conventional as well as biological) with glucocorticoids are necessary to draw scientific guidelines.

References

1. Salvarani C, Cantini F, Hunder GG (2008) Polymyalgia rheumatica and giant-cell arteritis. Lancet 372(9634):234–245
2. Weyand CM, Goronzy JJ (2014) Clinical practice. Giant-cell arteritis and polymyalgia rheumatica. N Engl J Med 371(1):50–57, 3 July 2014
3. Jennette JC, Falk RJ, Bacon PA, Basu N, Cid MC, Ferrario F, Flores-Suarez LF, Gross WL, Guillevin L, Hagen EC, Hoffman GS, Jayne DR, Kallenberg CG, Lamprecht P, Langford CA, Luqmani RA, Mahr AD, Matteson EL, Merkel PA, Ozen S, Pusey CD, Rasmussen N, Rees AJ, Scott DG, Specks U, Stone JH, Takahashi K, Watts RA (2013) 2012 revised international Chapel Hill consensus conference nomenclature of Vasculitides. Arthritis Rheum 65(1):1–11, January 2013
4. Kermani TA, Crowson CS, Muratore F, Schmidt J, Matteson EL, Warrington KJ (2015) Extracranial giant cell arteritis and Takayasu arteritis: how similar are they?. Semin Arthritis Rheum 44(6):724–728, June 2015
5. Chatterjee S, Flamm SD, Tan CD, Rodriguez ER (2014) Clinical diagnosis and management of large vessel vasculitis: giant cell arteritis. Curr Cardiol Rep 16(7):498, July 2014
6. Chacko JG, Chacko JA, Salter MW (2015) Review of giant cell arteritis. Saudi J Ophthalmol 29(1):48–52, January–March 2015
7. González-Gay MA, Pina T (2015) Giant cell arteritis and polymyalgia rheumatica: an update. Curr Rheumatol Rep 17(2):6, February 2015
8. Carmona FD, González-Gay MA, Martín J (2014) Genetic component of giant cell arteritis. Rheumatology (Oxford) 53(1):6–18, January 2014
9. Guida A, Tufano A, Perna P, Moscato P, De Donato MT, Finelli R, Caputo D, Di Minno MN (2014) The thromboembolic risk in giant cell arteritis: a critical review of the literature. Int J Rheumatol 2014, 806402
10. Deng J, Younge BR, Olshen RA, Goronzy JJ, Weyand CM (2010) Th17 and Th1 T-cell responses in giant cell arteritis. Circulation 121(7):906–915, 23 February 2010
11. Nesher G (2014) Autoimmune aspects of giant cell arteritis. Isr Med Assoc J 16(7):454–455
12. van der Geest KS, Abdulahad WH, Chalan P, Rutgers A, Horst G, Huitema MG, Roffel MP, Roozendaal C, Kluin PM, Bos NA, Boots AM, Brouwer E (2014) Disturbed B cell homeostasis in newly diagnosed giant cell arteritis and polymyalgia rheumatica. Arthritis Rheumatol 66(7):1927–1938, July 2014
13. Singh AG, Kermani TA, Crowson CS, Weyand CM, Matteson EL, Warrington KJ (2015) Visual manifestations in giant cell arteritis: trend over 5 decades in a population-based cohort. J Rheumatol 42(2):309–315, February 2015
14. Dasgupta B, Borg FA, Hassan N, Alexander L, Barraclough K, Bourke B, Fulcher J, Hollywood J, Hutchings A, James P, Kyle V, Nott J, Power M, Samanta A, BSR and BHPR Standards,

Guidelines and Audit Working Group (2010) BSR and BHPR guidelines for the management of giant cell arteritis. Rheumatology (Oxford) 49(8):1594–1597, August 2010
15. Hunder GG, Bloch DA, Michel BA, Stevens MB, Arend WP, Calabrese LH, Edworthy SM, Fauci AS, Leavitt RY, Lie JT et al (1990) The American college of Rheumatology 1990 criteria for the classification of giant cell arteritis. Arthritis Rheum 33(8):1122–1128, August 1990
16. Mukhtyar C, Guillevin L, Cid MC, Dasgupta B, de Groot K, Gross W, Hauser T, Hellmich B, Jayne D, Kallenberg CG, Merkel PA, Raspe H, Salvarani C, Scott DG, Stegeman C, Watts R, Westman K, Witter J, Yazici H, Luqmani R; European Vasculitis Study Group (2009) EULAR recommendations for the management of large vessel vasculitis. Ann Rheum Dis 68(3):318–323, March 2009
17. Santoro L, D'Onofrio F, Bernardi S, Gremese E, Ferraccioli G, Santoliquido A (2013) Temporal ultrasonography findings in temporal arteritis: early disappearance of halo sign after only 2 days of steroid treatment. Rheumatology (Oxford) 52(4):622, April 2013
18. Puppo C, Massollo M, Paparo F, Camellino D, Piccardo A, Shoushtari Zadeh Naseri M, Villavecchia G, Rollandi GA, Cimmino MA (2014) Giant cell arteritis: a systematic review of the qualitative and semiquantitative methods to assess vasculitis with 18F-fluorodeoxyglucose positron emission tomography. Biomed Res Int 2014, 574248
19. Tomasson G, Peloquin C, Mohammad A, Love TJ, Zhang Y, Choi HK, Merkel PA (2014) Risk for cardiovascular disease early and late after a diagnosis of giant-cell arteritis: a cohort study. Ann Intern Med 160(2):73–80, 21 January 2014
20. Mackie SL, Hensor EM, Morgan AW, Pease CT (2014) Should I send my patient with previous giant cell arteritis for imaging of the thoracic aorta? A systematic literature review and meta-analysis. Ann Rheum Dis 73(1):143–148, January 2014
21. Yates M, Loke YK, Watts RA, MacGregor AJ (2014) Prednisolone combined with adjunctive immunosuppression is not superior to prednisolone alone in terms of efficacy and safety in giant cell arteritis: meta-analysis. Clin Rheumatol 33(2):227–236, February 2014
22. Loricera J, Blanco R, Hernández JL, Castañeda S, Mera A, Pérez-Pampín E, Peiró E, Humbría A, Calvo-Alén J, Aurrecoechea E, Narváez J, Sánchez-Andrade A, Vela P, Díez E, Mata C, Lluch P, Moll C, Hernández Í, Calvo-Río V, Ortiz-Sanjuán F, González-Vela C, Pina T, González-Gay MÁ (2015) Tocilizumab in giant cell arteritis: multicenter open-label study of 22 patients. Semin Arthritis Rheum 44(6):717–723, June 2015

Chapter 9
HLA System and Giant Cell Arteritis

F. David Carmona and Javier Martín

Abstract Giant cell arteritis (GCA) is a complex condition in which many *loci* across the genome may be involved in its susceptibility and phenotypic expression. However, recent large-scale genetic data has shown that the HLA system exerts most of the genetic influence to disease risk, particularly class II genes. This is in contrast with that observed in Takayasu arteritis, the other large vessel vasculitis, in which the HLA association is mainly driven by class I haplotypes. The use of novel imputation methods has made possible an analysis of the HLA system at the amino acid level in GCA. In this context, three polymorphic amino acid positions (positions 13 and 56 of the class II molecules HLA-DRβ1 and HLA-DQα-1, respectively, and position 45 of the class I molecule HLA-B) have been proposed as the causative variants for the HLA association with this type of vasculitis. Although functional experiments may be carried out to confirm these findings, the current data clearly reinforces the idea of GCA as an antigen-driven disease with a major role of T cells.

9.1 Introduction

Giant cell arteritis (GCA) is an immune-mediated vasculitis characterised by inflammatory lesions in medium and large vessels. It shows a complex etiology, in which different factors (including genetic, epigenetic and environmental factors) may interact for its development and clinical manifestations [1]. Cumulated knowledge, based on the study of gene expression profiles and genetic markers of disease susceptibility, indicates that GCA is an antigen-driven condition [2]. In this regard, it has been described that different T cell populations are directly involved in the immunopathological mechanisms leading to GCA, including IFNγ-producing Th1 cells, Th17 cells, Treg cells and Th9 cells [3–6]. Consistent with the above, the main genetic associations with GCA are harbored within the human leukocyte antigen (HLA) region [7], as it occurs in most autoinflammatory and autoimmune diseases, emphasizing the central role of immunity in the pathogenesis of GCA.

F.D. Carmona (✉) • J. Martín
Instituto de Parasitología y Biomedicina "López-Neyra," CSIC, PTS Granada, Granada 18016, Spain
e-mail: dcarmona@ipb.csic.es

F. Dammacco et al. (eds.), *Systemic Vasculitides: Current Status and Perspectives*, DOI 10.1007/978-3-319-40136-2_9

The use of high-throughput genotyping platforms has allowed us to have a better perspective of the genetic background underlying GCA predisposition. Thanks to the recent large-scale genetic analysis performed in a well-powered GCA cohort from different populations of European origin [8], we currently know that HLA genes represent a considerably high proportion of the heritability of this type of vasculitis. Therefore, understanding how the HLA system influence GCA development may definitively help us in the challenging endeavour of designing more effective therapeutic strategies for this condition.

In this chapter, we will give an overview of the structure and function of the HLA system and we will summarise the recent findings on its contribution to GCA risk.

9.2 Structure and Function of HLA Class I and II Molecules

The acronym 'HLA' was first established more than six decades ago to list a group of serologically defined antigens, which varied from individual to individual and were associated with organ transplant rejection. Nowadays, we know that the HLA system is a complex genetic region involved in an important mechanism of the immune system aimed to differentiate self-components from potentially harmful non-self agents. Specifically, the HLA genes encode for surface glycoproteins that recognise and present antigens to the immune cells, thus having a major role in the control of the immune response. As this system operates not only in humans but also in most vertebrate species, the term 'Major Histocompatibility Complex' (MHC) was proposed to name the system in general. Therefore, the HLA genes are the human versions of the MHC [9].

The HLA region is located at the short arm of chromosome 6 and includes, amongst others, two major sets of genes: the HLA class I genes (HLA-A, -B, and -C) and the HLA class II genes (HLA-DR, -DQ, and -DP). Both classes of HLA genes have different biological functions.

The HLA class I proteins are expressed on the membrane of all nucleated cells in the body, and their main function is to display intracellular peptides to CD8+ T cells. They are composed of three α domains (encoded by the HLA class I genes) and are linked non-covalently to the β2-microglobulin, which is not polymorphic. In physiological conditions, CD8+ T cells are tuned during thymocyte maturation to the specific set of HLA class I and self proteins produced by the corresponding individual, and will not be activated in response to them in a process known as immune tolerance. However, if a foreign or strange peptide is presented by a HLA class I protein (because of a viral infection, for instance), an immediate immune response will be triggered against the infected cell [10].

On the other hand, HLA class II proteins consist of two homogenous peptides (designated as α and β chains), both of which are encoded by HLA class II genes. Their expression is generally restricted to some cell types known as antigen presenting cells (including dendritic cells, macrophages, and B cells), which do not present cytosolic antigens but those derived from extracellular components to CD4+

helper T cells. This is the mechanism that commonly operates in bacterial infections. In this case, extracellular pathogens are endocytosed, digested in lysosomes, loaded onto the antigen-binding groove of the HLA molecule, and recognised as non-self by CD4+ helper T cells, which, as a consequence, initiate an appropriate immune response consisting of monoclonal expansion, localised inflammation, release of chemoattractant cytokines to recruit phagocytes, and production of specific antibodies against the pathogen [11].

9.3 The Complexity of the HLA Genomic Region

The HLA region spans around 4-megabase pairs (Mbp) within the chromosome position 6p21.3 and it is characterised by three main features: (1) it contains a high gene density (more than 400 genes and pseudogenes have been annotated, many of them with related immune functions), (2) it shows an extreme sequence variation (more than 8000 alleles have been described for the classical HLA genes), and (3) there is an extensive linkage disequilibrium (LD) in the region [12]. These characteristics are a consequence of a unique evolutionary history that has shaped the genetic structure of this genomic region not only by recombination and gene conversion, but also by natural selection, which makes it difficult to tease apart effects of individual *loci* in disease association studies [13]. To facilitate this, a systematic nomenclature system based on the early serological studies was developed by the '*WHO Nomenclature Committee for Factors of the HLA System*' (http://hla.alleles.org/nomenclature/committee.html), which first met in 1968 and laid down the criteria for successive meetings [14]. At first, to define the different serotyped haplotypes (*i.e.* combinations of specific sets of amino acids of the HLA proteins responsible for transplant rejection), this nomenclature included names composed of the HLA gene that encoded the corresponding protein, followed by two-digit numbers (*e.g.* HLA-DRB1*01). Consequently, these two-digit alleles correlated with the variation of the protein epitopes to which the antibodies were bound. Later on, the use of the polymerase chain reaction (PCR) and DNA sequencing techniques allowed a better estimation of the sequence variation of the HLA genes, and names containing four-digits were established to define haplotypes including non-synonymous changes within exons (*e.g.* HLA-DRB1*01:01). Although successive digits have been added to improve the accuracy of the defined haplotypes (to consider synonymous changes, for example), it was accepted by consensus that the analysis of 4-digit types, known as classical HLA alleles, is an appropriate approach to obtain a good estimation of the HLA contribution to the studied phenotypes (Fig. 9.1).

However, it is important to note that the classical HLA alleles do not consider all genetic variants and polymorphic amino acid positions within the HLA region, but only specific haplotypes covering each of the HLA genes. As a consequence of the broad LD across many genes, the interpretation of the HLA associations with clinical phenotypes remains difficult. An optimal approach to identify causal variants for

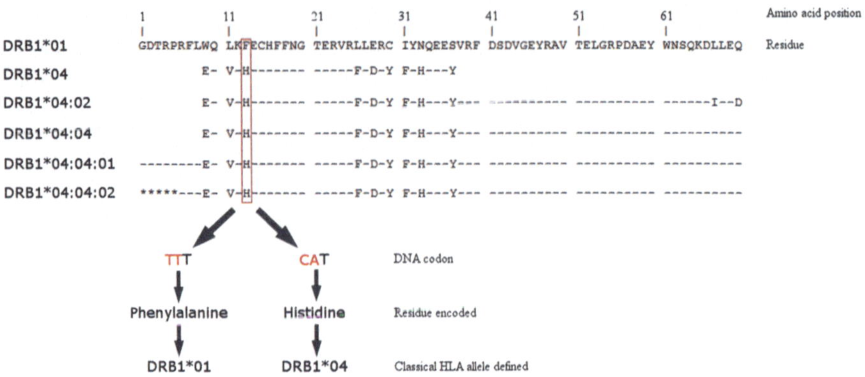

Fig. 9.1 Comparison of the amino acid sequence between different HLA-DRB1 classical alleles at two- and four-digit types. The reference sequence corresponds to that of the HLA-DRB1*01 allele. For the HLA-DRB1*01 sequences shown, a hyphen indicates identity to the reference sequence, whereas non-identity to the reference sequence is shown by displaying the appropriate residue at that position. The DNA codons for the two amino acids present in position 13 are also shown, with red bases indicating presence of different alleles of single-nucleotide polymorphisms

those associations would be to analyse the complete sequence data across the HLA region, but this represents a major challenge due to its complex genetic structure. Considering the above, researchers from the Broad Institute (Cambridge, Massachusetts, USA) proposed an alternative methodology consisting of testing single-nucleotide polymorphisms (SNP) and polymorphic amino acid positions individually, instead of the traditional approach of testing the established haplotypes [15]. To do that in a cost-effective manner, and taking advantage of the high-throughput genomics, they developed an imputation method that infers the variation within the HLA region of SNPs, classical HLA alleles, and amino acid polymorphisms at class I and II *loci*, using a reference panel of 5225 individuals of European origin with genotyping data of 8961 SNPs and indel polymorphisms across the HLA region, as well as four-digit genotyping data of the HLA class I and II molecules [16]. As discussed below, the imputation and association testing at the amino acid level have remarkably facilitated the fine-mapping of primary HLA association signals for many immune-mediated conditions, including GCA.

9.4 HLA and Human Disease

The HLA region has been associated with a wide spectrum of clinical conditions in humans, more than any other region of the genome, including autoimmune and autoinflammatory diseases, infectious diseases, cancer, graft-versus-host disease, and severe side-effects of drugs [17, 18].

With regards to autoimmunity, typical diseases associated with classical HLA class I alleles include ankylosing spondylitis (HLA-B*27), Behçet's disease (BD)

(HLA-B*51) and Takayasu disease (TAK) (HLA-B*51); whereas some examples of HLA class II diseases are type 1 diabetes (HLA-DRB1*04 and HLA-DRB1*03), rheumatoid arthritis (RA) (HLA-DRB1*04 and HLA-DQA1*03), celiac disease (CeD) (HLA-DQA1*05 and HLA-DQB1*02), multiple sclerosis (HLA-DRB1*15), systemic sclerosis (SSc) (HLA-DRB1*11 and HLA-DRB1*07), systemic lupus erythematosus (SLE) (HLA-DRB1*03), ulcerative colitis (HLA-DRB1*11), and GCA (HLA-DRB1*04), amongst others [18]. However, the extensive LD of this genomic region, together with the fact that disease predisposition concerns subtle effects of common alleles, has made it difficult to pinpoint causal coding or regulatory variants of those primary associations with conflicting results.

In any case, the novel imputation and fine-mapping approach described in the previous section have shed light into the pathological mechanisms underlying the HLA associations with autoimmune diseases. The first published study using this method was performed in RA in 2012 [19]. The authors proposed a model of three amino acid positions in HLA-DRβ1 (11, 71 and 74), one in HLA-DPβ1 (9), and one in HLA-B (9), that explained almost completely the HLA association with this rheumatic disease. All associated amino acids were located in peptide-binding pockets, implying a functional impact on antigenic peptide presentation to T cells.

Subsequent studies have also analysed the HLA region at the amino acid level in other autoimmune diseases such as SSc, SLE, BD, and GCA [8, 20–22], making a valuable contribution to the current knowledge about the complex HLA associations that account for most of their phenotypic variance.

In the following section we will summarise the recent advances achieved in the study of the HLA contribution to GCA susceptibility.

9.5 HLA Contribution to GCA Susceptibility

9.5.1 Early Studies

Despite being limited by a low statistical power, many studies from the early 1990s clearly pointed out the HLA class II as the most relevant genomic region for GCA pathogenesis. Specifically, HLA-DRB1*04 alleles (generally DRB1*0401 but also DRB1*0404) were directly involved in disease predisposition in almost all the independent candidate gene studies conducted in this type of vasculitis, which included populations of European ancestry from USA, Spain, Italy, France, and Denmark [23–30] (Table 9.1). Some studies also reported a correlation between these alleles and both resistance to corticosteroid treatment and the development of visual complications in GCA patients [25, 31, 32].

Regarding HLA class I, some classical alleles were also suggested as GCA risk factors, including HLA-A*31, HLA-B*8, HLA-Cw3 and HLA-Cw6, described in the early 1980s, and the more recent associations with HLA-B*15 and with the MHC class I polypeptide-related sequence A (*MICA*) gene [33–36]. Nevertheless,

Table 9.1 Described associations of giant cell arteritis with HLA-DRB1*04 alleles

Year of publication	Cohort origin	Sample size (case/control)	*P*-value	OR	Reference
1992	Rochester, Minnesota (USA)[a]	42/63	0.03	NA	Weyand et al. [29]
1994	Rochester, Minnesota (USA)[a]	52/72	1.00E-04	NA	Weyand et al. [26]
1998	Toulouse (France)	41/384	<1.00E-03	2.83[d]	Rauzy et al. [24]
1998	Montpellier (France)	42/1609	5.00E-04	3.10	Combe et al. [30]
1998	Lugo (Spain)	53/145	<0.05	NA	Dababneh et al. [27]
2002	Copenhagen (Denmark)	65/193	0.01	2.30	Jacobsen et al. [48]
2004	Cantabria (Spain)	44/99	0.04	1.90	Martínez-Taboda et al. [31]
2015	Spain	763/1517	5.75E-14	1.94	Carmona et al. [8]
2015	UK	251/8612	1.86E-12	2.00	Carmona et al. [8]
2015	Italy	238/1270	2.15E-03	1.68	Carmona et al. [8]
2015	North America[b]	205/1641	9.44E-09	2.02	Carmona et al. [8]
2015	Norway	99/374	9.29E-03	1.64	Carmona et al. [8]
2015	Germany	95/1892	2.31E-03	1.76	Carmona et al. [8]
2015	Combined[c]	1651/15,306	6.78E-38	1.92	Carmona et al. [8]

[a]Scandinavian descent
[b]USA and Canada
[c]Meta-analysis including Spain, UK, Italy, North America, Norway and Germany
[d]Risk ratio
NA, not available

the results of those studies were inconsistent in most cases and were not replicated in independent populations [37].

9.5.2 *Novel Associations Using High-Throughput Data*

The recently published large-scale genetic screening on GCA has represented a turning point in the elucidation of the HLA contribution to disease pathogenesis [8]. The study was possible thanks to the establishment of an exciting international collaborative effort, in which many research groups and hospitals worldwide, including the '*European Vasculitis Genetics Consortium*', the '*Spanish GCA Consortium*', the '*UKGCA Consortium*' and the '*Vasculitis Clinical Research Consortium*', contributed with more than 1,600 GCA samples (with a diagnosis confirmed either by temporal artery biopsy or imaging techniques) from seven countries (Spain, Italy, UK, USA, Canada, Germany and Norway). For the analysis, more than 15,000 matched controls were also included, thus representing the largest case–control cohort investigated in a genetic study on GCA so far.

Besides the high statistical power, what made this study of high relevance for the investigation of the HLA system in GCA, was the use of the 'Human Immuno DNA Analysis BeadChip Kit' (known as the Immunochip). This genotyping platform was designed by a consortium of leading groups covering all of the major autoimmune and seronegative diseases to identify immune-related risk variants [38]. The Immunochip includes probes to type almost 200,000 SNPs, rare variants, and insertion/deletion polymorphisms located within 186 known susceptibility *loci* for autoimmune and inflammatory disorders. The use of the Immunochip has been considerably helpful for the identification of novel specific and common genetic risk factors in multiple immune-mediated diseases, including TAK, CeD, RA and SSc amongst others [21, 39–42]. Remarkably, the chip has a dense coverage of polymorphisms within the HLA region, which can be used to impute classical alleles and amino acid variations with the method described in the third section of this chapter.

The study confirmed class II genes (*HLA-DRB1* and *HLA-DQA1*) as the main contributors to disease risk. Indeed, the considerably higher statistical significance observed within this region in comparison with the rest of immune genes analysed, suggested that most of the genetic component of GCA relies on HLA class II. This is consistent with the pathogen infection hypothesis proposed to explain the initial activation and expansion of local dendritic cells within the vessel wall of GCA patients [43]. Contrary to that observed in GCA, the strongest susceptibility markers for TAK, the other large vessel vasculitis, are harboured in the HLA class I region (specifically HLA-B*52). The different pattern of HLA associations observed between these two similar conditions is striking, and could reflect disease-specific mechanisms during the early development of both type of vasculitides [44].

However, the most relevant insight of the Immunochip study on GCA was the analysis of the HLA system at the amino acid level. The authors proposed a model of three amino acid positions that explained most of the differences in the HLA region between cases and controls. This model included the positions 13 and 56 of the HLA class II molecules HLA-DRβ1 and HLA-DQα1, respectively (representing the major contribution), and the position 45 of the HLA class I molecule HLA-B (which conferred a weaker but still significant disease risk) (Fig. 9.2). Remarkably, all three amino acid positions were located in the binding groove of their corresponding molecule, and they have been described to have a direct interaction with the bound antigen, which gives a functional implication to the model.

The amino acid with higher effect size amongst the six possible residues in the position 13 of HLA-DRβ1 was histidine (which was also the top signal in the whole Immunochip study). This is one of the polymorphic amino acids that defines the HLA alleles associated classically with GCA, HLA-DRB1*04:01 and HLA-

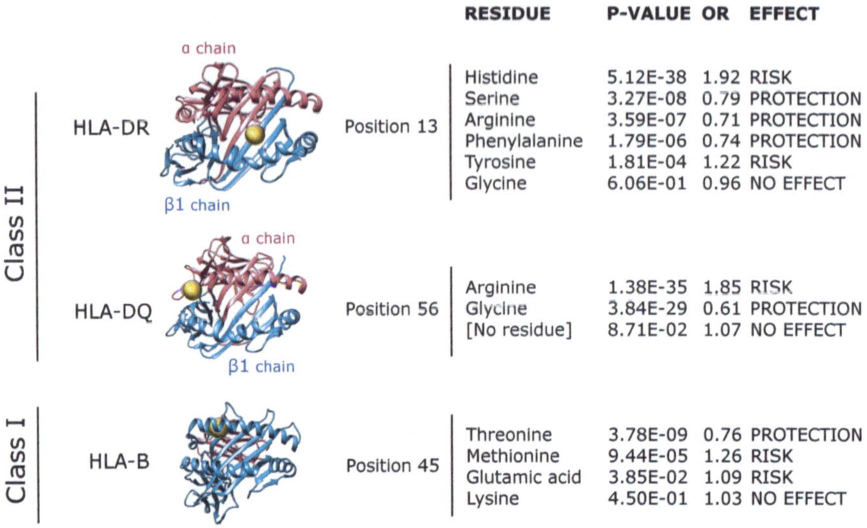

Fig. 9.2 Amino acid model that best explains the HLA association with giant cell arteritis susceptibility. The location of the associated amino acid positions is represented with a yellow sphere in the ribbon representation of each HLA molecule. P-values, odds ratios (OR) and effect conferred by the different residues at each position are also indicated (Data from Carmona et al. [8])

DRB1*04:04 (Fig. 9.1). Hence, it is likely that the presence of this histidine in the binding pocket of the HLA-DR molecule can predispose antigen-presenting cells to recognise self-antigens within the vascular walls.

It is interesting to note that the histidine in position 13 of HLA-DRβ1 (and, consequently, the HLA-DRB1*04 alleles that contain this residue) is one of the most associated variants with RA susceptibility [19]. In this context, old studies described an association between GCA and HLA-DRB1*04 alleles carrying the ‘shared epitope’ (a common region of the β chain of HLA-DR, comprising positions 67–74, that is commonly present in RA patients and could be involved in presenting autoimmunological peptides) [26, 27]. In addition, *PTPN22* (a central regulator of both B and T cell receptor signalling) also represent the strongest non-HLA marker for both diseases [45, 46], and a genetic score predictive for RA was shown to yield significantly higher values in GCA patients compared to controls [8]. Altogether, these evidences may suggest common pathological mechanisms between GCA and RA. However, these two diseases seldom co-occur and they clearly show a different phenotypic expression. It could be speculated that the risk HLA-DRB1 haplotypes (*i.e.* those including a histidine in position 13) act as major contributors to the loss of tolerance influencing the first stages of both conditions, with other genes acting as secondary ‘modifiers’ of the final phenotype leading to GCA or RA.

9.6 Future Perspectives

The establishment of an International GCA consortium, involving many groups and hospitals from Europe and North America, has allowed the fulfilment of the first large-scale genetic analysis on GCA, which has produced very exciting insights on the role of the HLA system in this type of vasculitis. However, the main conclusions are based on the assumption that all the effects are conferred on a log-additive scale, that is, the first and second copies of an allele from an associated variant multiplicatively increase risk by the same amount (so homozygosis for an associated allele would double the disease risk of heterozygosis). However, recent lines of evidence suggest that non-additive genetic effects of dominance and epistasis, as a consequence of differences in autoantigen-binding repertoires between a heterozygote's two expressed HLA variants, may modulate the risk of autoimmune diseases [47]. To continue shedding light into the influence of the HLA system in GCA, this approach should be considered, as it could explain moderate fractions of the phenotypic variance.

On the other hand, the ongoing collaboration of the GCA consortium will increase the current case–control cohort, and additional subphenotype analyses of the HLA system accordingly with the main clinical complications of the disease could be performed. These studies may have relevant therapeutic implications, as it could be possible that different HLA haplotypes are related with higher risk to develop severe complications like visual loss or with relapses after corticosteroid tapering [43].

9.7 Conclusion

Thanks to the advent of the new technologies for high-throughput genotyping, we have now a clearer overview of the genetic basis predisposing to complex traits such as the autoimmune diseases. Platforms like the GWAS or the Immunochip have helped us to make an accurate estimation of the contribution of the HLA system to the development of autoimmunity. We now know that the HLA region explains more disease risk than any other locus in the genome in most immune-mediated disorders. Elucidation of the functional implications of the autoimmune disease-associated HLA alleles is essential for a better understanding of the pathophysiology of these conditions, and may ultimately lead to more effective treatments. Regarding GCA, a comprehensive analysis of the HLA system has been possible taking advantage of the dense SNP coverage of the Immunochip for this genomic region and the use of novel imputation methods. The data indicated that certain amino acids located in the binding groove of the HLA-DR and HLA-DQ molecules confer the strongest risk for GCA development, with a weaker contribution of class I residues. These data support the hypothesis that GCA is an antigen-driven disease likely triggered by a pathogen infection.

References

1. Ly KH, Regent A, Tamby MC, Mouthon L (2010) Pathogenesis of giant cell arteritis: more than just an inflammatory condition? Autoimmun Rev 9:635–645
2. Weyand CM, Goronzy JJ (2013) Immune mechanisms in medium and large-vessel vasculitis. Nat Rev Rheumatol 9:731–740
3. Terrier B, Geri G, Chaara W, Allenbach Y, Rosenzwajg M, Costedoat-Chalumeau N et al (2012) Interleukin-21 modulates Th1 and Th17 responses in giant cell arteritis. Arthritis Rheum 64:2001–2011
4. Espigol-Frigole G, Corbera-Bellalta M, Planas-Rigol E, Lozano E, Segarra M, Garcia-Martinez A et al (2013) Increased IL-17A expression in temporal artery lesions is a predictor of sustained response to glucocorticoid treatment in patients with giant-cell arteritis. Ann Rheum Dis 72:1481–1487
5. Samson M, Audia S, Fraszczak J, Trad M, Ornetti P, Lakomy D et al (2012) Th1 and Th17 lymphocytes expressing CD161 are implicated in giant cell arteritis and polymyalgia rheumatica pathogenesis. Arthritis Rheum 64:3788–3798
6. Ciccia F, Rizzo A, Guggino G, Cavazza A, Alessandro R, Maugeri R et al (2015) Difference in the expression of IL-9 and IL-17 correlates with different histological pattern of vascular wall injury in giant cell arteritis. Rheumatology (Oxford) 54:1596–1604
7. Carmona FD, Gonzalez-Gay MA, Martin J (2014) Genetic component of giant cell arteritis. Rheumatology (Oxford) 53:6–18
8. Carmona FD, Mackie SL, Martin JE, Taylor JC, Vaglio A, Eyre S et al (2015) A large-scale genetic analysis reveals a strong contribution of the HLA class II region to giant cell arteritis susceptibility. Am J Hum Genet 96:565–580
9. Bodmer WF (1987) The HLA, system: structure and function. J Clin Pathol 40:948–958
10. Hewitt EW (2003) The MHC, class I antigen presentation pathway: strategies for viral immune evasion. Immunology 110:163–169
11. Cresswell P (1994) Assembly, transport, and function of MHC class II molecules. Annu Rev Immunol 12:259–293
12. Horton R, Wilming L, Rand V, Lovering RC, Bruford EA, Khodiyar VK et al (2004) Gene map of the extended human MHC. Nat Rev Genet 12:889–899
13. Traherne JA (2008) Human MHC, architecture and evolution: implications for disease association studies. Int J Immunogenet 35:179–192
14. Marsh SG, Albert ED, Bodmer WF, Bontrop RE, Dupont B, Erlich HA et al (2010) Nomenclature for factors of the HLA system, 2010. Tissue Antigens 75:291–455
15. de Bakker PI, Raychaudhuri S (2012) Interrogating the major histocompatibility complex with high-throughput genomics. Hum Mol Genet 21:R29–R36
16. Jia X, Han B, Onengut-Gumuscu S, Chen WM, Concannon PJ, Rich SS et al (2013) Imputing amino acid polymorphisms in human leukocyte antigens. PLoS One 8, e64683
17. Petersdorf EW (2013) The major histocompatibility complex: a model for understanding graft-versus-host disease. Blood 122:1863–1872
18. Trowsdale J, Knight JC (2013) Major histocompatibility complex genomics and human disease. Annu Rev Genomics Hum Genet 14:301–323
19. Raychaudhuri S, Sandor C, Stahl EA, Freudenberg J, Lee HS, Jia X et al (2012) Five amino acids in three HLA proteins explain most of the association between MHC and seropositive rheumatoid arthritis. Nat Genet 44:291–296
20. Kim K, Bang SY, Lee HS, Okada Y, Han B, Saw WY et al (2014) The HLA-DRbeta1 amino acid positions 11-13-26 explain the majority of SLE-MHC associations. Nat Commun 5:5902
21. Mayes MD, Bossini-Castillo L, Gorlova O, Martin JE, Zhou X, Chen WV et al (2014) Immunochip analysis identifies multiple susceptibility loci for systemic sclerosis. Am J Hum Genet 94:47–61

22. Ombrello MJ, Kirino Y, de Bakker PI, Gul A, Kastner DL, Remmers EF (2014) Behcet disease-associated MHC class I residues implicate antigen binding and regulation of cell-mediated cytotoxicity. Proc Natl Acad Sci U S A 111:8867–8872
23. Gonzalez-Gay MA, Amoli MM, Garcia-Porrua C, Ollier WE (2003) Genetic markers of disease susceptibility and severity in giant cell arteritis and polymyalgia rheumatica. Semin Arthritis Rheum 33:38–48
24. Rauzy O, Fort M, Nourhashemi F, Alric L, Juchet H, Ecoiffier M et al (1998) Relation between HLA DRB1 alleles and corticosteroid resistance in giant cell arteritis. Ann Rheum Dis 57:380–382
25. Gonzalez-Gay MA, Garcia-Porrua C, Llorca J, Hajeer AH, Branas F, Dababneh A et al (2000) Visual manifestations of giant cell arteritis. Trends and clinical spectrum in 161 patients. Medicine 79:283–292
26. Weyand CM, Hunder NN, Hicok KC, Hunder GG, Goronzy JJ (1994) HLA-DRB1 alleles in polymyalgia rheumatica, giant cell arteritis, and rheumatoid arthritis. Arthritis Rheum 37:514–520
27. Dababneh A, Gonzalez-Gay MA, Garcia-Porrua C, Hajeer A, Thomson W, Ollier W (1998) Giant cell arteritis and polymyalgia rheumatica can be differentiated by distinct patterns of HLA class II association. J Rheumatol 25:2140–2145
28. Cid MC, Ercilla G, Vilaseca J, Sanmarti R, Villalta J, Ingelmo M et al (1988) Polymyalgia rheumatica: a syndrome associated with HLA-DR4 antigen. Arthritis Rheum 31:678–682
29. Weyand CM, Hicok KC, Hunder GG, Goronzy JJ (1992) The HLA-DRB1 locus as a genetic component in giant cell arteritis. Mapping of a disease-linked sequence motif to the antigen binding site of the HLA-DR molecule. J Clin Invest 90:2355–2361
30. Combe B, Sany J, Le Quellec A, Clot J, Eliaou JF (1998) Distribution of HLA-DRB1 alleles of patients with polymyalgia rheumatica and giant cell arteritis in a Mediterranean population. J Rheumatol 25:94–98
31. Martinez-Taboda VM, Bartolome MJ, Lopez-Hoyos M, Blanco R, Mata C, Calvo J et al (2004) HLA-DRB1 allele distribution in polymyalgia rheumatica and giant cell arteritis: influence on clinical subgroups and prognosis. Semin Arthritis Rheum 34:454–464
32. Salvarani C, Boiardi L, Mantovani V, Ranzi A, Cantini F, Olivieri I et al (1999) HLA-DRB1, DQA1, and DQB1 alleles associated with giant cell arteritis in northern Italy. J Rheumatol 26:2395–2399
33. Armstrong RD, Behn A, Myles A, Panayi GS, Welsh KI (1983) Histocompatibility antigens in polymyalgia rheumatica and giant cell arteritis. J Rheumatol 10:659–661
34. Hansen JA, Healey LA, Wilske KR (1985) Association between giant cell (temporal) arteritis and HLA-Cw3. Hum Immunol 13:193–198
35. Kemp A, Marner K, Nissen SH, Heyn J, Kissmeyer-Nielsen F (1980) HLA antigens in cases of giant cell arteritis. Acta Ophthalmol 58:1000–1004
36. Gonzalez-Gay MA, Rueda B, Vilchez JR, Lopez-Nevot MA, Robledo G, Ruiz MP et al (2007) Contribution of MHC class I region to genetic susceptibility for giant cell arteritis. Rheumatology (Oxford) 46:431–434
37. Richardson JE, Gladman DD, Fam A, Keystone EC (1987) HLA-DR4 in giant cell arteritis: association with polymyalgia rheumatica syndrome. Arthritis Rheum 30:1293–1297
38. Cortes A, Brown MA (2011) Promise and pitfalls of the Immunochip. Arthritis Res Ther 13:101
39. Saruhan-Direskeneli G, Hughes T, Aksu K, Keser G, Coit P, Aydin SZ et al (2013) Identification of multiple genetic susceptibility loci in Takayasu arteritis. Am J Hum Genet 93:298–305
40. Trynka G, Hunt KA, Bockett NA, Romanos J, Mistry V, Szperl A et al (2011) Dense genotyping identifies and localizes multiple common and rare variant association signals in celiac disease. Nat Genet 43:1193–1201
41. Eyre S, Bowes J, Diogo D, Lee A, Barton A, Martin P et al (2012) High-density genetic mapping identifies new susceptibility loci for rheumatoid arthritis. Nat Genet 44:1336–1340

42. Parkes M, Cortes A, van Heel DA, Brown MA (2013) Genetic insights into common pathways and complex relationships among immune-mediated diseases. Nat Rev Genet 14:661–673
43. Carmona FD, Martin J, Gonzalez-Gay MA (2015) New insights into the pathogenesis of giant cell arteritis and hopes for the clinic. Expert Rev Clin Immunol 12:57–66
44. Carmona FD, Gonzalez-Gay MA, Martin J (2015) Genetic analysis of large vessel vasculitis. Nephron 129:3–5
45. Begovich AB, Carlton VE, Honigberg LA, Schrodi SJ, Chokkalingam AP, Alexander HC et al (2004) A missense single-nucleotide polymorphism in a gene encoding a protein tyrosine phosphatase (PTPN22) is associated with rheumatoid arthritis. Am J Hum Genet 75:330–337
46. Serrano A, Marquez A, Mackie SL, Carmona FD, Solans R, Miranda-Filloy JA et al (2013) Identification of the PTPN22 functional variant R620W as susceptibility genetic factor for giant cell arteritis. Ann Rheum Dis 72:1882–1886
47. Lenz TL, Deutsch AJ, Han B, Hu X, Okada Y, Eyre S et al (2015) Widespread non-additive and interaction effects within HLA loci modulate the risk of autoimmune diseases. Nat Genet 47:1085–1090
48. Jacobsen S, Baslund B, Madsen HO, Tvede N, Svejgaard A, Garred P (2002) Mannose-binding lectin variant alleles and HLA-DR4 alleles are associated with giant cell arteritis. J Rheumatol 29:2148–2153

Chapter 10
Microscopic Polyangiitis

Franco Dammacco and Angelo Vacca

Abstract Microscopic polyangiitis (MPA) is a necrotizing, systemic vasculitis usually affecting capillaries, venules and arterioles. Midium arteries can also be involved. Necrotizing glomerulonephritis and pulmonary capillaritis are the major pathological findings, but granulomatosis is regularly absent. Its prevalence is lower than that of granulomatosis with polyangiitis and of eosinophilic granulomatosis with polyangiitis but, similarly to these two conditions, it belongs to the group of vasculitides associated with anti-neutrophil cytoplasmic antibodies (ANCA), directed (in most though not all MPA patients) against myeloperoxidase rather than against proteinase-3. In addition to general constitutional symptoms, clinical features may range from a renal-restricted vasculitis usually consisting of idiopathic, necrotizing and crescentic glomerulonephritis to pulmonary capillaritis whose severity may reach the level of pulmonary hemorrhage. Purpuric eruptions and mononeuritis multiplex can also be detected with variable prevalence. At diagnosis, prompt and intensive treatment with the combination of corticosteroids and cyclophosphamide is the usual first-line approach, that is able to achieve a complete or partial response in over two-thirds of the patients. Maintenance therapy includes lower doses of corticosteroids, azathioprine and methotrexate. Rituximab has been shown to be effective in resistant and relapsing patients. End-stage renal disease, infections and cardiovascular failure are the most frequent causes of mortality.

Keywords Microscopic polyangiitis • Anti-neutrophil cytoplasmic antibodies • ANCA-associated vasculitides • Necrotizing glomerulonephritis • Pulmonary capillaritis

---

F. Dammacco (✉) • A. Vacca
Department of Biomedical Sciences and Human Oncology, Section of Internal Medicine, University of Bari Medical School, 70124 Bari, Italy
e-mail: francesco.dammacco@uniba.it

F. Dammacco et al. (eds.), *Systemic Vasculitides: Current Status and Perspectives*, DOI 10.1007/978-3-319-40136-2_10

10.1 Introduction

According to the nomenclature established by the 2012 revised international Chapel Hill Consensus Conference (CHCC) [1], microscopic polyangiitis (MPA) is a necrotizing vasculitis of unknown etiology belonging to the subgroup of small vessel vasculitides which predominantly affect small vessels such as intra-parenchymal small arteries, arterioles, capillaries and venules. Midium arteries and veins can, however, be affected to a much lower extent. The pathological cornerstones of the disease are necrotizing glomerulonephritis and pulmonary capillaritis. Based on the highly frequent (though not invariable) occurrence of serum anti-neutrophil cytoplasmic antibodies (ANCA) that MPA shares with granulomatosis with polyangiitis (GPA, formerly Wegener's disease) and eosinophilic granulomatosis with polyangiitis (EGPA, formerly Churg-Strauss syndrome), these three diseases are collectively indicated with the term ANCA-associated vasculitides (AAV).

10.2 Epidemiology

MPA is a relative rare systemic vasculitis whose prevalence is variable in different geographic areas but is increasing worldwide following the growing use of methods for the detection of ANCAs. In a study carried out in Norwich (United Kingdom), the overall annual incidence of MPA and, by comparison, of GPA was 5.9/million and 11.3/million, respectively. In addition, while GPA followed a cyclical pattern of occurrence with a periodicity of 7.6 years, no definite evidence of periodicity was detected for MPA. The cyclical pattern of occurrence for GPA suggests an infectious etiology and emphasizes its basic diversity from MPA [2]. Conversely, in spite of the substantial overlap in the incidence of AAV between Japan and the United Kingdom (UK), MPA was clearly prevalent in Japan (83 %), whereas GPA was more frequent in the UK (66 %) [3]. In more general terms, MPA with myeloperoxidase (MPO)-ANCA positivity is prevalent in Asian countries, whereas GPA with proteinase-3 (PR3)-ANCA positivity is prevalent in northern Europe and the United States [4].

10.3 Pathophysiology

Initially lumped together with polyarteritis nodosa, MPA is indeed distinctly different in terms of clinical manifestations and a high prevalence of ANCA that are instead consistently absent in patients with polyarteritis nodosa. As already mentioned above, no etiologic agent of MPA has so far been recognized. The necrotizing vasculitis that is the hallmark of the disease is characterized by the scarcity or the lack of immunoglobulin deposition along the vascular lesions, and this implies that the formation of immune complexes is likely not involved in the pathogenesis of MPA. In addition, at variance from GPA, extravascular granulomatous inflammation is regularly absent. Instead, growing evidence indicates that ANCA reactivity, prevalently directed to MPO, supports the autoimmune nature of MPA, although the persistence of several undefined issues casts shadows on a full comprehension of its pathogenetic mechanism(s). The following points should be emphasized:

(a) MPO-ANCA account for 58–70 % of cases [5, 6];
(b) approximately 1 out of 4 MPA patients is Pr3-ANCA-positive and MPO-ANCA-negative [7];
(c) a small percentage of patients may also result ANCA-negative [8], although this may reflect the lack of activity of the vasculitis process and/or the response to therapy. Furthermore, it should be kept in mind that the sensitivities and specificities of the methods for ANCA detection were formerly significantly different among the various commercial kits [9, 10], especially before reaching consensus on the standard, computer-based indirect immunofluorescence techniques and ANCA "capture" or "anchor" ELISA procedures [11, 12]: a long way since ANCA were first described in segmental necrotizing glomerulonephritis [13];
(d) studying pentraxin-3 (PTX3), a novel ANCA antigen that (similarly to MPO and Pr3) is stored in human neutrophil granules and is expressed on apoptotic neutrophil surface, anti-PTX3 autoantibodies with a specific fluorescence pattern have been detected in 37.3 % of a cohort of 150 AAV patients versus 5.3 % of healthy subjects ($p < 0.001$), including 7 of 14 patients who were both MPO and PR3 ANCA-negative. The titers of anti-PTX3 antibodies are higher in patients with active AAV [14]. Incidentally, search for anti-PTX3 antibodies might become a novel, useful test when MPO and PR3 ANCAs are negative;
(e) among the various animal models of MPA, those particularly suggestive of the pathogenicity of MPO-ANCA are described in chap. 2 of the present volume. Regardless of the different experimental procedures, the animal models consistently point to the pathogenic effect of MPO-ANCAs that are able to induce necrotizing glomerulonephritis and pulmonary capillaritis [15, 16];
(f) MPA and GPA are genetically distinct subsets of AAV, the strongest genetic associations being with the antigenic specificity of ANCA and not with the clinical syndrome [17]. It has also been reported that the three main AAV subtypes are associated with distinct HLA variants, namely GPA with HLA-DP1, MPA with HLA-DQ and EGPA with HLA-DRB4 [18].

Although these observations are to a large extent (though not univocally) consistent with a clear-cut autoimmune pathogenetic picture, it seems reasonable to hypothesize that, based on the ability of MPO-ANCA and PR3-ANCA to activate neutrophils, reactive oxygen radicals would be produced which would cause neutrophil degranulation and release of lytic enzymes. When this process takes place on the surface of endothelial cells, leukocyte-endothelial interactions would be enhanced, and this would result in detachment and lysis of the endothelial layer, and microvascular hemorrhage. Complement and effector T cells are also likely to be involved, but their role is still poorly defined [6, 19].

10.4 Clinical Features

The clinical spectrum of MPA is strikingly heterogeneous and may range from clinically mild renal and/or pulmonary signs and symptoms to rapidly progressing life-threatening renal-pulmonary vasculitis syndrome. There is an overwhelming literature on the clinical manifestations of MPA and more generally of AAV. For the sake of brevity, we refer to four comprehensive reviews [6, 20–22]. Usually, MPA

Table 10.1 The spectrum of the most common clinical manifestations in microscopic polyangiitis [5, 6, 20]

Organ involvement	Rough % range	Clinical manifestations
Kidney	80–100	Asymptomatic proteinuria, microscopic hematuria, urinary casts
		Idiopathic necrotizing and crescentic glomerulonephritis
		End-stage renal disease
Lung	25–80	Pulmonary capillaritis with hemoptysis, dyspnea, cough, pleuritis
		Interstitial pneumonia and pulmonary fibrosis
Skin	30–60	Palpable purpura
		Livedo reticularis
		Urticaria/angioedema
		Skin ulcers
Gastrointestinal tract	30–50	Abdominal pain
		Gastrointestinal hemorrhage
Cardiovascular system	10–25	Arterial hypertension
		Congestive heart failure
		Subclinical myocardial infarction
		Pericarditis
Peripheral nervous system	20–60	Mononeuritis multiplex
		Polyneuropathy
Central nervous system	15–30	Arachnoid hemorrhage
		Pachymeningitis
		Non-hemorrhagic cerebral infarction
Eye	10–20	Episcleritis, scleritis
		Iridocyclitis with hypopyon
		Optic neuropathy
Ear/Nose/Throat	9–15	Sinusitis
		Sensorineural hearing loss

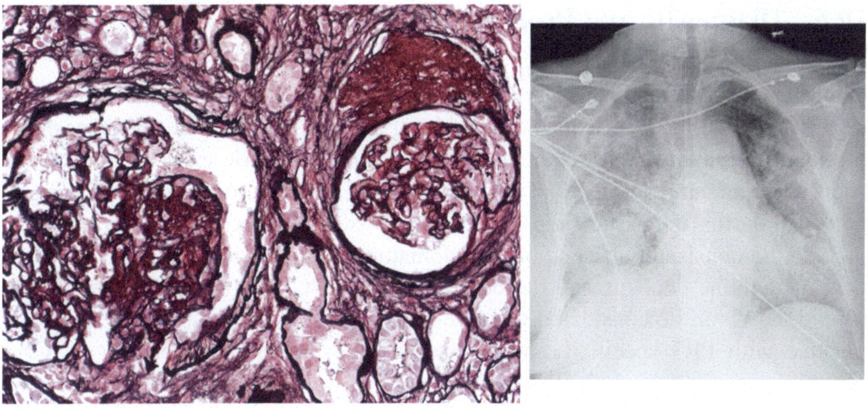

Fig. 10.1 Severe "kidney-lung syndrome" in a patient with microscopic polyangiitis

begins with insidious constitutional symptoms such as weight loss, irregular fever, anorexia, and arthralgia/myalgia that may erroneously suggest polymyalgia rheumatica. Table 10.1 summarizes the clinical manifestations of MPA in a decreasing order of frequency, with the understanding that large variations may occur in their combination, severity and evolution, and that, given its nature of systemic vasculitis, potentially every organ can be involved.

The so-called pulmonary-renal vasculitis syndrome (Fig. 10.1) deserves a special mention for its severe clinical course that, if not promptly treated with an aggressive therapeutic approach and, if possible, with the patient hospitalization in an intensive care unit, may result rapidly fatal [23].

10.5 Histopathology

Whenever possible, histopathological examination of bioptic samples should be performed to confirm the clinical diagnosis and to differentiate MPA from the other two common AAV such as GPA and EGPA. All three conditions share the characteristics of small vessel involvement and the necrotizing type of vasculitis.

In the most typical and severe MPA, renal biopsy shows crescentic glomerulonephritis. This finding is usually preceded by a stepwise pathological process which starts with patchy thrombosis of glomerular capillaries, then progresses to fibrinoid necrosis, discontinuation of the glomerular basement membrane, extracapillary proliferation and finally crescent formation. Pulmonary lesions usually include capillaritis with fibrinoid necrosis, whereas leukocytoclastic vasculitis is diagnosed when skin biopsies are examined [5, 6].

10.6 Diagnostic and Prognostic Features

Clinical involvement of the kidney and/or the lung, but sometimes also of other organs and systems, can be diagnosed with variable frequency in all AAV, thus making the differential diagnosis among them rather difficult on clinical grounds. In addition to the already mentioned focal crescentic glomerulonephritis and pulmonary capillaritis on tissue biopsy specimens, useful diagnostic criteria of MPA include the consistent absence of granulomatous inflammation, the paucity or the lack of immune deposits and the detection of p-ANCA with MPO specificity. It should, however, be considered that a minority of MPA patients may be c-ANCA positive with PR3 specificity, and an additional small group may test ANCA-negative, especially in clinically quiescent patients or in those with therapy-induced clinical and immunological remission. Anti-glomerular basement membrane antibodies are consistently absent.

In the absence of an early and suitable treatment, a severe prognosis can be anticipated for MPA as well as for the other AAV, the mortality being approximately 90 % at 1 year from diagnosis. The administration of the classical association of prednisone plus cyclophosphamide, and more recently of additional different combinations, has dramatically improved the prognosis. At 5 years, the overall survival rates are roughly 80 % even in patients with impaired renal function and pulmonary capillaritis [5]. Death within the first year is more frequently caused by infection and rapidly progressing renal failure, whereas in the following years the major causes of death still include infection but may also be ascribed to cardiovascular diseases and the onset of tumors [7].

10.7 Therapy

The first demonstration by Fauci and Wolff [24] of the strikingly effective results achieved with the association prednisone plus cyclophosphamide in patients with the previously termed Wegener's granulomatosis has marked a turning point in the therapy of AAV. For many years, this combination has indeed been considered the therapeutic gold standard for AAV. In the last several years, however, a number of drugs, given singly or in variable associations, have been proposed in the induction and maintenance of remission as well as in the treatment of relapses. Several reasons may account for the growing number of different therapeutic regimens: (a) given the large spectrum of clinical manifestations of AAV with variable grading of disease severity, it is reasonable to avoid aggressive induction treatments in patients with clinically limited forms of MPA; (b) unnecessary immunosuppression would result in higher toxicity and potential complications, especially in elderly, polypathologic patients; (c) conversely, large multicenter studies have shown that a small minority of patients are poorly responsive or refractory to current, established treatments and would therefore require newly conceived regimens; (d) although

Table 10.2 An incomplete list of the therapeutic regimens that have been used in patients with MPA [5, 6, 20, 25, 26]

Regimen	Indication	References
Remission induction (6–9 months)		
Corticosteroids alone	Clinically mild disease and lack of poor-prognosis factors. A higher relapse rate can be expected	[27]
Corticosteroids plus cyclophosphamide, oral or pulse	Conventional, gold standard for the large majority of patients	[28]
Corticosteroids plus methotrexate	Because of methotrexate tubular toxicity, this regimen should not be used in renal forms of MPA	[29]
Plasma exchange	To improve renal function in patients with crescentic glomerulonephritis, and possibly in patients with major alveolar hemorrhage	[30]
Anti-tumor necrosis factor-alpha agents (Infliximab or etanercept or adalimumab), in combination with conventional therapy	To treat patients with refractory disease. Treatment should be administered for a few weeks only, given the risk of serious complications. If possible, the use of these agents should be avoided.	[31]
Rituximab (anti-CD20 monoclonal antibody)	Highly effective and well tolerated in patients poorly responsive or unresponsive to conventional therapy	[32]
Alemtuzumab	Selectively depletes circulating T cells, monocytes and macrophages. Relapses and adverse events are common	[33]
Tocilizumab	The evidence for its clinical efficacy is so far circumstantial, although serum IL-6 levels are frequently elevated in AAV	[34]
Remission maintenance (18–24 months)		
Azathioprine or methotrexate	Add trimethoprim/sulfamethoxazole or aerosolized pentamidine to prevent Pneumocystis jiroveci pneumonia	[35]
Mycophenolate mofetil	Only in patients with azathioprine and/or methotrexate intolerance or side effects	[36]
Rituximab	It is a better maintenance therapy than azathioprine in patients with severe, refractory disease	[37]
Belimumab	In combination with azathioprine. Trial is under way	[26]
Relapse (length variable)		
Corticosteroids plus cyclophosphamide	Usually the conventional induction regimen is also effective in relapsing patients, but different regimens may be useful in refractory patients. However, so far their standardization has not been established	[20]
Intravenous immunoglobulins, alone or in addition to conventional therapy	In patients experiencing relapse during or after treatment or with persistently active (refractory) disease. Renal insufficiency is a contraindication.	[38]

receiving maintenance immunosuppression, a significant percentage of patients experience relapse of their disease, that not necessarily appears responsive to the previously administered regimens.

In Table 10.2 we have reported the main therapeutic protocols and biological drugs that have been used in patients with MPA and more generally with AAV, with the warning that only a few of these protocols have been the object of controlled clinical trials, and that individually tailoring, recommended regimens aimed at maximizing efficacy, avoid overtreatment and thus minimizing toxicity may only derive from international, multi-center, collaborative studies. Many of the regimens reported in Table 10.2 have indeed been studied in small uncontrolled trials. For the sake of simplicity, the table does not include the doses of the various drugs, that can be found in the reference mentioned for each regimen.

References

1. Jennette JC, Falk RJ, Bacon PA, Basu N, Cid MC, Ferrario F et al (2013) 2012 revised international Chapel Hill consensus conference nomenclature of vasculitides. Arthritis Rheum 65:1–11
2. Watts RA, Mooney J, Skinner J, Scott DG, Macgregor AJ (2012) The contrasting epidemiology of granulomatosis with polyangiitis (Wegener's) and microscopic polyangiitis. Rheumatology (Oxford) 51:926–931
3. Fujimoto S, Watts RA, Kobayashi S, Suzuki K, Jayne DR, Scott DG et al (2011) Comparison of the epidemiology of anti-neutrophil cytoplasmic antibody-associated vasculitis between Japan and the U.K. Rheumatology (Oxford) 50:1916–1920
4. Katsuyama T, Sada KE, Makino H (2014) Current concept and epidemiology of systemic vasculitides. Allergol Int 63:505–513
5. Kallenberg CG (2014) The diagnosis and classification of microscopic polyangiitis. J Autoimmun 48–49:90–93, February–March 2014
6. Greco A, De Virgilio A, Rizzo MI, Gallo A, Magliulo G, Fusconi M et al (2015) Microscopic polyangiitis: advances in diagnostic and therapeutic approaches. Autoimmun Rev 14:837–844
7. Flossmann O, Berden A, de Groot K, Hagen C, Harper L, Heijl C et al., European Vasculitis Study Group (2011) Long-term patient survival in ANCA-associated vasculitis. Ann Rheum Dis 70:488–494
8. Wang R, Yang XC, Zhou SJ, Sun GY (2015) Anti-neutrophil cytoplasmic antibody-negative microscopic polyangiitis: a case report and literature review. Exp Ther Med 10:749–752
9. Wang G, Csernok E, de Groot K, Gross WL (1997) Comparison of eight commercial kits for quantitation of antineutrophil cytoplasmic antibodies (ANCA). J Immunol Methods 208:203–211
10. Holle JU, Hellmich B, Backes M, Gross WL, Csernok E (2005) Variations in performance characteristics of commercial enzyme immunoassay kits for detection of antineutrophil cytoplasmic antibodies: what is the optimal cut off? Ann Rheum Dis 64:1773–1779
11. Csernok E, Moosig F (2014) Current and emerging techniques for ANCA detection in vasculitis. Nat Rev Rheumatol 10:494–501
12. Rasmussen N, Wiik A, Jayne DR (2015) A historical essay on detection of anti-neutrophil cytoplasmic antibodies. Nephrol Dial Transplant 30 Suppl 1:i8–i13
13. Davies DJ, Moran JE, Niall JF, Ryan GB (1982) Segmental necrotising glomerulonephritis with antineutrophil antibody: possible arbovirus aetiology? Br Med J 285:606

14. Simon A, Subra JF, Guilpain P, Jeannin P, Pignon P, Blanchard S et al (2016) Detection of anti-pentraxin-3 autoantibodies in ANCA-associated vasculitis. PLoS One 11, e0147091
15. Xiao H, Heeringa P, Hu P, Liu Z, Zhao M, Aratani Y et al (2002) Antineutrophil cytoplasmic autoantibodies specific for myeloperoxidase cause glomerulonephritis and vasculitis in mice. J Clin Invest 110:955–963
16. Little MA, Smyth CL, Yadav R, Ambrose L, Cook HT, Nourshargh S et al (2005) Antineutrophil cytoplasm antibodies directed against myeloperoxidase augment leukocyte-microvascular interactions in vivo. Blood 106:2050–2058
17. Lyons PA, Rayner TF, Trivedi S, Holle JU, Watts RA, Jayne DR et al (2012) Genetically distinct subsets within ANCA-associated vasculitis. N Engl J Med 367:214–223
18. Alberici F, Martorana D, Vaglio A (2015) Genetic aspects of anti-neutrophil cytoplasmic antibody-associated vasculitis. Nephrol Dial Transplant 30 Suppl 1:i37–i45
19. Chen M, Kallenberg CG (2010) ANCA-associated vasculitides – Advances in pathogenesis and treatment. Nat Rev Rheumatol 6:653–664
20. Villiger PM, Guillevin L (2010) Microscopic polyangiitis: clinical presentation. Autoimmun Rev 9:812–819
21. Lally L, Spiera RF (2015) Pulmonary vasculitis. Rheum Dis Clin North Am 41:315–331
22. Schirmer JH, Wright MN, Vonthein R, Herrmann K, Nölle B, Both M et al (2016) Clinical presentation and long-term outcome of 144 patients with microscopic polyangiitis in a monocentric German cohort. Rheumatology (Oxford) 55(1):71–79, January 2016
23. Lee RW, D'Cruz DP (2010) Pulmonary renal vasculitis syndromes. Autoimmun Rev 9:657–660
24. Fauci AS, Wolff SM (1973) Wegener's granulomatosis: studies in eighteen patients and a review of the literature. Medicine (Baltimore) 52:535–561
25. Lutalo PM, D'Cruz DP (2015) Biological drugs in ANCA-associated vasculitis. Int Immunopharmacol 27:209–212
26. Smith RM (2015) Update on the treatment of ANCA associated vasculitis. Presse Med 44:e241–e249
27. Ribi C, Cohen P, Pagnoux C, Mahr A, Arène JP, Puéchal X et al, French Vasculitis Study Group (2010) Treatment of polyarteritis nodosa and microscopic polyangiitis without poor-prognosis factors: a prospective randomized study of one hundred twenty-four patients. Arthritis Rheum 62:1186–1197
28. de Groot K, Harper L, Jayne DR, Flores Suarez LF, Gregorini G, Gross WL et al; EUVAS (European Vasculitis Study Group) (2009) Pulse versus daily oral cyclophosphamide for induction of remission in antineutrophil cytoplasmic antibody-associated vasculitis: a randomized trial. Ann Intern Med 150:670–680
29. de Groot K, Rasmussen N, Bacon PA, Tervaert JW, Feighery C, Gregorini G et al (2005) Randomized trial of cyclophosphamide versus methotrexate for induction of remission in early systemic antineutrophil cytoplasmic antibody-associated vasculitis. Arthritis Rheum 52:2461–2469
30. de Joode AA, Sanders JS, Smid WM, Stegeman CA (2014) Plasmapheresis rescue therapy in progressive systemic ANCA-associated vasculitis: single-center results of stepwise escalation of immunosuppression. J Clin Apher 29:266–272
31. Dharmapalaiah C, Watts RA (2012) The role of biologics in treatment of ANCA associated vasculitis. Mod Rheumatol 22:319–326
32. Stone JH, Merkel PA, Spiera R, Seo P, Langford CA, Hoffman GS et al, RAVE-ITN Research Group (2010) Rituximab versus cyclophosphamide for ANCA-associated vasculitis. N Engl J Med 363:221–232
33. Walsh M, Chaudhry A, Jayne D (2008) Long-term follow-up of relapsing/refractory anti-neutrophil cytoplasm antibody associated vasculitis treated with the lymphocyte depleting antibody alemtuzumab (CAMPATH-1H). Ann Rheum Dis 67:1322–1327

34. Berti A, Cavalli G, Campochiaro C, Guglielmi B, Baldissera E, Cappio S et al (2015) Interleukin-6 in ANCA-associated vasculitis: rationale for successful treatment with tocilizumab. Semin Arthritis Rheum 45:48–54
35. Puéchal X, Pagnoux C, Perrodeau É, Hamidou M, Boffa JJ, Kyndt X et al, French Vasculitis Study Group (2016) Long-term outcomes of the WEGENT trial on remission-maintenance for granulomatosis with polyangiitis or microscopic polyangiitis. Arthritis Rheumatol 68:690–701
36. Hiemstra TF, Walsh M, Mahr A, Savage CO, de Groot K Harper L et al, European Vasculitis Study Group (EUVAS) (2010) Mycophenolate mofetil vs azathioprine for remission maintenance in antineutrophil cytoplasmic antibody-associated vasculitis: a randomized controlled trial. JAMA 304:2381–2388
37. Charles P, Néel A, Tieulié N, Hot A, Pugnet G, Decaux O et al. French Vasculitis Study Group (2014) Rituximab for induction and maintenance treatment of ANCA-associated vasculitides: a multicentre retrospective study on 80 patients. Rheumatology (Oxford) 53:532–539
38. Crickx E, Machelart I, Lazaro E, Kahn JE, Cohen-Aubart F, Martin T et al, French Vasculitis Study Group (FVSG) (2016) Intravenous immunoglobulin as immunomodulating agent in ANCA-associated vasculitides: a French nationwide study of 92 patients. Arthritis Rheumatol 68:702–712

Chapter 11
Granulomatosis with Polyangiitis (Wegener's)

Franco Dammacco, Sebastiano Cicco, Domenico Ribatti, and Angelo Vacca

Abstract Granulomatosis with polyangiitis (GPA) is a relatively rare necrotizing vasculitis belonging to the so-called ANCA-associated vasculitides, in that circulating anti-proteinase-3 neutrophil cytoplasmic antibodies (c-ANCA) can be detected in the large majority of patients. The major clinical manifestations include necrotizing granulomatous lesions in the upper and/or lower respiratory tract, and glomerulonephritis. Although generalized necrotizing vasculitis involving both arteries and veins is a frequent indication of its systemic character, limited forms restricted to paranasal sinuses can also be observed. In the absence of suitable therapies, the disease can lead to death in a high percentage of patients. The introduction of glucocorticoids associated to cyclophosphamide has marked a milestone in the treatment of GPA. This combination has in fact been able to remarkably improve the previously ominous prognosis of these patients, resulting in a 10-years' survival rate of approximately 75 %. More recently, additional immunosuppressive drugs such as azathioprine, methotrexate, and rituximab have been employed, with comparable or even better results.

Keywords ANCA • Cyclophosphamide • Glomerulonephritis • Granulomatosis • Necrotizing vasculitis • Pulmonary infiltration

11.1 Definition and Epidemiology

Granulomatosis with polyangiitis (GPA), formerly named Wegener's granulomatosis, is a systemic, necrotizing vasculitis characterized by the involvement of the upper and lower respiratory tract, often associated with pauci-immune

F. Dammacco (✉) • S. Cicco • A. Vacca
Department of Biomedical Sciences and Human Oncology, Section of Internal Medicine, University of Bari Medical School, 70124 Bari, Italy
e-mail: francesco.dammacco@uniba.it

D. Ribatti
Department of Basic Medical Sciences, Neurosciences and Sensory Organs, University of Bari Medical School, Policlinico – Piazza G. Cesare, 11, 70124 Bari, Italy

National Cancer Institute, "Giovanni Paolo II", 70124 Bari, Italy

F. Dammacco et al. (eds.), *Systemic Vasculitides: Current Status and Perspectives*, DOI 10.1007/978-3-319-40136-2_11

glomerulonephritis. It is histologically characterized by granulomas within the medium and small vessel wall [1]. It is associated to the presence of antineutrophil cytoplasmic antibodies (ANCA), with a cytoplasmic staining pattern against proteinase-3 (PR3, c-ANCA) [2].

GPA is a rare disease with an estimated prevalence of 22–34 cases per million inhabitants [3, 4] and an incidence rate of 4–9 cases per million/year [4–6], with a higher incidence in Northern Europe. Mean age at diagnosis is between 50 and 65 years [4, 6], and males and females are equally affected. GPA is the most common of the pulmonary granulomatous vasculitides, involves the upper respiratory tract (e.g., sinuses, ears, nasopharynx, oropharynx, trachea), the lower respiratory tract (bronchi and lung), and the kidney, with varying degrees of disseminated vasculitis.

Although first described by Klinger in 1931 [7], GPA was established as a distinct clinico-pathologic entity by Wegener 5 years later [8]. The classic criteria for the diagnosis of GPA, described by Godman and Churg in 1954 [9], include the following triad features: (a) necrotizing granulomatous lesions in the upper and/or lower respiratory tract; (b) generalized necrotizing vasculitis involving both arteries and veins; (c) glomerulitis. Following the nomenclature system for Wegener's granulomatosis and other idiopathic vasculitides initially proposed by the Chapel Hill Consensus Conference in 1994 [10], the revised International Chapel Hill Consensus Conference Nomenclature of Vasculitides of 2012 [1] proposed a name change from "Wegener's granulomatosis" to "granulomatosis with polyangiitis (Wegener)", which is now accepted worldwide and has received the endorsement of the American College of Rheumatology, the American Society of Nephrology, and the European League Against Rheumatism.

The two distinguishing histopathologic features of the disease, namely the granuloma formation and the vasculitis process, obviously suggest that a hypersensitivity response plays an important role in the onset of the disease. However, so far no triggering or inciting antigen or pathogen has been identified. The discovery of ANCA [11] showed that pathogenic antibodies could be directly involved in the vasculitic process of GPA.

Under the effect of an unknown antigen and following the production of the tumor necrosis factor-alpha (TNFα) and interleukin-1beta (IL1β), neutrophils express PR3 on their surface, leading to the production of anti-PR3 antibodies, followed by leukocyte and monocyte adhesion to the vessel wall via adhesion molecules. The pathogenesis of GPA possibly involves the failure of alpha-1 antitrypsin (AAT), which *in vivo* is the primary inhibitor of PR3. Patients that have an AAT deficiency are at increased risk for GPA, suggesting a role for the increased presence of PR3 at inflammatory sites ([12]. The interaction of PR3-ANCA with TNF-α-primed mononuclear cells stimulates IL-8 release by cross-linking Fc gamma receptors (FcγR) and PR3 expressed on the monocyte cell surface ([13]. Moreover, these autoantibodies induce the release of monocyte chemoattractant protein-1 from mononuclear cells [14].

Under the influence of cytokines and ANCA, reactive oxygen species (ROS) are released and contribute to the aggression of the vessel wall. ANCA also directly attack the vessels while PR3 (Wegener's autoantigen) participates in these phenomena. Autoimmune response to PR3 plays a central role in disease development, according to *in vitro* and *in vivo* experimental data [15]. Immune complexes have been identified in serum [16] and skin [17] of patients with GPA, whereas these findings are generally absent in involved lung or renal tissues [18].

When granuloma formation occurs, lymphocytes and follicular dendritic cells form germinal center-like structures within the inflammatory lesions of granulomatous tissues. These follicular structures provide a microenvironment for the development of auto-reactive B cells. A large variability in the respective preponderance of the granulomatous or vasculitic features can be observed, depending on the stage of the disease.

11.2 Clinical Features

Constitutional signs (fever, asthenia and weight loss) are present in about 50 % of the patients [19]. In addition, ear, nose and throat (ENT) signs and symptoms occur in virtually all patients [20, 21], including crusting, rhinorrhea, sinusitis, nasal pain and chronic otitis media. Erosions of the facial cartilages may lead to perforation of the palate and/or pinna of the ear. When the nasal septum is involved, a relatively frequent outcome is nasal bridge collapse and a "saddle-nose" deformity. Sinusitis remains the most common hallmark of the disease and usually results in nasal obstruction, stuffiness, and hypo- or anosmia that are often the opening symptomatology. Localized forms of GPA may be characterized by ENT involvement only. When a hearing loss occurs, it is usually due to a Eustachian tube dysfunction secondary to naso-pharyngeal disease. In addition, inner ear disease can be associated to a sensorineural impairment and/or to a vestibular dysfunction [20].

The lower respiratory tract is involved in about 80 % of the patients. Hoarseness, cough, wheezing and stridor are symptoms of subglottic stenosis, a potentially fatal complication that is often asymptomatic in the initial phase of the disease and can affect roughly 15–20 % of the patients. The most common manifestations of lung involvement include alveolar hemorrhage of variable entity (from small to massive quantities, that may even lead to fulminant respiratory failure), pulmonary infiltrates and/or single or multiple parenchymal nodules which can undergo excavation. These signs can sometimes be misdiagnosed as mycobacterial or fungal infections [21].

In the systemic forms of GPA renal involvement can be detected in 50–100 % of the patients. The usual finding is glomerulonephritis with segmental necrosis, associated to extra-capillary proliferation and pauci-immune crescent formation. Kidney disease has a negative impact on the prognosis, the glomerular filtration rate (GFR) at diagnosis and the number of normal glomeruli at the renal biopsy being independent prognostic factors. The appearance of hematuria or proteinuria, and/or a significant increase in serum creatinine need an immediate evaluation and a prompt treatment. Additional urogenital manifestations (renal pseudotumor, prostatitis, orchitis, epididymitis, uretral stenosis or penis ulceration) have sometimes been described, but are rare complications [22].

Mucocutaneous manifestations are reported in about 50 % of patients as ulcerating, necrotizing lesions or widespread vascular purpura. Papules, subcutaneous nodules, raspberry-red gengivitis, and intraoral and/or genital ulcerations can less frequently be observed [19, 23, 24].

An ocular involvement is described in about half of the patients, usually in the form of scleritis or episcleritis, ranging in severity from an inflammatory process to a necrotizing disease that may even lead to blindness [25]. Other manifestations such as conjunctivitis, keratitis, uveitis, and nasolacrimal duct obstruction have been reported with variable frequency. Although of rare occurrence, the involvement of the eye socket is suggestive for GPA, especially if a granulomatous retro-orbital pseudo-tumor infiltration or a dacryoadenitis develop, clinically appearing as exophthalmos or ophthalmoplegia [25].

While peripheral neuropathy (sensorial and/or motor) occurs in about 30 % of patients, pachymeningitis is the most important involvement of central nervous system, is more rare (5–10 %), and can be caused by granulomatous deposits, intracerebral vascular lesions or an extension of sinus lesions.

Cardiovascular (conduction disorders, pericarditis or myocarditis ranging from subclinical manifestations to end-stage heart failure) [26] and gastrointestinal (ulcerative lesions as well as intestinal perforation) systems are quite unusual but possible targets in GPA.

11.3 Diagnostic Criteria and Follow-up

In 1990 the American College of Rheumatologists proposed four diagnostic criteria, of which at least two are needed for the diagnosis, with the aim of including homogeneous adult populations in research studies (Table 11.1A) [27]. In 2007 some surrogate diagnostic parameters were suggested [28] and in 2010 the European League Against Rheumatism worked out classification criteria for childhood GPA (Table 11.1B) [29]. The already mentioned 2012 Chapel Hill Consensus Conference [1] defined GPA as a necrotizing granulomatous inflammation of the upper and lower respiratory tract affecting small and medium vessel walls. According to this definition, glomerulonephritis is not necessary for the diagnosis, in spite of its relatively frequent occurrence. Moreover, this classification specifies that granulomas do not need to be confirmed histologically, but can be predicted by non-invasive techniques. The diagnosis can in fact be postulated by the combination of peculiar clinical features with the presence of PR3-cANCA [2]. However, histological findings (especially the results of renal biopsy) maintain a remarkable importance for both their diagnostic and prognostic implications.

Patients with suspected GPA should be evaluated by a combination of instrumental and laboratory findings. Complete blood count, creatinine, BUN, serum electrolytes, serum proteins with electrophoresis, erythrocyte sedimentation rate (ESR) and C-reactive protein (CRP) are mandatory tests in these patients. Increase of ESR, CRP and α2-globulins is obviously indicative of active disease. Urinalysis and

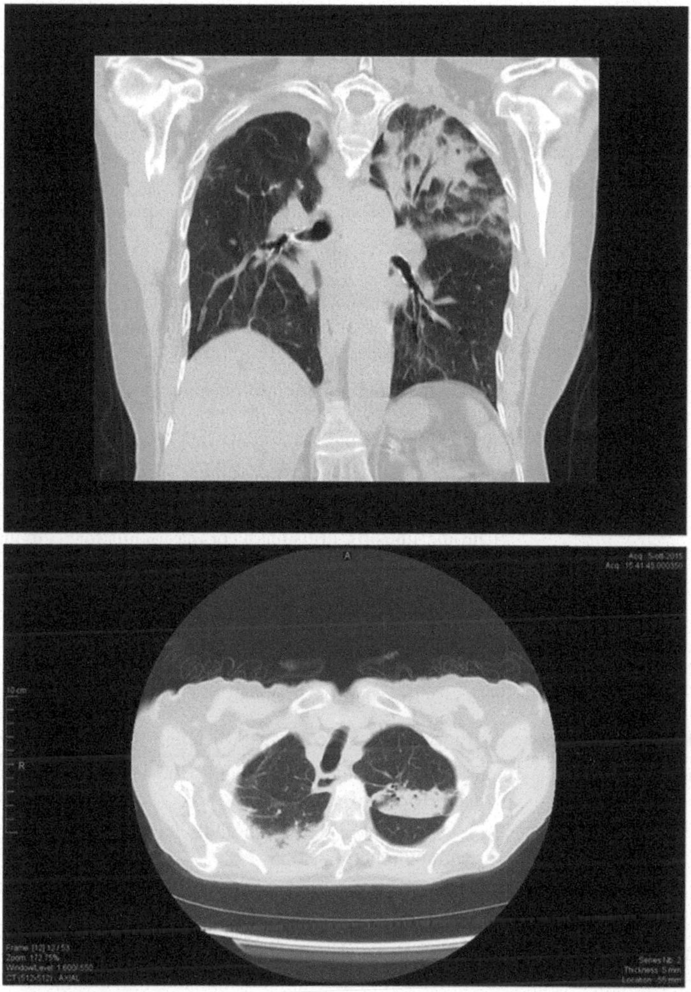

Fig. 11.1 Computed tomography scan of the thorax of a patient with granulomatosis with polyangiitis. Extensive infiltrates of the upper lobe of the left lung can be seen, in addition to ground-glass areas. Sub-pleural infiltrates also involve the right lung

24 h-proteinuria are useful to evaluate the kidney function. Approximately 90 % of GPA patients are positive for PR3 c-ANCA, whereas an MPO p-ANCA test is positive in about 5–10 %. Although ANCA titers are not related to disease activity, the reappearance of ANCA positivity in a previously negative patient usually precedes the clinical flare. Infections by fungi and Mycobacterium tuberculosis should be excluded before treatment.

Conventional chest and nasal-sinus X-ray is the first instrumental test to be performed in GPA: nasal sinuses obstruction and single/multiple nodules are the most common findings. CT scan is needed to better define the extension of inflammatory tissue (Fig. 11.1). Endoscopy of nose and larynx should be performed to exclude

Table 11.1 Diagnostic criteria of granulomatosis with polyangiitis (GPA)

(A) In adults according to the American College of Rheumatology [27, 28]	
Nasal or oral inflammation	Painful or painless oral ulcers or purulent or bloody nasal discharge
Abnormal chest radiograph	Nodules, fixed infiltrates, or cavities
Urinary sediment	Microhematuria or red cell casts
Granulomatous inflammation	Within the wall of an artery or in the perivascular area on biopsy specimens
Surrogate parameters	
Glomerulonephritis	Hematuria associated with red cell casts or ~10% dysmorphic erythrocytes or 2+ hematuria and 2+ proteinuria on urinalysis
Lower airways	
Upper airways	X-ray: fixed pulmonary infiltrates/nodules/cavitations for >1 month
	Bronchial stenosis
	Bloody nasal discharge and crusting for >1 month, or nasal ulceration
	Chronic sinusitis, otitis media or mastoiditis for >3 months
	Retro-orbital mass or inflammation (pseudotumor)
	Subglottic stenosis
	Saddle nose deformity/destructive sino-nasal disease
Diagnosis of GPA: at least 2 criteria (sensitivity 88.2%; specificity 92%)	
Only one surrogate marker is necessary to support a diagnosis of GPA	

(B) In childhood according to the European League Against Rheumatism (EULAR) [29]	
Histopathology	Granulomatous inflammation within the wall of an artery or in the perivascular or extravascular area
Upper airway involvement	Chronic purulent or bloody nasal discharge or recurrent epistaxis/crusts/granulomata
	Nasal septum perforation or saddle nose deformity
	Chronic or recurrent sinus inflammation
Laryngo-tracheo-bronchial involvement	Subglottic, tracheal or bronchial stenoses
Pulmonary involvement	Chest X-ray or CT showing the presence of nodules, cavities or fixed infiltrates
ANCA	ANCA positivity by immunofluorescence or by ELISA (PR3/c or MPO/p)
Renal involvement	Proteinuria >0.3 g/24 h or >30 mmol/mg of urine
	Albumin/creatinine ratio on a spot morning sample
	Hematuria or red blood cell casts: >5 red blood cells/high power field or red blood cell casts in the urinary sediment or ≥2+ on dipstick
	Necrotizing pauci-immune glomerulonephritis
GPA is diagnosed when at least three criteria are present (sensitivity 93.3%; specificity 99.2%)	

Table 11.2 Diseases to be considered in the differential diagnosis of granulomatosis with polyangiitis

Infections	Meningococcal sepsis
	Legionella pneumonia
	Lyme's disease
	Leptospirosis
	Tuberculosis
	Bacterial endocarditis
	Mycotic aneurysms
	Hemolytic-uremic syndrome
Autoimmune diseases	Systemic lupus erythematosus
	Rheumatoid arthritis
	Anti-phospholipid syndrome
	Sjögren's syndrome
	Cryoglobulinemic vasculitis
	Other ANCA vasculitides
	Goodpasture's disease
	Polyarteritis nodosa
	[Sarcoidosis][a]
Neoplastic diseases	Lymphoma/leukemia
	Paraneoplastic syndromes
Other	Thrombotic thrombocytopenic purpura

[a]The insertion of sarcoidosis among autoimmune diseases is questionable.

nasal septum erosions and to obtain histological specimens. Ultrasound echography is the best way to study kidney morphology and to guide a renal biopsy. Heart ultrasound and brain or abdomen CT scan are advisable when the involvement of the respective organs is suspected.

In addition to the already mentioned fungal and mycobacterial infections, other types of ANCA-associated vasculitides and Goodpasture's syndrome should be considered in the differential diagnosis (Table 11.2). Furthermore, the extent of disease activity should be assessed using an activity index (i.e. the widely used Birmingham Vasculitis Activity Score applied to GPA and indicated with the acronym BVAS/WG [30]) to perform a quantitative analysis that can be repeated in the course of the patient's follow-up.

11.4 Treatment

Until the 1950s GPA was an almost regularly fatal disease, only 10 % of the patients reaching a 2-years survival. Following the introduction of cyclophosphamide (CTX) and corticosteroids as the standard treatment [31], patients now experience a

1 year survival of 93 % and a 5-years and 10-years survival of 79 % and 75 % respectively [32].

Treatment must be personalized for each patient, according to disease severity and co-morbidities, in order to avoid either an excessive treatment that is associated with a significant risk of adverse events or side effects, or an insufficient treatment with a risk of failure and disease progression.

The therapeutic approach to GPA is divided into two phases: (a) "induction phase", that aims to achieve a disease remission in about 3–6 months, according to the clinical response; (b) "maintenance phase", that points to consolidate the disease control and reduce the risk of relapse. The maintenance phase lasts 18–24 months after the achievement of remission [33].

Steroids are the main drugs used in the induction phase. In limited GPA, oral prednisone at a daily starting dose of 1 mg/kg is the recommended treatment. After about 1 month, steroids should be gradually tapered to achieve a steroid-free treatment by 18–24 months. In severe or refractory disease (kidney involvement or hemorrhagic alveolitis), an intravenous bolus of methylprednisolone at 7.5–15 mg/kg/day should be given for 1–3 consecutive days before starting the full dose oral prednisone.

Immunosuppressive drugs are needed in the induction phase to achieve the remission (BVAS = 0). Azathioprine (AZA, 1–2 mg/kg/day) or methotrexate (MTX, 15–25 mg/week) are preferred options. These drugs may also be useful in the maintenance phase as steroid-sparing agents [34].

In severe disease, CTX remains the milestone. It must be given at 2 mg/kg/day for a maximum of 3–6 months, and then switched to AZA for maintenance. More recently, the monoclonal antibody Rituximab (RTX) has been found to be as effective as CTX to obtain a steroid-free remission. RTX is applied at the weekly dose of 375 mg/m^2 for 4 weeks [35] and its efficacy has also been shown in patients with kidney involvement [36, 37]. Additional evidence indicates that RTX may be useful in the maintenance phase as well: after achieving remission, RTX (500 mg on days +1, +15, and on months +6, +12, and +18) is more effective than AZA (2 mg/kg/day), resulting in a low risk of relapse at 44 months [38].

In life-threatening GPA or in patients admitted to intensive care units (especially those in ventilation therapy) intravenous boluses of CTX should be promptly administered. Doses of 600 mg/m^2 to a maximum of 1.2 g/day on days +1, +15, +30, followed by 700 mg/m^2 every three weeks for a maximum of 6–9 boluses are remarkably effective to control the disease activity [39]. Finally, plasma exchange can be considered as an additional treatment option in patients with severe and rapidly progressing GPA.

References

1. Jennette JC, Falk RJ, Bacon PA, Basu N, Cid MC, Ferrario F et al (2013) 2012 revised international Chapel Hill consensus conference nomenclature of vasculitides. Arthritis Rheum 65:1–11
2. Thai LH, Charles P, Resche-Rigon M, Desseaux K, Guillevin L (2014) Are anti-proteinase-3 ANCA a useful marker of granulomatosis with polyangiitis (Wegener's) relapses? Results of a retrospective study on 126 patients. Autoimmun Rev 13:313–318
3. Mahr A, Guillevin L, Poissonnet M, Aymé S (2004) Prevalences of polyarteritis nodosa, microscopic polyangiitis, Wegener's granulomatosis, and Churg-Strauss syndrome in a French urban multiethnic population in 2000: a capture-recapture estimate. Arthritis Rheum 51:92–99
4. Catanoso M, Macchioni P, Boiardi L, Manenti L, Tumiati B, Cavazza A et al (2014) Epidemiology of granulomatosis with polyangiitis (Wegener's granulomatosis) in Northern Italy: a 15-year population-based study. Semin Arthritis Rheum 44:202–207
5. Watts RA, Carruthers DM, Scott DG (1995) Epidemiology of systemic vasculitis: changing incidence or definition? Semin Arthritis Rheum 25:28–34
6. Mohammad AJ, Jacobsson LTH, Westman KWA, Sturfelt G, Segelmark M (2009) Incidence and survival rates in Wegener's granulomatosis, microscopic polyangiitis, Churg-Strauss syndrome and polyarteritis nodosa. Rheumatology (Oxford) 48:1560–1565
7. Klinger H (1931) Grenzformen der Pericarditis nodosa. Frankf Z Pathol 42:455–480
8. Wegener F (1936) Über generalisierte, septische GefäBerkrankungen [Regarding generalized, septic vascular diseases]. Verh Dtsch Pathol Gesellsch 26:202–210
9. Godman GC, Churg J (1954) Wegener's granulomatosis. Pathology and review of the literature. Arch Pathol 58:533–553
10. Jennette JC, Falk RJ, Andrassy K, Bacon PA, Churg J, Gross WL et al (1994) Nomenclature of systemic vasculitides: proposal of an international consensus conference. Arthritis Rheum 37:187–192
11. Davies DJ, Moran JE, Niall JF, GB (1982) Segmental necrotising glomerulonephritis with antineutrophil antibody: possible arbovirus aetiology? Br Med J 285:606
12. Elzouki AN, Segelmark M, Wieslander J, Eriksson S (1994) Strong link between the alpha 1-antitrypsin PiZ allele and Wegener's.granulomatosis. J Intern Med 236:543–548
13. Ralston DR, Marsh CB, Lowe MP, Wewers MD (1997) Antineutrophil cytoplasmic antibodies induce monocyte IL-8 release. Role of surface proteinase-3, alpha1 antitrypsin, and Fcgamma receptors. J Clin Invest 100:1416–1424
14. Casselman BL, Kilgore KS, Miller BF, Warren JS (1995) Antibodies to neutrophil cytoplasmic antigens induce monocyte chemoattractant protein-1 secretion from human monocytes. J Lab Clin Med 126:495–502
15. Kallenberg CG (2011) Pathogenesis of ANCA-associated vasculitides. Ann Rheum Dis 70(suppl 1):59–63
16. Pinching A, Lockwood C, Pussell B, Rees AJ, Sweny P, Evans DJ et al (1983) Wegener's granulomatosis: observations on 18 patients with severe renal disease. QJM 52:435–460
17. Brons RH, de Jong MC, de Boer NK, Stegeman CA, Kallenberg CG, Tervaert JW (2001) Detection of immune deposits in skin lesions of patients with Wegener's granulomatosis. Ann Rheum Dis 60:1097–1102
18. Horn RG, Fauci AS, Rosenthal AS, Wolff SM (1974) Renal biopsy pathology in Wegener's granulomatosis. Am J Pathol 74:423–440
19. Lane SE, Watts R, Shepstone L, Scott DGI (2005) Primary systemic vasculitis: clinical features and mortality. QJM 98:97–111
20. Trimarchi M, Sinico RA, Teggi R, Bussi M, Specks U, Meroni PL et al (2013) Otorhinolaryngological manifestations in granulomatosis with polyangiitis (Wegener's). Autoimmun Rev 12:501–505

21. Gómez-Gómez A, Martínez-Martínez MU, Cuevas-Orta E, Bernal-Blanco JM, Cervantes-Ramírez D, Martínez-Martínez R (2014) Pulmonary manifestations of granulomatosis with polyangiitis. Reum Clin 10:288–293
22. Chang DY, Wu LH, Liu G, Chen M, Kallenberg CG, Zhao MH (2012) Re-evaluation of the histopathologic classification of ANCA-associated glomerulonephritis: a study of 121 patients in a single center. Nephrol Dial Transplant 27:2343–2349
23. Stone JH (2003) Limited versus severe Wegener's granulomatosis: baseline data on patients in the Wegener's granulomatosis etanercept trial. Arthritis Rheum 48:2299–2309
24. Sung IY, Kim YM, Cho YC, Son JH (2015) Role of gingival manifestation in diagnosis of granulomatosis with polyangiitis (Wegener's granulomatosis). J Periodontal Implant Sci 45:247–251
25. Tarabishy AB, Schulte M, Papaliodis GN, Hoffman GS (2010) Wegener's granulomatosis: clinical manifestations, differential diagnosis, and management of ocular and systemic disease. Surv Ophthalmol 55:429–444
26. McGeoch L, Carette S, Cuthbertson D, Hoffman GS, Khalidi N, Koening CL et al (2015) Cardiac involvement in granulomatosis with polyangiitis. J Rheumatol 42:1209–1212
27. Leavitt RY, Fauci AS, Bloch DA, Michel BA, Hunder GG, Arend WP et al (1990) The American College of Rheumatology 1990 criteria for the classification of Wegener's granulomatosis. Arthritis Rheum 33:1101–1107
28. Watts R, Lane S, Hanslik T, Hauser T, Hellmich B, Koldingsnes W et al (2007) Development and validation of a consensus methodology for the classification of the ANCA-associated vasculitides and polyarteritis nodosa for epidemiological studies. Ann Rheum Dis 66:222–227
29. Ozen S, Pistorio A, Iusan SM, Bakkaloglu A, Herlin T, Brik R et al (2010) EULAR/PRINTO/PRES criteria for Henoch-Schönlein purpura, childhood polyarteritis nodosa, childhood Wegener granulomatosis and childhood Takayasu arteritis: Ankara 2008. Part II: Final classification criteria. Ann Rheum Dis 69:798–806
30. Mukhtyar C, Lee R, Brown D, Carruthers D, Dasgupta B, Dubey S et al (2009) Modification and validation of the birmingham vasculitis activity score (version 3). Ann Rheum Dis 68:1827–1832
31. Fauci AS, Haynes BF, Katz P, Wolff SM (1983) Wegener's granulomatosis: prospective clinical and therapeutic experience with 85 patients for 21 years. Ann Intern Med 1:76–85
32. Koldingsnes W, Nossent H (2002) Predictors of survival and organ damage in Wegener's granulomatosis. Rheumatology (Oxford) 41:572–581
33. Springer J, Nutter B, Langford CA, Hoffman GS, Villa-Forte A (2014) Granulomatosis with polyangiitis (Wegener's). Medicine (Baltimore) 93:82–90
34. Pagnoux C, Mahr A, Hamidou MA, Boffa JJ, Ruivard M, Ducroix JP et al (2008) French Vasculitis Study Group. Azathioprine or methotrexate maintenance for ANCA-associated vasculitis. N Engl J Med 359:2790–2803
35. Stone JH, Merkel PA, Spiera R, Seo P, Langford CA, Hoffman GS et al (2010) RAVE-ITN research group. Rituximab versus cyclophosphamide for ANCA-associated vasculitis. N Engl J Med 363:221–232
36. Jones RB, Tervaert JW, Hauser T, Luqmani R, Morgan MD, Peh CA et al (2010) European vasculitis study group. Rituximab versus cyclophosphamide in ANCA-associated renal vasculitis. N Engl J Med 363:211–220
37. Geetha D, Specks U, Stone JH, Merkel PA, Seo P, Spiera R et al (2015) Rituximab for ANCA-associated vasculitis immune tolerance network research group. Rituximab versus cyclophosphamide for ANCA-associated vasculitis with renal involvement. J Am Soc Nephrol 26:976–985
38. Guillevin L, Pagnoux C, Karras A, Khouatra C, Aumaître O, Cohen P et al (2014) French vasculitis study group. Rituximab versus azathioprine for maintenance in ANCA-associated vasculitis. N Engl J Med 371:1771–1780
39. Kimmoun A, Baux E, Das V, Terzi N, Talec P, Asfar P et al (2016) Outcomes of patients admitted to intensive care units for acute manifestation of small-vessel vasculitis: a multicenter, retrospective study. Crit Care 20:27

Chapter 12
Eosinophilic Granulomatosis with Polyangiitis (Churg-Straus Syndrome)

Renato Alberto Sinico and Paolo Bottero

Abstract Eosinophilic Granulomatosis with Polyangiitis (EGPA) is defined as an eosinophil-rich and granulomatous inflammation involving the respiratory tract, and necrotizing vasculitis affecting small to medium-sized vessels, and associated with asthma and eosinophilia. It is usually classified among the so-called anti-neutrophil antibody (ANCA)-associated systemic vasculitides (AASVs) because of its clinical and pathologic features that overlap with those of the other AASVs. However, recent studies on large cohorts of patients have found that ANCA, usually P-ANCA/MPO-ANCA, were present in less than 40 % of patients. Moreover, ANCA status was shown to segregate with clinical phenotype. Preliminary results suggest that ANCA-positive and ANCA-negative patients also might have a different genetic background. Corticosteroids remain the cornerstone of the initial treatment of EGPA. The addition of cyclophosphamide is indicated in patients with poor-prognosis factors or in patients without poor-prognosis factors that relapse early. How long should maintenance therapy be continued remains to be established. However, the vast majority of patients require long-term corticosteroids treatment to control asthma.

Keywords Churg-Strauss syndrome • Eosinophilic granulomatosis with polyangiitis (EGPA) • Asthma • Eosinophilia • Anti-neutrophil cytoplasmic antibody (ANCA) • Vasculitis • ANCA associated vasculitis

R.A. Sinico (✉)
Clinical Immunology Unit and Renal Unit, Azienda Ospedaliera Ospedale San Carlo Borromeo, Milano, Italy
e-mail: renatoalberto.sinico@fastwebnet.it

P. Bottero
Poliambulatorio Santa Crescenzia, Magenta, Milano, Italy

F. Dammacco et al. (eds.), *Systemic Vasculitides: Current Status and Perspectives*, DOI 10.1007/978-3-319-40136-2_12

12.1 Introduction

Eosinophilic Granulomatosis with Polyangiitis, EGPA, (formerly known as Churg-Strauss Syndrome, from the names of the pathologists who first described the syndrome in 1951) is defined as an eosinophil-rich and granulomatous inflammation involving the respiratory tract, with necrotizing vasculitis affecting small to medium-sized vessels, and associated with asthma and eosinophilia [1, 2]. EGPA is a rare disorder affecting 1.3 and 6.8 cases per 1,000,000 people per year and with an overall prevalence of 10.7–13 per 1,000,000 adults [3–5].

EGPA is considered one of the so-called anti-neutrophil cytoplasmic antibody (ANCA) associated systemic vasculitis because of its clinical and pathologic features that overlap with those of the other anti-neutrophil cytoplasmic antibody-associated vasculitis (AASV), Wegener's granulomatosis (WG) and microscopic polyangiitis (MPA) [2, 5–8], although the prevalence of ANCA in EGPA is much lower (about 40 % of patients) [9, 10].

12.2 Clinical Features

Usually, three different phases can be distinguished in EGPA [11]. A prodromal phase with asthma and rhino-sinusitis/nasal polyposis [12], that may precede by months, and sometimes by several years, the development of an eosinophilic infiltrative disease with eosinophilic pneumonia or gastro-enteritis, followed by the vasculitic phase [11–18]. Mean age at diagnosis is around 50 years with no gender prevalence.

Constitutional symptoms such as fever, myalgias, arthralgias and weight loss are common whereas asthma may be silent at vasculitic onset.

Pulmonary. Non-fixed pulmonary infiltrates are seen in most patients. Pleural effusions with a large number of eosinophils occur in 22 % of cases [10] and rarely cavitating parenchymal nodules may occur [9, 16–19]. Alveolar haemorrhage is not common but [20] can be severe and massive [9, 10].

Extra pulmonary. Skin, peripheral nerves, gastrointestinal tract, heart and kidney may be involved.

Peripheral nerves (65–76 %), with lung (51–65 %), and skin (52–57 %) are the most frequently involved organ/systems [9, 10, 20]. Mononeuritis multiplex is the usual sign of peripheral nervous system involvement but symmetrical peripheral neuropathy can be seen [14–18]. Skin involvement may include rash, palpable purpura (Fig. 12.1), infarction, and nodules [14, 16, 21–23]. Heart involvement has been documented in 16–50 % of cases and includes myocarditis, coronary vasculitis, valvular heart abnormalities, congestive heart failure, and pericarditis [9, 10, 14–19]. Coronary vasculitis and myocarditis are the main causes of morbidity and mortality. Moreover, venous thrombo-embolic events such as deep venous thrombosis and/or pulmonary embolism can occur with relatively high frequency (8 %) [24].

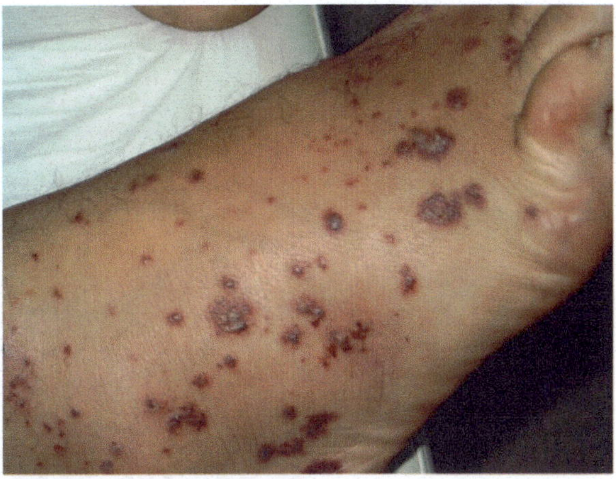

Fig. 12.1 Purpura in a patient with Eosinophilic Granulomatosis with Polyangiitis (EGPA)

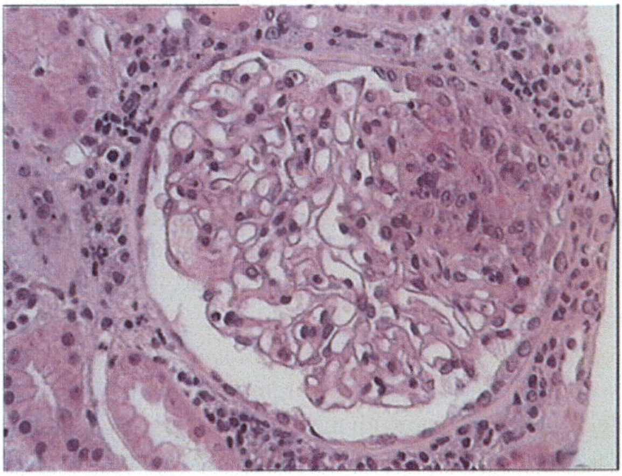

Fig. 12.2 Necrotizing crescentic glomerulonephritis in a patient with Eosinophilic Granulomatosis with Polyangiitis (EGPA)

Abdominal pain, ascites, diarrhea, even bloody, are present in 22–32 % of cases as signs of gastrointestinal involvement. Eosinophilic gastroenteritis and mesenteric vasculitis are the main pathological pictures but pancreatitis and cholecystitis have also been reported [14–18]. Kidney is affected in about 25 % of patients [25]. The main picture is necrotizing crescentic glomerulonephritis (Fig. 12.2) but other forms of nephropathy may also occur such as eosinophilic tubulo-interstitial nephritis (Fig. 12.3), focal mesangial proliferative glomerulonephritis, and focal segmental glomerulosclerosis [25]. Obstructive uropathy caused by vasculitis and/or granulomatous inflammation of the ureter or prostate may be present [25]. Central nervous system (CNS) involvement is not frequent (9–14 %) but it can have a great impact on mortality and morbidity [1, 7–18] because include palsies of the cranial nerves, cerebral hemorrhage or infarction.

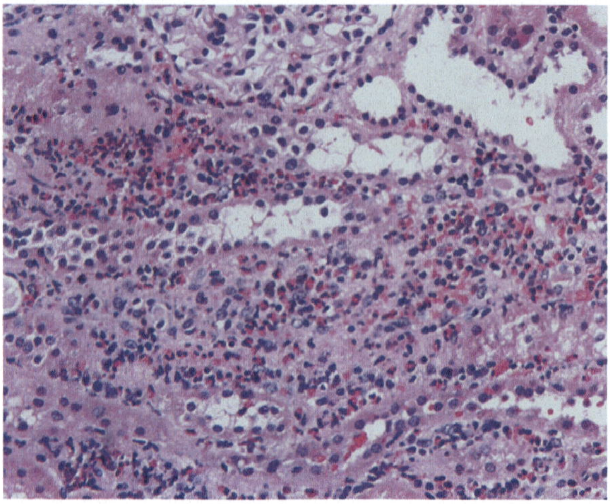

Fig. 12.3 Eosinophilic tubulo-interstitial nephritis in a patient with Eosinophilic Granulomatosis with Polyangiitis (EGPA) (Courtesy of Dr. G. Gregorini, Brescia, Italy)

12.3 Laboratory, Immunological Findings and Pathology

Most patients show blood eosinophilia (more than 1500 cells/mm^3 or more than 10 %), but eosinophils count can be normal in patients previously treated with systemic glucocorticoids for asthma [9, 11, 13]. ANCA (usually P-ANCA with specificity for myeloperoxidase) are present in less than 40 % of patients [9, 10]. Serum creatinine may be increased when kidney is involved and can represent a negative prognostic factor [26]. Total immunoglobulins E are often elevated but evidence of allergy, considered as the demonstration of specific IgE consistent with the clinical history, is rarely found [12]. Serum immunoglobulins G4 can be elevated during active EGPA and correlate with the number of disease manifestations and the Birmingham Vasculitis Activity Score (BVAS) [27]. Apart from ANCA, several other biomarkers have been proposed to differentiate EGPA from other diseases. CCL17, eotaxin-1 and IL-5 levels cannot differentiate EGPA from other hypereosinophilic diseases, but eotaxin-3 and IL-25 levels might be increased in EGPA only [28].

A significant prevalence of HLA-DRB4 gene is present in EGPA patients mainly in those ANCA positive [29, 30].

The presence of extravascular granulomas in association with necrosis and predominant extravascular eosinophils, as well as necrotizing vasculitis are the pathological findings [1, 2] but all of these lesions are rarely found together in biopsy specimens [11, 16–18].

12.4 Triggers

Various possible triggers have been reported in the development of EGPA: infections, vaccinations, allergic hyposensitizations and drugs, mainly leukotriene-receptors antagonists (LTRA), even if, in spite of the number of case reports, it is not possible to determine in individual cases whether the association between EGPA and LTRA therapy is causal, coincidental or directly related to other patterns of disease presentation or medication use [31].

12.5 Subgrouping EGPA Patients by ANCA Status

In 2005 two independent studies [9, 10] suggested that EGPA might include two clinical subsets: an ANCA-positive phenotype associated with a higher frequency of renal involvement, peripheral neuropathy, alveolar hemorrhage and purpura, and an ANCA-negative phenotype associated with heart and lung disease (other than alveolar hemorrhage). Vasculitis was documented less frequently in histological specimens from ANCA-negative patients in comparison with ANCA-positive ones. These findings have led to postulate the predominance of distinct pathogenetic mechanisms in the two subsets of patients: an ANCA-mediated process in ANCA-positive patients and tissue infiltration by eosinophils with subsequent release of toxic product in ANCA-negative cases. Similar conclusions have been reported in a larger survey more recently [32].

12.6 Diagnosis and Classification

Diagnosis of EGPA may be difficult for the phasic nature of the disease, the absence of specific symptoms and signs, the possibility of "formes frustes" [33] in which the clinical manifestations and histological findings may be partially or totally suppressed by corticosteroid therapy for asthma. Therefore, classification criteria have been used as diagnostic surrogates. Churg and Strauss originally described 13 asthmatic patients with blood and tissue eosinophilia, necrotizing vasculitis, and necrotizing granulomas centred on necrotic eosinophils [1], but all these pathological criteria are present in a minority of patients. In 1984 Lanham et al. [11] suggested that diagnosis could be made on clinical ground in patients with history of asthma, eosinophilia higher than 1500 cells/mm^3 and clinical/histological vasculitis involving two or more extrapulmonary organs. Asthma, eosinophilia greater than 10 % on differential white blood cell count, mononeuropathy (including multiplex) or polyneuropathy, migratory or transient pulmonary infiltrates, paranasal sinus abnormality, biopsy containing a blood vessel with extravascular eosinophils were proposed as 6 criteria for classifying CSS, with 4 being necessary for diagnosis, with a sensitivity of 85 % and a specificity of 99.7 %, by the American College of Rheumatology (ACR) in 1990 [34]. The ACR criteria have been used widely as diagnostic surrogates even if it is important to underline that these criteria, in the absence of histologically or clinically proven vasculitis, are insensitive and poorly specific [15] because 4 criteria for the diagnosis of CSS can be fitted by other diseases [9, 34–36]. In 2007 a stepwise algorithm was developed and validated for the classification of patients with a clinical diagnosis of ANCA-associated vasculitis and polyarteritis nodosa in which EGPA could be diagnosed if the ACR and/or the Lanham criteria are fulfilled in a patient with either histological proof of vasculitis or surrogate markers for vasculitis [37]. In the same year, diagnostic criteria for AASVs, including EGPA, have been elaborated by the Japanese Research Group of Intractable Vasculitis, but they have not yet been validated and compared with ACR and Lanham criteria in the English Literature [38].

12.7 Differential Diagnosis

A number of different diseases can share several clinical and/or histological features of EGPA such as other forms of AASVs (microscopic polyangiitis and Wegener's granulomatosis), in which however, asthma and eosinophilia (especially higher than 1500 cells/mm^3) are not usually present [2, 7, 38]. Also in idiopathic hypereosinophilic syndrome (HES), defined as a sustained peripheral blood eosinophilia of unknown origin, exceeding 1500 cells/mm^3 for more than six consecutive months [39] the organs involved are similar and cardiac disease is the major cause of death in both. However, asthma is usually absent in this condition and signs of vasculitis are not found on biopsy specimens. The diagnosis of HES will be facilitated by the use of molecular biology techniques since specific mutations have been identified in some subsets of this syndrome [40]. Asthma, eosinophilia, sinusitis and lung infiltrates are usually present in allergic bronchopulmonary aspergillosis and chronic eosinophilic pneumonia, but they lack the extrapulmonary involvement [35, 36, 41]. Finally, parasites such as toxocara and strongyloides stercoralis should be excluded for their possible expanded clinical spectrum [42, 43].

12.8 Prognosis and Treatment

Corticosteroids dramatically improve the prognosis of EGPA [1, 9, 10, 13, 19, 44–46] and are considered the cornerstone of initial EGPA treatment. Cardiomyopathy and older age are independent risk factors for death [32]. Prednisone, 1 mg/kg/day, or its equivalents, are given orally but methylprednisolone pulses are required in the most severe cases. Steroids are slowly tapered in some weeks [44] but residual asthma can prevent from steroids withdrawal [43].

The addition of immunosuppressive treatment (oral azathioprine or cyclophosphamide pulses, preferred to oral administration because of the lower cumulative dosage) is required in case of treatment failure or relapse [44] and in patients with poor prognostic factor.

Five prognostic factors, the so-called Five Factor Score (FFS), (elevated serum creatinine levels > 1.58 mg/dl, proteinuria > 1 g/day, gastrointestinal tract involvement, cardiomyopathy, central nervous system involvement) have been identified in patients with necrotizing vasculitis including EGPA [26]. Corticosteroids alone should be the treatment of choice in EGPA patients without poor-prognosis factors (FFS of 0) [44] and additional immunosuppressive treatment (azathioprine or cyclophosphamide pulses) should be reserved to the patients with treatment failure or relapse [44]. Patients with poor-prognosis factors (FFS = or > 1) should be treated with 3 consecutive methylprednisolone pulses on days 1–3 followed by oral prednisone 1 mg/kg daily for 3 weeks, tapering 5 mg every 10 days to 0.5 mg/kg and, afterwards, tapering 2.5 mg every 10 days to the minimal effective dosage or, when possible until definitive withdrawal plus 12 cyclophosphamide pulses (600 mg/m^2) every 2 weeks for 1 month, then every 4 weeks thereafter or short-course of cyclophosphamide (oral 2 mg/kg for 3 months or 6 cyclophosphamide pulse [600 mg/m^2] every 2 weeks for 1 month, then every 4 weeks thereafter), followed by azathioprine 2 mg/kg for 1 year or more [44, 47]. Plasma exchange should be considered (in addition to corticosteroids and cyclophosphamide) in patients with rapidly progressive glomerulonephritis and/or alveolar hemorrhage [20, 48]. Methotrexate, cyclosporin-A and Azathioprine have been proposed as drugs for maintenance of remission [28]. Different treatments as intravenous immunoglobulins, interferon-alpha, anti-TNF alpha agents, rituximab, mepolizumab and omalizumab have been used in refractory or frequently relapsing patients with promising results [28]. Finally, the treatment of residual asthma should be optimized in accordance with modern guidelines of asthma management, with the goal of reducing the use of oral corticosteroids.

Recently, an international task force of experts from different specialties has published disease specific recommendations for the diagnosis and management of EGPA [49]. These recommendations aim to give physicians tools for effective and individual management of EGPA patients and to provide guidance for future targeted research (Table 12.1).

Table 12.1 Recommendations for the diagnosis, follow-up and management of EGPA with the corresponding level of evidence established by the International Task Force (49)

The EGPA consensus task force recommendations		Level of evidence
1.	EGPA should be managed in (or in collaboration with) centers with established expertise	NA
2.	Minimal initial differential diagnosis work-up should include testing for toxocariasis, HIV, aspergillus, triptase, vit. B12, peripheral blood smear, chest CT-scan	NA
3.	Obtaining biopsies is encouraged	NA
4.	ANCA testing (indirect immunofluorescence and PR3/MPO specific immunoassay) should be performed	NA
5.	No reliable biomarker to measure disease activity is available	NA
6.	Once EGPA is diagnosed, evaluation of organ/system involvement is indicated	NA
7.	Remission: the absence of a clinical systemic manifestation (excluding asthma and/or ENT)	NA
8.	Relapse: the new appearance or recurrence or worsening of clinical EGPA manifestations (excluding asthma and/or ENT) requiring treatment	NA
9.	Glucocorticoids (prednisone 1 mg/kg) are indicated to induce remission	A
10.	Patients with life and/or organ threatening manifestations should be treated with a combination of glucocorticoids and immunosuppressant (e.g. cyclophosphamide)	B
11.	Maintenance therapy (azathioprine or methotrexate) is recommended for patients with life and/or organ threatening disease	C
12.	Glucocorticoids alone may be suitable for patients without life and/or organ threatening manifestations	C
13.	Plasma-exchange is usually not effective but can be considered for patients with rapidly progressive glomerulonephritis and/or alveolar hemorrhage	D
14.	Rituximab can be considered in selected ANCA positive cases	C
15.	IVIg can be considered as a second line treatment	C
16.	Interferon-alpha may be considered as a second/third line treatment for selected patients	C
17.	Leukotriene-receptor antagonists can be prescribed	B
18.	Vaccination with inactivated vaccines (influenza and pneumococci) should be encouraged	D
19.	Implementation of patient educational program is encouraged	D
20.	Patients with peripheral nerve involvement and motor deficit(s) should be referred to a physiotherapist	D
21.	Patients should avoid tobacco smoking and irritants	D
22.	Venous thromboembolic events should be treated according to general guidelines	D

EGPA Eosinophilic Granulomatosis with Polyangiitis, *ANCA* anti-neutrophil cytoplasmic antibody, *CT* computed tomography, *ENT* ear, nose and throat, *HIV* human immunodeficiency virus, *IVIg* intravenous immunoglobulins, *NA* not applicable

References

1. Churg J, Strauss L (1951) Allergic granulomatosis angiitis and periarteritis nodosa. Am J Pathol 27:277–294
2. Jennette JC, Falk RJ, Andrassy K, Bacon PA, Churg J, Gross WL et al (1994) Nomenclature of systemic vasculitides: proposal of an international consensus conference. Arthritis Rheum 37:187–192
3. Watts RA, Lane SE, Bentham G, Scott DG (2000) Epidemiology of systemic vasculitis: a ten-year study in the United Kingdom. Arthritis Rheum 43:414–419
4. Martin RM, Wilton LV, Mann RD (1999) Prevalence of Churg-Strauss syndrome, vasculitis, eosinophilia and associated conditions: retrospective analysis of 58 prescription event monitoring cohort studies. Pharmacoepidemiol Drug Saf 8:179–189
5. Mahr A, Guillevin L, Poissonnet M, Ayme S (2004) Prevalence of polyarteritis nodosa, microscopic polyangiitis, Wegener's granulomatosis, and Churg-Strauss syndrome in a French urban multiethnic population in 2000: a capture-recapture estimate. Arthritis Rheum 51:92–99
6. Jennette JC, Falk RJ (1997) Small-vessel vasculitis. N Engl J Med 337:1512–1523
7. Savage CO, Harper L, Adu D (1997) Primary systemic vasculitis. Lancet 349:553–558
8. Kallenberg CG, Brouwer E, Weening JJ, Tervaert JW (1994) Anti-neutrophil cytoplasmic antibodies: current diagnostic and pathophysiological potential. Kidney Int 46:1–15
9. Sinico RA, Di Toma L, Maggiore U, Bottero P, Radice A, Tosoni C et al (2005) Prevalence and clinical significance of antinutrophil cytoplasmic antibodies in Churg-Strauss syndrome. Arthritis Rheum 52:2926–2936
10. Sable-Fourtassou R, Cohen P, Mahr A, Pagnoux C, Mouthon L, Jayne D et al (2005) Antineutrophil cytoplasmic antibodies and the Churg-Strauss syndrome. Ann Intern Med 143:632–638
11. Lanham JG, Elkon KB, Pusey CD, Hughes GR (1984) Systemic vasculitis with asthma and eosinophilia: a clinical approach to the Churg-Strauss syndrome. Medicine (Baltimore) 63:65–81
12. Bottero P, Bonini M, Vecchio F, Grittini A, Patruno GM, Colombo B et al (2007) The common allergens in the Churg–Strauss syndrome. Allergy 62(11):1288–1294
13. Guillevin L, Cohen P, Gayraud M, Lhote F, Jarrousse B, Casassus P (1999) Churg Strauss syndrome: clinical study and long term follow-up of 96 patients. Medicine (Baltimore) 78:26–37
14. Eustace JA, Nadasdy T, Choi M (1999) The Churg Strauss syndrome. J Am Soc Nephrol 10:2048–2055
15. Conron M, Beynon HL (2000) Churg-Strauss syndrome. Thorax 55:870–877
16. Churg A (2001) Recent advances in the diagnosis of Churg-Strauss syndrome. Mod Pathol 14:1284–1293
17. Noth I, Strek ME, Leff AR (2003) Churg Strauss syndrome. Lancet 361:587–594
18. Abril A, Calamia KT, Cohen MD (2003) The Churg Strauss syndrome (allergic granulomatous angiitis): review and update. Semin Arthritis Rheum 33:106–114
19. Keogh KA, Specks U (2003) Churg Strauss syndrome. Clinical presentation, antineutrophil cytoplasmic antibodies, and leukotriene receptor antagonists. Am J Med 115:284–290
20. Specks U (2001) Diffuse alveolar hemorrhage syndromes. Curr Opin Rheumatol 13:12–17
21. Reid AJC, Harrison BDW, Watts RA, Watkin SW, McCann BG, Scott DG (1998) Churg Strauss syndrome in a district hospital. QJM 91:219–229
22. Della Rossa A, Baldini C, Tavoni A, Tognetti A, Neglia D, Sambuceti G et al (2002) Churg-Strauss syndrome: clinical and serological features of 19 patients from a single Italian centre. Rheumatology 41:1286–1294
23. Solans R, Bosch JA, Perez-Bocanegra C, Selva A, Huguet P, Alijotas J et al (2001) Churg-Strauss syndrome: outcome and long-term follow-up of 32 patients. Rheumatology 40:763–771

24. Allenbach Y, Seror R, Pagnoux C, Teixeira L, Guilpain P, Guillevin L (2009) High frequencyof venous thromboembolic events in Churg–Strauss syndrome, Wegener's granulomatosis and microscopic polyangiitis but not polyarteritis nodosa: a systematic retrospective study on 1130 patients. Ann Rheum Dis 68:564–567
25. Sinico RA, Di Toma L, Maggiore U, Tosoni C, Bottero P, Sabadini E et al (2006) Renal involvement in Churg-Strauss syndrome. Am J Kidney Dis 47:770–779
26. Guillevin L, Lhote F, Gayraud M, Cohen P, Jarrousse B, Lortholary O et al (1996) Prognostic factors in polyarteritis nodosa and Churg-Strauss syndrome. A prospective study in 342 patients. Medicine (Baltimore) 75(1):17–28
27. Vaglio A, Strehl JD, Manger B, Maritati F, Alberici F, Beyer C et al (2012) IgG4 immune response in Churg-Strauss syndrome. Ann Rheum Dis 71(3):390–393
28. Mahr A, Moosig F, Neumann T, Szczeklik W, Taillé C, Vaglio A et al (2014) Eosinophilic granulomatosis with polyangiitis (Churg-Strauss): evolutions inclassification, etiopathogenesis, assessment and management. Curr Opin Rheumatol 26(1):16–23
29. Vaglio D, Martorana UM, Grasselli C, Zanetti A, Pesci A et al (2007) HLA-DRB4 as a genetic risk factor for Churg-Strauss syndrome. Arthritis Rheum 56:3159–3166
30. Wieczorek S, Hellmich B, Gross WL, Epplen JT (2008) Associations of Churg-Strauss syndrome with the HLA-DRB1 locus, and relationship to the genetics of antineutrophil cytoplasmic antibody-associated vasculitides. Arthritis Rheum 58:329–330
31. Bibby S, Healy B, Steele R, Kumareswaran K, Nelson H, Beasley R (2010) Association between leukotriene receptor antagonist therapy and Churg-Strauss syndrome: an analysis of the FDA AERS database. Thorax 65(2):132–138
32. Comarmond C, Pagnoux C, Khellaf M, Cordier JF, Hamidou M, Viallard JF et al (2013) Eosinophilic granulomatosis with polyangiitis (Churg-Strauss): clinical characteristics and long termfollowup of the 383 patients enrolled in the French Vasculitis Study Group cohort. Arthritis Rheum 65(1):270–281
33. Churg A, Brallas M, Cronin SR, Churg J (1995) Formesfrustes of Churg-Strauss syndrome. Chest 108:320–323
34. Masi AT, Hunder GG, Lie JT, Michel BA, Bloch DA, Arend WP et al (1990) The American College of Rheumatology 1990 criteria for the classification of Churg-Strauss syndrome (allergic granulomatosis and angiitis). Arthritis Rheum 33:1094–1100
35. Alberts WM (2007) Pulmonary manifestations of the Churg-Strauss syndrome and related idiopathic small vessel vasculitis syndromes. Curr Opin Pulm Med 13:445–450
36. Katzenstein AL (2000) Diagnostic features and differential diagnosis of Churg-Strauss syndrome in the lung. A review. Am J Clin Pathol 114:767–772
37. Watts R, Lane S, Hanslik T, Hauser T, Hellmich B, Koldingsnes W et al (2007) Development and validation of a consensus methodology for the classification of the ANCA-associated vasculitides and polyartetritis nodosa for epidemiology studies. Ann Rheum Dis 66:222–227
38. Saleh A, Stone JH (2005) Classification and diagnostic criteria in systemic vasculitis. Best Pract Res Clin Rheumatol 19:209–221
39. Roufosse FE, Goldam M, Cogan E (2007) Hypereosinophilic syndromes. Orphanet J Rare Dis 2:37
40. Mukhtyar C, Guillevin L, Cid MC, Dasgupta B, de Groot K, Gross W et al (2009) EULAR recommendations for the management of primary small and medium vessel vasculitis. Ann Rheum Dis 68(3):318–323
41. Wechsler ME (2007) Pulmonary eosinophilic syndromes. Immunol Allergy Clin N Am 27:477–492
42. Taylor MR, Keane CT, O'Connor P, Muvihill E, Holland C (1988) The expanded spectrum of toxocaral disease. Lancet 1:692–695
43. Siddiqui AA, Berk SL (2001) Diagnosis of *Strongyloides stercoralis* infection. Clin Infect Dis 33:1040–1047
44. Pagnoux C, Guilpain P, Guillevin L (2007) Churg-Strauss syndrome. Curr Opin Rheumatol 19:25–32

45. Guillevin L, Jarrousse B, Lok C, Lhote F, Jais JP, Le ThiHuong DD et al (1991) Longterm followup after treatment of polyarteritis nodosa and Churg-Strauss angiitis with comparison of steroids, plasma exchange and cyclophosphamide to steroids and plasma exchange. A prospective randomized trial of 71 patients. The Cooperative Study Group for Polyarteritis Nodosa. J Rheumatol 18(4):567–574
46. Gayraud M, Guillevin L, Cohen P, Lhote F, Cacoub P, Deblois P et al (1997) Treatment of good-prognosis polyarteritis nodosa and Churg–Strauss syndrome: comparison of steroids and oral or pulse cyclophosphamide in 25 patients. French Cooperative Study Group for Vasculitides. Br J Rheumatol 36:1290–1297
47. Cohen P, Pagnoux C, Mahr A, Arène JP, Mouthon L, Le Guern V et al (2007) Churg-Strauss syndrome with poor-prognosis factors: a prospective multicenter trial comparing glucocorticoids and six or twelve cyclophosphamide pulses in forty-eight patients. Arthritis Rheum 57:686–693
48. Jayne DR, Gaskin G, Rasmussen N, Abramowicz D, Ferrario F, Guillevin L et al (2007) Randomized trial of plasma exchange or high-dosage methylprednisolone as adjunctive therapy for severe renal vasculitis. J Am Soc Nephrol 18:2180–2188
49. Groh M, Pagnoux G, Baldini C, Bel E, Bottero P, Cottin V et al (2015) Eosinophilic granulomatous with polyangiitis (Churg-Strauss) (EGPA) consensus task force recommendations for evaluation and management. Eur J Intern Med 26(7):545–553

Chapter 13
ANCA-Associated Vasculitis and the Mechanisms of Tissue Injury

Adrian Schreiber and Mira Choi

Abstract ANCA associated vasculitides (AAVs) comprise four different disease entities: I) granulomatosis with polyangiitis (GPA, formerly Wegener's granulomatosis), II) microscopic polyangiitis (MPA), III) eosinophilic granulomatosis with polyangiitis (EGPA, formerly Churg-Strauss syndrome), and IV) renal-limited vasculitis or isolated pauci-immune necrotizing and crescentic glomerulonephritis (NCGN). Experimental data support the notion that ANCA-induced activation of both neutrophils and monocytes is one of the main pathogenic mechanisms involved in disease induction. Binding of ANCA IgG to surface expressed ANCA antigens on myeloid cells leads to generation of reactive oxygen species (ROS), degranulation and activation of proteases, and formation of neutrophil extracellular traps (NET). Finally, activation of the complement system in AAVs by ANCA stimulated neutrophils leads to generation of C5a, which plays an important role in an amplifying inflammatory loop.

13.1 Introduction

Anti-neutrophil cytoplasmic autoantibodies (ANCA) associated vasculitides (AAV) comprise systemic diseases with the characteristics of small and medium vessel inflammation which might target almost every organ with the risk of fatal outcomes. AAV comprise four different disease entities: I) granulomatosis with polyangiitis (GPA, formerly Wegener's granulomatosis), mainly associated with ANCA against the proteinase 3 (PR3), II) microscopic polyangiitis (MPA), mainly associated with ANCA against myeloperoxidase (MPO), III) eosinophilic granulomatosis with polyangiitis (EGPA, formerly Churg-Strauss syndrome), only in 50 % presenting

A. Schreiber (✉) • M. Choi
Experimental and Clinical Research Center (ECRC) at the MDC Berlin, Charité Berlin, Lindenberger Weg 80, 13125 Berlin, Germany

Department of Nephrology and Intensive Care Medicine, Campus Virchow Clinic, Medical Faculty of the Charité, Berlin, Germany
e-mail: adrian.schreiber@charite.de

F. Dammacco et al. (eds.), *Systemic Vasculitides: Current Status and Perspectives*, DOI 10.1007/978-3-319-40136-2_13

with ANCA against MPO, and IV) renal-limited vasculitis or isolated pauci-immune necrotizing and crescentic glomerulonephritis (NCGN) [1].

ANCA are directed against the granule proteins MPO or PR3, and there is evidence from in vitro and from animal studies that ANCA are pathogenic and cause effector functions in neutrophils and monocytes, leading to subsequent tissue damage via interaction with endothelial cells as one hallmark of the disease [2, 3]. In addition, another antigen called lysosomal-associated membrane protein-2 (hLAMP-2) has been described [4–7].

The dominant histological finding is characterized by granuloma formation, leukocyte accumulation and endothelial necrosis within the vessel wall.

A broad range of clinical manifestations are common to all types of AAV, such as fever, malaise, weight loss, and arthralgia. GPA often presents with pulmonary and renal involvement, vasculitis of the skin and eye- and ear-nose-throat involvement, they often present with coughing/hemoptysis, a skin rash, laryngitis, recurrent rhinitis, mastoiditis, episcleritis and sometimes with oliguria/anuria if renal involvement is severe. MPA shows similarities to GPA, but without the formation of granulomatous inflammation and, more common, with limited renal involvement [8]. EGPA depicts features of eosinophilic asthma and vasculitic inflammation within the skin, and often presents with nervous system and cardiac involvement.

In the following pages we will discuss genetic implications of ANCA disease, the contribution of different cell types to the initiation and progress of AAV and different tissue injury mechanisms crucial to disease activity in vivo.

Finally, future directions to ongoing research and novel therapeutic options will be given according to recently published work in understanding the pathogenic mechanisms in AAV.

13.2 Genetic Regulation

For a long time it has been proposed that AAV patients have a genetic predisposition to the disease. In different autoimmune diseases such as type 1 diabetes and rheumatoid arthritis, the HLA region is central for autoimmunity. In AAV the strongest evidence for HLA association is with HLA-DPB1. This association was demonstrated in several candidate gene studies. However, our understanding of the genetic regulation of AAV has dramatically increased with introduction of genome-wide association studies (GWAS). Two studies in AAV patients have been published so far: the first conducted by the European Vasculitis Genetic Consortium (EVGC; 2,687 cases of GPA and MPA, 6,858 controls) [9] and the second by the US Vasculitis Clinical Research Consortium (VCRC; 987 GPA cases, 2,731 controls) [10]. In the European study, patients with PR3-ANCA showed a significant association both with HLA-DPB1, the PR3 inhibitor alpha1-antytrypsin (SERPINA1) and PR3 (PRTN3). In contrast, patients with MPO-ANCA showed an association with HLA-DQ. Interestingly, the study found these differences in genetic association rather in respect to ANCA serotype than to clinical phenotype of AAV. The US

American study confirmed in its cohort the association with HLA-DPB1, in addition an association with SEM6A6 (coding for semaphorin 6A, a protein with largely unknown function) was found. However, no statistically significant difference for this gene could be found in a recent study in a cohort of European patients [11].

In summary, the strongest data for a genetic background of AAV is for the HLA region and different studies have found a strong signal for HLA-DPB1 in AAV patients from different genetic background. Very interestingly, patients with PR3-ANCA show a signal both in PR3 and the PR3-inhibitor alpha1-antytrypsin; it is conceivable that a dysregulated balance between PR3 and its inhibitor leads through insufficient clearance to stimulation of the immune system with development of PR3-ANCA.

13.3 ANCA Autoantigens

13.3.1 PR3 Membrane Expression by CD177

The ANCA antigens PR3 and MPO are expressed by both neutrophils and monocytes. PR3 is stored mainly in azurophil granules, and is exposed in variable amounts on the surface of resting neutrophils. Interestingly, expression of surface PR3 is bimodal, with expression on the neutrophil membrane only in a proportion of cells (defined as $mPR3^{high}$) but not in the others (defined as $mPR3^{low}$). Expression varies from 0 to 100 % of neutrophils from one individual and remains very stable over longer time periods. It appears that PR3 is presented on the membrane by a unique interaction with CD177, where CD177 is only expressed on a subset of neutrophils [4, 5, 12, 13]. Patients with AAV display a significantly higher percentage of PR3-expressing neutrophils and patients with higher membrane PR3 expression do worse [14–16]. Interestingly, patients with severe bacterial sepsis displayed dynamic changes in the percentage of CD177-positive (and $mPR3^{high}$) neutrophils and these changes affected clinical outcome which makes regulation of the expression of CD177 and mPR3 important for diseases far beyond AAV [17].

Several explanations for a correlation between membrane PR3 expression and disease have been proposed: I) A direct effect by stronger activation of PR3-expressing neutrophils with PR3-ANCA. The higher expression of PR3 by CD177 on the surface of neutrophils enables stronger ANCA IgG binding [14] and subsequent activation mediated by a signaling complex of PR3/CD177 together with Mac-1 (CD11b/CD18) [18]. II) A rather indirect effect by promoting trans-endothelial migration of CD177 positive neutrophils. It has been described that CD177 acts as a counter-receptor for the endothelial junctional protein PECAM-1 (platelet endothelial cell adhesion molecule 1, CD31). The heterophilic CD177/PECAM-1 interaction facilitates neutrophil transmigration through suppression of PECAM-1 ITIM phosphorylation, leading to a decrease in endothelial cell junctional integrity and finally a migration advantage [19, 20]. In addition, the interaction between the CD177-presented PR3 and PECAM-1 facilitates trans-endothelial migration and reestablish vascular integrity after leukocyte transmigration [21]. Both processes – stronger activation by ANCA IgG and support of trans-endothelial migration – could finally lead to a stronger vascular damage in AAV.

13.3.2 Alternative ANCA Antigens

With respect to ANCA target autoantigens other than MPO and PR3, controversy persists. In 2008 Kain et al. reported the presence of autoantibodies against hLAMP-2 in the majority of patients with AAV [5]. hLAMP-2 is a membrane glycoprotein expressed mainly in lysosomes but also in late endosomes and on the plasma membrane of neutrophils. In addition, hLAMP-2 is expressed by glomerular endothelial cells. Kain et al. demonstrated anti-hLAMP-2 autoantibodies in patients with AAV, in addition an epitope within hLAMP-2 was shown to have 100 % homology to the bacterial protein FimH. Furthermore, immunization of WKY rats with FimH induced autoantibodies to both rat and human hLAMP-2 and induced NCGN [5], suggesting molecular mimicry as cause of AAV. The same group later published that anti-hLAMP-2 antibodies correlated very well with disease activity, disappeared rapidly after disease induction therapy and could be detected in 73 % of patients with ANCA-negative NCGN [7]. Therefore, antibodies to hLAMP-2 could potentially serve as a better activity marker than conventional ANCA tests. However, Roth et al. were not able to reproduce these findings as they detected anti-hLAMP-2 autoantibodies in only 21 % of sera from patients with ANCA-associated vasculitides [22]. Thus, the relevance of hLAMP-2 and anti-hLAMP-2 antibodies in ANCA-associated vasculitis requires further investigation.

13.3.3 Epigenetics

Transcription of PR3 and MPO is silenced in healthy mature circulating neutrophils and monocytes. However, by microarray analyses an aberrant expression of both PR3 (PRTN3) and MPO was identified in AAV patients. In addition to both genes, a global granulopoiesis signature was identified in mature PMN and monocytes [23]. MPO and PR3 transcriptional upregulation correlated with clinical disease activity and different laboratory markers of disease activity. Since PRTN3 and MPO genes locate on different chromosomes, this coordinated expression was postulated to be caused by epigenetic modifications. Ciavatta et al. demonstrated that active transcription of PRTN3 and MPO results from defective epigenetic silencing [24]. The authors found depletion of histone methylation H3K27me3 at PRTN3 and MPO due to increased expression of Jumonji domain-containing 3 (JMJD3) demethylase as well as failure of the transcription factor RUNX3 to recruit EZH2, which is responsible for H3K27me3 methylation. In a recent follow-up study the same authors showed that the alternative PR3 transcript lacked exon 1 as a consequence of a transcription start in intron 1 encoding a 24-kD protein (p24PR3/MBN), a transcript known as myeloblastin. Importantly, active synthesis of PR3, MPO, and p24PR3/MBN protein was demonstrated in patient's neutrophils and binding of ANCA IgG to p24PR3/MBN. As a result of this aberrant antigen expression, different pathogenic settings are conceivable: First, synthesis of PR3 or MPO in mature neutrophils could lead to mistargeting followed by immunogenic presentation, thereby supporting autoantibody generation. Second, the alternative transcript in itself could break tolerance by serving as a foreign antigen leading to autoantibody generation.

13.3.4 Epitope Analysis

An elegant approach for identification of pathogenic epitopes of ANCA antigens has been followed recently in a multicenter study by Roth et al. [25]. Here, a novel assay termed "epitope excision" was used to identify the specific target epitopes of ANCA IgG, which involves analysis of ANCA-binded MPO-epitopes by MALDI-TOF/TOF mass spectrometry. This novel powerful method allows both detection of low-titer antibodies and identification of conformational epitopes. The authors found substantial MPO-ANCA epitope diversity by detection of epitopes exclusive for active disease, distinct epitopes not specific for active disease, and epitopes not associated with disease (so called natural occurring autoantibodies). Interestingly, the authors could detect autoantibodies to MPO in patients with ANCA-negative pauci-immune NCGN, which appeared to have been masked by a circulating fragment of the endogenous MPO-inhibitor ceruloplasmin in the sera of the patients. It appears that epitope specificity determines the pathogenicity of autoantibodies. Interestingly, Olsen et al. recently detected PR3-ANCA IgG in GPA patients prior to disease onset [26]. It is conceivable that in these patients over time epitope spreading occurs towards pathogenic epitopes, finally inducing active disease.

13.4 ANCA Interaction with Neutrophils and Monocytes

The ANCA target antigens PR3 and MPO reside within the azurophilic granules of neutrophils and within the lysosomes of monocytes. To interact with ANCA PR3 and MPO must be presented to the cell surface, a movement which is called *translocation*. The trigger for translocation is mediated by cytokine stimulation, so called *priming*, induced by low-dose TNFα, IL-1ß, IL-18 or G-CSF, by C5a, or even polymers of the PR3 inhibitor a1-antitrypsin [27–33]. In patients with AAV viral and bacterial infections are observed during the prodromal stage of the disease and, like *in-vivo priming*, it is known that colonization with staphylococcus aureus contributes significantly to recurring disease activity [34]. While priming itself does not result in full stimulation, primed neutrophils and monocytes respond very rapidly to stimulation with ANCA IgG with the release of oxygen radicals and proteolytically active granule proteins as neutrophil serine proteases, MPO, cytotoxic mediators or even microparticles [27, 28, 35, 36].

The contribution of neutrophils to ANCA-induced activation is multifaceted and complex. Proof of the central and essential role of neutrophils as the primary effector cell in AAV comes from murine animal studies by Xiao et al. where, using an anti-MPO IgG-induced NCGN model, neutrophil depletion protected from disease [37].

13.4.1 Role of Fcγ Receptors in ANCA Induced NCGN

ANCA activate cytokine-primed neutrophils through both, binding of ANCA IgG to surface-expressed antigens with their $F(ab)_2$ portions and through Fc-part mediated Fcγ receptor (FcγR) engagement [38–40]. FcγRs are present on neutrophils, macrophages, and dendritic cells. The Fcγ RIIB is also present on B cells containing an inhibitory signaling motif [41, 42]. Indeed, it could be shown that RIIB limits autoimmunity to MPO and absence of FcγRIIB lead to disease aggravation by increasing autoreactive T- and B-cell responses, activation of neutrophils by ANCA, and the recruitment of macrophages and neutrophils by effector T cells [43].

By genotyping it has been described that specifically the NA1 allele of FcγRIIIB correlates with the severity of renal involvement [44].

ANCA are predominantly IgG and are found in all four human IgG subclasses. There is still ongoing research about the predominance of IgGs in AAV patients and its pathogenic distribution to vasculitic inflammation. Apparently the ANCA IgG3 subclass interacts most strikingly with Fcγ receptor, is most potent for neutrophil capture from flow and induction of adhesion and has more pronounced capacity to induce neutrophil burst [45].

IgG molecules require glycosylation in the CH2 domains of both heavy chains of the Fc fragment for effective Fcγ receptor interaction. Noteworthy, modulation of ANCA IgG glycosylation through treatment with the bacterial enzyme endoglycosidase S inhibited neutrophil activation in vitro and reduced anti-MPO IgG-induced glomerulonephritis in mice [46].

13.4.2 Signaling Pathways in Neutrophil Activation

Multiple signaling pathways are involved in the processes of neutrophil priming and ANCA-induced activation. The mitogen-activated protein kinases (MAPK) p38 and extracellular signal-regulated kinase (ERK) mediate priming of TNF-α-stimulated neutrophils enabling subsequent ANCA-induced respiratory burst, while p38MAPK only mediates translocation of ANCA antigens to the cell surface [47]. Importance for p38MAPK has been demonstrated in vivo, where p38MAPK inhibition partly reduced crescent formation in mice with anti-MPO NCGN [48]. Another crucial signaling pathway involved in neutrophil activation is the phosphoinositide-3 kinase (PI3-K) pathway. Inhibition of the PI3-K pathway leads to decreased ANCA-induced superoxide production in neutrophils [49]. In vivo, the inhibition of the PI3-Kγ subform either by specific pharmacologic blockade or genetic deficiency demonstrated protection from ANCA-associated necrotizing crescentic glomerulonephritis (NCGN) [50].

The alternative pathway of complement is involved as well. Release of factor B, properdin and C3 from activated neutrophils results in activation of the complement system with the generation of C5a which is a very strong chemoattractant for neutrophils and has also the ability to prime neutrophils. Responsible mediators of C5a-mediated priming of human neutrophils with subsequent ANCA-induced respiratory burst and degranulation are regulated by p38MAPK, ERK and PI3-K signaling pathways [51].

13.4.3 Role of Monocytes in ANCA-Induced Activation

While most studies described a crucial role of neutrophils in ANCA-induced effector functions, monocytes also express both ANCA antigens, PR3 and MPO, and react with ANCA IgG. ANCA IgG induced in vitro up-regulation of CD14 and CD18 on monocytes, triggered ROS generation and lead to the release of pro-inflammatory cytokines, such as IL-8 [52, 53].

Renal biopsies from patients with ANCA GN revealed that monocytes/macrophages were the predominant invading cells in glomeruli, followed by neutrophils, but to a lesser extent [54, 55]. Monocytes transform into macrophages and reside within different phases of inflammation from acute to chronic, while neutrophils are short lived and undergo apoptosis or necrosis.

Recently, monocytes were classified into 3 subsets based on their expression of the Fcγ III receptor, CD16 ($CD16^-$ and $CD16^+$ monocytes), and based on their surface expression of the lipopolysaccharide (LPS) co-receptor, CD14, on $CD16^+$ monocytes [56]. Systematic investigations in patients with AAV elucidated that PR3 and MPO are differentially expressed on distinct monocyte subsets, with the highest expression on intermediate cells ($CD14^+/CD16^+$) [57]. Interestingly, monocyte subsets react in a different manner to anti-MPO antibodies, with intermediate monocytes producing the highest amount of IL-1β and IL-6 after stimulation. Furthermore, the percentage of intermediate monocytes was higher in AAV patients compared to healthy controls, supporting the hypothesis that expansion of this monocyte subset in AAV has an impact in disease pathogenesis.

The production of IL-1β plays a critical role in the pathogenesis of ANCA vasculitis presenting an important inflammatory mediator. The observation that IL-1 receptor antagonist treatment abrogated anti-MPO antibody–induced NCGN confirmed that IL-1β is indeed crucial for disease development [58]. It could be demonstrated that anti-MPO antibody–activated monocytes and neutrophils generate IL-1β, which in turn stimulated the release of additional inflammatory cytokines (including TNFα and IL-6), causing local inflammation and renal injury leading to NCGN.

Soluble fms-like tyrosine kinase-1 (sFlt1), a circulating form of VEGF-R1 that lacks the transmembrane domain of the intact protein, is mainly secreted by monocytes, endothelial cells, and placental cytotrophoblasts and displays potent inhibitory effects on VEGF. Noteworthy, anti-PR3 mAb and serum containing PR3-ANCA from patients with active vasculitis were found to induce sFLt1 release from monocytes, whereas anti-MPO mAb did not [59]. As a consequence, the increase might impair endothelial repair during active vasculitis culminating in pronounced vasculitic inflammation and more severe organ damage.

But still, the true in vivo importance of monocytes in the pathogenesis of AAV disease development has to be defined in animal models.

13.5 Interaction of ANCA-Activated Neutrophils with Endothelial Cells in Vitro

In locally restricted inflammation neutrophils upregulate adhesion molecules (ß2 integrins) on their surface and adhere to the endothelium, which itself upregulates intercellular adhesion molecule-1 (ICAM-1) or vascular cell adhesion molecule-1 (VCAM-1), transforming themselves into a pro-adhesive state. In the absence of ANCA, neutrophils start to transmigrate through the endothelial layer and cell activation does not take place until they reach the site of inflammation. Thus, release of toxic contents and oxidative burst do not lead to vascular damage. In contrast, in the presence of circulating ANCA binding to primed neutrophils results in more enhanced inflammatory responses and cell activation. Subsequently, enhanced inflammatory responses of adherent neutrophils promote vascular damage with increased vascular permeability and detachment of the endothelium. In renal histologic stains of patients with AAV this is depicted as fibrin depositions (necrosis) within the vessel wall. Thus, the interaction of ANCA-activated neutrophils with the endothelial cell surface is a crucial step in the initiation of the development of necrotizing vasculitis.

Studies from cremasteric microvasculature in mice demonstrated that primed neutrophils showed reduced rolling in the presence of anti-MPO antibody stimulation and an increase of adhesion to the endothelium [60]. These steps were dependent on β2-integrins, such as CD11b and FcγR. In a flow based assay, TNF-α-primed neutrophils tend to adhere more firmly to endothelial cells in the presence of anti-MPO and the number of transmigrating neutrophils increased.

As a proof of principle, enhanced neutrophil infiltration within glomeruli has been demonstrated in renal biopsies from patients with AAV and renal involvement and in animal models with NCGN. In ANCA-induced NCGN neutrophils undergo rapid destruction by mechanisms such as apoptosis and necrosis and could be observed as fragmentation of nuclei, also called leukocytoclasia, at sites of acute fibrinoid necrosis [61]. This is apparently due to gross neutrophil infiltration or reduced clearance of neutrophil fragments. Furthermore, neutrophil extracellular traps (NETs) were detected at sites of inflammation and acute necrosis [62].

The important role of an inflammatory environment to enhance neutrophilic functions in close proximity to the endothelium has been observed in a mice model of anti-MPO NCGN, where additional LPS treatment in addition to anti-MPO IgG increased neutrophil influx into glomeruli compared to anti-MPO IgG only [63]. Evidence that the anti-MPO antibody itself increases neutrophil accumulation within glomeruli comes from a study with the use of relatively high dose of anti-MPO antibody [64] and in vitro using isolated mouse glomerular endothelial cells anti-MPO antibody induces upregulation of adhesion molecules such as ICAM-1, VCAM-1 and E-selectin. Furthermore, the in vitro data showed that these glomerular EC also secreted chemokines such as KC and MIP-2 and they identified a molecule called moesin as a probable autoantigen for endothelial activation.

Thorough attention has been drawn to the role of chemokines during neutrophil endothelial cell interactions in ANCA settings. Neutrophils express the chemokine receptor CXCR1 and CXCR2 for IL-8, the most potent member of the CXC family. In a flow model, inhibition of CXCR2 was associated with an increase in rolling and significantly reduced the level of migration [65]. Similar observations were made in vivo in patients with AAV in active disease, showing a decreased CXCR1 and 2 surface expression on circulating neutrophils compared to patients in remission [66]. In the same line, in vitro blockade of CXCR1 and CXCR2 significantly increased neutrophil adhesion and inhibited migration through glomerular EC monolayers, which supports the observation of ongoing inflammation at the site of the vessel wall.

13.6 Mechanisms of Tissue Injury in AAV

13.6.1 Degranulation of Toxic Granule Contents

Primed neutrophils and monocytes respond very rapidly to stimulation with ANCA IgG with release of proteolytically active granule proteins as neutrophil serine proteases (NSPs), MPO or MMP9. Recent evidence shows that neutrophil-derived degranulated extracellular MPO is deposited in glomeruli of AAV patients [67] and furthermore that MPO can potentially attenuate the development of adaptive immunity and autoimmunity by inhibition of antigen presenting cells [68]. We have recently focused on the specific role of NSPs in the pathogenesis of AAV. NSPs, namely PR3, neutrophil elastase (NE), cathepsin G (CG), and NSP4, reside in neutrophil granules and monocytic lysosomes [69]. NSPs are generated as inactive proenzymes that require proteolytic proform cleavage by the lysosomal cysteine protease dipeptidyl peptidase I (DPPI) [70]. The functional role of NSP was tested by using DPPI-deficient mice lacking functional active serine proteases in an animal model of anti-MPO induced NCGN, which is induced by immunization of MPO-deficient mice with murine MPO, followed by subsequent bone marrow (BM) transplantation with MPO-positive BM. Compared with WT BM, DPPI-deficient BM protected from anti-MPO induced NCGN [58]. The protection was mediated by massively reduced ANCA-stimulated monocytic and neutrophil IL-1β generation and release in DPPI-deficient cells, an effect much more pronounced in monocytes than in neutrophils. In line with this, DPPI-deficient mice showed strongly reduced renal IL-1β levels. Moreover, PR3/NE-double-deficient mice showed diminished IL-1β production to anti-MPO antibodies in vitro and were protected in vivo from anti-MPO induced NCGN. Finally, specific IL-1β blockade by Anakinra reduced ANCA-induced NCGN. These observations suggest that NSP-mediated IL-1β processing is fundamental for induction of ANCA-NCGN. Generation of IL-1β is usually regulated by a multiprotein complex, called the inflammasome. Inflammasome stimulation leads to processing of inactive procaspase-1 to active caspase-1, which then cleaves pro-IL-1β into the active IL-1β form. However, it appears that in the setting of ANCA-stimulation of monocytes this classical pathway seems to be rather inactivated as inhibition of caspase-1 had no influence on IL-1β generation. We will discuss the mechanism involved herein in the next paragraph.

13.6.2 Generation of Reactive Oxygen Species

Multiple in vitro studies have demonstrated that ANCA IgG stimulate a robust respiratory burst in myeloid cells by activation of the phagosomal NADPH-oxidase (NOX2), leading to the generation of potential damaging ROS. Surprisingly, even though the in vitro effect of ANCA-stimulated ROS generation is well established, the net in vivo effect has not been demonstrated yet. Therefore, anti-MPO NCGN was induced in two different NOX-deficient mouse strains (namely gp91phox-deficient and p47phox-deficient mice) and compared with wild-type mice. Surprisingly, both NADPH-oxidase deficient mouse strains showed a strongly aggravated phenotype of NCGN, which was paralleled by increased renal IL-1β level [71]. In different experimental in vitro settings we could demonstrate that ANCA-stimulated ROS generation by NOX2 inhibits the inflammasome-mediated caspase-1-dependent IL-1β generation. We could finally prove this process by inducing anti-MPO NCGN in caspase-1/gp91phox-double deficient mice, which were rescued from the aggravated phenotype observed in gp91phox-deficient mice. Interestingly, the ANCA-induced ROS-mediated caspase-1 inhibition was observed in both neutrophils and monocytes, but monocyte were still the main producer of IL-1β.

In summary, it appears that in ANCA-stimulated myeloid cells two different pathways controlling the generation of the potent pro-inflammatory cytokine IL-1β act in parallel: ANCA-induced activation of the NSPs PR3 and elastase lead to inflammasome-independent proteolytic IL-1β generation, whereas at the same time

NADPH-oxidase dependent ROS negatively regulates caspase-1 activity by oxidative inhibition, thereby limiting IL-1β generation.

13.6.3 Generation of Neutrophil Extracellular Traps

A formerly unrecognized neutrophil defense mechanism was described by Brinkmann et al., who showed release of decondensed chromatin fibers together with histones and neutrophil granule proteins such as neutrophil elastase, MPO, the antimicrobial peptide cathelicidin from PMA-stimulated neutrophils [72]. This process of generation of neutrophil extracellular traps (NET) is now widely denoted as "NETosis" and was subsequently found to be involved in bacterial defense by trapping of bacteria, but even more in the pathogenesis of a variety of different diseases such as thrombosis formation, TRALI, SLE, psoriasis, or hantavirus-disease. However, even an anti-inflammatory effect of NETs was recently described by demonstrating degradation of cytokines and chemokines with consecutive resolution of neutrophilic inflammation [73]. A role of NETosis in ANCA NCGN was first demonstrated by Kessenbrock et al.: the authors showed generation of NETs by ANCA-stimulated neutrophils in vitro, local kidney NET deposition in patients with NCGN, and detected circulating NETs in patients with active disease by ELISA [62]. This study confirm previous findings of increased levels of nucleosomes in the peripheral blood of patients with AAV [74]. It is conceivable that, through local enrichment of active proteases and histones through NETosis, glomerular endothelial cell damage and vasculitis could be induced. Furthermore, the presentation of ANCA antigens together with dsDNA and LL-37 could break tolerance and induce autoantibody generation. It was demonstrated that these complexes could stimulate Tcells both indirectly and directly; indirectly by stimulation of INF-γ secretion from plasmacytoid dendritic cells (pDCs) or by priming of macrophage cytokine release finally activating Tcells [75], and directly by reduction of the Tcell activation threshold [76]. A potential role for MPO-ANCA induction was recently demonstrated in rats [77]: NETs generated in the presence of PTU and PMA induced MPO-ANCA formation and pulmonary capillaritis. Furthermore, treatment of rats with PTU and PMA administration led to MPO-ANCA generation and subsequently to NCGN. The authors demonstrated that NETs present the ANCA antigens PR3 and MPO and transfer these to APCs, such as myeloid DCs. Immunization of mice with NET-treated myeloid DCs induced both MPO- and PR3-ANCA, as well as anti-dsDNA autoantibodies, leading to autoimmune vasculitis. In a recent follow-up study the same authors demonstrated that MPO-ANCA sera induced NETosis and impaired NET degradation by inhibition of DNase I [78]. These observations suggest that sera from ANCA patients both induce NETosis and inhibit NET degradation with an overall effect of strongly upregulated NET formation. However, it remains to be formally demonstrated that NETosis is more than a marker of disease activity and is actively contributing to tissue injury.

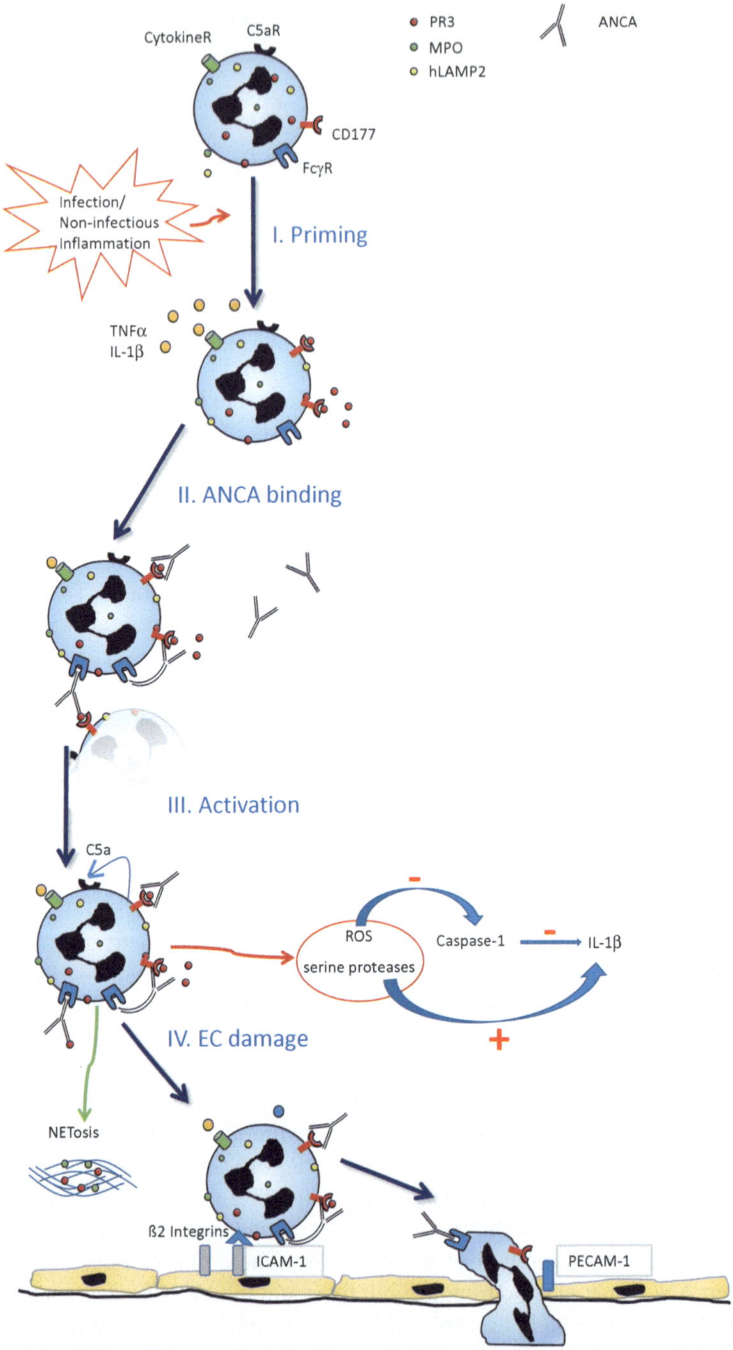

Fig. 13.1 Mechanisms of ANCA-induced vasculitis. In the resting state the neutrophil granulocyte is not activated by ANCA IgG. (I) Priming of neutrophil granulocyte by different inflammatory stimuli (TNFα, IL-ß, C5a) results in up-regulated membrane antigen expression. (II) ANCA

13.6.4 *Activation of the Pathway of Alternative Complement Activation*

Increasing data predominantly from animal studies demonstrated a crucial role of the complement system in the pathogenesis of AAV. In a passive antibody transfer model, mice deficient in the complement factors B and C5 were protected from disease, whereas factor C4-deficient mice developed disease comparable to wild-type mice, proving a central role of the alternative pathway of complement activation, whereas the classical or lectin binding pathway were not involved [79]. In a different study, anti-MPO NCGN could be blocked by treating mice with a C5-inhibiting monoclonal antibody [80]. Finally, C5aR-deficient mice were protected from anti-MPO induced disease and ANCA-stimulated neutrophils activated the complement system with generation of C5a [32]. Recently, this work was confirmed in transgenic mice expressing the human C5aR, where a specific small molecule blocker of the human C5aR (CCX168) reduced disease in an anti-MPO model [81]. A clinical trial testing the safety and efficacy of CCX168 in AAV patients has now been performed. In addition, human data demonstrated activation of the complement system in urine and biopsies from patients with AAV, thereby confirming the animal data [31, 82, 83] (Fig. 13.1).

References

1. Jennette JC, Falk RJ, Bacon PA, Basu N, Cid MC, Ferrario F, Flores-Suarez LF, Gross WL, Guillevin L, Hagen EC, Hoffman GS, Jayne DR, Kallenberg CGM, Lamprecht P, Langford CA, Luqmani RA, Mahr AD, Matteson EL, Merkel PA, Ozen S, Pusey CD, Rasmussen N, Rees AJ, Scott DGI, Specks U, Stone JH, Takahashi K, Watts RA (2013) 2012 revised international Chapel Hill consensus conference nomenclature of vasculitides. Arthritis Rheum 65:1–11
2. Falk RJ, Jennette JC (2002) ANCA are pathogenic – oh yes they are! J Am Soc Nephrol 13:1977–1979
3. Jennette JC, Falk RJ, Gasim AH (2011) Pathogenesis of antineutrophil cytoplasmic autoantibody vasculitis. Curr Opin Nephrol Hypertens 20:263–270
4. Kain R, Matsui K, Exner M, Binder S, Schaffner G, Sommer EM, Kerjaschki D (1995) A novel class of autoantigens of anti-neutrophil cytoplasmic antibodies in necrotizing and crescentic glomerulonephritis: the lysosomal membrane glycoprotein h-lamp-2 in neutrophil granulocytes and a related membrane protein in glomerular endothelial cells. J Exp Med 181:585–597

Fig. 13.1 (continued) can now bind to surface expressed antigens. Neutrophils are activated by binding of $F(ab')_2$ to membrane expressed PR3 or MPO and through FcγR activation. (III) Binding results in activation with generation of NETs, degranulation of neutrophil serine proteases (NSPs) and generation of reactive oxygen species. ROS inhibit caspase-1 and thereby reduce IL-1ß generation, whereas NSPs cleave and activate IL-1ß independent from caspase-1. (IV) These processes result in PMN/endothelial cell interaction followed by PMN transmigration

5. Kain R, Exner M, Brandes R, Ziebermayr R, Cunningham D, Alderson CA, Davidovits A, Raab I, Jahn R, Ashour O, Spitzauer S, Sunder-Plassmann G, Fukuda M, Klemm P, Rees AJ, Kerjaschki D (2008) Molecular mimicry in pauci-immune focal necrotizing glomerulonephritis. Nat Med 14:1088–1096
6. Kain R, Tadema H, McKinney EF, Benharkou A, Brandes R, Peschel A, Hubert V, Feenstra T, Sengölge G, Stegeman C, Heeringa P, Lyons PA, Smith KGC, Kallenberg C, Rees AJ (2012) High prevalence of autoantibodies to hLAMP-2 in anti-neutrophil cytoplasmic antibody-associated vasculitis. J Am Soc Nephrol 23:556–566
7. Peschel A, Basu N, Benharkou A, Brandes R, Brown M, Dieckmann R, Rees AJ, Kain R (2014) Autoantibodies to hLAMP-2 in ANCA-negative pauci-immune focal necrotizing GN. J Am Soc Nephrol 25:455–463
8. Mueller A, Holl-Ulrich K, Gross WL (2013) Granuloma in ANCA-associated vasculitides: another reason to distinguish between syndromes? Curr Rheumatol Rep 15:376
9. Lyons PA, Rayner TF, Trivedi S, Holle JU, Watts RA, Jayne DRW, Baslund B, Brenchley P, Bruchfeld A, Chaudhry AN, Cohen Tervaert JW, Deloukas P, Feighery C, Gross WL, Guillevin L, Gunnarsson I, Harper L, Hrušková Z, Little MA, Martorana D, Neumann T, Ohlsson S, Padmanabhan S, Pusey CD, Salama AD, Sanders J-SF, Savage CO, Segelmark M, Stegeman CA, Tesař V, Vaglio A, Wieczorek S, Wilde B, Zwerina J, Rees AJ, Clayton DG, Smith KGC (2012) Genetically distinct subsets within ANCA-associated vasculitis. N Engl J Med 367:214–223
10. Xie G, Roshandel D, Sherva R, Monach PA, Lu EY, Kung T, Carrington K, Zhang SS, Pulit SL, Ripke S, Carette S, Dellaripa PF, Edberg JC, Hoffman GS, Khalidi N, Langford CA, Mahr AD, St Clair EW, Seo P, Specks U, Spiera RF, Stone JH, Ytterberg SR, Raychaudhuri S, de Bakker PIW, Farrer LA, Amos CI, Merkel PA, Siminovitch KA (2013) Association of granulomatosis with polyangiitis (Wegener's) with HLA-DPB1*04 and SEMA6A gene variants: evidence from genome-wide analysis. Arthritis Rheum 65:2457–2468
11. Wieczorek S, Holle JU, Cohen Tervaert JW, Harper L, Moosig F, Gross WL, Epplen JT (2014) The SEM6A6 locus is not associated with granulomatosis with polyangiitis or other forms of antineutrophil cytoplasmic antibody-associated vasculitides in Europeans: comment on the article by Xie et al. Arthritis Rheumatol (Hoboken) 66:1400–1401
12. Bauer S, Abdgawad M, Gunnarsson L, Segelmark M, Tapper H, Hellmark T (2007) Proteinase 3 and CD177 are expressed on the plasma membrane of the same subset of neutrophils. J Leukoc Biol 81:458–464
13. Von Vietinghoff S, Tunnemann G, Eulenberg C, Wellner M, Cristina Cardoso M, Luft FC, Kettritz R (2007) NB1 mediates surface expression of the ANCA antigen proteinase 3 on human neutrophils. Blood 109:4487–4493
14. Schreiber A, Luft FC, Kettritz R (2004) Membrane proteinase 3 expression and ANCA-induced neutrophil activation. Kidney Int 65:2172–2183
15. Rarok AA, Stegeman CA, Limburg PC, Kallenberg CGM (2002) Neutrophil membrane expression of proteinase 3 (PR3) is related to relapse in PR3-ANCA-associated vasculitis. J Am Soc Nephrol 13:2232–2238
16. Witko-Sarsat V, Lesavre P, Lopez S, Bessou G, Hieblot C, Prum B, Noël LH, Guillevin L, Ravaud P, Sermet-Gaudelus I, Timsit J, Grünfeld JP, Halbwachs-Mecarelli L (1999) A large subset of neutrophils expressing membrane proteinase 3 is a risk factor for vasculitis and rheumatoid arthritis. J Am Soc Nephrol 10:1224–1233
17. Schreiber A (2015) CD177/NB1 receptor expression is dynamically regulated in sepsis patients. Immunohematol Am Red Cross 31:128–129
18. Jerke U, Rolle S, Dittmar G, Bayat B, Santoso S, Sporbert A, Luft F, Kettritz R (2011) Complement receptor Mac-1 is an adaptor for NB1 (CD177)-mediated PR3-ANCA neutrophil activation. J Biol Chem 286:7070–7081
19. Sachs UJH, Andrei-Selmer CL, Maniar A, Weiss T, Paddock C, Orlova VV, Choi EY, Newman PJ, Preissner KT, Chavakis T, Santoso S (2007) The neutrophil-specific antigen CD177 is a counter-receptor for platelet endothelial cell adhesion molecule-1 (CD31). J Biol Chem 282:23603–23612

20. Bayat B, Werth S, Sachs UJH, Newman DK, Newman PJ, Santoso S (2010) Neutrophil transmigration mediated by the neutrophil-specific antigen CD177 is influenced by the endothelial S536N dimorphism of platelet endothelial cell adhesion molecule-1. J Immunol (Baltimore 1950) 184:3889–3896
21. Kuckleburg CJ, Tilkens SB, Santoso S, Newman PJ (2012) Proteinase 3 contributes to transendothelial migration of NB1-positive neutrophils. J Immunol (Baltimore 1950) 188:2419–2426
22. Roth AJ, Brown MC, Smith RN, Badhwar AK, Parente O, Chung H chul, Bunch DO, McGregor JG, Hogan SL, Hu Y, Yang J-J, Berg EA, Niles J, Jennette JC, Preston GA, Falk RJ (2012) Anti-LAMP-2 antibodies are not prevalent in patients with antineutrophil cytoplasmic autoantibody glomerulonephritis. J Am Soc Nephrol 23:545–555
23. Yang JJ, Pendergraft WF, Alcorta DA, Nachman PH, Hogan SL, Thomas RP, Sullivan P, Jennette JC, Falk RJ, Preston GA (2004) Circumvention of normal constraints on granule protein gene expression in peripheral blood neutrophils and monocytes of patients with antineutrophil cytoplasmic autoantibody-associated glomerulonephritis. J Am Soc Nephrol 15:2103–2114
24. Ciavatta DJ, Yang J, Preston GA, Badhwar AK, Xiao H, Hewins P, Nester CM, Pendergraft WF, Magnuson TR, Jennette JC, Falk RJ (2010) Epigenetic basis for aberrant upregulation of autoantigen genes in humans with ANCA vasculitis. J Clin Invest 120:3209–3219
25. Roth AJ, Ooi JD, Hess JJ, van Timmeren MM, Berg EA, Poulton CE, McGregor J, Burkart M, Hogan SL, Hu Y, Winnik W, Nachman PH, Stegeman CA, Niles J, Heeringa P, Kitching AR, Holdsworth S, Jennette JC, Preston GA, Falk RJ (2013) Epitope specificity determines pathogenicity and detectability in ANCA-associated vasculitis. J Clin Invest 123:1773–1783
26. Olson SW, Owshalimpur D, Yuan CM, Arbogast C, Baker TP, Oliver D, Abbott KC (2013) Relation between asymptomatic proteinase 3 antibodies and future granulomatosis with polyangiitis. Clin J Am Soc Nephrol 8:1312–1318
27. Falk RJ, Terrell RS, Charles LA, Jennette JC (1990) Anti-neutrophil cytoplasmic autoantibodies induce neutrophils to degranulate and produce oxygen radicals in vitro. Proc Natl Acad Sci U S A 87:4115–4119
28. Charles LA, Caldas ML, Falk RJ, Terrell RS, Jennette JC (1991) Antibodies against granule proteins activate neutrophils in vitro. J Leukoc Biol 50:539–546
29. Morris H, Morgan MD, Wood AM, Smith SW, Ekeowa UI, Herrmann K, Holle JU, Guillevin L, Lomas DA, Perez J, Pusey CD, Salama AD, Stockley R, Wieczorek S, McKnight AJ, Maxwell AP, Miranda E, Williams J, Savage CO, Harper L (2011) ANCA-associated vasculitis is linked to carriage of the Z allele of α_1 antitrypsin and its polymers. Ann Rheum Dis 70:1851–1856
30. Freeley SJ, Coughlan AM, Popat RJ, Dunn-Walters DK, Robson MG (2013) Granulocyte colony stimulating factor exacerbates antineutrophil cytoplasmic antibody vasculitis. Ann Rheum Dis 72:1053–1058
31. Gou S-J, Yuan J, Wang C, Zhao M-H, Chen M (2013) Alternative complement pathway activation products in urine and kidneys of patients with ANCA-associated GN. Clin J Am Soc Nephrol 8:1884–1891
32. Schreiber A, Xiao H, Jennette JC, Schneider W, Luft FC, Kettritz R (2009) C5a receptor mediates neutrophil activation and ANCA-induced glomerulonephritis. J Am Soc Nephrol 20:289–298
33. Hewins P, Morgan MD, Holden N, Neil D, Williams JM, Savage COS, Harper L (2006) IL-18 is upregulated in the kidney and primes neutrophil responsiveness in ANCA-associated vasculitis. Kidney Int 69:605–615
34. Stegeman CA, Tervaert JW, Sluiter WJ, Manson WL, de Jong PE, Kallenberg CG (1994) Association of chronic nasal carriage of Staphylococcus aureus and higher relapse rates in Wegener granulomatosis. Ann Intern Med 120:12–17

35. Hong Y, Eleftheriou D, Hussain AAK, Price-Kuehne FE, Savage CO, Jayne D, Little MA, Salama AD, Klein NJ, Brogan PA (2012) Anti-neutrophil cytoplasmic antibodies stimulate release of neutrophil microparticles. J Am Soc Nephrol 23:49–62
36. Harper L, Radford D, Plant T, Drayson M, Adu D, Savage CO (2001) IgG from myeloperoxidase-antineutrophil cytoplasmic antibody-positive patients stimulates greater activation of primed neutrophils than IgG from proteinase 3-antineutrophil cytosplasmic antibody-positive patients. Arthritis Rheum 44:921–930
37. Xiao H, Heeringa P, Liu Z, Huugen D, Hu P, Maeda N, Falk RJ, Jennette JC (2005) The role of neutrophils in the induction of glomerulonephritis by anti-myeloperoxidase antibodies. Am J Pathol 167:39–45
38. Porges AJ, Redecha PB, Kimberly WT, Csernok E, Gross WL, Kimberly RP (1994) Anti-neutrophil cytoplasmic antibodies engage and activate human neutrophils via Fc gamma RIIa. J Immunol (Baltimore 1950) 153:1271–1280
39. Kocher M, Edberg JC, Fleit HB, Kimberly RP (1998) Antineutrophil cytoplasmic antibodies preferentially engage Fc gammaRIIIb on human neutrophils. J Immunol (Baltimore 1950) 161:6909–6914
40. Kettritz R, Jennette JC, Falk RJ (1997) Crosslinking of ANCA-antigens stimulates superoxide release by human neutrophils. J Am Soc Nephrol 8:386–394
41. Nimmerjahn F, Ravetch JV (2011) FcγRs in health and disease. Curr Top Microbiol Immunol 350:105–125
42. Smith KGC, Clatworthy MR (2010) FcgammaRIIB in autoimmunity and infection: evolutionary and therapeutic implications. Nat Rev Immunol 10:328–343
43. Ooi JD, Gan P-Y, Chen T, Eggenhuizen PJ, Chang J, Alikhan MA, Odobasic D, Holdsworth SR, Kitching AR (2014) FcγRIIB regulates T-cell autoreactivity, ANCA production, and neutrophil activation to suppress anti-myeloperoxidase glomerulonephritis. Kidney Int 86:1140–1149
44. Kelley JM, Monach PA, Ji C, Zhou Y, Wu J, Tanaka S, Mahr AD, Johnson S, McAlear C, Cuthbertson D, Carette S, Davis JC, Dellaripa PF, Hoffman GS, Khalidi N, Langford CA, Seo P, St Clair EW, Specks U, Stone JH, Spiera RF, Ytterberg SR, Merkel PA, Edberg JC, Kimberly RP (2011) IgA and IgG antineutrophil cytoplasmic antibody engagement of Fc receptor genetic variants influences granulomatosis with polyangiitis. Proc Natl Acad Sci U S A 108:20736–20741
45. Pankhurst T, Nash G, Williams J, Colman R, Hussain A, Savage C (2011) Immunoglobulin subclass determines ability of immunoglobulin (Ig)G to capture and activate neutrophils presented as normal human IgG or disease-associated anti-neutrophil cytoplasm antibody (ANCA)-IgG. Clin Exp Immunol 164:218–226
46. Van Timmeren MM, van der Veen BS, Stegeman CA, Petersen AH, Hellmark T, Collin M, Heeringa P (2010) IgG glycan hydrolysis attenuates ANCA-mediated glomerulonephritis. J Am Soc Nephrol 21:1103–1114
47. Kettritz R, Schreiber A, Luft FC, Haller H (2001) Role of mitogen-activated protein kinases in activation of human neutrophils by antineutrophil cytoplasmic antibodies. J Am Soc Nephrol 12:37–46
48. Van der Veen BS, Chen M, Müller R, van Timmeren MM, Petersen AH, Lee PA, Satchell SC, Mathieson PW, Saleem MA, Stegeman CA, Zwerina J, Molema G, Heeringa P (2011) Effects of p38 mitogen-activated protein kinase inhibition on anti-neutrophil cytoplasmic autoantibody pathogenicity in vitro and in vivo. Ann Rheum Dis 70:356–365
49. Kettritz R, Choi M, Butt W, Rane M, Rolle S, Luft FC, Klein JB (2002) Phosphatidylinositol 3-kinase controls antineutrophil cytoplasmic antibodies-induced respiratory burst in human neutrophils. J Am Soc Nephrol 13:1740–1749
50. Schreiber A, Rolle S, Peripelittchenko L, Rademann J, Schneider W, Luft FC, Kettritz R (2010) Phosphoinositol 3-kinase-gamma mediates antineutrophil cytoplasmic autoantibody-induced glomerulonephritis. Kidney Int 77:118–128

51. Hao J, Meng L-Q, Xu P-C, Chen M, Zhao M-H (2012) p38MAPK, ERK and PI3K signaling pathways are involved in C5a-primed neutrophils for ANCA-mediated activation. PLoS One 7:e38317
52. Weidner S, Neupert W, Goppelt-Struebe M, Rupprecht HD (2001) Antineutrophil cytoplasmic antibodies induce human monocytes to produce oxygen radicals in vitro. Arthritis Rheum 44:1698–1706
53. Ralston DR, Marsh CB, Lowe MP (1997) Antineutrophil cytoplasmic antibodies induce monocyte IL-8 release. Role of surface proteinase-3, alpha1-antitrypsin, and Fcgamma receptors. J Clin Invest 100:1416–1424
54. Weidner S, Carl M, Riess R, Rupprecht HD (2004) Histologic analysis of renal leukocyte infiltration in antineutrophil cytoplasmic antibody-associated vasculitis: importance of monocyte and neutrophil infiltration in tissue damage. Arthritis Rheum 50:3651–3657
55. Zhao L, David MZ, Hyjek E, Chang A, Meehan SM (2015) M2 macrophage infiltrates in the early stages of ANCA-associated pauci-immune necrotizing GN. Clin J Am Soc Nephrol 10:54–62
56. Ziegler-Heitbrock L, Ancuta P, Crowe S, Dalod M, Grau V, Hart DN, Leenen PJM, Liu Y-J, MacPherson G, Randolph GJ, Scherberich J, Schmitz J, Shortman K, Sozzani S, Strobl H, Zembala M, Austyn JM, Lutz MB (2010) Nomenclature of monocytes and dendritic cells in blood. Blood 116:e74–e80
57. O'Brien EC, Abdulahad WH, Rutgers A, Huitema MG, O'Reilly VP, Coughlan AM, Harrington M, Heeringa P, Little MA, Hickey FB (2015) Intermediate monocytes in ANCA vasculitis: increased surface expression of ANCA autoantigens and IL-1β secretion in response to anti-MPO antibodies. Sci Rep 5:11888
58. Schreiber A, Pham CTN, Hu Y, Schneider W, Luft FC, Kettritz R (2012) Neutrophil serine proteases promote IL-1β generation and injury in necrotizing crescentic glomerulonephritis. J Am Soc Nephrol 23:470–482
59. Le Roux S, Pepper RJ, Dufay A, Néel M, Meffray E, Lamandé N, Rimbert M, Josien R, Hamidou M, Hourmant M, Cook HT, Charreau B, Larger E, Salama AD, Fakhouri F (2012) Elevated soluble Flt1 inhibits endothelial repair in PR3-ANCA-associated vasculitis. J Am Soc Nephrol 23:155–164
60. Nolan SL, Kalia N, Nash GB, Kamel D, Heeringa P, Savage COS (2008) Mechanisms of ANCA-mediated leukocyte-endothelial cell interactions in vivo. J Am Soc Nephrol 19:973–984
61. Bajema IM, Hagen EC, de Heer E, van der Woude FJ, Bruijn JA (2001) Colocalization of ANCA-antigens and fibrinoid necrosis in ANCA-associated vasculitis. Kidney Int 60:2025–2030
62. Kessenbrock K, Krumbholz M, Schönermarck U, Back W, Gross WL, Werb Z, Gröne H-J, Brinkmann V, Jenne DE (2009) Netting neutrophils in autoimmune small-vessel vasculitis. Nat Med 15:623–625
63. Huugen D, Xiao H, van Esch A, Falk RJ, Peutz-Kootstra CJ, Buurman WA, Tervaert JWC, Jennette JC, Heeringa P (2005) Aggravation of anti-myeloperoxidase antibody-induced glomerulonephritis by bacterial lipopolysaccharide: role of tumor necrosis factor-alpha. Am J Pathol 167:47–58
64. Kuligowski MP, Kwan RYQ, Lo C, Wong C, James WG, Bourges D, Ooi JD, Abeynaike LD, Hall P, Kitching AR, Hickey MJ (2009) Antimyeloperoxidase antibodies rapidly induce alpha-4-integrin-dependent glomerular neutrophil adhesion. Blood 113:6485–6494
65. Calderwood JW, Williams JM, Morgan MD, Nash GB, Savage COS (2005) ANCA induces beta2 integrin and CXC chemokine-dependent neutrophil-endothelial cell interactions that mimic those of highly cytokine-activated endothelium. J Leukoc Biol 77:33–43
66. Hu N, Westra J, Rutgers A, Doornbos-Van der Meer B, Huitema MG, Stegeman CA, Abdulahad WH, Satchell SC, Mathieson PW, Heeringa P, Kallenberg CGM (2011) Decreased CXCR1 and CXCR2 expression on neutrophils in anti-neutrophil cytoplasmic autoantibody-associated vasculitides potentially increases neutrophil adhesion and impairs migration. Arthritis Res Ther 13:R201

67. O'Sullivan KM, Lo CY, Summers SA, Elgass KD, McMillan PJ, Longano A, Ford SL, Gan P-Y, Kerr PG, Kitching AR, Holdsworth SR (2015) Renal participation of myeloperoxidase in antineutrophil cytoplasmic antibody (ANCA)-associated glomerulonephritis. Kidney Int 88:1030–1046
68. Odobasic D, Kitching AR, Yang Y, O'Sullivan KM, Muljadi RCM, Edgtton KL, Tan DSY, Summers SA, Morand EF, Holdsworth SR (2013) Neutrophil myeloperoxidase regulates T-cell-driven tissue inflammation in mice by inhibiting dendritic cell function. Blood 121:4195–4204
69. Perera NC, Wiesmüller K-H, Larsen MT, Schacher B, Eickholz P, Borregaard N, Jenne DE (2013) NSP4 is stored in azurophil granules and released by activated neutrophils as active endoprotease with restricted specificity. J Immunol (Baltimore 1950) 191:2700–2707
70. Pham CTN (2008) Neutrophil serine proteases fine-tune the inflammatory response. Int J Biochem Cell Biol 40:1317–1333
71. Schreiber A, Luft FC, Kettritz R (2015) Phagocyte NADPH oxidase restrains the inflammasome in ANCA-induced GN. J Am Soc Nephrol 26:411–424
72. Brinkmann V, Reichard U, Goosmann C, Fauler B, Uhlemann Y, Weiss DS, Weinrauch Y, Zychlinsky A (2004) Neutrophil extracellular traps kill bacteria. Science 303:1532–1535
73. Schauer C, Janko C, Munoz LE, Zhao Y, Kienhöfer D, Frey B, Lell M, Manger B, Rech J, Naschberger E, Holmdahl R, Krenn V, Harrer T, Jeremic I, Bilyy R, Schett G, Hoffmann M, Herrmann M (2014) Aggregated neutrophil extracellular traps limit inflammation by degrading cytokines and chemokines. Nat Med 20:511–517
74. Holdenrieder S, Eichhorn P, Beuers U, Samtleben W, Schoenermarck U, Zachoval R, Nagel D, Stieber P (2006) Nucleosomal DNA fragments in autoimmune diseases. Ann N Y Acad Sci 1075:318–327
75. Warnatsch A, Ioannou M, Wang Q, Papayannopoulos V (2015) Inflammation. Neutrophil extracellular traps license macrophages for cytokine production in atherosclerosis. Science 349:316–320
76. Tillack K, Breiden P, Martin R, Sospedra M (2012) T lymphocyte priming by neutrophil extracellular traps links innate and adaptive immune responses. J Immunol (Baltimore 1950) 188:3150–3159
77. Nakazawa D, Tomaru U, Suzuki A, Masuda S, Hasegawa R, Kobayashi T, Nishio S, Kasahara M, Ishizu A (2012) Abnormal conformation and impaired degradation of propylthiouracil-induced neutrophil extracellular traps: implications of disordered neutrophil extracellular traps in a rat model of myeloperoxidase antineutrophil cytoplasmic antibody-associated vasculitis. Arthritis Rheum 64:3779–3787
78. Nakazawa D, Shida H, Tomaru U, Yoshida M, Nishio S, Atsumi T, Ishizu A (2014) Enhanced formation and disordered regulation of NETs in myeloperoxidase-ANCA-associated microscopic polyangiitis. J Am Soc Nephrol 25:990–997
79. Xiao H, Schreiber A, Heeringa P, Falk RJ, Jennette JC (2007) Alternative complement pathway in the pathogenesis of disease mediated by anti-neutrophil cytoplasmic autoantibodies. Am J Pathol 170:52–64
80. Huugen D, van Esch A, Xiao H, Peutz-Kootstra CJ, Buurman WA, Tervaert JWC, Jennette JC, Heeringa P (2007) Inhibition of complement factor C5 protects against anti-myeloperoxidase antibody-mediated glomerulonephritis in mice. Kidney Int 71:646–654
81. Xiao H, Dairaghi DJ, Powers JP, Ertl LS, Baumgart T, Wang Y, Seitz LC, Penfold MET, Gan L, Hu P, Lu B, Gerard NP, Gerard C, Schall TJ, Jaen JC, Falk RJ, Jennette JC (2014) C5a receptor (CD88) blockade protects against MPO-ANCA GN. J Am Soc Nephrol 25:225–231
82. Gou S-J, Yuan J, Chen M, Yu F, Zhao M-H (2013) Circulating complement activation in patients with anti-neutrophil cytoplasmic antibody-associated vasculitis. Kidney Int 83:129–137
83. Manenti L, Vaglio A, Gnappi E, Maggiore U, Allegri L, Allinovi M, Urban ML, Delsante M, Galetti M, Nicastro M, Pilato FP, Buzio C (2015) Association of serum C3 concentration and histologic signs of thrombotic microangiopathy with outcomes among patients with ANCA-associated renal vasculitis. Clin J Am Soc Nephrol 10:2143–2151

Chapter 14
Long-Term Outcome of ANCA-Associated Systemic Vasculitis

James Ritchie, Timothy Reynolds, and Joanna C. Robson

14.1 Introduction

Long-term outcomes for patients with ANCA associated vasculitis have been transformed by treatment regimens including high dose glucocorticoids and cyclophosphamide; from a fatal disease in 80 % of patients in the 1950s, to a chronic disease with periods of remission and relapse in the majority in the last decade [1]. The focus now is on refining the regimens used, including the use of cyclophosphamide and glucocorticoid reduction strategies, to maximise effectiveness of remission induction and maintenance, and reduce irreversible damage accumulation, whilst minimising adverse effects including infection, cardiovascular disease and malignancy. A key element of the analysis of long-term outcomes has been the identification of prognostic markers for poorer outcomes (see Table 14.1) which may respond to targeted treatment regimens. One example is the response to Rituximab, particularly in relapsing disease which is PR3 ANCA positive [2]. The Outcome Measures in Rheumatology (OMERACT) Vasculitis Working group has spearheaded the development and use of outcome measurement within ANCA-associated vasculitis and defined a core set for use in clinical trials [3]. Outcomes including measures of disease activity and irreversible damage have enabled standardised disease assessments within clinical trials and enabled the description and analysis of long-term outcomes in a way that was previously difficult to do in what is a relatively rare disease. The OMERACT Vasculitis group is now also looking beyond the use of physician based outcome tools to the development of disease specific patient reported outcome measures to enable future treatment regimens to be assessed via outcomes that measure what is important to patients [4].

J. Ritchie • T. Reynolds • J.C. Robson (✉)
Academic Rheumatology Unit, The Courtyard, Bristol Royal Infirmary,
BS2 8HW Bristol, UK
e-mail: Jo.Robson@uwe.ac.uk

F. Dammacco et al. (eds.), *Systemic Vasculitides: Current Status and Perspectives*, DOI 10.1007/978-3-319-40136-2_14

Table 14.1 Overview of risk factors for mortality and relapse

Risk factors for mortality	Risk factors for relapse
Alveolar haemorrhage	PR3 ANCA positivity
Renal impairment (eGFR <15 ml/min or dialysis)	Upper respiratory tract involvement
Baseline severity of disease activity measured via the BVAS or FFS (≥1)	Respiratory involvement
Older age	Renal impairment associated with a decreased risk
Accumulation of damage on the VDI (≥5 items)	Switch from ANCA negativity to positivity
Hypoalbuminaemia	If using treatment with Rituximab, B cell return
Cardiomyopathy	
Pre-existing co-morbidities	

BVAS Birmingham vasculitis activity score, *FFS* five factor score, *VDI* vasculitis damage index

14.2 Mortality

The European Vasculitis Study Group (EUVAS) long-term follow up of 535 newly diagnosed patients from four randomised controlled trials has demonstrated that patients with ANCA-associated vasculitis (AAV) still have a 2.6 fold (95 % CI 2.2–3.1) increased risk of death compared with the general population, when matched for age, sex and country of origin [1]. Survival rates are 88 % (95 % CI 86 to 91 %), 85 % (95 % CI 82 to 88 %) and 78 % (95 % CI 75 to 82 %) at 1, 2 and 5 years from diagnosis [1], although patients with severe pulmonary haemorrhage were excluded from the original trials [1]. A retrospective cohort of 53 patients with AAV with alveolar haemorrhage, has demonstrated lower survival rates of 83 % at 3 months and 58 % at 49 months [5].

There has however been a definite improvement in mortality rates over the last 30 years, probably due to the advent of more effective and targeted treatment regimens, including reduced cyclophosphamide exposure through use of a pulsed regimen [6, 7], and use of alternatives such as methotrexate for non-organ or life threatening involvement [6, 8–10]. An inception cohort of 181 patients with AAV from The Netherlands demonstrated a difference in mortality between groups of patients diagnosed at different time points; with a Hazard ratio (HR) for death of 2.9, for those diagnosed between1990–2000 and 3.9 for 1979–1989, in comparison with a group diagnosed later, between 2001 and 2009 [8]. A single cohort of 445 German patients with Granulomatosis with Polyangiitis (GPA) also reported a decline in standardized mortality rates over the last four decades, and described a shorter interval between symptoms and diagnosis (from 8 months to 4 months), and reduction in cyclophosphamide cumulative dosages (from 67 g diagnosed in 1966–1993, to 24 g in those diagnosed between 1999 and 2002) over time in their cohort [11].

A general practice cohort study of 255 patients with newly diagnosed GPA demonstrated a peak in mortality during the first year following diagnosis (HR 9.0) [12]. Mortality then dropped but remained higher than in the general population, gradually rising to a further peak at around 10–15 years post diagnosis (HR 4.0) [12]. Data from the EUVAS long-term follow up [1] and The Netherlands cohorts [8] has highlighted that early deaths within the first year are commonly due to infections or active vasculitis [1], whilst malignancies and cardiovascular events are responsible for later deaths [1].

Baseline clinical features associated with a later increase in mortality, include worse renal function (either eGFR <15 ml/min or needing dialysis) [1, 8, 13], older age [1, 14], pulmonary haemorrhage [5, 8, 13, 14], low albumin [10, 15], high levels of proteinase 3- anti-neutrophil cytoplasmic antibody (PR3 ANCA) compared to perinuclear ANCA in patients with renal involvement [13, 16], and cardiomyopathy [17]. Older patients (>65 years) with renal vasculitis are more likely to present with severe renal disease and have an increased risk of infection due to leucopenia [15] and may have worse pulmonary involvement than younger patients [14]. Patients with GPA and pre-existing illnesses have an increased mortality risk than those without any co-morbidities [18].

The Birmingham Vasculitis Activity Score (BVAS) is an index of systemic symptoms and signs of abnormalities in eight different organ systems: skin, mucus membranes and eyes, ears nose and throat, chest, cardiovascular system, gastrointestinal tract, kidney and nervous system [19] which has been endorsed by the Outcome Measures in Rheumatology (OMERACT) initiative as one of the key measure of disease activity within clinical trials [3]. Predictors of increased mortality include baseline severity of disease, as measured by the Birmingham Vasculitis Activity Score (BVAS) and the Five-Factor Score (FFS) [20].

The French Vasculitis Study Group (FVSG) has developed the Five Factor Score of prognostic markers including classes of serum creatinine levels (≤1.58 mg/dl and ≥1.58 mg/dl) and proteinuria (≤1 g/day and ≥1 g/day), presence of severe gastrointestinal (GI) tract involvement, cardiomyopathy, and/or central nervous system involvement [21]. The presence of each factor is accorded 1 point, and the score is defined as follows: 0 represents no factors, 1 represents 1 factor, 2 represents ≥2 factors [21]. Patients with microscopic polyangiitis and polyarteritis nodosa with a FFS of 0 have a 5 and 8 year survival rate of 93 % and 86 % respectively [22]. A long-term cohort of 101 patients with eosinophilic granulomatosis with polyangiitis (EGPA) demonstrated that no one factor predicted mortality, but those over 65 years old with cardiomyopathy or ANCA positivity were more likely to develop chronic kidney disease and chronic neurological disability [23]. The Vasculitis Damage Index is the OMERACT endorsed measure of irreversible damage or scarring [3]; patients who score more than 5 items of damage have a 6.4-fold increase in mortality risk (95 % CI 2.1 to 19) [24].

14.3 Renal Outcomes

Between 75 % and 90 % of patients with ANCA associated vasculitis (AAV) will develop renal involvement at some point in their disease course [25]. Patients with severe renal impairment and dialysis-dependence at baseline have worse overall patient and renal survival [26]. In a Swedish population-based study of 201 newly diagnosed patients with AAV, the percentages with renal involvement at baseline were 85 % and 68 % for those with MPO-ANCA and PR3-ANCA respectively [27]. Patients with MPO-ANCA positivity appear to be more at risk of progression to end stage renal failure (38 % vs 15 %) [27]. In a EUVAS study comparing plasma exchange to methylprednisolone in 137 patients presenting with dialysis dependence or creatinine >500 μmol/L, patient and renal survival were 49 % and 61 % respectively after a median of 3.95 years [26]. The Glomerular Disease Collaboration Network (GDCN) AAV registry, an inception cohort based in the United States, consists of 544 patients diagnosed with AAV with renal involvement since 1985 [28]. Data from this registry has demonstrated that rates of patient and renal survival at 5 years have improved over time (64 % and 51 % in 1985–1989, increasing to 83 % and 82 %, respectively, in 2005–2009, $P < 0.001$ for both) [28]. Analysis of patients presenting with severe renal disease (eGFR <45 ml/min) showed that survival had also improved over time in this group [28]. Elevated serum creatinine at baseline was the only factor associated with worse patient and renal survival (HR 1.11 per 1 mg/dL of serum creatinine [95 % CI 1.04 to 1.18], $p = 0.002$) [28]. Age, ANCA type and duration of cyclophosphamide treatment were not found to be significant predictors of death, progression to dialysis or relapse in this cohort [28].

A histological classification system has been proposed to predict outcomes in AAV based on renal histology; 5 year renal survival rates are 93 % for focal, 76 % for crescentic, 61 % for mixed and 50 % for sclerotic categories [29, 30]. Separate cohorts have produced similar data, although a recent retrospective cohort of 93 Italian patients with ANCA-associated glomerulonephritis, demonstrated differing outcomes, with worse outcomes in the crescentic and sclerotic classes [31]. In this cohort independent predictors of end-stage renal disease were high serum creatinine and hypertension at presentation and less than 20 % of normal glomeruli at kidney biopsy [31]. An analysis of long-term outcomes in 26 patients with biopsy-proven ANCA-associated glomerulonephritis but normal eGFR at diagnosis demonstrated similar outcomes at 3 years in relation to remission, relapse rates, disease activity and damage levels as patients with abnormal eGFR at diagnosis, suggesting that both groups may need intensive treatment to prevent adverse renal outcomes [32].

Outcomes for patients with AAV who receive a renal transplant are generally good, with relatively low relapse rates at 0.01 per patient per year [30]. There is a suggestion that patients who are PR3 positive at the time of transplant are more likely to relapse, but more work is needed to determine predictors of graft rejection, relapse and death in transplant patients [33].

14.4 Relapse

Relapse rates in the French Vasculitis Study Group cohort (FVSG, n = 434) and the Glomerular Disease Collaborative Network (GDCN, n = 347) cohort in the United States were 54 % and 42 % respectively, after a median follow up of 44 months [34]. Predictors of relapse in both cohorts were PR3 ANCA positivity and lung involvement, with upper respiratory tract involvement a predictor in the GDCN but not the FVSG [34]. In the long-term follow up from the European Vasculitis Study Group (EUVAS) trials, which combined patients who had received a range of induction and maintenance regimes; PR3 positivity was again independently associated with an increased risk of relapse, as was cardiovascular disease, whilst impaired renal function was associated with lower relapse rates [35]. A small study of 16 patients with AAV who received a renal transplant suggested that PR3 positive patients were two times (OR 2.19, $p = 0.71$) more likely to experience a relapse compared to MPO positive patients, although this did not reach statistical significance [33].

The long-term follow up of the EUVAS trials, which investigated cyclophosphamide-sparing treatment strategies at induction, has found an increased risk of relapse in those receiving pulsed IV cyclophosphamide or methotrexate as induction therapy, compared with oral cyclophosphamide[7, 36]. It appears that although there may be benefits in terms of adverse events with less intensive immunosuppression at induction, the use of these regimens may be associated with earlier relapse (see Fig. 14.1) [37].

Long term data from the Wegener's Granulomatosis-Entretien (WEGENT) study [38] showed that, at a median follow up of 11.9 years, Azathioprine and Methotrexate as maintenance therapy, were comparable in terms of survival (10 year survival 75.1 % vs 79.9 %, $p = 0.56$) and relapse rates (relapse-free survival 26.3 % vs 33.9 %,

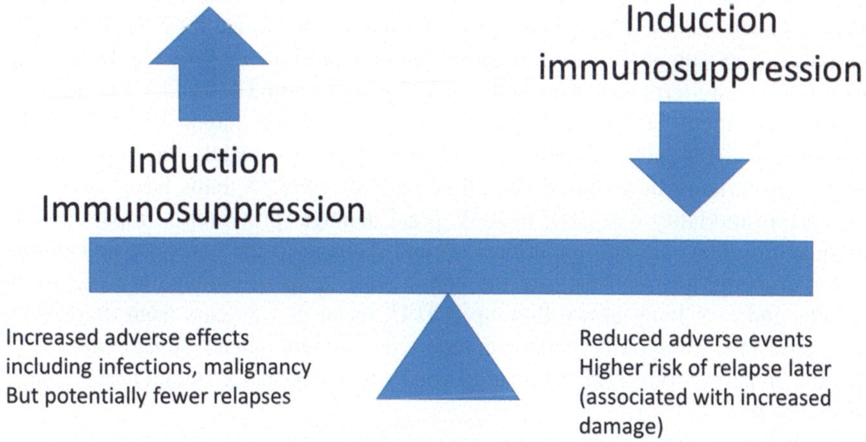

Fig. 14.1 Balance between risks and benefits of differing intensities of induction immunosuppression regimens in ANCA associated vasculitis: initial adverse events versus long-term relapse risk

p = 0.29). PR3-ANCA positivity, but not disease subtype, was again associated with relapse [38].

The French Maintenance of Remission using Rituximab in Systemic ANCA-associated Vasculitis (MAINRITSAN) study of 115 patients with newly diagnosed or relapsing GPA, MPA or renal-limited ANCA associated vasculitis, who were in complete remission after induction with cyclophosphamide and glucocorticoids, demonstrated that more patients had sustained remission at month 28 with Rituximab than with Azathioprine (relapses seen in 29 % of Azathioprine arm versus 5 % in Rituximab arm; HR for relapse, 6.61; 95 % CI 1.56 to 27.96) [39]. Using data from the Rituximab in ANCA-associated Vasculitis (RAVE) trial; PR3 positive AAV patients with relapsing disease achieved complete remission more often following treatment with Rituximab treatment versus cyclophosphamide plus azathioprine at 6 months (OR 3.57; 95 % CI 1.43 to 8.93), 12 months (OR 4.32; 95 % CI 1.53 to 12,15) and 18 months (OR 3.06; 95 % CI 1.05 to 8.97). There was no association between treatment and remission in the MPO positive patients, or those divided according to diagnosis [2]. Use of a Rituximab protocol including induction and maintenance phases (6-monthly 1 g repeated infusions over 24 months) in an open label report of 69 patients, demonstrated all were in remission at the end of 24 months, with 28 patients relapsing a median of 34.4 months after their last infusion [40]. PR3-ANCA positivity (P = 0.039), B cell return within 12 months of the last RTX infusion (P = 0.0038) and switch from ANCA negativity to positivity (P = 0.0046) were predictors of relapse [40].

14.5 Damage

The Vasculitis Damage Index (VDI) is a validated index of 54 items of irreversible aspects of vasculitis or its treatment, accrued since the onset of the disease [41]. The VDI has been endorsed by the Outcome Measures in Rheumatology (OMERACT) initiative as an outcome measurement within clinical trials in AAV [3]. Data from the EUVAS long-term follow-up study of 735 patients with GPA and MPA, demonstrated that patients commonly present with damage even at diagnosis; 34.5 % have ≥1 item and 5.1 % have ≥5 items at baseline [42]. The total VDI score is predictive of future mortality in systemic vasculitis; patients with ≥5 items have a 6.4-fold increase in mortality risk [24]. In AAV, renal damage (proteinuria, impaired glomerular filtration rate and hypertension) and damage to the ear, nose and throat (nasal crusting, hearing loss) are the commonest items of damage accrued by 6 months and over long-term follow up (LTFU, mean of 7.3 years from diagnosis) [42]. By LTFU, two-thirds of patients reported treatment related damage, most frequently hypertension (41.5 %), osteoporosis (14.1 %), malignancy (12.6 %) and diabetes (10.4 %) [42].

Higher baseline BVAS scores, lower glomerular filtration rate and increasing age were baseline predictors of high levels of subsequent damage; the number of relapses and cumulative dosages of glucocorticoids (GCs) were also associated with increasing VDI scores [43]. A total VDI score of ≥5 was more likely to occur with every year of GC treatment [Odds Ratio 1.26 per 12 months of GC use (95% CI 1.03 to 1.53), P = 0.022] [42,43].

14.6 Infection

Adverse events within the first year of treatment can be significant in AAV, causing 59 % of deaths compared with 14 % of deaths caused by active vasculitis [44]. Severity of leucopenia, infection, GFR and other adverse events are all independently associated with mortality [44]. Within a cohort of 489 patients with AAV followed for a median of 2.8 years, the 1, 2 and 5 year cumulative incidence of infection was 51, 58 and 65 % and severe infection (defined as requiring intravenous antibiotics, admittance to intensive care) was 22, 23 and 26 % [45]. Severe infection increased the risk of death within the first year fourfold (95% CI 2.0–8.7; P = 0.001) [45].

14.7 Cardiovascular Disease

From the EUVAS long-term follow-up data, 14 % of patients with GPA and MPA had a recorded cardiovascular event within 5 years from diagnosis [46]. In patients with no previous history of cardiovascular disease, predictors were age and raised diastolic blood pressure at diagnosis, whilst PR3 ANCA positivity was associated with a reduced risk compared to MPO ANCA or negative ANCA status [46]. Patients with AAV also have increased prevalence of cardiovascular risk factors including renal dysfunction, and hypertension, diabetes, increased BMI and hyper lipidaemia secondary to glucocorticoid treatment [42, 43]. A Danish study of 293 patients with GPA also demonstrated an increased risk of early (<5 years) and late (>5 years from diagnosis) cardiovascular disease, including an observed: expected ratio for ischaemic heart disease (IHD) of 1.9 (95% CI 1.4 to 2.4) [47]. In a prospective cohort study of 91 patients with AAV, 62 % of EGPA and 46 % of GPA patients vs 20 % of controls, were found to have cardiac abnormalities on baseline ECG or echocardiography [48]. These cardiac abnormalities were associated with an increased all-cause and cardiovascular mortality (Log-rank P = 0.015 and Log-rank P = 0.021, respectively) by a mean of 53 months post-diagnosis [48]. The significance of late gadolinium enhancement on cardiac magnetic resonance imaging (CMRI) is not yet confirmed in patients with AAV; in patients diagnosed with cardiomyopathy independently of CMRI, LGE may be a poor prognostic sign, but as an isolated abnormality on CMRI alone, it may prove to be clinically unimportant [49].

14.8 Malignancy

With improvements in long-term survival rates data relating to the development of malignancies amongst those treated for ANCA-associated vasculitis is now available.

Retrospective data from the French Vasculitis Study Group followed 383 patients with EGPA over a mean period of 66.8 months. 14 malignancies were reported amongst 13 patients. These consisted of 11 solid-organ cancers (of which 5 were ultimately fatal) and 3 haematological neoplasms. This did not represent an increase beyond the background risk for age and sex-matched individuals (SIR 1.15 [95 % CI 0.63–1.94]) [17].

A Danish cohort consisting of 293 patients diagnosed with GPA between 1973 and 1999 were followed for a longer median period of 9.7 years [50]. Amongst this group 73 cancers were reported. This represented an overall SIR for all cancers of 1.9 (95 % CI 1.5–2.4) [50]. Higher rates of non-melanoma skin cancers (SIR 4.0 [95 % CI 2.7–5.7]), bladder cancers (SIR 5.5 [95 % CI 2.7–9.8]) and myeloid leukaemia (SIR 13.3 [95 % CI 3.6–34]) were found, especially amongst those treated with higher cumulative doses of cyclophosphamide (above 36 g in total) [50]. Nonetheless the data were not adjusted for cumulative doses of other immunosuppressive drugs, smoking status, co-morbidities or the use of 2-mercaptoetane sulfonate Na (mesna).

Analysis of follow-up data from 535 patients in the EUVAS study group followed over a mean period of 4.95 years revealed 50 new cancers during the follow-up period. There was an increase in the rate of cancers at all sites (SIR 1.58 [95% CI 1.17 to 2.08]) but this was influenced significantly by the increased rate of non-melanoma skin cancers (SIR 2.78 [1.56 to 4.59]). When these were excluded the SIR was 1.30 (95 % CI 0.90 to 1.80) [51]. It was not clear whether this lower rate of cancer was related to the relatively short follow-up period or the lower cumulative doses of immunosuppression typically used in more modern treatment regimens [51].

Further long-term data is needed to determine the risk of developing malignancies later in life as there can be a significant latency period before these arise. The available literature highlights the importance of good quality skin care and surveillance amongst patients on long term immunosuppression. The data also reflects the importance of hydration and the prevention of urothelial toxicity associated with higher cumulative cyclophosphamide doses to reduce the risk of bladder cancer later in life. This is reflected in the 2009 European League against Rheumatism (EULAR) guidelines [52] and the subsequent 2014 British Society for Rheumatology (BSR) guidelines [53] which support the use of mesna together with careful urine monitoring during follow-up. This is largely based on expert opinion as there are no vasculitis trials that have focused on mesna use in reducing the urothelial toxicity of cyclophosphamide.

The BSR guidelines also support limiting the cumulative lifetime doses of cyclophosphamide to less than 25 g. This recommendation is based on the findings of a data-driven review article by Monach et al. in 2010 [54] which highlighted the risks

Table 14.2 Summary of risk reduction strategies to reduce adverse long-term outcomes

Adverse outcomes	Risk reduction strategies
Infection	Immunisations against influenza and Pneumocystis pneumonia (PCP)
Cancer	Reduction in cumulative dose of cyclophosphamide
	Use of mesna and oral hydration during cyclophosphamide
Cardiovascular disease	Address modifiable cardiovascular risk factors (smoking, hypercholesterolaemia, hypertension)
Relapse	Consider continuing immunosuppression in patients with ongoing PR3 ANCA positivity and or upper respiratory tract disease

of increased dose and duration of cyclophosphamide treatment in the subsequent development of bladder cancer. See Table 14.2 for further risk prevention strategies.

14.9 Pregnancy Outcomes

With the advent of better and targeted treatments which aim to spare fertility whilst gaining tighter control of the underlying disease, there is now emerging evidence on a range of pregnancy outcomes. A matched case control study of 51 pregnancies in 29 patients with mainly ANCA associated vasculitis demonstrated that patients had lower gestational age at delivery than controls, 36 weeks versus 40 weeks ($P < 0.003$), but there was no difference in birth centiles once the gestational age difference was adjusted for [55]. Flares of vasculitis do occur during pregnancy and postpartum with figures ranging from 10 % up to a third of patients in different cohorts [55, 56]. Three TIAs were reported in one cohort, suggesting that thrombotic complications may be an issue [56].

14.10 Health Related Quality of Life

Health related quality of life is impaired in patient with ANCA association at diagnosis. Use of the Short-Form 36 to measure HRQoL in 346 patients enrolled in the EUVAS trials, demonstrated physical functioning scores were the most affected, specifically in those with neurologic involvement (−8.48 points [95 % CI −12.90, −4.06]; $P < 0.001$) and older age (−0.25 points/year [95 % CI −0.38, −0.11]; $P < 0.001$) [57]. Patients rank fatigue and reduced energy levels as having the greatest impact or relevance to their HRQoL, rather than specific organ manifestations, in contrast to physicians who rank dialysis or oxygen dependence as more important [58]. A case control study from Scotland demonstrated that patients with AAV are twice as likely to have mild or moderate fatigue [odds ratio (OR) 2.0; 95 % CI 1.1, 3.8] or severe fatigue (OR 2.5; 95 % CI 1.4, 4.5) than age, gender and location matched controls, and this is strongly associated with impaired physical health [59]. Survey data suggests that AAV-related fatigue is multifactorial, and associated with pain, sleep disturbance and levels of inflammation [60]. An international collaboration to develop a disease specific patient reported outcome for AAV is underway as part of the OMERACT initiative to expand and refine the core-set of patient-specific outcome measures for use in clinical trials [4].

14.11 Summary

In summary, the long-term outcomes for patients with ANCA-associated disease have significantly improved in the last few decades. The challenge now is to further refine our management regimens to develop better targeted treatments which have high efficacy but low toxicity, whilst continuing to address any modifiable risk factors for adverse outcomes including infection and malignancy.

References

1. Flossmann O, Berden A, de Groot K, Hagen C, Harper L, Heijl C et al (2011) Long-term patient survival in ANCA-associated vasculitis. Ann Rheum Dis 70(3):488–494
2. Unizony S, Villarreal M, Miloslavsky EM, Lu N, Merkel PA, Spiera R et al (2015) Clinical outcomes of treatment of anti-neutrophil cytoplasmic antibody (ANCA)-associated vasculitis based on ANCA type. Ann Rheum Dis 75(6):1166–1169
3. Merkel PA, Aydin SZ, Boers M, Direskeneli H, Herlyn K, Seo P et al (2011) The OMERACT core set of outcome measures for use in clinical trials of ANCA-associated vasculitis. J Rheumatol 38(7):1480–1486
4. Robson JC, Milman N, Tomasson G, Dawson J, Cronholm PF, Kellom K et al (2015) Exploration, development, and validation of patient-reported outcomes in antineutrophil cytoplasmic antibody-associated vasculitis using the OMERACT process. J Rheumatol 42(11):2204–2209
5. Hruskova Z, Casian AL, Konopasek P, Svobodova B, Frausova D, Lanska V et al (2013) Long-term outcome of severe alveolar haemorrhage in ANCA-associated vasculitis: a retrospective cohort study. Scand J Rheumatol 42(3):211–214
6. Guillevin L, Cordier JF, Lhote F, Cohen P, Jarrousse B, Royer I et al (1997) A prospective, multicenter, randomized trial comparing steroids and pulse cyclophosphamide versus steroids and oral cyclophosphamide in the treatment of generalized Wegener's granulomatosis. Arthritis Rheum 40(12):2187–2198
7. Harper L, Morgan MD, Walsh M, Hoglund P, Westman K, Flossmann O et al (2012) Pulse versus daily oral cyclophosphamide for induction of remission in ANCA-associated vasculitis: long-term follow-up. Ann Rheum Dis 71(6):955–960
8. Hilhorst M, Wilde B, van Paassen P, Winkens B, van Breda VP, Cohen Tervaert JW et al (2013) Improved outcome in anti-neutrophil cytoplasmic antibody (ANCA)-associated glomerulonephritis: a 30-year follow-up study. Nephrol Dial Transplant 28(2):373–379
9. De Groot K, Rasmussen N, Bacon PA, Tervaert JW, Feighery C, Gregorini G et al (2005) Randomized trial of cyclophosphamide versus methotrexate for induction of remission in early systemic antineutrophil cytoplasmic antibody-associated vasculitis. Arthritis Rheum 52(8):2461–2469

10. Aasarod K, Iversen BM, Hammerstrom J, Bostad L, Vatten L, Jorstad S (2000) Wegener's granulomatosis: clinical course in 108 patients with renal involvement. Nephrol Dial Transplant 15(5):611–618
11. Holle JU, Gross WL, Latza U, Nolle B, Ambrosch P, Heller M et al (2011) Improved outcome in 445 patients with Wegener's granulomatosis in a German vasculitis center over four decades. Arthritis Rheum 63(1):257–266
12. Luqmani R, Suppiah R, Edwards CJ, Phillip R, Maskell J, Culliford D et al (2011) Mortality in Wegener's granulomatosis: a bimodal pattern. Rheumatology 50(4):697–702
13. Hogan SL, Falk RJ, Chin H, Cai J, Jennette CE, Jennette JC et al (2005) Predictors of relapse and treatment resistance in antineutrophil cytoplasmic antibody-associated small-vessel vasculitis. Ann Intern Med 143(9):621–631
14. Chen M, Yu F, Zhang Y, Zhao MH (2008) Antineutrophil cytoplasmic autoantibody-associated vasculitis in older patients. Medicine 87(4):203–209
15. Harper L, Savage CO (2005) ANCA-associated renal vasculitis at the end of the twentieth century–a disease of older patients. Rheumatology 44(4):495–501
16. Westman KW, Selga D, Isberg PE, Bladstrom A, Olsson H (2003) High proteinase 3-anti-neutrophil cytoplasmic antibody (ANCA) level measured by the capture enzyme-linked immunosorbent assay method is associated with decreased patient survival in ANCA-associated vasculitis with renal involvement. J Am Soc Nephrol 14(11):2926–2933
17. Comarmond C, Pagnoux C, Khellaf M, Cordier JF, Hamidou M, Viallard JF et al (2013) Eosinophilic granulomatosis with polyangiitis (Churg-Strauss): clinical characteristics and long-term followup of the 383 patients enrolled in the French Vasculitis Study Group cohort. Arthritis Rheum 65(1):270–281
18. Faurschou M, Ahlstrom MG, Lindhardsen J, Baslund B, Obel N (2015) Impact of pre-existing co-morbidities on mortality in granulomatosis with polyangiitis: a cohort study. Rheumatology (Oxford) 55(4):649–653
19. Luqmani RA, Bacon PA, Moots RJ, Janssen BA, Pall A, Emery P et al (1994) Birmingham Vasculitis Activity Score (BVAS) in systemic necrotizing vasculitis. QJM 87(11):671–678
20. Gayraud M, Guillevin L, le Toumelin P, Cohen P, Lhote F, Casassus P et al (2001) Long-term followup of polyarteritis nodosa, microscopic polyangiitis, and Churg-Strauss syndrome: analysis of four prospective trials including 278 patients. Arthritis Rheum 44(3):666–675
21. Guillevin L, Lhote F, Gayraud M, Cohen P, Jarrousse B, Lortholary O et al (1996) Prognostic factors in polyarteritis nodosa and Churg-Strauss syndrome. A prospective study in 342 patients. Medicine 75(1):17–28
22. Samson M, Puechal X, Devilliers H, Ribi C, Cohen P, Bienvenu B et al (2014) Long-term follow up of a randomized trial on 118 patients with polyarteritis nodosa or microscopic polyangiitis without poor-prognosis factors. Autoimmun Rev 13(2):197–205
23. Durel CA, Berthiller J, Caboni S, Jayne D, Ninet J, Hot A (2015) Long-term follow-up of a multicentre cohort of 101 patients with eosinophilic granulomatosis with polyangiitis (EGPA). Arthritis Care Res 68(3):374–387
24. Exley AR, Carruthers DM, Luqmani RA, Kitas GD, Gordon C, Janssen BA et al (1997) Damage occurs early in systemic vasculitis and is an index of outcome. QJM 90(6):391–399
25. Franssen CF, Stegeman CA, Kallenberg CG, Gans RO, De Jong PE, Hoorntje SJ et al (2000) Antiproteinase 3- and antimyeloperoxidase-associated vasculitis. Kidney Int 57(6):2195–2206
26. Walsh M, Casian A, Flossmann O, Westman K, Hoglund P, Pusey C et al (2013) Long-term follow-up of patients with severe ANCA-associated vasculitis comparing plasma exchange to intravenous methylprednisolone treatment is unclear. Kidney Int 84(2):397–402
27. Mohammad AJ, Segelmark M (2014) A population-based study showing better renal prognosis for proteinase 3 antineutrophil cytoplasmic antibody (ANCA)-associated nephritis versus myeloperoxidase ANCA-associated nephritis. J Rheumatol 41(7):1366–1373
28. Rhee RL, Hogan SL, Poulton CJ, McGregor JA, Landis JR, Falk RJ, et al (2016) Trends in long-term outcomes among patients with ANCA-associated vasculitis with renal disease. Arthritis Rheumatol 68(7):1711–1720

29. Berden AE, Ferrario F, Hagen EC, Jayne DR, Jennette JC, Joh K et al (2010) Histopathologic classification of ANCA-associated glomerulonephritis. J Am Soc Nephrol 21(10):1628–1636
30. Marco H, Mirapeix E, Arcos E, Comas J, Ara J, Gil-Vernet S et al (2013) Long-term outcome of antineutrophil cytoplasmic antibody-associated small vessel vasculitis after renal transplantation. Clin Transplant 27(3):338–347
31. Moroni G, Binda V, Leoni A, Raffiotta F, Quaglini S, Banfi G et al (2015) Predictors of renal survival in ANCA-associated vasculitis. Validation of a histopatological classification schema and review of the literature. Clin Exp Rheumatol 33(2 Suppl 89):S-56–63
32. Hanaoka H, Ota Y, Takeuchi T, Kuwana M (2016) Poor renal outcomes in patients with antineutrophil cytoplasmic antibody-associated crescentic glomerulonephritis and normal renal function at diagnosis. Clin Rheumatol 35(2):495–500
33. Geetha D, Lee SM, Shah S, Rahman HM (2015) Relevance of ANCA positivity at the time of renal transplantation in ANCA associated vasculitis. J Nephrol
34. Pagnoux C, Hogan SL, Chin H, Jennette JC, Falk RJ, Guillevin L et al (2008) Predictors of treatment resistance and relapse in antineutrophil cytoplasmic antibody-associated small-vessel vasculitis: comparison of two independent cohorts. Arthritis Rheum 58(9):2908–2918
35. Walsh M, Flossmann O, Berden A, Westman K, Hoglund P, Stegeman C et al (2012) Risk factors for relapse of antineutrophil cytoplasmic antibody-associated vasculitis. Arthritis Rheum 64(2):542–548
36. Faurschou M, Westman K, Rasmussen N, de Groot K, Flossmann O, Hoglund P et al (2012) Brief report: long-term outcome of a randomized clinical trial comparing methotrexate to cyclophosphamide for remission induction in early systemic antineutrophil cytoplasmic antibody-associated vasculitis. Arthritis Rheum 64(10):3472–3477
37. Weidner S, Geuss S, Hafezi-Rachti S, Wonka A, Rupprecht HD (2004) ANCA-associated vasculitis with renal involvement: an outcome analysis. Nephrol Dial Transplant 19(6):1403–1411
38. Puechal X, Pagnoux C, Perrodeau E, Hamidou M, Boffa JJ, Kyndt X et al (2015) Long-term outcomes of the WEGENT trial on remission-maintenance for granulomatosis with polyangiitis or microscopic polyangiitis. Arthritis Rheumatol 67(4):1117–1127
39. Guillevin L, Pagnoux C, Karras A, Khouatra C, Aumaitre O, Cohen P et al (2014) Rituximab versus azathioprine for maintenance in ANCA-associated vasculitis. N Engl J Med 371(19):1771–1780
40. Alberici F, Smith RM, Jones RB, Roberts DM, Willcocks LC, Chaudhry A et al (2015) Long-term follow-up of patients who received repeat-dose rituximab as maintenance therapy for ANCA-associated vasculitis. Rheumatology (Oxford) 54(7):1153–1160
41. Exley AR, Bacon PA, Luqmani RA, Kitas GD, Gordon C, Savage CO et al (1997) Development and initial validation of the vasculitis damage index for the standardized clinical assessment of damage in the systemic vasculitides. Arthritis Rheum 40(2):371–380
42. Robson J, Doll H, Suppiah R, Flossmann O, Harper L, Hoglund P et al (2015) Damage in the anca-associated vasculitides: long-term data from the European vasculitis study group (EUVAS) therapeutic trials. Ann Rheum Dis 74(1):177–184
43. Robson J, Doll H, Suppiah R, Flossmann O, Harper L, Hoglund P et al (2015) Glucocorticoid treatment and damage in the anti-neutrophil cytoplasm antibody-associated vasculitides: long-term data from the European Vasculitis Study Group trials. Rheumatology 54(3):471–481
44. Little MA, Nightingale P, Verburgh CA, Hauser T, De Groot K, Savage C et al (2010) Early mortality in systemic vasculitis: relative contribution of adverse events and active vasculitis. Ann Rheum Dis 69(6):1036–1043
45. McGregor JG, Negrete-Lopez R, Poulton CJ, Kidd JM, Katsanos SL, Goetz L et al (2015) Adverse events and infectious burden, microbes and temporal outline from immunosuppressive therapy in antineutrophil cytoplasmic antibody-associated vasculitis with native renal function. Nephrol Dial Transplant 30(Suppl 1):i171–i181
46. Suppiah R, Judge A, Batra R, Flossmann O, Harper L, Hoglund P et al (2011) A model to predict cardiovascular events in patients with newly diagnosed Wegener's granulomatosis and microscopic polyangiitis. Arthritis Care Res (Hoboken) 63(4):588–596

47. Faurschou M, Mellemkjaer L, Sorensen IJ, Svalgaard Thomsen B, Dreyer L, Baslund B (2009) Increased morbidity from ischemic heart disease in patients with Wegener's granulomatosis. Arthritis Rheum 60(4):1187–1192
48. Hazebroek MR, Kemna MJ, Schalla S, Sanders-van Wijk S, Gerretsen SC, Dennert R et al (2015) Prevalence and prognostic relevance of cardiac involvement in ANCA-associated vasculitis: eosinophilic granulomatosis with polyangiitis and granulomatosis with polyangiitis. Int J Cardiol 199:170–179
49. Dunogue B, Terrier B, Cohen P, Marmursztejn J, Legmann P, Mouthon L et al (2015) Impact of cardiac magnetic resonance imaging on eosinophilic granulomatosis with polyangiitis outcomes: A long-term retrospective study on 42 patients. Autoimmun Rev 14(9):774–780
50. Faurschou M, Mellemkjaer L, Voss A, Keller KK, Hansen IT, Baslund B (2015) Prolonged risk of specific malignancies following cyclophosphamide therapy among patients with granulomatosis with polyangiitis. Rheumatology 54(8):1345–1350
51. Heijl C, Harper L, Flossmann O, Stucker I, Scott DG, Watts RA et al (2011) Incidence of malignancy in patients treated for antineutrophil cytoplasm antibody-associated vasculitis: follow-up data from European Vasculitis Study Group clinical trials. Ann Rheum Dis 70(8):1415–1421
52. Mukhtyar C, Guillevin L, Cid MC, Dasgupta B, de Groot K, Gross W et al (2009) EULAR recommendations for the management of primary small and medium vessel vasculitis. Ann Rheum Dis 68(3):310–317
53. Ntatsaki E, Carruthers D, Chakravarty K, D'Cruz D, Harper L, Jayne D et al (2014) BSR and BHPR guideline for the management of adults with ANCA-associated vasculitis. Rheumatology 53(12):2306–2309
54. Monach PA, Arnold LM, Merkel PA (2010) Incidence and prevention of bladder toxicity from cyclophosphamide in the treatment of rheumatic diseases: a data-driven review. Arthritis Rheum 62(1):9–21
55. Sangle SR, Vounotrypidis P, Briley A, Nel L, Lutalo PM, Sanchez-Fernandez S et al (2015) Pregnancy outcome in patients with systemic vasculitis: a single-centre matched case-control study. Rheumatology 54(9):1582–1586
56. Fredi M, Lazzaroni MG, Tani C, Ramoni V, Gerosa M, Inverardi F et al (2015) Systemic vasculitis and pregnancy: a multicenter study on maternal and neonatal outcome of 65 prospectively followed pregnancies. Autoimmun Rev 14(8):686–691
57. Walsh M, Mukhtyar C, Mahr A, Herlyn K, Luqmani R, Merkel PA et al (2011) Health-related quality of life in patients with newly diagnosed antineutrophil cytoplasmic antibody-associated vasculitis. Arthritis Care Res (Hoboken) 63(7):1055–1061
58. Herlyn K, Hellmich B, Seo P, Merkel PA (2010) Patient-reported outcome assessment in vasculitis may provide important data and a unique perspective. Arthritis Care Res (Hoboken) 62(11):1639–1645
59. Basu N, Jones GT, Fluck N, MacDonald AG, Pang D, Dospinescu P et al (2010) Fatigue: a principal contributor to impaired quality of life in ANCA-associated vasculitis. Rheumatology (Oxford) 49(7):1383–1390
60. Basu N, McClean A, Harper L, Amft EN, Dhaun N, Luqmani RA et al (2013) Explaining fatigue in ANCA-associated vasculitis. Rheumatology 52(9):1680–1685

Chapter 15
Kawasaki Disease: Past, Present and Future

Fernanda Falcini and Gemma Lepri

Abstract Forty eight years ago Doctor Tomisaku Kawasaki firstly described a disease called "*Acute febrile muco-cutaneous lymph node syndrome*", after labeled as Kawasaki disease (KD). It still remains a mysterious illness affecting coronaries in a quarter of untreated patients and is the most common cause of childhood-acquired heart disease in industrialized countries. Many gaps exist in our knowledge in the etiology and pathogenesis of KD. Numerous KD features still demand further efforts to achieve a better understanding of the disease. Some of these issues include coronaries' injuries in children not fulfilling the classical diagnostic criteria, genetic predisposition, unpredictable ineffectiveness of current therapy in some cases, vascular dysfunction in patients not showing echocardiographic evidence of coronaries abnormalities in the acute phase and risk of potential premature atherosclerosis. The lack of specific laboratory tests for early identification of the atypical and incomplete cases, especially in infants, is one of the obstacles to treat early patients in order to reduce cardiovascular involvement. Transthoracic echocardiography remains the gold-standard for evaluation of coronaries in the acute phase and follow-up. In patients with severe vascular complications, more expensive and invasive investigations, such as coronary CT angiography and MRI, may be required. As children with KD grow-up, the acknowledgment and treatment of the potential sequelae become critical, requiring rheumatologists, cardiologists and infectious disease specialists cooperate to develop guidelines for a proper evaluation and management of patients. Deep education is recommended for physicians and other professionals about how to recognize the long-term impact of systemic problems related to KD.

Keywords Kawasaki disease • Coronary damage • Kawasaki therapy • Incomplete Kawasaki • Atypical Kawasaki

---

F. Falcini (✉) • G. Lepri
Department of Clinical and Experimental Medicine, Section of Rheumatology, Transition Clinic, University of Florence, Florence, Italy
e-mail: falcini@unifi.it

F. Dammacco et al. (eds.), *Systemic Vasculitides: Current Status and Perspectives*, DOI 10.1007/978-3-319-40136-2_15

15.1 Introduction

This is the first description of Kawasaki Disease by Doctor Tomisaku Kawasaki: "*In January1961, as I look back now, I saw the first case of what it is known as a typical Kawasaki disease case which I had not experienced in my ten year career as a pediatrician, with this kind of unique symptom complex. It was a 4-year and 3-month-old boy. There was high fever which have lasted for 2 weeks, bilateral conjunctival hyperemia, dried reddish, fissured, bleeding lips, diffuse erythematosus of the oral cavity mucous membrane and strawberry tongue. There was left cervical lymphoadenopathy. There was polymorphous erythema all over the body. Red palms and soles were seen. Indurative oedema was on the hands and feet. After 10 to 14 days, there was membranous desquamation of hands and feet. I presented this case to which I could not give diagnosis at a pediatric department meeting in my hospital. One of the my colleagues suggested atypical scarlet fever, another suggested mild form of Stevens-Johnson Syndrome. I could not agree with either of this opinions. I released the case with" diagnosis unknown". In February 1962, a 2 – year-old boy with suspected sepsis was admitted at the Emergency room, when I saw him I immediately remembered the case I had seen the previous year. After that I experienced a total of 50 cases which fell into the above category by the end of 1966, and I reported on these cases under the title" Acute febrile muco-cutaneous lymph node syndrome: clinical observation of 50 cases" published in The Japanese Journal Allergy in 1964"* [1].

Since then, Kawasaki Disease (KD) was recognized in America, Europe and all over the world.

KD formerly known as febrile mucocutaneous lymph node syndrome is the most common systemic vasculitis in childhood after Henoch-Schoenlein purpura, the main cause of acquired heart disease among children living in industrialized countries where it has surpassed Rheumatic fever, and an important cause of long-term cardiac disease in adult life [2, 3]. Following the initial reports of the disease, it became apparent that KD was not benign, since a number of children were reported to die of this illness, mainly due to the cardiovascular complications [4]. KD is an acute, self-limited necrotizing vasculitis of unknown origin affecting medium and small-sized vessels that predominantly occurs in young children, aged 6 months to 5 years, with a male-to-female ratio of 1.8–1. It may be also observed in neonates, adolescents or young adults. The peak age in Japan is 5–11 months while in the United States it is between 1 and 2 years. KD is more prevalent in Asian countries, especially in Japan, with an annual incidence of 216.9 per 100,000 children aged 0–4 years but it is universally distributed and can affect children of any ethnicity. In US, where it is reported a hospitalization of 17.1 per 100,000 children, it is more common in Americans of Asian and Pacific Island descent. In Europe, it is reported an incidence ranging from 4.9 per 100, 000 children under 5 years in Denmark to 8.1 per 100,000 in United Kingdom, and to 9 per 100,000 in France [5, 6].

The etiology of KD is still unknown but current hypothesis suggests an unusual inflammatory response to one or more yet unidentified pathogen(s) occurring in genetically predisposed individuals. The discovery of viral-like cytoplasmic inclusion bodies in ciliated bronchial epithelial cells supports the hypothesis that KD might arise from a previously unidentified ubiquitous RNA virus inducing a state of persistent viral infection [7].

The immune response in KD encompasses both aspects of innate and adaptive immunity with a significant overproduction of different cytokines and activation of endothelial cells. Activation of both B and T cells has been detected together with increased proinflammatory cytokine production, including tumor necrosis factor(TNF)-α, interleukin-1 and interleukin-6 [8, 9].

As no diagnostic tests are available, the diagnosis relies on recognition of the clinical criteria, which include fever for at least 5 days and four or more of the following five features: (i) Bilateral conjunctival injection; (ii) Cervical lymphadenopathy (Fig. 15.1); (iii) Oral mucosal changes (Fig. 15.2); (iv) Polymorphous rashes(Fig. 15.3); (v) Swelling or redness of the extremities, and the exclusion of other diagnoses. Fever is typically hectic and remittent, with peak temperatures frequently exceeding 39 °C or higher, unresponsive to antibiotics but partially responsive to antipyretics. For untreated children, the febrile period lasts for a mean of 11 days. Bilateral vascular injection of the bulbar conjunctivae is generally seen in the first week of illness, not associated with exudate. Cervical lymphadenopathy is usually unilateral, restricted to the anterior cervical lymph nodes that are enlarged more than 1.5 cm, non-fluctuant, and moderately tender (Fig. 15.1). Changes in the mouth are characterized by diffuse erythema, dryness, fissuring, cracking, and bleeding of the lips (Fig. 15.2). Fever lasting at least 5 days and four of the five typical features are required to meet the diagnosis [10]. Among the clinical manifestations, mucous membrane changes are present in more than 90 % of patients while lymphadenopathy is less frequently observed. An additional sign that distinguishes KD from other febrile diseases in infants and very young children is an extreme irritability due to aseptic meningitis. The presence of fever is mandatory and its onset is considered the first day of the disease. It is important to remember that the

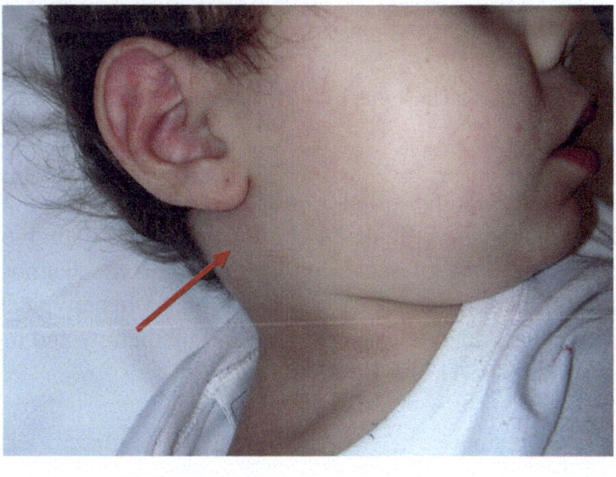

Fig. 15.1 Lymph node enlargement

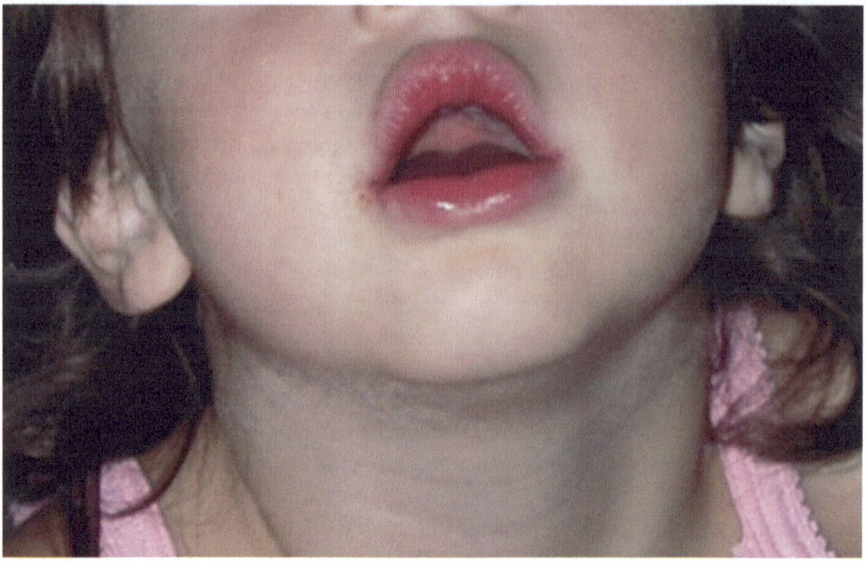

Fig. 15.2 Redding, swelling and vertical cracking of the lips

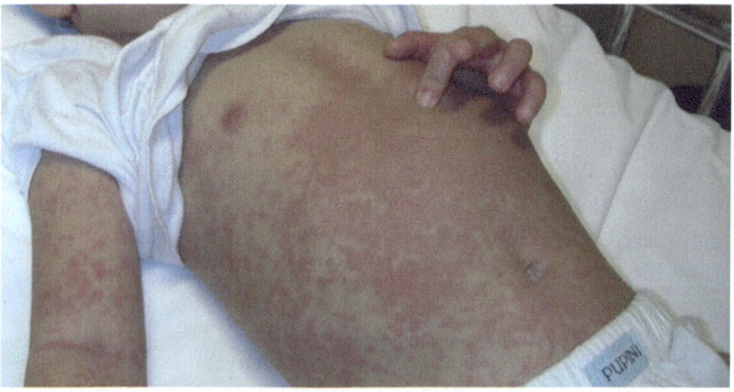

Fig. 15.3 Rash on the trunk

diagnostic features often appear consecutively and may be transient, making the early diagnosis difficult. In children aged less than 3–6 months as well in school-aged children, adolescents, and young adults the diagnosis is a challenge for general practitioners as the disease presentation is often atypical and the course incomplete with consequent delay in the appropriate therapy and high risk of cardiac coronary aneurysms (CA), see Table 15.1.

Table 15.1 Criteria for the diagnosis of Kawasaki disease

Fever persisting for more than 5 days plus at least four of the following clinical signs not explained by another disease
Bilateral bulbar conjunctival non purulent injection
Changes in extremities and perineum: acute phase: redding of palms and soles, indurative edema (Fig. 15.4) convalescent phase: membranous desquamation from fingertips (Figs. 15.5, 15.6, 15.7, 15.8 and 15.9)
Polymorphous exanthaem primarily on the trunk
Changes in lips and oral cavity
Acute non purulent cervical lymphadenopathy (>1.5 cm diameter), usually unilateral
Exclusion of other febrile diseases with similar features or echocardiographic evidence of coronary artery abnormalities.
Evidence of other organ involvement
Gastrointestinal tract (abdominal pain, hepatic dysfunction, hydrops of gallbladder)
Lungs and respiratory tract (pneumonia, et al.)
Central nervous system (aseptic meningitis, peripheral facial palsy, sensorineural hearing loss, et al.)
Kidneys and genitourinary system: (haematuria, urethritis, etc.)
Musculoskeletal system (arthritides).

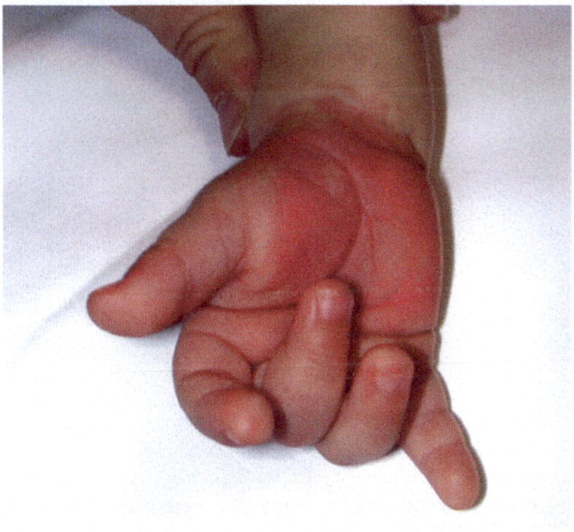

Fig. 15.4 Palmar erythema

15.2 Incomplete Kawasaki Disease

Children with at least 5 days of fever and two or three of the major clinical criteria are labeled as 'incomplete Kawasaki disease' and account for 15–20 % of children with the illness. Children with incomplete KD may develop CA abnormalities in a higher percentage as the treatment is delayed or omitted. If coronary artery dilatation is detected on echocardiography, the diagnosis of KD is confirmed, even if less

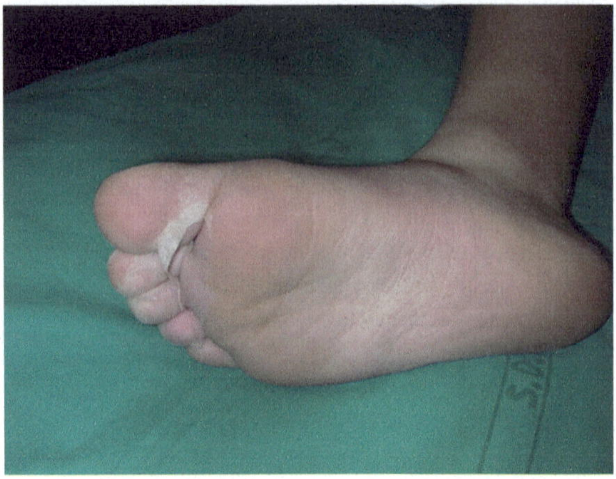

Fig. 15.5 Skin peeling of foot

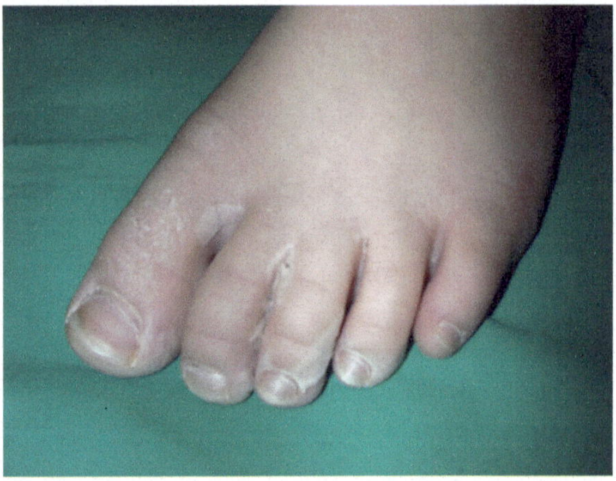

Fig. 15.6 Skin peeling of foot

than four principal features are present. The diagnosis of incomplete KD is particularly challenging, and more common in infants, below 6–12 months, and in children older than 5 years of age. It is strongly recommended to suspect incomplete KD in any infant who has a fever for more than 7 days with laboratory evidence of inflammation without a reasonable cause. The American Heart Association has formulated an algorithm to provide a structured approach to the child with suspected incomplete KD: children over 6 months of age with prolonged fever of at least 5 days and two or three principal diagnostic criteria need to be evaluated for the presence of additional clinical and laboratory features, which may lead to echocardiographic assessment and/or treatment in those who have other suggestive findings [11, 12].

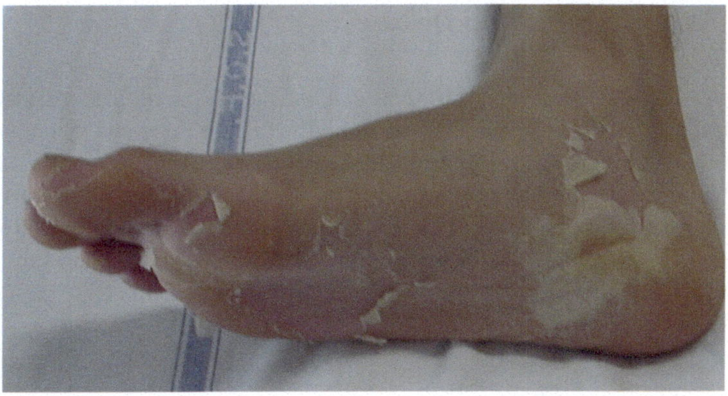

Fig. 15.7 Sheet-like desquamation

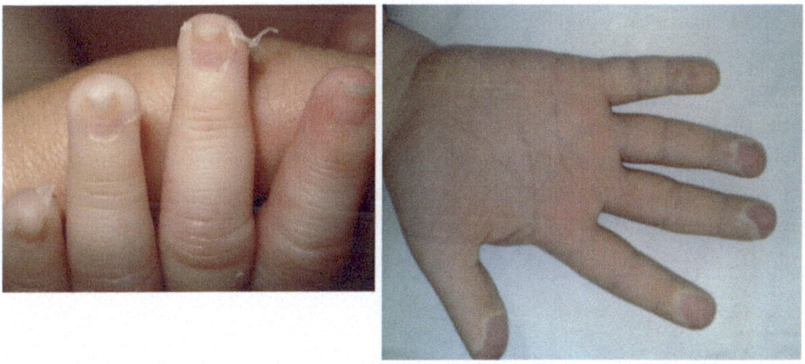

Figs. 15.8 Fingertip peeling of hands

Fig. 15.9 Bowl sign

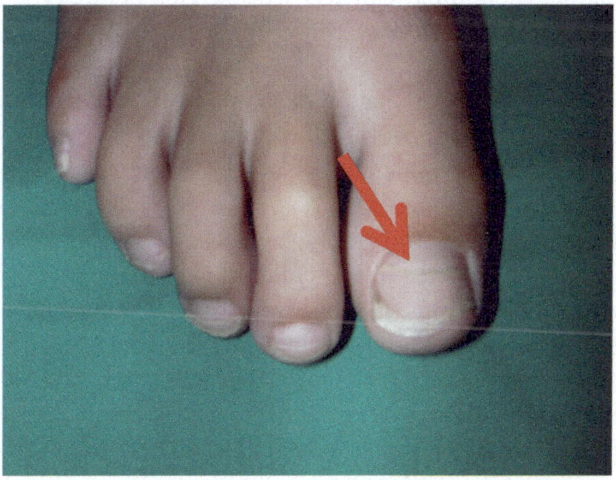

15.3 Atypical Kawasaki Disease

According to the American Heart Association and the American Academy of Pediatrics, "atypical" KD" is defined as a child with high fever and symptoms and signs at disease onset that are not included in the clinical major criteria, such as acute abdominal pain, pulmonary or gastrointestinal signs, facial nerve palsy, sensorineural hearing loss. Extreme irritability is an additional symptom present in most KD patients and may help on suspecting the diagnosis in patients with incomplete and atypical disease [13, 14].

15.4 Laboratory Tests

Laboratory findings show elevation of erythrocyte sedimentation rate, C-reactive protein, and white cell count. At onset, not all inflammatory markers are present and, if KD is deeply suspected, it is recommended to repeat the exams. Thrombocytosis occurs through the end of the second week of the illness and, therefore, may not help the diagnosis in the early stages. In a few cases, acute thrombocytopenia or low platelet count may occur and may be associated with a poorer prognosis. Abnormal liver function may be observed and some patients present with jaundice and elevated transaminases. Hypoalbuminemia is common; sterile pyuria and cerebrospinal fluid lymphocytosis representing aseptic meningitis may also be detected [10, 15].

15.5 Cardiac Complications

KD displays a high risk of subsequent cardiac abnormalities. Fifteen to 25 % of untreated children with KD will develop coronary aneurysm or coronary dilation and the cardiac disease can also include myocardial infarction, sudden death and heart ischemia.

Coronary inflammation develops within the first week of the disease with infiltration by inflammatory cells of intima and adventitia rapidly leading to panarteritis and widespread inflammation of whole artery edge. Inflammatory process continues and subsides around day 40 after onset. Thus, a prompt treatment is recommended before the echocardiographic evidence of coronary damage. Despite the occurrence of cardiac sequelae in KD children is becoming less prevalent due to the improvement in the therapy, in Japan more than 10,000 adults with history of KD in infancy are affected by cardiovascular complications. Several methods are available to detect coronary artery injury. Transthoracic echocardiography is the main tool in monitoring the coronary involvement, others more expensive and invasive encompass Magnetic resonance imaging (MRI), multislice CT (MSCT), single-photon emission CT (SPECT), and X-ray coronary angiography.

Coronary artery rupture, a rare event, is mainly the result of progressive aneurysm dilation in the acute phase. Moreover, larger aneurysms are at greater risk of developing stenosis and subsequent myocardial ischemia or infarction. Consequently, precise aneurysm measurements at onset are important to close follow-up of their evolution. Severe stenotic lesions or giant aneurysms developing late after Kawasaki disease are often associated with coronary artery calcification, responsible for irreversible change in the coronary artery wall after the acute vasculitis and can strongly predict future endothelial dysfunction of the coronary artery. Thus, early discovery of coronary injury and prompt treatment are extremely important and can help to predict prognosis [16, 17].

15.6 Present Situation and New Challenges for Medical Treatment of Kawasaki Disease

Standard treatment for acute KD consists of high dose intravenous immunoglobulin (IVIG)2 g/kg, administered within the day 10 from the onset of fever. A single infusion generally reduces the risk of CA lesions from 20–25 % to 2–4 %. IVIG should also be given to children presenting after day 10 of illness if fever and signs of inflammation persist, although the prognosis worsens with the delayed therapy. Conversely, administration of IVIG before day 5 does not appear to improve coronary artery outcomes and may increase the need for retreatment. Treatment with IVIG should also be given if incomplete KD is suspected, regardless of the echocardiographic findings. Aspirin at a dosage of 50–80 mg/kg/day is recommended during the acute phase of the illness; this dosage may be better tolerated than higher doses as regards the gastrointestinal and other side effects. Then, it should be reduced to an antiplatelet dosage of 3–5 mg/Kg once fever and inflammation have receded. There is no evidence that aspirin improves the prevalence and the outcome of coronary artery abnormalities that are highly dependent on gamma globulin dose but independent of salicylate dose [18]. Eleven to 38 % of timely IVIG treated patients continue to have persistent or recurrent fever at least 36 h after infusion. IVIG-resistant patients are at higher risk for the development of CA damage. The optimal therapy for these patients remains controversial [19].

15.7 A Concern About the Use of Steroids in KD is Still Unsolved. Who Should be Treated?

Agents used for secondary or "rescue" therapy include additional doses of IVIG, intravenous methylprednisolone (IVMP), oral corticosteroids, cyclophosphamide, cyclosporine, methotrexate, and plasma exchange. IVMP, 30 mg/kg, maximum 1 g is usually given because of its powerful and rapid immunosuppressive effect. In a randomized control trial for all patients with KD, IVMP plus initial IVIG, compared with IVIG and placebo, did not decrease the incidence of coronary artery lesions [18]. Conversely, it has been reported that suspected IVIG-resistant patients who received initial IVIG plus IVMP, compared with IVIG alone, had earlier defervescence and a significantly lower rate of coronary damage. For patients resistant to initial or additional IVIG, some studies show that IVMP was effective for rapid defervescence and prevention of CA injury [20–23]. In IVIG-resistant patients, the efficacy and safety of anti-cytokine therapy with infliximab (Remicade), a chimera type anti TNF-α agent, has been tested. After May 2005, Remicade has been used in more than 500 pediatric patients IVIG and IVMP resistant. The efficacy and safety has been observed, though 10–20 % of patients resulted Remicade-resistant. Re-treatment with IVIG or steroids was also effective. The efficacy of Remicade for reducing the fever duration, CRP, WBC counts was promising, but reduction of the incidence of coronary aneurysm has not been confirmed. In a retrospective study, patients with IVIG-resistant KD whose first re-treatment was with infliximab, compared with IVIG, had faster resolution of fever and fewer days of hospitalization. Coronary artery outcomes and adverse events were comparable [24–30]. Nearly 2–3 % of untreated children die as a consequence of CA thrombosis, myocardial infarction, or seldom aneurysm rupture. Patients with giant CA aneurysms (8 mm or more) are at long-term risk of developing aneurysms thrombosis, CA stenosis and myocardial infarction even years after the acute phase of KD. Several scoring systems have been developed to identify children at highest risk of IVIG resistance and, hence, highest risk of developing CAA. Kobayashi et al. developed a model to predict unresponsiveness to IVIG in Japanese children with KD [23, 31, 32]. In Japan, the Kobayashi score appears to identify these patients, but outside Japan, it seems reasonable to offer steroid treatment in addition to IVIG to patients with features of the most severe disease at higher risk of developing CAA, including the very young, and those with markers of severe disease, including intense inflammation, liver dysfunction, hypoalbuminemia, anemia and organ dysfunction. Steroids should also be given to patients who do not respond promptly to initial IVIG. All patients treated with steroids should be followed to identify adverse effects including osteonecrosis and intercurrent infection.

Few data are available about the treatment with Anakinra, an IL-1 receptor antagonist. The first reports in a child with severe relapsing KD, and recently in a 11-week-old female with severe KD resistant to 3 IVIG infusions and high dose corticosteroids, and complicated by macrophage activation syndrome, indicated a dramatic improvement [33, 34].

15.8 Vaccinations in Kawasaki Disease Children

A concern with the use of IVIG is that the passively acquired antibodies may interfere with the serologic response to active immunization. Current guidelines recommend postponing the measles, mumps, and rubella (MMR) vaccination to at least

6 months after IVIG treatment, but there is no consensus about the time interval. In Japan, an interval of 6–9 months is recommended. A study was conducted in The Netherlands where children receive this vaccine at the age of 14 months and 9 years. MMR vaccination response was evaluated in patients treated with IVIG for KD, in comparison with healthy controls, and no difference was detected in the two groups in terms of immunological response. However, in IVIG KD patients the immunogenic response after vaccination is reduced, especially in the measles component that seems to be less immunogenic than mumps and rubella.

In patients who received one additional dose of IVIG because they did not respond to the first infusion, a few data are available. Miura et al. studied the persistence of measles antibody titers in six episodes of KD in 5 patients aged 4–28 months without history of measles infection or vaccination who received additional infusion of IVIG with a total dose of 4 g/kg in five episodes and 6 g/kg in one episode. Enzyme immunoassay antibody titers against measles tested 3 months after the second IVIG administration were still positive but negative in all 9 months after infusion. So, the authors suggest that an appropriate interval between infusion of 4 g/kg of immunoglobulin and measles vaccination should be 9 months. The interval of 11 months recommended in the United States in all KD children who received 2 g/kg is too longer than necessary. No studies have been performed in KD patients treated with steroids in addition to IVIG.

Patients who require long-term aspirin for persistent CAA should be considered for immunization with varicella zoster virus (VZV) vaccine in view of the association of VZV and aspirin with Reye syndrome [35, 36].

15.9 Kawasaki Disease and Atherosclerotic Risk

One of the most conceivable issues for long-term prognosis of KD is that the disease may represent a risk factor for atherosclerosis. KD is a severe systemic vasculitis and post inflammatory vasculature may not return to normal. As atherosclerosis is an inflammatory process, many similarities have been observed in post KD patients.

Numerous studies have focused on atherosclerosis in KD with conflicting results, and a direct evidence demonstrating that the disease induces atherosclerotic lesions is still lacking.

Ross et al. advocated the hypothesis that endothelial injury triggers endothelial dysfunction and both are responsible for early atherosclerosis in young adults with KD in infancy [37, 38].

Fibroblast Growth Factor (FGF) 23 has been reported to influence endothelial integrity. A recent study detected that the intact serum FGF23 levels in children with KD were significantly higher than in healthy controls, in particular in those with coronary artery injury, suggesting FGF23 as a marker evocative of cardiac complications. In addition, genetic variation in the FGF23 gene has been reported in a group of KD children with cardiac disease, and its correlation with the higher serum FGF23 levels that promote coronary artery damage [39, 40].

15.10 Kawasaki Disease and Vitamin D

Twenty-five-hydroxyvitamin D (25(OH)-vitamin D) is crucial in the regulation of immunologic processes, but although its deficiency has been reported in patients with different rheumatologic disorders, no data were available for Kawasaki disease (KD). Recently, serum levels of 25(OH)-vitamin D in children with KD and the relationship with the eventual occurrence of KD-related vascular abnormalities have been assessed. Twenty (OH)-vitamin D levels were measured in 79 children with KD (21 females, 58 males, median age 4.9 years, range 1.4–7.5 years) and compared with healthy sex-/age-matched controls. A significantly higher percentage of KD patients (98.7 %) showed reduced 25(OH)-vitamin D levels (<30 ng/mL) in comparison with controls (78.6 %, $p < 0.0001$). Furthermore, KD patients had severely low levels of 25(OH)-vitamin D than controls (9.17 ± 4.94 vs 23.3 ± 10.6 ng/mL, $p < 0.0001$), especially the subgroup who developed coronary artery abnormalities (4.92 ± 1.36 vs 9.41 ± 4.95 ng/mL, $p < 0.0001$). In addition, serum 25(OH)-vitamin D levels correlated not only with erythrosedimentation rate ($p < 0.0001$), C-reactive protein ($p < 0.0001$), hemoglobin level at KD diagnosis ($p < 0.0001$) but also with both coronary artery aneurysms ($p = 0.005$) and non-aneurysmatic cardiovascular lesions ($p < 0.05$), supporting the hypothesis that the VITD may play a role in the development of vascular damage [41].

15.11 KD and Adult Life

Among the big cohort of KD survivors, it is important to determine which patients have residual coronary abnormalities that are associated with late cardiovascular events. The grade level of initial coronary involvement must be considered: patients with giant aneurysms (8 mm), large aneurysms (6 to 8 mm), smaller aneurysms (persistent or regressed), and transient coronary dilation and without evident coronary involvement at diagnosis. Persistent giant aneurysms are associated with a high risk for late complications, including thrombosis, stenosis, and calcification, possibly leading to myocardial infarction and major late mortality. Patients with large aneurysms may present coronary findings later in life. There is general agreement that these require follow-up, testing, and management. Fortunately, individuals with residual giant or large aneurysms represent roughly 1 % of patients who have had KD. Transthoracic echocardiography remains the gold-standard for evaluation of coronary arteries in the acute phase and follow-up. In KD patients with severe vascular complications, more costly and potentially invasive investigations such as coronary CT angiography and MRI maybe necessary. As children with KD with or without heart involvement become adolescents and adults, the recognition and treatment of the potential long term sequelae become crucial, requiring that rheumatologists, infectious disease specialists, and cardiologists cooperate to develop specific guidelines for a proper evaluation and management of these patients. More education is needed for physicians and other professionals about how to recognize the long-term impact of systemic problems related to KD [42–48].

15.12 Future

A genetic contribution to the risk of KD is suggested by the much higher risk of the disease in Asian children, particularly the Japanese and Koreans, which persists when patients of these ethnicities migrate to other countries, from the increased relative risk to siblings of index cases compared with the general population, from twin studies and from well documented multicase families. In the future, studies need to be addressed to candidate genes, either as susceptibility genes for developing KD, or increasing risk of CAA [49–51].

References

1. Burns JC (2002) Commentary: translation of Dr. Tomisaku Kawasaki's original report of fifty patients in 1967. Pediatr Infect Dis J 21:993–1005
2. Kawasaki T, Kosaki T, Okawa S, Shigematsu I, Yanagawa H (1974) A new infantile acute febrile mucocutaneous lymph node syndrome (MLNS) prevailing in Japan. Pediatrics 54:271–276
3. Gardner-Medwin JM, Dolezalova P, Cummins C, Southwood TR (2002) Incidence of Henoch-Schonlein purpura, Kawasaki disease, and rare vasculitides in children of different ethnic origins. Lancet 360:1197–1202
4. Kato H, Koike S, Yamamoto M, Ito Y, Yano E (1975) Coronary aneurysms in infants and young children with acute febrile mucocutaneous lymph node syndrome. J Pediatr 86:892–898
5. Harnden A, Alves B, Sheikh A (2002) Rising incidence of Kawasaki disease in England: analysis of hospital admission data. BMJ 324:1424–1425
6. Nakamura Y, Yashiro M, Uehara R, Sadakane A, Tsuboi S, Aoyama Y et al (2012) Epidemiologic features of Kawasaki disease in Japan: results of the 2009–2010 nationwide survey. J Epidemiol 22:216–221
7. Rowley AH, Baker SC, Shulman ST, Fox LM, Takahashi K, Garcia FL et al (2005) Cytoplasmic inclusion bodies are detected by synthetic antibody in ciliated bronchial epithelium during acute Kawasaki disease. J Infect Dis 192:1757–1766
8. Hirao J, Hibi S, Andoh T, Ichimura T (1997) High levels of circulating interleukin- 4 and interleukin-10 in Kawasaki disease. Int Arch Allergy Immunol 112:152–156
9. Chan WC, Duong TT, Yeung RS (2001) Presence of IFN-gamma does not indicate its necessity for induction of coronary arteritis in an animal model of Kawasaki disease. J Immunol 173:3492–3503
10. Newburger JW, Takahashi M, Gerber MA, Gewitz MH, Tani LY, Burns JC et al (2004) Diagnosis, treatment, and long- term management of Kawasaki disease. Circulation 110:2747–2771
11. Yeo Y, Kim T, Ha K, Jang G, Lee J, Lee K, Son C, Lee J (2009) Incomplete Kawasaki disease in patients younger than 1 year of age: a possible inherent risk factor. Eur J Pediatr 168:157–162

12. Sonobe T, Kiyosawa N, Tsuchiya K, Aso S, Imada Y, Imai Y et al (2007) Prevalence of coronary artery abnormality in incomplete Kawasaki disease. Pediatr Int 49:421–426
13. Genizi J, Miron D, Spiegel R, Fink D, Horovitz Y (2003) Kawasaki disease in very young infants: high prevalence of atypical presentation and coronary arteritis. Clin Pediatr (Phila) 42:263–267
14. Falcini F, Ozen S, Magni-Manzoni S, Candelli M, Ricci L, Martini G et al (2012) Discrimination between incomplete and atypical Kawasaki syndrome versus other febrile diseases in childhood: results from an international registry-based survey. Clin Exp Rheumatol 30:799–804
15. Huang MY, Gupta-Malhotra M, Huang JJ, Syu FK, Huang TY (2010) Acute-phase reactants and a supplemental diagnostic aid for Kawasaki disease. Pediatr Cardiol 31:1209–1213
16. Tacke CE, Romeih S, Kuipers IM, Spijkerboer AM, Groenink M, Kuijpers TW (2013) Evaluation of cardiac function by magnetic resonance imaging during the follow-up of patients with Kawasaki disease. Circ Cardiovasc Imaging 6:67–73
17. Dietz SM, Tacke CE, Kuipers IM, Viegman A, de Winter RJ, Burns C et al (2015) Cardiovascular imaging in children and adults following Kawasaki disease. Insight Imaging 6:697–705
18. Eleftheriou D, Levin M, ShingadiaD TR, Klein NJ, Brogan PA (2014) Management of Kawasaki disease. Arch Dis Child 99:74–83
19. Newburger JW, Sleeper LA, McCrindle BW, McCrindle DW, Minich LU, Gersony W et al (2007) Randomized trial of pulse corticosteroid therapy for primary treatment of Kawasaki disease. N Eng J Med 356:663–675
20. Tremoulet AH, Best BM, Song S, Wang S, Corinaldesi E, Eichenfield JR et al (2008) Resistance to intravenous immunoglobulin in children with Kawasaki disease. J Pediatr 153:117–121
21. Kobayashi T, Saji T, Otani T, Kobayashi T, Saji T, Otani T (2012) Efficacy of immunoglobulin plus prednisolone for prevention of coronary artery abnormalities in severe Kawasaki disease (RAISE study): a randomised, open-label, blinded-endpoints trial. Lancet 379:1613–1620
22. Chen S, Dong Y, Yin Y, Krucoff MW (2013) Intravenous immunoglobulin plus corticosteroid to prevent coronary artery abnormalities in Kawasaki disease: a meta-analysis. Heart 99:76–82
23. Sato S, Kawashima H, Kashiwagi Y, Oshika A (2013) Inflammatory cytokines as predictors of resistance to intravenous immunoglobulin therapy in Kawasaki disease patients. Int J Rheum Dis 16:168–172
24. Burns JC, Mason WH, Hauger SB, Janai H, Bastian JF, Wohrley JD et al (2005) Infliximab treatment for refractory Kawasaki syndrome. J Pediatr 146:662–667
25. Burns JC, Best BM, Mejias A, Mahony L, Fixler DE, Jafri HS et al (2008) Infliximabtreatment of intravenous immunoglobulin-resistant kawasaki disease. J Pediatr 153:833–838
26. Son MB, Gauvreau K, Burns JC, Corinaldesi E, Tremoulet AH, Watson VE et al (2011) Infliximab for intravenous immunoglobulin resistance in Kawasaki disease: a retrospective study. J Pediatr 158:644–649
27. Hirono K, Kemmotsu Y, Wittkowski H, Foell D, Saito K, Ibuki K et al (2009) Infliximab reduces the cytokine-mediated inflammation but does not suppress cellular infiltration of the vessel wall in refractory Kawasaki disease. Pediatr Res 65:696–701
28. Levin M, Burgner D (2014) Treatment of Kawasaki disease with anti-TNF antibodies. Lancet 383:1700–1703
29. Sonoda K, Mori M, Hokosaki T, Yokota S (2014) Infliximab plus plasma exchange rescue therapy in Kawasaki disease. J Pediatr 164:1128–1132
30. Davies S, Gold-von SG (2013) Should Infliximab be used as an adjuvant to IVIG in the treatment of children with Kawasaki disease who are at high risk for resistance to conventional therapy? Pediatr Cardiol 34:1756
31. Kobayashi T, Inoue Y, Takeuchi K, Okada Y, Tamura K, Tomomasa T et al (2006) Prediction of intravenous immunoglobulin unresponsiveness in patients with Kawasaki disease. Circulation 113:2606–2612
32. Sleeper LA, Minich LL, McCrindle BM, Lee JS, Mason W, Colan SD, Atz AM et al (2011) Evaluation of Kawasaki disease risk-scoring systems for intravenous immunoglobulin resistance. J Pediatr 158:831–835

33. Cohen S, Tacke CE, Straver B, Meijer N, Kuipers IM, Kuijpers TW (2012) A child with severe relapsing Kawasaki disease rescued by IL-1 receptor blockade and extracorporeal membrane oxygenation. Ann Rheum Dis 71:2059–2061
34. Shafferman A, Birmingham JD, Randy QC (2014) High dose anakinra for treatment of severe neonatal Kawasaki disease: a case report. Pediatr Rheumatol Online 12:26
35. Miura M, Katada Y, Ishihara J (2004) Time interval of measles vaccination in patients with Kawasaki disease treated with additional intravenous immunoglobulin. Eur J Pediatr 163:25–29
36. Tacke CE, Smits GP, van der Klis FRM, Kuipers IM, Zaaijer HL, Kuijpers TW (2014) Reduced serologic response to mumps, measles, and rubella vaccination in patients treated with intravenous immunoglobulin for Kawasaki disease. Allergy Clin Immunol 131:1701
37. Cheung YF, Yung TC, Tam SC, Ho MH, Chau AK (2004) Novel and traditional cardiovascular risk factors in children after Kawasaki disease: implications for premature atherosclerosis. J Am Coll Cardiol 43:120–124
38. Noto N, Okada T, Karasawa K, Ayusawa M, Sumitomo N, Harada K et al (2009) Age-related acceleration of endothelial dysfunction and subclinical atherosclerosis in subjects with coronary artery lesions after Kawasaki disease. Pediatr Cardiol 30:262–268
39. Masi L, Franceschelli F, Leoncini G (2013) Can Fibroblast growth factor (FGF)-23 circulating levels suggest coronary artery abnormalities in children with Kawasaki disease? Clin Exper Rheumatol 31:149–153
40. Falcini F, Rigante D, Masi L, Covino M, Franceschetti F, Leoncincini G, Tarantino G, MatucciCerinic M, Brandi ML (2013) Fibroblast growth factor (FGF23) gene polymorphism in children with Kawasaki syndrome (KS) and susceptibility to cardiac abnormalities. Italian J Pediatr 39:69–74
41. Stagi S, Rigante D, Lepri G, MatucciCerinic M, Falcini F, (2016) Severe Vitamin D deficiency in patients with Kawasaki a potential role in the risk to develop heart vascular abnormalities?. Clin Rheumatol 35(7):1865–1872
42. Manlhiot C, Niedra E, McCrindle BW (2013) Long-term management of Kawasaki disease: implications for the adult patient. Pediatr Neonatol 54:12–21
43. JCS Joint Working Group (2014) Guidelines for diagnosis and management of cardiovascular Sequelae in Kawasaki disease (JCS 2013). Dig VersionCirc J 78:2521–2562
44. Levin M, Burns JC, Gordon JB (2014) Warfarin plus Aspirin or Aspirin alone for patient with giant coronary artery Aneurysms secondary to Kawasaki disease? Cardiology 129:174–177
45. Saij T, Ayusawa M, Miura M, Gobayashi T, Suzuki H, Mori M et al (2014) Guidelines for medical treatment of acute Kawasaki disease: report of the research committee of the Japanese Society of Pediatric Cardiology and Cardiac Surgery (2012 revised version). Pediatr Int 56:135–158
46. Maric LS, Knezovic I, Papic N, Mise B, Roglic S, Markovinovic L et al (2014) Risk factors for coronary artery abnormalities in childrenwith Kawasaki disease: a 10-year experience. RheumatolInt. Epubahead of print
47. Ishikawa T, Iwashima S (2013) Endothelial dysfunction in children within 5 years after onset of Kawasaki disease. Pediatrics 163(4):1117–1121
48. Rizk SR, El Said G, Daniels LB, Burns JC, El Said H, Sorour KA et al (2015) Acute myocardial ischemia in adults secondary to missed Kawasaki disease in childhood. Am J Cardiol 115:423–427
49. Onouchi Y (2012) Genetics of Kawasaki disease. Circul J 76:1581–1586
50. Burgner D, Davila S, Breunis WB, Ng SB, Li Y, Bonnard C, Ling L et al (2009) A genome-wide association study identifies novel and functionally related susceptibility Loci for Kawasaki disease. PLoS Genet 5:e1000319
51. Onouchi Y, Ozaki K, Burns JC, Shimizu C, Terai M, Hamada H et al (2012) A genome-wide association study identifies three new risk loci for Kawasaki disease. Nat Genet 44:517–521

Chapter 16
Polyarteritis Nodosa

Nicolò Pipitone and Carlo Salvarani

Abstract Polyarteritis nodosa (PAN) is a rare vasculitis affecting middle-sized arteries. Its annual incidence in Europe is estimated to range between 0 and 1.6 cases per million, while the prevalence is about 31 cases per million. The frequency of hepatitis B virus (HBV)-related PAN has declined in developed countries since vaccination against HBV has been implemented. Specifically, before vaccination against HBV was implemented on a large scale, more than one-third of adults with PAN were infected by HBV, whereas currently only 5 % of European adults with PAN are infected by HBV. PAN is usually considered an immune-complex-driven vasculitis. However, the evidence of abundant CD4+ T cells in vascular inflammatory infiltrates suggests that PAN may also be induced by a T-cell response. Clinically, PAN usually presents with constitutional manifestations as well as symptoms and signs related to the organs affected. Organ ischemia is thought to be due to vascular stenoses, while ruptured aneurysms can result in tissue hemorrhage. The most frequent clinical features include constitutional manifestations, myalgia, arthralgia, peripheral neuropathy and mononeuritis multiplex. There are no specific blood tests to diagnose PAN, but inflammatory markers are typically elevated. Therefore, the diagnosis rests on histological changes in affected organs, showing a transmural vessel wall infiltrate, or angiographic findings, including small saccular or fusiform aneurysms and stenoses. Treatment includes glucocorticoids in patients without poor prognostic factors or cyclophosphamide if the disease is life- or organ-threatening.

N. Pipitone
Unità Operativa di Reumatologia, Dipartimento di Medicina Interna, Azienda Ospedaliera ASMN, Istituto di Ricovero e Cura a Carattere Scientifico, Reggio Emilia, Italy

C. Salvarani (✉)
Unit of Rheumatology, Arcispedale S.Maria Nuova,
V.le Risorgimento N80, 42100 Reggio Emilia, Italy
e-mail: Salvarani.carlo@asmn.re.it

F. Dammacco et al. (eds.), *Systemic Vasculitides: Current Status and Perspectives*, DOI 10.1007/978-3-319-40136-2_16

16.1 History

The credit for describing the first case of polyarteritis nodosa (PAN) is usually attributed to Kußmaul (a physician) and Maier (a pathologist), who reported a young man who looked severely ill and died shortly after his admission to the hospital [1]. Autopsy revealed nodular arterial aneurysms and inflammation of the adventitia, which prompted the authors to name the disease "periarteritis nodosa". The term PAN was coined only two decades later by Ferrari [2], who demonstrated the involvement of multiple ("poly") organs by nodular arterial aneurysms, while the first mention of the term PAN in the English literature dates back to 1908 [3]. However, despite the relatively early recognition of PAN as a specific disease entity, over the following decades PAN was jumbled together with what we now call "microscopic polyangiitis" (MPA), which was considered a variant ("the microscopic form") of PAN [4, 5].

16.2 Classification Criteria

In 1990, the American College of Rheumatology (ACR) developed classification criteria for various vasculitides including PAN [6] (Table 16.1). These criteria have often been used in clinical practice to support the diagnosis of PAN, but were actually designed to discriminate between different types of vasculitides, but not to distinguish vasculitis from other disorders. Therefore, their performance crucially hinges on a high pre-test probability of patients having a vasculitis in the first place. Further, the ACR panel did not consider at that time MPA as a condition different from PAN. The differentiation of PAN from MPA was made instead by the Chapel Hill consensus criteria (CHCC), which identified ten distinct vasculitis entities and classified them into three groups on the basis of the size of the vessels involved (small, medium, and large) [7]. According to the CHCC, the term PAN should be restricted to an arteritis involving medium-sized and small arteries without involvement of smaller vessels. Therefore, patients with vasculitis affecting arterioles, venules, or capillaries, including glomerular capillaries (i.e., with glomerulonephritis), were excluded from this diagnostic category. In contrast, MPA was defined as pauci-immune (i.e., few or no immune deposits) necrotizing vasculitis affecting small vessels, with or without involvement of medium-sized arteries. The CHCC definition had the merit of clearly separating out PAN from MPA on the basis of the absence of small vessels (arterioles, venules, and capillaries) in PAN, and small-vessel sparing has been retained as a key differential feature in the revised CHCC nomenclature published in 2013 [8]. On the other hand, the CHCC nomenclature is essentially based on histological features, and thus does not easily lend itself to classify PAN for clinical purposes. These shortcomings have led to subsequent efforts to provide more robust classification criteria for both pediatric and adult PAN. On this line, in 2007, Watts et al proposed a diagnostic algorithm [9] which has later

Table 16.1 1990 ACR criteria for the classification of polyarteritis nodosa

1. Weight loss >4 kg
Loss of 4 kg or more of body weight since illness began, not due to dieting or other factors
2. Livedo reticularis
Mottled reticular pattern over the skin or portions of the extremities or torso
3. Testicular pain or tenderness
Pain or tenderness of the testicles, not due to infection, trauma, or other causes
4. Myalgias, weakness or leg tenderness
Diffuse myalgias (excluding shoulder and hip girdle) or weakness of muscles or tenderness of leg muscles
5. Mononeuropathy or polyneuropathy
Development of mononeuropathy, multiple mononeuropathies, or polyneuropathy
6. Diastolic BP >90 mm Hg
Development of hypertension with diastolic BP higher than 90 mm Hg
7. Elevated BUN or creatinine
Elevation of BUN >40 mg/dl or creatinine >1.5 mg/dl, not due to dehydration or obstruction
8. Hepatitis B virus
Presence of hepatitis B surface antigen or antibody in serum
9. Arteriographic abnormalities
Arteriogram showing aneurysms or occlusions of the visceral arteries, not due to arteriosclerosis, fibromuscular dysplasia, or other non-inflammatory causes
10. Biopsy of small or medium-sized artery containing PMN
Histologic changes showing the presence of granulocytes or granulocytes and mononuclear leukocytes in the artery wall

Source: 1990 ACR Criteria For the Classification of Polyarteritis Nodosa [6]
For classification purposes, a patient shall be said to have polyarteritis nodosa if at least 3 of these 10 criteria are present. The presence of any 3 or more criteria yields a sensitivity of 82.2 % and a specificity of 86.6 %

been officially endorsed by the EMA (European Medicine Agency) to diagnose adult PAN (Table 16.2), while in 2010, the EULAR (European League Against Rheumatism), PRINTO (Pediatric Rheumatology International Trials Organization) and the PRES (Pediatric Rheumatology European Society) jointly published classification criteria for childhood PAN [10] (Table 16.3). These criteria have undergone validation, and in particular the EMA algorithm has been shown to perform satisfactorily in clinical practice [11].

Attempts have also been made to define the so-called cutaneous form of PAN (shorthand cPAN, although in other contexts, quite confusingly, the abbreviation cPAN is sometimes used to denote childhood PAN or classical PAN, respectively). There is a broad agreement that cPAN is characterized by lesions limited to the skin without internal organ involvement, but it is often accepted that there may be concomitant involvement of muscles and peripheral nerves, if such an involvement is mild and transient [12], or if it occurs in the same areas affected by skin lesions [13]. Formal criteria to classify cPAN have been proposed, but not validated [13].

Table 16.2 EMA algorithm to diagnose vasculitis including PAN [9]

Three criteria must be fulfilled to use this algorithm: (1) compatible clinical features, (2) at least one of the following: (a) histological evidence of vasculitis or of granuloma formation (within the wall of an artery or in the perivascular or extravascular area of an artery or arteriole), (b) positive ANCA, (c) eosinophilia (>10 % or >1.5 × 10^9), (d) specific investigations strongly suggestive of vasculitis and/or granuloma (e.g. EMG showing mononeuritis multiplex), (3) exclusion of other conditions.
If fulfills the ACR or Lanham criteria for Churg-Strauss syndrome (CSS), diagnose as CSS. If not, proceed further.
Diagnose as Wegener granulomatosis (WG) alias granulomatosis with polyangiitis if: fulfills the ACR criteria for WG, has histology consistent with WG (CHCC criteria), has histology consistent with CHCC MPA and WG surrogate markers, or if there is no histology but are positive WG surrogate markers plus anti-PR3 or anti-MPO, If not, proceed further.
Diagnose as MPA if clinical features and histology are consistent with small-vessel vasculitis and WG surrogate markers are negative, or if there is no histology, WG surrogate markers are negative but are positive surrogate markers for renal vasculitis and positive anti-PR3 or anti-MPO (includes renal limited vasculitis). If not, proceed further.
Diagnose as PAN if histology is compatible with CHCC PAN or there are typical angiographic features of PAN. If not, diagnose as "unclassifiable"

Table 16.3 EULAR/PRINTO/PRES classification criteria for childhood PAN [10]

Histopathology (necrotizing vasculitis in medium or small sized arteries) or angiographic abnormalities (aneurysm, stenosis or occlusion of medium or small sized arteries) (mandatory) plus one of the five following criteria:
Skin involvement
Myalgia/muscle tenderness
Hypertension
Peripheral neuropathy
Renal involvement

For further details, please refer to [10]

It is also recognized that some patients may present with isolated organ involvement due to vasculitis histologically indistinguishable from PAN, eg of the calf muscles [14] or of the testis [15]; such forms are sometimes referred to as "isolated PAN". Single-organ vasculitic involvement (eg of the gastrointestinal tract) has also been considered a limited form of PAN in the presence of the characteristic vascular changes of PAN (ie microaneurysms) on imaging [16]; however, in such cases the diagnosis of limited PAN requires exclusion of other disorders, because microaneurysms *per se* are not pathognomonic for PAN.

16.3 Epidemiology

PAN is a very rare disease. Its annual incidence in Europe is estimated to range between 0 and 1.6 cases per million, while the prevalence is about 31 cases per million [17, 18]. The aggressive campaign of vaccination against hepatitis B virus

(HBV), a known trigger of PAN, may have contributed to lower the frequency of PAN [19]. Before vaccination against HBV was implemented on a large scale, more than one-third of adults with PAN were infected by HBV, whereas currently only 5 % of European adults with PAN are infected by HBV [17].

PAN can affect individuals of any age, gender, and race, with a peak occurrence in the fifth to sixth decade of life (3996}.

16.4 Pathogenesis

PAN has often been regarded as an immune-complex-driven vasculitis, not least because of the association with active HBV infection, which was supposed to result in the formation of pathogenic circulating HBsAg-antibody complexes [20]. However, the absence of glomerulonephritis (a typical immune-complex-driven condition) and the lack of complement consumption appear to militate against this hypothesis. An alternative hypothesis, based on the demonstration of abundant CD4+ T cells in vascular inflammatory infiltrates, posits a role for a T cell-mediated immune response [21]. HBV is the most commonly identified trigger of PAN, but other microorganisms, such as cytomegalovirus, hepatitis C virus (HCV), human immunodeficiency virus (HIV) and Parvovirus B19 have occasionally been implicated as etiological agents of PAN [21].

16.5 Clinical Manifestations

PAN usually presents with constitutional manifestations as well as symptoms and signs related to the organs affected. Organ ischemia is thought to be due to vascular stenoses, while ruptured aneurysms can result in tissue hemorrhage [21]. The main clinical features of PAN are summarized in Table 16.4.

Table 16.4 Main clinical manifestations of PAN (% of patients) [22]

Constitutional manifestations	94 %
Myalgia	53 %
Arthralgia	52 %
Peripheral neuropathy	85 %
Mononeuritis multiplex	82 %
Skin nodules	6 %
Skin purpura	18 %
Livedo	11 %
Abdominal pain	50 %
Cardiomyopathy	13 %

16.6 Investigations

Laboratory tests usually reveal a systemic inflammatory response (raised inflammatory markers, chronic anemia, leukocytosis...) in PAN [22]. Anti-neutrophil cytoplasmic antibodies (ANCA) are typically negative and are helpful in differentiating PAN from MPA, where ANCA (usually with a p-ANCA pattern) are detectable in about 75 % of patients. HBV or, less frequently, other microorganisms may be associated with PAN.

Histology of active lesions characteristically shows a segmental necrotizing vasculitis with mixed inflammatory infiltrates (consisting in variable proportions of lymphocytes, macrophages, neutrophils and eosinophils) [23]. On the other hand, in chronic lesions, neoangiogenesis, fibrotic changes, and thrombosis may predominate [23]. Active and chronic lesions may at times coexist in the same specimen [23].

Imaging is a key part of the work-up of PAN with suspected internal organ involvement, and is sometimes used as a surrogate of biopsy to support the diagnosis when biopsy is unfeasible or negative. Specifically, angiography is still considered the gold standard to demonstrate vascular changes in patients with manifestations suggestive of abdominal, renal or cardiac involvement. The typical vascular lesions of PAN are small (1–5 mm in diameter) saccular or fusiform aneurysms, but stenoses are also common. The kidney, mesenteric and hepatic arteries are prevalently affected. However, these lesions are not exclusive to PAN and must be interpreted in the broader clinical context. Imaging can also be used for monitoring response to therapy; regression of vascular lesions has been reported after successful medical treatment [24].

16.7 Treatment

As a rule, the treatment of PAN should be tailored to its severity. In patients without poor prognostic factors (five-factors score [FFS] of 0) [25] (Table 16.5) glucocorticoids (GC) alone (usually prednisone 1 mg/kg/day tapering) may be used; however,

Table 16.5 Five-factors score [25]

Proteinuria >1 g/24 h
Serum creatinine > 140 μmol/L (1.58 mg/dl)
Cardiomyopathy, specific
GI involvement (specific): bleeding, perforation, closely spaced episodes of abdominal pain, and pancreatitis
CNS involvement, specific

Five-year mortality rate is: 12 % when the FFS = 0, when the FFS = 1 is 26 %, when the FFS ≥ 2 is 46 %

relapsing patients may require additional immunosuppressive therapy (azathioprine or cyclophosphamide) [26]. In contrast, patients with poor prognostic factors (FFS ≥1) should receive immunosuppressive therapy (usually cyclophosphamide) on top of GC [27]. Immunosuppressive therapy may also be warranted in patients with a FFS of 0, but with severe disease manifestations (eg severe peripheral neuropathy) not included in the FFS [28]. Whether biological agents may be of benefit in severe or relapsing PAN remains at the present hypothetical. Another controversial issue is whether agents less toxic than cyclophosphamide (such as methotrexate and azathioprine) might be effective in maintaining remission [29].

HBV-related PAN should be treated with antiviral agents [30]: among virological responders, PAN virtually never relapses [31]. A short course of glucocorticoids (GC) with or without plasma exchanges is also empirically advocated to curb inflammation and clear immune-complexes [29, 30, 32].

cPAN may not require aggressive treatment as it only exceptionally progresses to systemic ("classical") PAN [29]. Non-steroidal anti-inflammatory drugs and/or colchicine could suffice in mild cases, whereas a more aggressive approach may be required in relapsing cases [29] and probably in those with severe peripheral nerve involvement.

References

1. Kußmaul A, Maier R (1886) Über eine bisher nicht beschriebene eigenthümliche Arterienerkrankung (Periarteritis nodosa), die mit Morbus Brightii und rapid fortschreitender allgemeiner Muskellähmung einhergeht. Dtsch Arch Klin Med 1:484–516
2. Ferrari E (1903) Über polyarteritis acuta nodosa (sogenannte Periarteritis nodosa) und ihre Beziehungen zur Polymyositis and Polyneuritis acuta. Beitr Pathol Anat 34:350–386
3. Dickson WEC (1908) Polyarteritis acuta nodosa and periarteritis nodosa. J Pathol 12:31–47
4. Wainwright J, Davson J (1950) The renal appearances in the microscopic form of periarteritis nodosa. J Pathol 62:189–196
5. Wohlwill F (1923) Über die nur mikroskopisch erkennbare Form der Periarteritis nodosa. Virchows Arch 246:377–411
6. Lightfoot RW Jr, Michel BA, Bloch DA et al (1990) The American College of Rheumatology 1990 criteria for the classification of polyarteritis nodosa. Arthritis Rheum 33:1088–1093
7. Jennette JC, Falk RJ, Andrassy K et al (1994) Nomenclature of systemic vasculitides. Proposal of an international consensus conference. Arthritis Rheum 37:187–192
8. Jennette JC, Falk RJ, Bacon PA et al (2013) 2012 revised international Chapel Hill consensus conference nomenclature of vasculitides. Arthritis Rheum 65:1–11
9. Watts R, Lane S, Hanslik T et al (2007) Development and validation of a consensus methodology for the classification of the ANCA-associated vasculitides and polyarteritis nodosa for epidemiological studies. Ann Rheum Dis 66:222–227
10. Ozen S, Pistorio A, Iusan SM et al (2010) EULAR/PRINTO/PRES criteria for Henoch-Schonlein purpura, childhood polyarteritis nodosa, childhood Wegener granulomatosis and childhood Takayasu arteritis: Ankara 2008. Part II: final classification criteria. Ann Rheum Dis 69:798–806
11. Flores-Suarez LF, Contreras-Rodriguez FJ (2014) Critical appraisal of classification criteria for vasculitides. Curr Rheumatol Rep 16:422

12. Ishiguro N, Kawashima M (2010) Cutaneous polyarteritis nodosa: a report of 16 cases with clinical and histopathological analysis and a review of the published work. J Dermatol 37:85–93
13. Nakamura T, Kanazawa N, Ikeda T et al (2009) Cutaneous polyarteritis nodosa: revisiting its definition and diagnostic criteria. Arch Dermatol Res 301:117–121
14. Nakamura T, Tomoda K, Yamamura Y, Tsukano M, Honda I, Iyama K (2003) Polyarteritis nodosa limited to calf muscles: a case report and review of the literature. Clin Rheumatol 22:149–153
15. Dotan ZA, Laufer M, Heldenberg E et al (2003) Isolated testicular polyarteritis nodosa mimicking testicular neoplasm-long-term follow-up. Urology 62:352
16. Ebert EC, Hagspiel KD, Nagar M, Schlesinger N (2008) Gastrointestinal involvement in polyarteritis nodosa. Clin Gastroenterol Hepatol 6:960–966
17. Mahr A, Guillevin L, Poissonnet M, Ayme S (2004) Prevalences of polyarteritis nodosa, microscopic polyangiitis, Wegener's granulomatosis, and Churg-Strauss syndrome in a French urban multiethnic population in 2000: a capture-recapture estimate. Arthritis Rheum 51:92–99
18. Mohammad AJ, Jacobsson LT, Mahr AD, Sturfelt G, Segelmark M (2007) Prevalence of Wegener's granulomatosis, microscopic polyangiitis, polyarteritis nodosa and Churg-Strauss syndrome within a defined population in southern Sweden. Rheumatology (Oxford) 46:1329–1337
19. Pagnoux C, Cohen P, Guillevin L (2006) Vasculitides secondary to infections. Clin Exp Rheumatol 24:S71–S81
20. Pipitone N, Salvarani C (2008) The role of infectious agents in the pathogenesis of vasculitis. Best Pract Res Clin Rheumatol 22:897–911
21. Hernandez-Rodriguez J, Alba MA, Prieto-Gonzalez S, Cid MC (2014) Diagnosis and classification of polyarteritis nodosa. J Autoimmun 48–49:84–89
22. Pagnoux C, Seror R, Henegar C et al (2010) Clinical features and outcomes in 348 patients with polyarteritis nodosa: a systematic retrospective study of patients diagnosed between 1963 and 2005 and entered into the French Vasculitis Study Group Database. Arthritis Rheum 62:616–626
23. Bonsib SM (2001) Polyarteritis nodosa. Semin Diagn Pathol 18:14–23
24. Pagnoux C, Mahr A, Cohen P, Guillevin L (2005) Presentation and outcome of gastrointestinal involvement in systemic necrotizing vasculitides: analysis of 62 patients with polyarteritis nodosa, microscopic polyangiitis, Wegener granulomatosis, Churg-Strauss syndrome, or rheumatoid arthritis-associated vasculitis. Medicine (Baltimore) 84:115–128
25. Guillevin L, Lhote F, Gayraud M et al (1996) Prognostic factors in polyarteritis nodosa and Churg-Strauss syndrome. A prospective study in 342 patients. Medicine (Baltimore) 75:17–28
26. Ribi C, Cohen P, Pagnoux C et al (2010) Treatment of polyarteritis nodosa and microscopic polyangiitis without poor-prognosis factors: a prospective randomized study of one hundred twenty-four patients. Arthritis Rheum 62:1186–1197
27. Guillevin L, Cohen P, Mahr A et al (2003) Treatment of polyarteritis nodosa and microscopic polyangiitis with poor prognosis factors: a prospective trial comparing glucocorticoids and six or twelve cyclophosphamide pulses in sixty-five patients. Arthritis Rheum 49:93–100
28. Langford CA (2001) Treatment of polyarteritis nodosa, microscopic polyangiitis, and Churg-Strauss syndrome: where do we stand? Arthritis Rheum 44:508–512
29. de MM, Mahr A (2011) Treating polyarteritis nodosa: current state of the art. Clin Exp Rheumatol 29:S110–S116
30. Guillevin L, Mahr A, Cohen P et al (2004) Short-term corticosteroids then lamivudine and plasma exchanges to treat hepatitis B virus-related polyarteritis nodosa. Arthritis Rheum 51:482–487
31. Guillevin L, Mahr A, Callard P et al (2005) Hepatitis B virus-associated polyarteritis nodosa: clinical characteristics, outcome, and impact of treatment in 115 patients. Medicine (Baltimore) 84:313–322
32. Segelmark M, Selga D (2007) The challenge of managing patients with polyarteritis nodosa. Curr Opin Rheumatol 19:33–38

Chapter 17
Anti-Glomerular Basement Membrane Disease

Michele Rossini, Annamaria Di Palma, Vito Racanelli, Franco Dammacco, and Loreto Gesualdo

Abstract Anti-glomerular basement membrane disease is a severe autoimmune disease caused by autoantibodies against epitopes on alpha 3 chains of type IV collagen, found in the alveolar and glomerular basement membrane. It presents with renal failure often requiring dialysis and/or pulmonary symptoms due to alveolar hemorrhage that may be life threatening when massive. Diagnosis rely on recognition of anti-GBM antibodies at tissue level (often kidneys) and/or in the blood. Treatment is based on plasma exchange in order to remove autoantibodies from circulation and on immunosuppressive agents in order to reduce antibodies production.

17.1 Introduction

Anti-glomerular basement membrane (GBM) disease is a rare, life threatening, severe autoimmune disease caused by autoantibodies that react with the non-collagenous domain 1 (NC1) of alpha 3 chain of type IV collagen (Goodpasture's antigen) [1, 2]. In the 2012 revised international Chapel Hill consensus conference nomenclature of vasculitides, anti-GBM disease is included in the small vessel vasculitides and is defined as a "vasculitis affecting glomerular capillaries, pulmonary capillaries, or both with GBM deposition of anti-GBM autoantibodies" [3]. The antigen is found in the basement membrane of lung (alveoli) and kidney (glomeruli). It is also expressed in the inner ear, eye, and choroid plexus. Patients affected by this condition often present with pulmonary symptoms, renal failure, or both. When the disease affects both lung and kidneys, the term Goodpasture's syndrome is used since E. Goodasture was the first, in 1919, to report a case of a young man

M. Rossini • A. Di Palma • L. Gesualdo (✉)
Nephrology, Dialysis and Transplantation Unit, Department of Emergency and Organ Transplantation, University of Bari "A. Moro", Policlinico, Bari, Italy
e-mail: loreto.gesualdo@uniba.it

V. Racanelli • F. Dammacco
Department of Biomedical Sciences and Human Oncology, Section of Internal Medicine, University of Bari Medical School, Policlinico, 70124 Bari, Italy

F. Dammacco et al. (eds.), *Systemic Vasculitides: Current Status and Perspectives*, DOI 10.1007/978-3-319-40136-2_17

who died after developing hemoptysis and renal failure with the autopsy findings of pulmonary hemorrhage and glomerulonephritis (GN) with features of crescentic necrotizing GN [4]. However, since pulmonary-renal syndrome can recognize different etiologies, the eponym should be used only when this is caused by anti-GBM antibodies. The incidence of the disease ranges from 0.5 to 1 pmp per year [5] and genetic association with HLA- DRB*1501 and 1502 has been reported [6, 7].

17.2 Pathogenesis

The pathogenicity of anti-GBM antibodies is well documented in animal experimental models [8-10] and in humans as well. Sera from patients with anti-GBM disease cause glomerulonehritis when injected in animals [11]. On the other hand, patients with Alport syndrome can experience an anti-GBM glomerulonephritis when receive a kidney transplant [12, 13]. Indeed, these patients lack normal type IV collagen since mutations in alpha 3, alpha 4 or alpha 5 chains of type IV collagen found in Alport patients prevent normal collagen type IV assembly. Therefore the normal type IV collagen expressed in the transplanted kidney is regarded as a "not self" antigen by the immune system of the recipients and can trigger an immune response with anti-GBM production that can cause glomeulonephritis.

However, epitopes of the Goodpasture's antigen are cryptic suggesting that other synergistic conditions that modify the basement membrane structures are necessary for the development of the disease. Exposure to cigarette smoking, inhaled cocaine, metal dusts, organic solvents, infections might damage alveolar capillary basement membrane resulting in the exposure of cryptic epitopes that are pathogenic [14]. Similarly, damage in glomerular basement membrane could result in the exposure of pathogenic epitopes in kidney. Indeed an association between shock wave lithotripsy and anti-GBM have been described [15, 16]. Presence of pre-existing antibodies against GBM and afterwards development of disease has been documented [17]. In a study on sera obtained from 30 patients prior to the onset of disease from the Department of Defense Serum Repository and compared with 30 matched healthy controls, low detectable anti-GBM antibody levels in a single serum sample before diagnosis were found more frequently in cases than controls (70 % versus 17 %, $p < 0.001$). Only patients had detectable anti-GBM levels in multiple samples before diagnosis (50 % versus 0 %, $P < 0.001$).

A genetic background appears to play a role as predisposing factor for anti-GBM disease. This is supported by the strong association between certain alleles of HLA DR and DQ antigens and the disease [18].

The deposited anti-GBM antibodies are predominantly IgG1, complement fixing and activate complement mainly by the classical pathway triggering a neutrophil dependent inflammation. There is, however, emerging evidence that also T-cells might play a role in the pathogenesis of the disease. It has been shown in animal models that immunization with T-cell Goodpasture's antigen epitope can cause an active glomerulonephritis in rats [19].

17.3 Clinical Presentation

Renal involvement is usually characterized by rapidly progressive renal failure with oligoanuria often requiring hemodialysis [20]. Oligoanuria at the time of presentation has been shown the strongest predictor of patient survival, followed by older age, in a recently published case series [20]. Macroscopic hematuria is also frequent. Microscopic hematuria is present in almost all cases with a small amount of proteinuria. Nephrotic range proteinuria is rare. Pulmonary involvement is present in the 60 % of cases with alveolar hemorrhage. Pulmonary symptoms include hemoptysis, cough, hypoxia, and dyspnea. Chest X-ray can usually show diffuse infiltrates. When pulmonary massive hemorrhage occurs, respiratory failure can lead to death in a short period of time. Lung and kidney involvement are not necessarily present together or at the same time. Other symptoms include anemia, fatigue, weight loss and may be related to hemorrhage, renal failure, and inflammation.

17.4 Diagnosis

The renal biopsy shows a picture of crescentic necrotizing glomerulonephritis. The glomerular involvement is usually diffuse involving frequently more than 80 % of glomeruli sampled (Fig. 17.1). Direct immunofluorescence testing using anti-IgG sera shows a linear IgG staining along glomerular and, sometimes, tubular basement membranes (Fig. 17.2). Electron microscopy of renal tissues sample does not show deposits. Renal biopsy plays a crucial role not only in the diagnosis, given that the percentage of glomeruli affected predict renal response to treatment strategies.

Anti-GBM antibodies can be detected in sera of patients with different methodologies. Indirect immunofluorescence can be used to detect anti-GBM in patient's

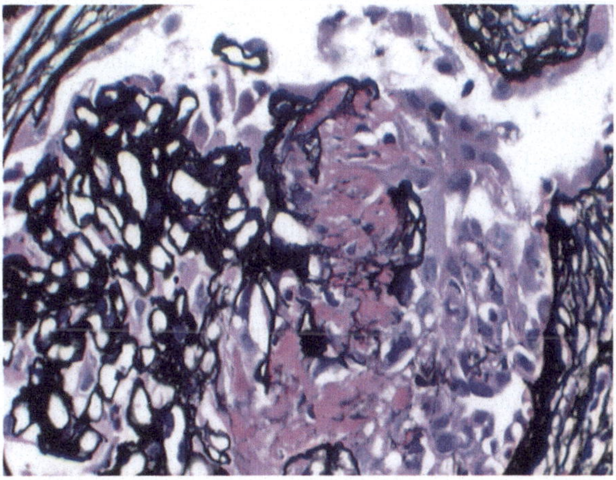

Fig. 17.1 A glomerulus showing fibrinoid necrosis and a cellular crescent (Jones' silver stain x400)

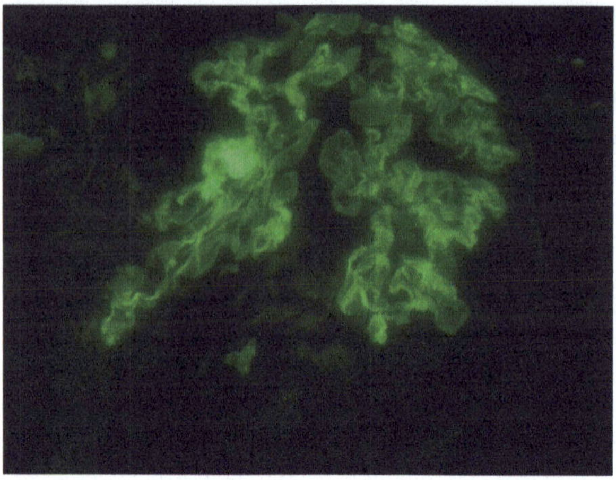

Fig. 17.2 Strong linear IgG staining along glomerular basement membrane (anti-IgG x400)

serum utilizing sections of normal kidney as substrate. However, there are several ELISA kits commercially available that are widely used. Anti-GBM should always be tested along with c/p ANCA and anti-PR3 and ANTI-MPO antibodies since there is a 20–30 % of patients who might also be ANCA-positive.

17.5 Treatment

Given the aggressive nature of the disease, an adequate treatment has to be established as soon as possible.

Anti-GBM disease is treated with a variable combination of plasma exchange, high doses of corticosteroids and immunosuppressive drugs [21–23].

Plasmapheresis is able to remove from circulation anti-GBM antibodies. In the only controlled clinical trial available [21], 17 patients were divided into two groups of 9 patients receiving prednisone and cyclophosphamide alone and 8 patients treated with prednisone, cyclophosphamide and plasma exchange. In the first group 6 of the 9 patients remained on dialysis while only 2 of the 8 patients treated with plasmapheresis were dialysis dependent after treatment. According to the American Society For Apheresis (ASFA) guidelines, all patients with diffuse alveolar hemorrhage and those that are not dialysis dependent should receive plasma exchange for at least 14 days or until antibodies become undetectable [22]. Volume treated in every session should be 1–1.5 the total plasma volume, with a daily or every other day frequency. Replacement fluid can be albumin or plasma, especially if hemorrhage is present or invasive diagnostic procedure (e.g. a renal biopsy) are planned. With the introduction of plasma exchange an improvement in mortality and renal survival rates compared to historical series has been observed.

While plasma exchange assures fast removal of antibodies from blood, immunossupressive agents reduce antibody formation. Steroids are given initially intravenously at high doses (usually methylprednisolone 1,000 mg/day for 3 consecutive days followed by 1 mg/kg/day up to 60–80 mg/day). Cyclophosphamide is given orally at doses of 2 mg/kg/day or intravenously with pulses of 500–1,000 mg every 2–3 weeks. In most of the reports with long follow-up period after an induction phase of 2–3 weeks plasmapheresis and 3 months treatment with steroids and cyclophosphamide, steroids alone were administered for 6–9 months [23].

Several case reports show that anti-CD20 therapy with rituximab can be successfully used in patients with anti-GBM disease [24]. This could represent a therapeutic option in patients who develop side effects for or that are ineligible to cyclophosphamide.

The degree of renal injury and the levels of serum creatinine at the time of treatment are best predictors of outcome [23]. In a study of 71 patients treated with corticosteroids, cyclophosphamide and plasma exchange, patients who presented with a creatinine concentration less than 5.7 mg/dL had 100 % patient survival and 95 % renal survival at 1 year. Patients presenting with a creatinine concentration higher than 5.7 mg/dL not requiring immediate dialysis, had patient and renal survival of 83 % and 82 % at 1 year respectively. In patients who presented with dialysis-dependent renal failure patient and renal survival were 65 % and 8 % at 1 year.

After remission of the first acute episode, recurrence is uncommon but might happen even many years later [25]. Anti-GBM disease can recur after renal transplant although the recurrence is uncommon if the transplant is performed after the disappearance of antibodies from circulation [26].

References

1. Kalluri R, Wilson CB, Weber M, Gunwar S, Chonko AM, Neilson EG, Hudson BG (1995) Identification of the alpha 3 chain of type IV collagen as the common autoantigen in antibasement membrane disease and Goodpasture syndrome. J Am Soc Nephrol 6:1178–1185
2. Dammacco F, Battaglia S, Gesualdo L, Racanelli V (2013) Goodpasture's disease: a report of ten cases and a review of the literature. Autoimmun Rev 12:1101–1108
3. Jennette JC, Falk RJ, Bacon PA, Basu N, Cid MC, Ferrario F et al (2013) 2012 revised international Chapel Hill consensus conference nomenclature of vasculitides. Arthritis Rheum 65:1–11
4. Goodpasture WE (1919) The significance of certain pulmonary lesions in relation to the etiology of influenza. Am J Med Sci 158:863–870
5. Tang W, McDonald SP, Hawley CM et al (2013) Anti-glomerular basement membrane antibody disease is an uncommon cause of end-stage renal disease. Kidney Int 83:503–510
6. Zhou XJ, Lv JC, Zhao MH, Zhang H (2010) Advances in the genetics of anti-glomerular basement membrane disease. Am J Nephrol 32:482–490
7. Ooi JD, Chang J, O'Sullivan KM, Pedchenko V, Hudson BG, Vanderbark AA (2013) The HLA-DRB1*15:01–Restricted Goodpasture's T Cell Epitope Induces GN. J Am Soc Nephrol 24:419–431
8. Germuth FG Jr, Choi IJ, Taylor JJ, Rodriguez E (1972) Antibasement membrane disease. I. The glomerular lesions of Goodpasture's disease and experimental disease in sheep. Johns Hopkins Med 131:367–384

9. Steblay RW (1962) Glomerulonephritis induced in sheep by injections of heterologous glomerular basement membrane and Freund's complete adjuvant. J Exp Med 116:253–272
10. Ohnuki T (1975) Crescentic glomeulonephritis induced in the goat by immunization with homologous or heterologous glomerular basement membrane antigen. Acta Pathol Jpn 25:319–331
11. Lerner RA, Glassock RJ, Dixon FJ (1999) The role of anti-glomerular basement membrane antibody in the pathogenesis of human glomerulonephritis. J Am Soc Nephrol 10:1389–1404
12. Shah B, Fisrt MR, Mendoza NC, Clyne DH (1988) Alport's syndrome: risk of glomeulonephritis induced by anti glomerular basement membrane antibody after renal transplantation. Nephron 50:34–38
13. Oliver TB, Glouldesbrough DR, Swainson CP (1991) Acute crescentic glomerulonephritis associated with antiglomerular basement membrane antibody in Alport's syndrome after second transplantation. Nephrol Dial Transplant 6:893–895
14. Kobayashi S, Inokuma S (2009) Intrapulmonary hemorrhage in collagen-vascular diseases includes a spectrum of underlying conditions. Intern Med 48:891–897
15. Iwamoto I, Yonekawa S, Takeda T et al (1998) Anti- glomerular basement membrane nephrithis after extracorporeal shock wave lithotripsy. Am J Nephrol 18:534–537
16. Xenocostas A, Jothy S, Collins B, Loertscher R, Levy M (1999) Anti- glomerular basement membrane glomerulo nephrithis after extracorporeal shock wave lithotripsy. Am J Kidney Dis 33:128–132
17. Olson SW, Arbogast CB, Baker TP et al (2011) Asymptomatic autoantibodies associate with future anti-glomerular basement membrane disease. J Am Soc Nephrol 22:1946–1952
18. Huey B, McCormick K, Capper J et al (1993) Associations of HLA-DR and HLA-DQ types with anti-GBM nephritis by sequence-specific oligonucleotide probe hybridization. Kidney Int 44:307–312
19. Bolton WK, Chen L, Hellmark T, Wieslander J, Fox JW (2005) Epitope spreading and autoimmune glomerulonephritis in rats induced by a T cell epitope of Goodpasture's antigen. J Am Soc Nephrol 16:2657–2666
20. Alchi B, Griffiths M, Sivalingam M, Jayne D, Farrington K (2015) Predictors of renal and patient outcomes in anti-GBM disease: clinicopathologic analysis of a two-centre cohort. Nephrol Dial Transplant 30:814–821
21. Johnson JP, Moore J Jr, Austin HA 3rd, Balow JE, Antonovych TT, Wilson CB (1985) Therapy of anti-glomerular basement membrane antibody disease: analysis of prognostic significance of clinical, pathologic and treatment factors. Medicine (Baltimore) 64:219–227
22. Schwartz J, Winters JL, Padmanabhan A, Balogun RA, Delaney M, Linenberger ML et al (2013) Guidelines on the use of therapeutic apheresis in clinical practice-evidence-based approach from the Writing Committee of the American Society for Apheresis: the sixth special issue. J Clin Apher 28:145–284
23. Levy JB, Turner AN, Rees AJ, Pusey CD (2001) Long-term outcome of anti-glomerular basement membrane antibody disease treated with plasma exchange and immunosuppression. Ann Intern Med 134:1033–1042
24. Syeda UA, Singer NG, Magrey M (2013) Anti-glomerular basement membrane antibody disease treated with rituximab: a case-based review. Semin Arthritis Rheum 42:567–572
25. Klasa RJ, Abboud RT, Ballon HS, Grossman L (1988) Goodpasture's syndrome: recurrence after a five-year remission. Case report and review of the literature. Am J Med 84:751–755
26. Almkuist RD, Buckalew VM Jr, Hirszel P, Maher JF, James PM, Wilson CB (1981) Recurrence of anti-glomerular basement membrane antibody mediated glomerulonephrithis in as isograft. Clin Immunol Immunopathol 18:54

Chapter 18
IgA Vasculitis

Roberta Fenoglio and Dario Roccatello

Abstract Henoch-Schonlein purpura (HSP), recently re-named IgA vasculitis, is a systemic small vessel vasculitis with immunoglobulin (Ig) A1-dominant immune deposits. The disease is characterized by a tetrad of clinical manifestations: palpable purpuric rash, arthralgia/arthritis, gastrointestinal symptoms (i.e., abdominal pain, gastrointestinal bleeding), but any organ may be involved. It is considered a form of immune complex-mediated vasculitis, but the etiopathogenesis is not yet completely understood.

The goals of treating HSP are typically to ameliorate acute symptoms, mitigate short-term morbidity and prevent chronic renal insufficiency. Corticosteroids and immunosuppressive agents have been postulated to be effective, but their role remains controversial. The efficacy of Rituximab in adult with IgA vasculitis has been reported in few cases.

Keywords IgA vasculitis • Leukocytoclastic vasculitis • Henoch-Schoenlein purpura nephritis • Henoch-Schoenlein treatment • Rituximab

18.1 Introduction and Epidemiology

Henoch-Schonlein purpura (HSP), also called IgA vasculitis (IgAV) is a systemic small vessel vasculitis with immunoglobulin (Ig) A1-dominant immune deposits. The disease is characterized by a tetrad of clinical manifestations: palpable purpuric rash, arthralgia/arthritis, gastrointestinal symptoms (i.e., abdominal pain, gastrointestinal bleeding) and renal disease. However, any organ may be involved as a result of systemic leukocytoclastic vasculitis.

Unlike many other forms of vasculitis, IgAV is self-limited in most cases (generally < 5 weeks). HSP primarily affects children; 90 % of patients are under 10 years of age. The annual incidence of HSP in children is estimated to be 15–20

R. Fenoglio • D. Roccatello (✉)
Center of Immunopathology and Rare Diseases and Nephrology and Dialysis Unit, San G. Bosco Hospital and University of Turin, Turin, Italy
e-mail: dario.roccatello@unito.it

F. Dammacco et al. (eds.), *Systemic Vasculitides: Current Status and Perspectives*, DOI 10.1007/978-3-319-40136-2_18

cases/100,000/year, while the disease is less common in adults and has an incidence of 1.3 cases/100,000/year. There is a male predominance (2 M:1 F). Previous epidemiological studies have found striking seasonal variations in HSP, with a prevalence of cases in the autumn and winter.

18.2 Etiopathogenesis

Although a variety of chemical and infectious triggers are well-known, the detailed pathogenetic mechanisms of IgAV have not been completely defined. The most common scenario comprises an abnormal inflammatory process deriving from immune reactions to various antigenic stimuli, which may be bacterial, viral, or caused by parasitic agents in a genetically prone individual. Then, a peculiar immune complex deposition in the vascular walls and overproduction of various proinflammatory molecules elicit clinical signs, which may be differentiated according to either a specific trigger or a specific genetic make-up. Genetic, immunologic and environmental factors all seem to play a role [1].

Acquired factors appear to be determinant, and several lines of evidence support that HSP may have an infectious origin. No consistent causative agent has been identified, and IgAV has been linked to a wide array of pathogens, including bacteria, viruses and parasites. Therefore, there may not be one specific microorganism triggering HSP, and it may not simply be limited to an agent penetrating the body through the respiratory or gastrointestinal mucosal surfaces [2]. IgAV may also be related to exposure to pharmacological agents (antibiotics, chemotherapeutics) or vaccine antigens.

Genetic Susceptibility Several predisposing or protective susceptibility loci within and outside the human leukocyte antigen (HLA) region have been identified, thus supporting the role played by genetics in IgAV. HLA B35 has been associated with an increased risk of nephritis in patients with IgAV. HLA DRB1*01, DRB1*11 and HLA DRB1*14 also seem to increase susceptibility to glomerulonephritis. HLA-DRB1*07 was found to have a protective effect in some populations. Whether HLA polymorphisms are also involved in the severity of the disease remains controversial. Recent studies have focused on polymorphisms of genes encoding proinflammatory adhesion molecules associated with endothelial cell activation, i.e., TNFα, IL1β, IL8, TGFβ and VEGF [3]. Patients with HSP Nephritis (HSPN) show higher blood and urinary leukotriene B_4 (LTB_4) levels and lower lipoxin A_4 (LXA_4) levels as compared to patients without HSPN, which is consistent with this association [4]. Indeed, both recruitment of neutrophils and chemotaxis are activated by LTB_4 and are inhibited by LXA_4. Some other studies established a link between IgAV and mutations in the familial Mediterranean fever (MEFV) gene.

Abnormalities in IgA_1 Glycosylation An increased plasma IgA level alone is not enough to produce mesangial IgA deposits. Patients with IgAV must produce circulating IgA molecules with special characteristics. Over 15 years ago an excess of poorly galactosylated IgA1 was found to be present both in the serum and in the glomerular immune deposits of patients with IgA-Nephropathy (IgAN) and HSPN. Galactose-deficient IgA1 (Gd-IgA1) is currently believed to play a pivotal role in the pathogenesis of both IgA nephropathy and Henoch–Schönlein purpura nephritis.

IgA1 contains a 17 amino-acid hinge region which undergoes co/post translational modification by the addition of up to six O-glycan chains. These chains comprise N-acetylgalactosamine (GAlNAc) in O-linkage with either serine or threonine residues. Galactose may be β1,3-linked to Ga1NAc by the enzyme C1GalT1. Both the galactose residue and the GalNAc may be sialylated [5]. Studies on immortalized circulating B cells from patients with IgAN have suggested decreased β1,3-galactosyltransferase enzyme activity that could result in impaired galactosylation of the core GalNaC residues [6] and an aberrantly exposed GalNac moiety. Both genetic abnormalities in enzyme structure or function, and acquired perturbations due to microenvironmental influences on $IgA1_1$ maturation might contribute to the defective enzymatic glycosylation of GalNac. An alternative explanation for the apparent excess of undergalactosylated IgA1in the circulation of IgAN patients could be that, unlike the response to systemic antigen challenge, the immune response to mucosal antigens is characterized by mucosally derived, relatively galactose-deficient IgA1[7].

Underglycosylation makes the IgA1 molecule susceptible to autoaggregation, thus forming molecular species whose properties resemble "immune complexes" in the circulation. Galactose-deficient IgA1 is also the target of an "autoimmune" response carried out both by IgA and IgG antibodies, thus forming galactose-deficient IgA1/IgA1 and galactose-deficient IgA1/IgG immune complexes [8]. An alternative concept has been proposed which states the increase in plasma levels of galactose deficient IgA1 is not the result of an aberrant B cell production of IgA1 but rather, is due to an aberrant distribution. Galactose-deficient IgA1 has a tendency to bind to a variety of glycoproteins including constituents of normal glomeruli thus leading to "planted" antigens which promote *in situ* immune complex formation. Moreover, desialysation enables IgA to activate complement through a C4-independent, alternative complement pathway, thus promoting deposition and assembly of the membrane attack complex.

Removal of immune complexes from circulation is delayed in HSP. Hepatic Ashwell Morell receptors, originally named asialoglycoprotein receptors (ASGPRs), represent a major pathway of IgA catabolism. Undergalactosylated IgA1 could make the normally expressed ASGPRs unable to remove IgA1 aggregates or IgA1 immune complexes. Aberrantly glycosylated IgA1 has also been thought to form immune complexes that are too large to enter the space of Disse and reach the ASGPRs [9]. Other removal pathways include macrophage Fc alfa receptors (FcαR). FcαR expression on phagocytes is decreased in IgAN-patients, possibly due to continuous receptor occupation resulting in receptor down-regulation or shedding [10].

In adults, drugs, toxins, vaccinations, insect bites and food allergies may be the primary predisposing factors for HSP. Infection may trigger autoimmune responses favoring the appearance of circulating IgA1 immune complexes [8]. A mechanism of molecular mimicry has also been postulated [11] The conformation of GalNac, which is exposed in galactose-deficient IgA1, is similar to some bacterial and viral epitopes. This reaction could be broadened through intramolecular and intermolecular epitope "spreading" where the primary response against the dominant initiating epitope extends to other epitopes within the same molecule or among different molecules. Currently, one particular area of interest is the role of Toll-like receptors (TLRs) in driving IgA synthesis and perhaps even modifying glycosyltransferase activity. Expression of TLR 4, whose activation is associated with bacterial lipopolysaccharides, is enhanced in IgAN.

These data suggest that abnormal IgA1 aggregates and IgA1 immune complexes can escape the physiologic removal systems. Immune complex deposition occurs through a combination of mesangial trapping and increased affinity of poorly galactosylated IgA1 to extracellular matrix components including fibronectin and type IV collagen. Subsequent events include complement activation via the alternative and mannose binding lectin pathways and release cytokines that lead to glomerular hypercellularity, matrix production, podocyte injury and scarring.

18.3 Diagnostic Criteria of HSP and Clinical Manifestations

There are no disease-specific laboratory abnormalities; the diagnosis of IgA vasculitis is currently based on symptoms, signs and histopathological findings. IgAV was first described in 1837 by Schonlein who defined the disease as a triad of purpuric rash, arthritis and abnormalities of the urinary sediment. Later (1874), Henoch described the association of purpuric rash, abdominal pain with bloody diarrhea, and proteinuria. More recently, several sets of diagnostic criteria for IgAV have been proposed (Table 18.1). The diagnostic criteria have high sensitivity/specificity and allow a diagnosis to be made in most patients. However, making a diagnosis can be difficult in the presence of atypical presentation [12].

In adults the clinical features do not completely overlap those of children (Table 18.2). At symptom onset, adults have a lower frequency of abdominal pain and fever, and a higher frequency of joint symptoms. During the clinical course, one of the main differences is an increased risk for developing significant renal involvement and end-stage kidney disease. In a previous retrospective study, a cohort of 250 adults suffering from HSP was analyzed (follow-up 14.8 years). Eleven percent of patients reached end-stage kidney disease, 13 % exhibited severe renal failure and 14 % showed moderate renal failure. Clinical remission was achieved in only 20 % of subjects. Renal function impairment and proteinuria levels at presentation, a significant degree of interstitial fibrosis, the presence of glomeruli with fibrinoid necrosis and a high percentage of sclerotic glomeruli were all associated with a poor renal prognosis [13].

Table 18.1 Classification criteria for HSP diagnosis

Classification	Diagnostic criteria
ACR 1990	≥2 of the following:
	(1) Palpable purpura, not thrombocytopenic
	(2) Bowel angina
	(3) Wall granulocytes on biopsy
	(4) Age ≤20 years at disease onset
Michel et al. 1992	≥ 3 of the following: HSP; ≤2 of the following HV:
	(1) Palpable purpura, not thrombocytopenic
	(2) Bowel angina
	(3) Gastrointestinal bleeding
	(4) Hematuria
	(5) Age ≤20 years at disease onset
	(6) No history of medication intake at disease onset
CHCC 1994	Vasculitis, with IgA-dominant immune deposits, affecting small vessels (i.e., capillaries, venules, or arterioles); typically involves skin, gut, and glomeruli and is associated with arthralgias or arthritis
Helander et al. 1995	Palpable purpura not thrombocytopenic with LCV + ≥ 3 of the following:
	(1) IgA vascular deposition
	(2) Age ≤20 years at disease onset
	(3) Gastrointestinal involvement
	(4) Upper respiratory tract infection prodrome
	(5) Mesangioproliferative glomerulonephritis with or without IgA deposition
EULAR/PRINTO/PRES (2010)	Palpable purpura not thrombocytopenic/petechiae (mandatory) +1 of the following:
	(1) Diffuse abdominal pain
	(2) Histopathology: typical LCV with predominant IgA deposits or proliferative glomerulonrphritis with predominant IgA deposits
	(3) Arthritis or Arthralgias
	(4) Renal involvement (Pto: >0.3 g/24 h or >30 mmol/mg of urine albumin to creatinine ratio on a spot morning sample; and/or hematuria, red blood cell casts: >5 red cells per high power field or ≥2 + on dipstick or red blood cell casts in the urinary sediment)

The diagnosis is straightforward when there is a classical presentation, especially palpable purpura of the lower extremities, but a skin biopsy is often needed in order to make a certain diagnosis. Maculopapular, urticarial or vescicular lesions may precede or occur together with palpable purpura. The characteristic rash which appears in nearly all patients is the expression of leukocytoclastic vasculitis. Skin biopsy shows fibrinoid necrosis of the wall vessels and perivascular accumulation of inflammatory infiltrate consisting predominantly of neutrophils and eosinophils

Table 18.2 Main clinical manifestations in children and adults

	Children (1)	Adults (2–3–4)
Purpura	100 %	96 %
Arthritis	75 %	61 %
Abdominal pain	65 %	65 %
Gastrointestinal bleeding	35 %	
Glomerulonephritis	40 %	40 %
Microhematuria	40 %	85 %
Macrohematuria	10 %	10 %
Proteinuria	25 %	–
Nephrotic proteinuria	–	28 %
CKD	–	32 %
End stage kidney disease	1 %	11 %

surrounding the capillaries and the postcapillary venules of the dermis (Fig. 18.1). Direct immunofluorescence showing dermal granular IgA deposition in the wall of the superficial dermis arterioles is pathognomonic of IgAV [14]. The biopsy should be performed within 24 h of the onset of the rash because in chronic lesions, vessel damage leads to nonspecific leakage of all isotypes of immunoglobulin.

Gastrointestinal involvement is a typical feature of IgAV and occurs in about 48 % of patients [15]. The most common presentation is colicky pain or bleeding. Leukocytoclastic vasculitis accompanying IgA deposits (even in the digestive tract) has been reported in HSP.

Renal impairment, which is the most severe complication, ranges from microscopic hematuria to nephrotic syndrome. The most frequently observed pathology is mesangial or endocapillary proliferative glomerulonephritis. The classification of pathologic glomerular changes in HSPN is based on endocapillary and extracapillary inflammation of the glomerulus. Crescents are present in more than 50 % of patients. Predominant IgA deposits are observed in the mesangium of all glomeruli in HSPN (Fig. 18.2). Capillary wall staining for IgA is more frequently found in HSPN and may even predominate on mesangial IgA (Fig. 18.2).

Unusual presentations include pulmonary involvement (alveolar hemorrhage, interstitial pneumonia or fibrosis) and central and/or peripheral nervous system involvement.

18.4 Laboratory Test Biomarkers for HSP Diagnosis

There are currently no existing diagnostic laboratory tests for HSP. Serum IgA levels are reportedly elevated in 50–70 % of patients with IgAV. Findings on routine blood tests are non specific and often reflect the triggering condition. Patients may have normochromic anemia because of gastrointestinal bleeding. Serum creatinine

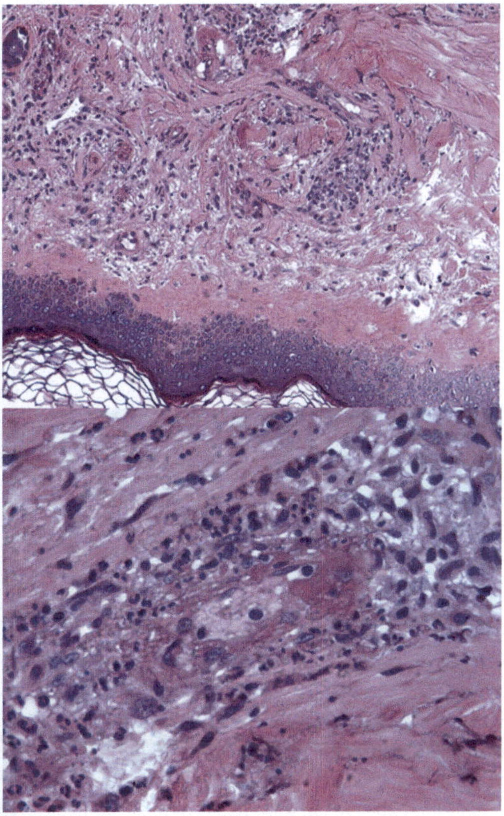

Fig. 18.1 Skin biopsy. Superficial dermal vessels showing inflammatory infiltrate consisting predominantly of neutrophils (*upper panel*). Destruction of the vessel wall and dissemination of *nuclear debris* from leukocytes which confers the *leukocytoclastic* characteristics to the vasculitis (*lower panel*)

should be evaluated in adults because of the increased risk of renal disease. Although some proinflammatory cytokines and chemokines are elevated in the acute phase [16], these laboratory abnormalities are not specific for IgAV. Urinalysis should be carried out in all patients because the results reflect the degree of renal involvement; the findings include evidence of red-white cell cellular casts and proteinuria. Urinary screening is recommended for 3–6 months after the onset (renal involvement decreases significantly after 3 months).

18.5 Therapy

The goals of treating HSP are typically to ameliorate acute symptoms, mitigate short-term morbidity and prevent chronic renal insufficiency. However, some patients receive only supportive therapy. Since IgAV is characterized by leukocyte infiltration of the blood vessel walls along with immunoglobulin A deposition (resulting in vascular injury and necrosis), and since corticosteroids inhibit inflammatory processes, early treatment with corticosteroids has been postulated to be

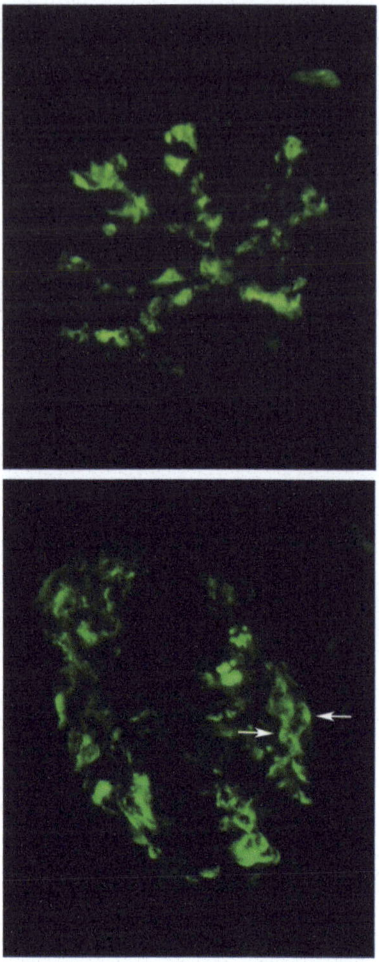

Fig. 18.2 Kidney biopsy. Deposition of IgA immunoglobulin in the mesangium (*upper panel*) and within mesangium and (*lower panel*) along capillary membrane (original magnification 400×)

effective for all three therapeutic goals, but much controversy still remains. There is suggestive evidence that glucocorticoids reduce the severity and enhance the rate of resolution of extrarenal symptoms, most notably arthritis and abdominal pain and swelling. Though this therapy provides symptomatic relief, there would appear to be no long-term benefits in using steroids in terms of shortening the overall length of the illness, or in reducing recurrences and progression of HSN or preventing renal involvement. Immunosuppressive agents (including Azathioprine, Cyclophosphamide, Cyclosporine, Mycophenolate) have been used in combination with corticosteroids with variable results.

Specific treatment of HSP nephritis should be considered in patients with marked proteinuria and/or impaired renal function during the acute event. Although nephritis is the most serious long-term complication of IgAV, few data are available to determine the best treatment. There are not robust evidence from controlled trials that therapy with conventional doses of glucocorticoids has any beneficial effects in patients with renal involvement [17]. Limited data suggest that cyclosporine may be beneficial in patients with HSP and severe proteinuria [18]. Plasmapheresis, alone or in conjunction with immunosuppressive agents, has also been used in patients with severe, usually crescentic, renal involvement. The data on the beneficial effects of mycophenolate mofetil and rituximab are encouraging, but limited. We also observed promising clinical results in a monocentric experience on the use of Rituximab in 5 adult IgA-vasculitis with biopsy-proven nephritis. The patients achieved a complete remission of nephritis and syndromic manifestations, without clinically relevant adverse reactions (paper submitted).

References

1. Rigante D, Castellazzi L, Bosco A et al (2013) Is there a crossroad between infections, genetics and Shonlein-Henoch Purpura? Autoimmun Rev 12:1016
2. Piram M, Mahr A (2013) Epidemiology of immunoglobulin A vasculitis (Henoch-Schonlein): current state of knowledge. Curr Opin Rheumatol 25:171–178
3. Saulsbury FT (2010) Shonlein-Henoch purpura. Curr Opin Rheumatol 22:598–602
4. WU SH, Liao PY, Yin PY et al (2009) Inverse temporal changes of lipoxin A_4, and leukotrienes in children with Shonlein-Henoch Purpura. Prostaglandins Leukot Essent Fatty Acids 80:177–183
5. Boyd JK, Cheung CK, Molyneux K et al (2012) An update on the pathogenesisis and treatment of IgA nephropathy. Kidney Int 81(9):833–843
6. Suzuki H, Moldoveanu Z, Hall S et al (2008) IgA1 secreting cell lines from patients with IgA nephropathy produce aberrantly glycosylated IgA1. Clin Invest 118:629–639
7. Barratt J, Eitner F, Feehally J et al (2009) Immune complex formation in IgA nephropathy: a case of the right antibodies in the wrong place at the wrong time. Nephrol Dial Transplant 24:3620–3623
8. Tomana M, Novak J, Julian BA et al (1999) Circulating immune complexes in IgA nephropathy consist of IgA1 with galactose deficient hinge region and antiglycan antibodies. J Clin Invest 10(4):73–81
9. Mestecky J, Hashim OH, Tomana M et al (1995) Alterations in the IgA carbohydrate chains influence the cellular distribution of IgA_1. Contrib Nephrol 111:66–71
10. Grossetete B, Launay P, Leheun A et al (1998) Downregulation of Fc α receptors on blood cells of IgA nephropathy patients: evidence for a negative regulatory role of serum IgA. Kidney Int 53:1321–1335
11. Glassock RJ (2011) The pathogenesis of IgA nephropathy. Curr Opin Nephrol Hypertens 20:153–160
12. Yang Y-H et al (2014) The diagnosis and classification of Henoch-Shonlein purpura: an update review. Autoimmuty Rev 13:355–358
13. Pillebout E, Thervet E, Hill G et al (2002) Henoch-Shonlein Purpura in adults: outcome and prognostic factors. J Am Soc Nephrol 13(5):1271–1278
14. Pillebout E, Verine J (2014) Henoch-Shonlein Purpura in the adult. Rev Med Interne 35(6):372–381
15. Jithpratuck W, Elshenawy Y, Saleh H et al (2011) The clinical implications of adult-onset Henoch-Shonlein Purpura. Clin Mol 9(1):9
16. Yang YH, Chuang Y, Wang LC et al (2008) The immunobiology of Henoch-Shonlein Purpura. Autoimmun Rev 7:179–184

17. Dudley J, Smith G, Llewellyn-Edwards A et al (2007) Randomized placebo controlled trial to asses the role of early prednisone on the development and progression of Henoch-Shonlein purpura nephrithis. Pediatr Nephrol 22:1457
18. Park JM, Won SC, Shin JI et al (2011) Cyclosporin A therapy for Henoch-Schonlein nephritis with nephrotic range proteinuria. Pediatr Nephrol 26:411

Chapter 19
Systemic Vasculitis and Pregnancy

Maria Grazia Lazzaroni, Micaela Fredi, Sonia Zatti, Andrea Lojacono, and Angela Tincani

Abstract Systemic Vasculitides (SV) are rare diseases and, unlike other rheumatic diseases, do not preferentially affect women during their childbearing age (with the exception of Takayasu arteritis and Behcet's disease), so that pregnancy in these patients is still an uncommon event. In the last decades, the improvement in diagnosis and management of SV determined a reduction in mortality and morbidity of these patients; the consequent improvement in quality of life includes for young women the possibility to carry out one or more pregnancies. Some general guidelines for the management of patients with vasculitis during pregnancy can be suggested: (1) pregnancy in such patients must be followed by a multispecialistic team; (2) pregnancies should be planned when the disease is in remission; (3) flares of maternal disease during pregnancy should be promptly and aggressively treated, because active disease can cause pregnancy complications; (4) pregnancies complicated by the onset of vasculitis have a particularly severe prognosis, therefore efficient management and close clinical surveillance are indicated; (5) a revision of therapy must be performed for all female patients that ask for a pregnancy.

Systemic Vasculitides (SV) are rare diseases and, unlike other rheumatic diseases, do not preferentially affect women during their childbearing age (with the exception of Takayasu arteritis and Behcet's disease), so that pregnancy in these patients is still an uncommon event. In the last decades, the improvement in diagnosis and management of SV determined a reduction in mortality and morbidity of these patients; the consequent improvement in quality of life could include for young women the possibility to carry out one or more pregnancies. Gestation in women with vasculitis requires a constant and careful management by Rheumatologists/

M.G. Lazzaroni • A. Tincani (✉)
Rheumathology and Clinical Immunology, University and Spedali Civili of Brescia, Brescia, Italy
e-mail: tincani@med.unibs.it

M. Fredi
Rheumathology and Clinical immunology, Spedali Civili of Brescia, Brescia, Italy

S. Zatti • A. Lojacono
Obstetric and Gynecology Unit, University and Spedali Civili of Brescia, Brescia, Italy

F. Dammacco et al. (eds.), *Systemic Vasculitides: Current Status and Perspectives*, DOI 10.1007/978-3-319-40136-2_19

Internists and Obstetricians, Neonatologists and also other medical specialties (i.e. nephrology) when indicated [1] to increase the possibilities of success. In fact, pregnancy-related modifications (immunological, cardiovascular and thrombophilic) can affect the course of the disease and the pregnancy outcome.

Preconception counseling is crucial for any patient with connective tissue disease in order to estimate the chance of both fetal and maternal problems, related to either disease activity or organ involvement, as contraindication to pregnancy. In particular, special attention should be given to cardiac or renal impairment, arterial hypertension, refractory asthma and tracheal stenosis, which can affect the course of the pregnancy and/or delivery [2]. Careful consideration must also be given to other common epidemiological risk factors that are relevant to any pregnancy as age, tobacco use, low or excess weight, arterial hypertension, diabetes mellitus, previous obstetric history.

When talking about her productive plans, any patient will ask questions concerning their health and offsprings' health. Therefore, the following issues must be discussed with patients who desire a pregnancy:

- Disease activity during pregnancy and post-partum period and the possible effects on fetal outcome;
- Therapies compatible or not with pregnancy, and which therapy should be used during pregnancy if needed.

Patients with a recent onset of vasculitis, with uncontrolled active disease or with irreversible damage should be discouraged on conception.

In fact there is a general agreement that a pregnancy in patients with SV, as well as for the other systemic autoimmune diseases, should be planned only when the maternal disease is in sustained remission (at least 6 months) or minimal disease activity.

One of the most critical issue in the management of a pregnancy is to choose or modify the medication in order to treat the mother without harming the fetus. In fact, women who plan a pregnancy often stop taking medications out of fear of harming the fetus. However, the withdrawal of effective therapy may result in a flare of maternal disease. Therefore adjustment or change of medication must be discussed with any patient that asks for pregnancy.

Most of the medications regularly used to treat vasculitis are contraindicated during pregnancy, with the exception of corticosteroids, antimalarial drugs, cyclosporine, azathioprine and intravenous immunoglobulins [3]. Other drugs commonly used, such as Cyclophosphamide (CYC), Methotrexate or Mycophenolate Mofetile are contraindicated because mutagenic and/or embriotoxic and must be suspended 3–6 months before conception.

Special attention must be given to patients currently or previously treated with CYC: this drug is associated with high risk of inducing infertility or subfertility, a risk that is related to the cumulative dose received and to the age of the patient [4, 5].

Protection of ovarian function may be provided by treatment with analogues of gonadotropin-releasing hormone as leuprolide acetate for depot suspension; in this way menstrual cycles could be blocked and CYC diffusion into ovaries limited, preserving ovarian function [6].

Concerning biological agents, data on the use of TNF blockers and Rituximab during pregnancy are limited and the majority come from single reports or limited case series; however it is reassuring to note that the number of reported successful pregnancies with such treatment is increasing [7].

As we explained above, SV women who start a pregnancy in a period of stable remission or minimal disease activity are less likely to flare during gestation. However, there is still the risk of a new flare, that can be treated according to the severity of the manifestations. Patients who experienced a mild reactivation can be treated with low-medium dose oral steroids. On the other side, in case of severe or life-treathening manifestations more potent immunosuppressants may be carefully used, as pulse intravenous steroids, intravenous immunoglobulins, azathioprine, cyclosporine, and in the second and third trimester also intravenous CYC [1, 8]. Intravenous corticosteroids are fluorinate compounds that reach the fetus and are capable of potential severe side effects as sudden hypertension and potential placental ischemia.

19.1 Large-Vessel Vasculitis

19.1.1 Takayasu Arteritis

Takayasu arteritis (TA) is a chronic, inflammatory, SV that predominantly involves large vessels, in particular the aorta and its main branches. Inflammation can cause wall thickening with stenosis or occlusion of the lumen, while the impairment of the media can determine aneurysm formation.

This type of vasculitis preferentially affects young women in their child-bearing age, therefore is not uncommon to observe a pregnancy in a woman diagnosed with TA. In fact the literature about pregnancy and TA accounts for a large number of cases in the field of pregnancies in SV, reaching now 388 pregnancies [2, 9-17].

According to the current literature, the most frequent obstetrical complications are related with hypertension, including new onset or worsening of pre-existent hypertension and/or pre-eclampsia. In a systematic review [2] and in the largest single-center series [12] the rate of these complications was around 45 % of pregnant women with TA.

Consequently, other frequent complications observed are intrauterine growth restriction (IUGR) and neonatal low birth weight, that have been reported with a frequency that varies from 0 to 20 % [2, 9, 10, 12, 16]. The higher rate of low birth weight is consistent with the increase of preterm delivery, that has been observed in variable rates from 8 to 50 % [9, 10, 12, 16], and was estimated about 23 % in a systematic review [2].

Notably some authors [10, 18, 19] have described a relationship between these complications and the involvement of renal arteries. Some Authors have suggested that an angioplastic surgery of renal arteries stenosis before pregnancy should be considered to improve the subsequent pregnancy outcome [20].

Disease flare was described as very infrequent (<5 %) in the review of by Gatto et al. [2], but in all the latest series published the rate of flare was higher, ranging from 10 to 40 % [9-12, 16]. Notably, in the study with the highest rate of flare (40 %, [12]) a large number of patients had an active disease at the beginning of pregnancy or were diagnosed with TA during pregnancy.

Given that a flare of TA does not occur in the majority of patients, it is now common opinion that cardiovascular complications are not only related to an active vessel inflammation, but instead to the hemodynamic changes related to pregnancy, which determine an overload on a vascular system that has already been damaged by the disease [10, 12].

Therefore the main issue in the management of TA during pregnancy is the treatment of hypertension. This goal should be obtained through the use of calcium channel blockers, alfa-metildopa or hydralazine and avoiding fetotoxic agents such as ACE inhibitors. On the other hand, the risk of hypoperfusion due to excessive hypotension in regions vascularized by stenotic vessels, should be considered and avoided [2, 21]. The management of hypertension should also include the delivery and the post-partum period. In fact during delivery, epidural analgesia is suggested to avoid the fluctuations in blood pressure that physiologically accompany the vaginal delivery [22]. Cesarean section should instead be limited to the cases of obstetric complications. The post-partum period should also be monitored (24–48 h after delivery) because of hemodynamical changes that could determine aortic dissection [23].

19.2 Medium-Vessel Vasculitis

19.2.1 Polyarteritis Nodosa

Polyarteritis Nodosa (PAN) is a disorder characterized by necrotizing inflammation of small to medium-sized arteries.

In patients with PAN prevalent features are constitutional symptoms, cutaneous, musculoskeletal, gastrointestinal and neurological involvement, especially peripheral neuropathy with mononeuritis multiplex.

There is a well-known association between active hepatitis B virus (HBV) and vasculitis, in fact cohort studies have found an association in one-third of PAN cases, thus all patients with PAN should be tested for HBV [24].

PAN is a vasculitis with a slight male predominance and with a late peak incidence (in the sixth decade of life); therefore there are few case reported of pregnancy; actually twenty-three cases of pregnancy are reported in these patients [2, 9].

In general, patients who conceive during disease remission seem to have a favorable outcome with uneventful pregnancies, rare disease relapse and, in the majority of cases, delivery of healthy babies. Preterm delivery with low birth weight babies are the most common obstetrical complications, occurring in more than 50 % of pregnancies [2, 25].

On the other side, women with an onset of PAN during pregnancy had a poor maternal outcome. High rates of maternal mortality have been reported, 7 out of 9 de novo PAN died [2] during pregnancy or puerperium, with the death due to aneurysm rupture, uncontrolled hypertension and renal failure. The other two pregnancies both ended prematurely due to persistent active disease, with severe preterm delivery [9, 25]. It has to be underlined, however, that all the cases with maternal death occurred in the '80, while the 2 cases with favorable outcome happened in the last 10 years [9, 25].

19.3 Small Vessel Vasculitis

19.3.1 EGPA

Eosinophilic Granulomatosis with Polyangiitis (EGPA) is a multi-system disorder characterized by small vessel necrotizing vasculitis, extravascular granulomas and peripheral and tissue hypereosinophilia, occurring in patients with asthma and allergic rhinitis. Cardiac, gastrointestinal, and central nervous system involvement are linked with poor prognosis [26].

There are approximately 50 reports in the literature of pregnancy in women with EGPA [2, 9, 11, 15]. The vast majority of EGPA pregnancies resulted in uncomplicated pregnancy with good maternal and neonatal outcome. The majority of cases reported occurred in women with EGPA prior to pregnancy and inactive disease at the time of conception; EGPA flare during pregnancy seems to occur in around 20 % of women in remission at the time of conception. Skin rash and worsening of asthma are the most common flare during pregnancy, even if also more severe disease manifestations such as mononeuritis multiplex and cardiac involvement with cardiac failure have been reported [25, 27]. There are two cases of maternal death both related to cardiac and respiratory involvement [28, 29].

Less than ten cases of disease onset during pregnancy or postpartum are reported, with a very poor prognosis and the severity is strictly related with the patients' outcome, especially if there was a cardiac involvement [27, 28, 30].

Few cases of fetal loss have been reported [2], while many studies reported high rates of preterm birth, with a range around 20–25 %, frequently associated with low birth weight [11].

Rates of cesarean section were very different, with the highest prevalence (70 %) recently reported by a multicentric Italian study [9].

19.4 MPA

Microscopic Polyangiitis (MPA) is a systemic, pauci-immune, necrotizing, small-vessel vasculitis. MPA was initially recognized as a subset of PAN from which it may be difficult to distinguish, therefore it is also possible that in many series reported in the literature MPA patients had been grouped with PAN.

Seven pregnancies in patients with MPA are available in the literature [31-36] and more recently, an internet self administered survey of patients with vasculitis and self-enrolled in the Vasculitis Clinical Research Consortium (VCRC) reported other 6 pregnancies in 2 patients [15].

Excluding the patients reported in the paper of Clowse at al (not available data concerning disease activity for each patient), MPA relapsed in two out of five cases with disease onset before pregnancy [31, 35] with worsening in respiratory function, constitutional symptoms and serological activity. In the two onsets during pregnancy, one patient died owing to pulmonary infection [32] while the other survived [34].

In two pregnancies that [31, 35] ended with a live birth, the authors described the occurrence of symptoms suggestive for MPA in the newborns, which could potentially be due to the transfer of maternal MPO antibodies through the placenta. One newborn presented skin purpuric rash that spontaneously disappeared few days later. The other baby developed renal involvement and pulmonary hemorrhage and the infant was successfully treated with corticosteroids and supportive care. Neither fetal losses nor medical terminations of pregnancy were reported.

19.5 GPA

Granulomatosis with Polyangiitis (GPA) is a rare small-vessel, necrotizing vasculitis which commonly affects the kidneys and the upper and lower respiratory tract. The onset of the disease is often in the forth to sixth decade of life, thus pregnancies are uncommonly observed. The limited number of publications concerning pregnancies in these patients report nearly 90 cases [2, 9, 11, 36, 37]. Onset of GPA during gestation occurred in 15 cases, and in the majority of the patients disease was considered inactive at conception.

Women with persistent activity or those with onset during gestation are at high risk of adverse outcome because of disease complications and maternal and/or fetal death. Poor fetal outcome, with either fetal loss or elective termination of pregnancy [38, 39] maternal death [40] and severe relapses have been reported, in fact the rate of flare in patients with active disease approached 100 % [2, 9].

Cesarean section and preterm delivery are the most frequent obstetrical complications. Obviously the rate of both events are increased in patients not in remission (nearly 40 % of in both cases) [2, 25].

In women who conceived in remission, a flare has been reported in 10–40 % of pregnancies [2, 25, 36, 37]; it can occur at any time during gestation and also during post-partum [9]. The most frequent manifestations during re-activations consist of constitutional symptoms, respiratory exacerbation with nasal inflammation and/or subglottic stenosis, arthralgias or arthritis and renal involvement. Worsening of the renal function is challenging, because it could be due to vasculitis flares or pre-eclampsia. Analysis of the urinary sediment and arterial hypertension could be useful. A very recent publication reported the case of systemic relapses during the second trimester with a progressive worsening of renal function, and the renal involvement was confirmed by renal biopsy [36].

19.6 Variable Vessels Vasculitis

19.6.1 Behçet Disease

Behçet disease (BD) is a multisystem inflammatory chronic disorder, characterized by recurrent oral or genital ulcerations; cutaneous, ocular, neurological and gastrointestinal involvement are also possible. Other important manifestations are thrombosis, that can occur in vessels of all sizes and both in venous and arterial district, even if venous involvement is more common. The thrombotic process is mediated by active vascular inflammation, which determines the activation of coagulation.

BD is more prevalent in young adults aged 20–40 years, with no difference in male to female ratio in patients from endemic areas, which comprise the areas located on the Ancient Silk Road. Reports from USA and Northern Europe instead, reveal an increased frequency of BD in women [41]. As a consequence, there are quite large series in literature about BD during pregnancy; there are 396 pregnancies in BD published so far [2, 9, 11, 15, 42, 43].

The majority of the studies report a rate of obstetrical complications similar to that of general healthy population [2, 42, 43]. One large case-control study [44] found an increased frequency of miscarriage, hypertension, gestational diabetes, premature delivery and cesarean section in BD patients compared to matched healthy women (26 % vs. 2 %). Another recent series also confirmed an increased rate of gestational diabetes and hypertension in BD women [9]. In terms of risk factors for obstetric complications, Noel et al. [43] demonstrated an association between previous venous involvement and miscarriage or cesarean delivery, as previously suggested by Jadaon [44]. Furthermore, one study examined the placental hystopathologic alterations from two BD women (with both positive and negative outcome) and identified necrotizing villitis and decidual vasculitis, that are similar to those found in other tissues involved. The authors supposed that this could be responsible for an increased rate of miscarriage observed in BD patients and could also be the demonstration of a possible intrauterine transmission of the disease [45]. In fact rare cases of neonatal BD are reported, all born from mothers with active disease during pregnancy and most of these babies with transient, mild clinical manifestations (oral ulcers or skin lesions).

Disease flares are reported in 30–36 % of pregnancies in a systematic review [2] and in a large recent series [43], but there are also reports of lower (8 %, [42]) and higher rates (45 %, [9]). In general, the most frequent manifestations during the reactivations of disease are oral or genital ulcers.

Another important issue is thrombosis: pregnancy and the post-partum increase the risk of thrombotic complications in these patients, that are per se at higher risk. There are several reports of vascular events during pregnancies [44, 46-48], including superior vena cava thrombosis [49], cerebral venous thrombosis [50] and transient ischemic attacks [9].

The management of these patients includes the use of corticosteroids at low dose, azathioprine and also colchicine, that has been demonstrated to be safe during pregnancy [51, 52]. Besides this, an effective prophylaxis of thrombosis during pregnancy and the post-partum period should be considered, using low dose aspirin or low molecular weight heparin. In the case of a previous thrombotic event, an anticoagulation therapy should be undertaken.

Table 19.1 shows a synthesis of the most frequent complications during pregnancy in each type of vasculitis.

Table 19.1 Systemic Vasculitis and Pregnancy: description of obstetrical complications and risk of disease flares, according to different types of vasculitis

Type of vasculitis	Number of cases reported (references)	Main obstetrical complications and pregnancy outcomes	Risk of vasculitis flares
Takayasu Arteritis (TA)	350 [2, 9-16]	Hypertensive complications (about 45 %)	Variable rates of flare described: <5 % in a systematic review; from 10 to 40 % in latest series.
		IUGR (0–20 %)	The highest rate (40 %) recorded in a group with a large number of patients with active disease/onset during pregnancy
		Low birth weight (8–50 %; 23 % in a systematic review)	
Panarteritis nodosa	23 [2, 9]	Preterm delivery (>50 %)	Rare if disease in remission
			Poor maternal outcome with high risk of maternal death if onset during pregnancy.
Eosinophilic granulomatosis with polyangiitis (EGPA)	51 [2, 9, 11, 15]	Preterm delivery (20–25 %)	Around 20 % of women in remission at the time of conception (most frequent manifestations are skin rash and asthma)
		Cesarean section (10–70 %)	
		Few cases of fetal loss (<10 %)	Rare case of maternal death.

(continued)

Table 19.1 (continued)

Type of vasculitis	Number of cases reported (references)	Main obstetrical complications and pregnancy outcomes	Risk of vasculitis flares
Microscopic Polyangiitis (MPA)	13 [15, 30-35]	Two cases reported of occurrence of symptoms suggestive for MPA in the newborns	Poor outcome if onset during pregnancy (one case of maternal death)
			Relapse in 40 % of established cases
		No cases of fetal losses nor medical terminations of pregnancy.	
Granulomatosis with Polyangiitis (GPA)	90 [2, 9, 11, 35, 36]	High risk of adverse outcome in patients with active disease at conception or onset during pregnancy:	Flare rate of 10–40 % with disease in remission;
			High rate in patients with active disease (nearly 100 %)
		cesarean section 40 %	
		preterm delivery 40 %	
Behçet's disease	396 [2, 9, 11, 15, 41, 42]	Rates comparable to general obstetric population in the majority of the studies; higher frequency of gestational diabetes and hypertension in two studies	Variable rates from 8 to 45 %, ranging around 30–36 % in the two largest studies
			The most frequent manifestations during re-activations are oral or genital ulcers
			A higher risk of thrombosis during pregnancy and the puerperium is described.

References

1. Østensen M, Andreoli L, Brucato A, Cetin I, Chambers C, Clowse ME et al (2015) State of the art: reproduction and pregnancy in rheumatic diseases. Autoimmun Rev 14(5):376–386
2. Gatto M, Iaccarino L, Canova M, Zen M, Nalotto L, Ramonda R et al (2012) Pregnancy and vasculitis: a systematic review of the literature. Autoimmun Rev 11(6):A447–A459
3. Østensen M, Förger F (2013) How safe are anti-rheumatic drugs during pregnancy? Curr Opin Pharmacol 13(3):470–475
4. Clowse ME, Copland SC, Hsieh TC et al (2011) Ovarian reserve is diminished by oral cyclophosphamide therapy for granulomatosis with polyangiitis (Wegener's). Arthritis Care Res 63(12):1777–1781
5. Ioannidis JP, Katsifis GE, Tzioufas AG et al (2002) Predictors of sustained amenorrhea from pulsed intravenous cyclophosphamide in premenopausal women with systemic lupus erythematosus. J Rheumatol 29:2129–2135

6. Blumenfeld Z, Mischari O, Schultz N, Boulman N, Balbir-Gurman A (2011) Gonadotropin releasing hormone agonists may minimize cyclophosphamide associated gonadotoxicity in SLE and autoimmune diseases. Semin Arthritis Rheum 41(3):346–352
7. Bazzani C, Scrivo R, Andreoli L, Baldissera E, Biggioggero M, Canti V (2015) Prospectively-followed pregnancies in patients with inflammatory arthritis taking biological drugs: an Italian multicentre study. Clin Exp Rheumatol 33(5):688–693
8. Le Thi Huong D, Amoura Z, Duhaut P, Sbai A, Costedoat N, Wechsler B et al (2002) Risk of ovarian failure and fertility after intravenous cyclophosphamide. A study in 84 patients. J Rheumatol 29(12):2571–2576
9. Fredi M, Lazzaroni MG, Tani C, Ramoni V, Gerosa M, Inverardi F et al (2015) Systemic vasculitis and pregnancy: a multicenter study on maternal and neonatal outcome of 65 prospectively followed pregnancies. Autoimmun Rev 14(8):686–691
10. Tanaka H, Tanaka K, Kamiya C, Iwanaga N, Yoshimatsu J (2014) Analysis of pregnancies in women with Takayasu arteritis: complication of Takayasu arteritis involving obstetric or cardiovascular events. J Obstet Gynaecol Res 40(9):2031–2036
11. Sangle SR, Vounotrypidis P, Briley A, Nel L, Lutalo PMK, Sanchez-Fernandez S, Chaib A et al (2015) Pregnancy outcome in patients with systemic vasculitis: a single-centre matched case-control study. Rheumatology (Oxford) 54(9):1582–1586
12. Comarmond C, Mirault T, Biard L, Nizard J, Lambert M, Wechsler B et al (2015) Takayasu's arteritis and pregnancy. Arthritis Rheumatol 67:3262–3269
13. Nalini S, Santa SA (2015) Takayasu arteritis with bilateral renal artery stenosis and left subclavian artery stenosis in pregnancy. J Clin Diagn Res 9(9):7–8
14. Chourasia U, Modi JN, Jain M, Biswas R (2015) Pulseless and ambulatory pregnant woman: an obstetric and medical challenge. J Obstet Gynecol India 65(2):125–128
15. Clowse ME, Richeson RL, Pieper C, Merkel PA (2013) Pregnancy outcomes among patients with vasculitis. Arthritis Care Res 65(8):1370–1374
16. Alpay-Kanitez N, Omma A, Erer B, Artim-Esen B, Gül A, Inanç M, Öcal L, Kamali S (2015) Favourable pregnancy outcome in Takayasu arteritis: a single-centre experience. Clin Exp Rheumatol 33(2 Suppl 89):S-7-10
17. Assad APL, da Silva TF, Bonfa E, Pereira RMR (2015) Maternal and neonatal outcomes in 89 patients with Takayasu Arteritis (TA): comparison before and after the TA diagnosis. J Rheumatol 42(10):1861–1864
18. Suri V, Aggarwal N, Keepanasseril A, Chopra S, Vijayvergiya R, Jain S (2010) Pregnancy and Takayasu arteritis: a single centre experience from North India. J Obstet Gynaecol Res 36(3):519–524
19. Chaudhry MA, Latif F (2013) Takayasu's arteritis and its role in causing renal artery stenosis. Am J Med Sci 346(4):314–318
20. Manika A, Ananya D (2010) Pregnancy in Takayasu's arteritis with bilateral renal artery and abdominal aorta stents. Aust N Z J Obstet Gynaecol 50(2):203–204
21. Shafi NA, Malik A, Silverman DI (2009) Management of Takayasu arteritis during pregnancy. J Clin Hypertens 11(7):383–385
22. Ioscovich A, Gislason R, Fadeev A, Grisaru-Granovsky S, Halpern S (2008) Peripartum anesthetic management of patients with Takayasu's arteritis: case series and review. Int J Obstet Anesth 17(4):358–364
23. Lakhi NA, Johannes J (2010) Takayasu's arteritis in pregnancy complicated by peripartum aortic dissection. Arch Gynecol Obstet 282(1):103–106
24. Pagnoux C, Seror R, Henegar C et al (2010) Clinical features and outcomes in 348 patients with polyarteritis nodosa: a systematic retrospective study of patients diagnosed between 1963 and 2005 and entered into the French Vasculitis Study Group Database. Arthritis Rheum 62(2):616
25. Pagnoux C, Le Guern V, Goffinet F, Diot E, Limal N, Pannier E et al (2011) Pregnancies in systemic necrotizing vasculitides: report on 12 women and their 20 pregnancies. Rheumatology 50(5):953–961

26. Comarmond C, Pagnoux C, Khellaf M, Cordier JF, Hamidou M, Viallard JF et al (2013) Eosinophilic granulomatosis with polyangiitis (Churg–Strauss): clinical characteristics and long-term followup of the 383 patients enrolled in the French Vasculitis Study Group cohort. Arthritis Rheum 65(1):270–281
27. Corradi D, Maestri R, Facchetti F (2009) Postpartum Churg–Strauss syndrome with severe cardiac involvement: description of a case and review of the literature. Clin Rheumatol 28(6):739–743
28. Connolly JO, Lanham JG, Partrige MR (1994) Fulminant pregnancy related Churg–Strauss syndrome. Br J Rheumatol 33(8):776–777
29. Abul-Haj SK, Flanagan P (1961) Asthma associated with disseminated necrotizing granulomas vasculitis, the Churg Strauss syndrome. Report of a case. Med Ann Dist Columbia 30:670–676
30. Edwards MH, Curtis EM, Ledingham JM (2015) Postpartum onset and subsequent relapse of eosinophilic granulomatosis with polyangiitis. BMJ Case Rep 2015:pii
31. Bansal PJ, Tobin MC (2004) Neonatal microscopic polyangiitis secondary to transfer of maternal myeloperoxidase antibody resulting in neonatal pulmonary haemorrhage and renal involvement. Ann Allergy Asthma Immunol 93:398–401
32. Silva F, Specks U, Sethi S et al (2009) Successful pregnancy and delivery of a healthy newborn despite transplacental transfer of antimyeloperoxidase antibodies from a mother with microscopic polyangiitis. Am J Kidney Dis 54:542–545
33. Cetinkaya R, Odabas AR, Gursan N et al (2002) Microscopic polyangiitis in a pregnant woman. South Med J 95(12):1441–1443
34. Milne KL, Stanley KP, Temple RC et al (2004) Microscopic polyangiitis: first report of a case with onset during pregnancy. Nephrol Dial Transplant 19(1):234–237
35. Morton MR (1998) Hypersensitivity vasculitis (microscopic polyangiitis) in pregnancy with transmission to the neonate. Br J Obstet Gynaecol 105:928–930
36. Croft AP, Smith SW, Carr S, Youssouf S, Salama AD, Burns A et al (2015) Successful outcome of pregnancy in patients with anti-neutrophil cytoplasm antibody–associated small vessel vasculitis. Kidney Int 87(4):807–811
37. Tuin J, Sanders JSF, de Joode AAE et al (2012) Pregnancy in women diagnosed with Antineutrophil cytoplasmic antibody-associated vasculitis: outcome for the mother and the child. Arthritis Care Res 64(4):539–545
38. Kumar A, Mohan A, Gupta R et al (1998) Relapse of Wegener's granulomatosis in the first trimester of pregnancy: a case report. Br J Rheumatol 37(3):331–333
39. M'Rad S, Moalla M, Ben MK et al (1989) Wegener's granulomatosis in pregnancy. One case report. Rev Med Interne 10(1):69–72
40. Pauzner R, Mayan H, Hershko E et al (1994) Exacerbation of Wegener's granulomatosis during pregnancy: report of a case with tracheal stenosis and literature review. J Rheumatol 21:1153–1156
41. Sakane T, Takeno M, Suzuki N, Inaba G (1999) Behçet's disease. N Engl J Med 341:1284
42. Iskender C, Yasar O, Kaymak O, Yaman ST, Uygur D, Danisman N (2014) Behçet's disease and pregnancy: a retrospective analysis of course of disease and pregnancy outcome. J Obstet Gynaecol Res 40(6):1598–1602
43. Noel N, Wechsler B, Nizard J, Costedoat-Chalumeau N, Boutin DLTH, Dommergues M et al (2013) Behcet's disease and pregnancy. Arthritis Rheum 65(9):2450–2456
44. Jadaon J, Shushan A, Ezra Y, Sela HY, Ozcan C, Rojansky N (2005) Behçet's disease and pregnancy. Acta Obstet Gynecol Scand 84(10):939–944
45. Hwang I, Lee CK, Yoo B, Lee I (2009) Necrotizing villitis and decidual vasculitis in the placentas of mothers with Behçet disease. Hum Pathol 40(1):135–138
46. Komaba H, Takeda Y, Fukagawa M (2007) Extensive deep vein thrombosis in a postpartum woman with Behcet's disease associated with nephrotic syndrome. Kidney Int 71(1):6
47. Hiwarkar P, Stasi R, Sutherland G, Shannon M (2010) Deep vein and intracardiac thrombosis during the post-partum period in Behçet's disease. Int J Hematol 91(4):679–686

48. Marsal S, Falga C, Simeon CP, Vilardell M, Bosch JA (1997) Behçet's disease and pregnancy relationship study. Rheumatology 36(2):234–238
49. Kale A, Akyildiz L, Akdeniz N, Kale E (2006) Pregnancy complicated by superior vena cava thrombosis and pulmonary embolism in a patient with Behcet disease and the use of heparin for treatment. Saudi Med J 27(1):95–97
50. Wechsler B, Généreau T, Biousse V, Vauthier-Brouzes D, Seebacher J, Dormont D, Godeau P (1995) Pregnancy complicated by cerebral venous thrombosis in Behçet's disease. Am J Obstet Gynecol 173(5):1627–1629
51. Diav-Citrin O, Shechtman S, Schwartz V, Avgil-Tsadok M, Finkel-Pekarsky V, Wajnberg R et al (2010) Pregnancy outcome after in utero exposure to colchicine. Am J Obstet Gynecol 203(2):144.e1–144.e6
52. Nabil H, Zayed A, State O, Badawy A (2012) Pregnancy outcome in women with familial Mediterranean fever. J Obstet Gynaecol 32(8):756–759

Chapter 20
Behçet Disease

Rosaria Talarico and Stefano Bombardieri

Abstract Behçet's disease (BD) is globally characterized by a variable spectrum of disease profile: while prevalent muco-cutaneous lesions and arthritis represent the only clinical features in patients with a benign disease subset, there are other patients who develop potentially sight or life-threatening manifestations, due to ocular, neurological or major vascular involvement. Beside the organ involvement, demographic factors could considerably influence the long-term and short-term outcomes of BD: age at disease onset, duration of disease, gender and sex. Younger men patients are more suitable to have a more severe disease, due to an increased frequency both of morbidity and mortality, related to ocular, vascular and neurological involvement. Eye involvement represents one of the most serious manifestations of BD and occurs in half of the patients. It seems more frequent and more severe among young males and, unluckily, it still remains one of the most significant causes of morbidity. Usually, ocular disease develops within the first years of disease onset and it seems to be more severe in this period; moreover, growing data suggest that the prognosis of BD patients with ocular involvement is mainly dependent upon the severity of visual acuity at presentation. Although not included in the International Study Group criteria for BD, neurological involvement represents the second main cause of mortality, preceded by large vessel disease. Parenchymal CNS involvement represents a serious morbidity of disease, often leading to disability and, if not treated early, to mortality, while dural sinus thrombosis is associated with a more favourable outcome than parenchymal involvement. The prevalence of vascular involvement varies from 20 to 35 % of BD patients and it may involve all types of vessels within the arterial and venous system. It is characterized by a clear male preponderance. Vascular involvement in BD represents a serious risk for multiple vessel-related complications, including thromboses, stenoses, occlusions, and aneurysms. Since there are no established laboratory findings to define BD and the diagnosis remains mainly dependent on the identification of the typical clinical pictures, to a certain extent, there are no optimal measures that would simplify the evaluation of the disease. Unluckily, there are no validated biomarkers that could reflect disease activity over time. Management of serious organ involvement such as ocular,

R. Talarico (✉) • S. Bombardieri
Rheumatology Unit, Department of Internal Medicine, University of Pisa, 56126 Pisa, Italy
e-mail: sara.talarico76@gmail.com

F. Dammacco et al. (eds.), *Systemic Vasculitides: Current Status and Perspectives*, DOI 10.1007/978-3-319-40136-2_20

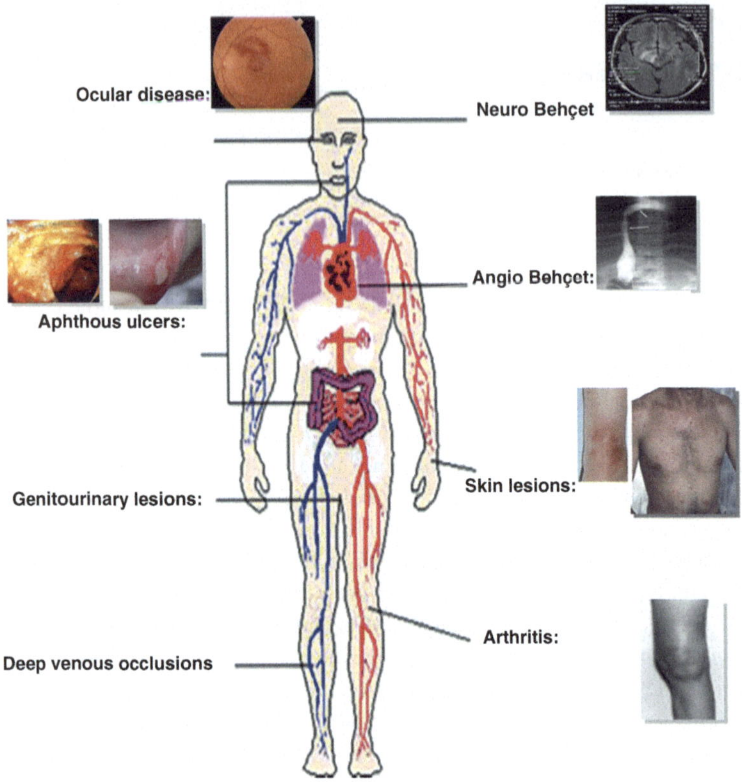

Fig. 20.1 Clinical manifestations of BD

vascular, gastrointestinal, and neurologic disease is with corticosteroids and immunosuppressives, which should be started early to prevent irreversible damage. In particular, TNF-alpha antagonists are exceptionally promising for treatment of resistant cases and their efficacy and safety have been proven in recalcitrant ocular, vascular, gastrointestinal and neurologic involvements.

20.1 Introduction

Behçet's disease (BD) is a systemic, chronic-relapsing vasculitis, typically characterised by recurrent oro-genital ulcers, ocular inflammation and skin manifestations (Fig. 20.1); articular, vascular, gastro-enteric and neurological involvement may also occur [1, 2]. The onset of disease typically occurs in patients in the late third and early fourth decades of life. Despite BD has a worldwide distribution, it is most commonly seen in the Middle East, Far East and the Mediterranean basin (Fig. 20.2), a particular trend that reminds of the ancient Silk Road [3, 4]. Moreover,

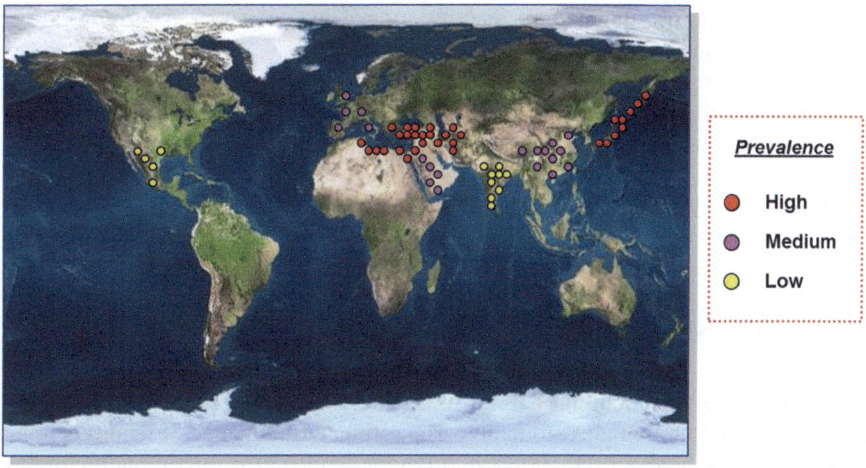

Fig. 20.2 Geographical distribution of BD

Table 20.1 International Study Group Classification Criteria for BD

Recurring oral ulcerations	Minor aphthous, major aphthous, or herpetiform ulcerations observed by physician or patient recurring at least 3 times in one 12-month period
Plus 2 of the following	
Recurring genital ulcerations	Aphthous ulcerations or scarring observed by physician or patient
Eye lesions	Anterior uveitis, posterior uveitis, cells in the vitreous on slit-lamp examination or retinal vasculitis observed by ophthalmologist
Skin lesions	Erythema nodosum observed by the physician or patient, pseudofolliculitis, papulopustular lesions, or acneiform nodules observed by physician in post-adolescent patients not receiving corticosteroid treatment
Pathergy	Test results read by physician at 24–48 h

the prevalence rates of BD in the endemic areas are strongly correlated to the prevalence of human leukocyte antigen (HLA)-B51. It is believed that a complex background with both genetic and environmental factors contributes to the disease development. Since there are no established laboratory findings to define BD, the diagnosis remains mainly dependent on the identification of the typical clinical pictures. In 1990 the International Study Group (ISG) for BD has proposed the validated classification criteria; to fulfill these criteria a *conditio sine qua non* for the diagnosis must be the presence of recurrent oral ulcers, together with two or more of the following: recurrent genital ulcerations, eye lesions, skin lesions or a positive pathergy test (Table 20.1) [5].

However, articular, vascular, gastro-enteric and neurological involvement may also occur. Globally, BD is characterized by a variable spectrum of disease profile: while prevalent muco-cutaneous lesions and arthritis represent the only clinical feature in patients with a benign disease subset, there are other patients who potentially develop sight or life-threatening manifestations, due to ocular, neurological or major vascular involvement [6–8].

Beside the organ involvement, a number of demographic factors could considerably influence the long-term and short-term outcomes of BD: age at disease onset, duration of disease, gender and sex [9]. Younger men patients are more suitable to have a more severe disease, due to an increased frequency both of morbidity and mortality, related to ocular, vascular and neurological involvement [10].

20.2 Mucocutaneous Involvement

The first sign of presentation of BD is usually represented by mucocutaneous features and recurrent aphthous stomatitis. The estimated frequency of recurrent aphthous stomatitis is about 95–100 % in different countries; it represents the initial manifestation of the disease in 90 % of the cases and is characterized by painful, shallow, round or oval ulcers covered with a yellowish pseudomembrane surrounded by a red border. Oral ulcers can occur anywhere in the oral cavity, but they are more frequent in buccal mucosae, tongue and mucosal surface of the lips. Notably, oral traumatic lesions have been implicated as a predisposing factor in oral aphthae [9, 10]. Moreover, growing data have shown an association between a lower prevalence of recurrent aphthous stomatitis and smoking.

Genital ulcers represent another major and specific cardinal manifestation of BD. Their frequency is about 50–85 % and they usually begin as a papule, pustule or circumscribed necrosis that ulcerates within a short period. In male, genital ulcers occur mostly on scrotum, penis and femoro-inguinal regions. In females, ulcers are commonly found on both major and minor labiae [9–11].

The skin lesions of BD are characterized by: (1) Nodular lesions (i.e. erythema nodosum-like lesions); (2) Papulopustular and acneiform lesions; (3) Other lesions (i.e. Sweet's syndrome) [11].

20.3 Ocular Involvement

Eye involvement represents one of the most serious manifestations of BD and occurs in half of the patients. It seems more frequent and severe among young males and, unluckily, it still remains one of the most significant causes of morbidity [10]. Usually, ocular disease develops within the first years of the disease onset and seems to be more severe in this period; moreover, growing data have suggested that the prognosis of BD patients with ocular involvement is mainly dependent upon the severity of visual acuity at presentation [11–13]. Posterior or pan-uveitis associated with retinal vasculitis are the most common ocular manifestations; they can be extremely severe, causing hemorrhages, retinal exudates, venous thrombosis, papilledema and macular disease. Structural changes, such as synechiae and retinal scars, are the consequences of its relapse course, which may lead in many patients to permanent visual impairment or blindness. The involvement of the anterior chamber with severe inflammation (hypopyion) indicates a poor outcome and is generally associated with severe retinal vasculitis. Isolated anterior uveitis is less frequent; moreover, conjunctivitis is sporadic. The anatomical classification of ocular involvement in BD has important therapeutic and prognostic implications: while attacks restricted to the anterior segment can be sufficiently managed with topical treatments, inflammation localized to the posterior segment always needs treatment with glucocorticoids and immunosuppressive therapies. Moreover, the specific site of inflammation, associated to the clinical course, represents a relevant reference point for the choice and duration of the therapy. Globally, the main determinant of visual prognosis is represented by the number of ocular attacks during the follow-up period; in this scenario, visual acuity represents the best marker of damage in ocular involvement.

20.4 Neurological Involvement

Although not included in the ISG criteria for BD, neurological involvement represents the second main cause of mortality, preceded by large vessel disease [14]. There have been many studies describing the prevalence of neuro-BD in different countries that varies from 2 to 50 % [14–17]. Despite immunosuppressive therapy, neurological involvement is still considered a worrying complication of the disease, representing an important cause of morbidity and mortality. Although neuro-BD may present with different neurological problems, directly or indirectly related to the systemic disease, it is usually categorised into two main groups: parenchymal brain involvement (more frequent, 80 % of cases) and non-parenchymal or vascular disease. Parenchymal central nervous system (CNS) involvement (Fig. 20.3), mainly affecting the brainstem, occurs with pyramidal signs, cerebellar symptoms, sphincter disturbance and behavioural changes. Vascular disease is generally due to intracranial hypertension secondary to dural sinus thrombosis. Headache undoubtedly represents the most common neurological symptom observed in patients with neuro-BD, and can be associated to different etiologies (i.e. parenchymal involvement, cerebral venous sinus thrombosis, ocular inflammation, co-existing primary headache). Moreover, it seems relatively frequent that patients with BD may develop a neuro-behavioral syndrome, characterized by bipolar disorders and paranoid attitudes, called "neuro-psycho-BD". So far, it is not clear which is the pathogenetic mechanism underlying this syndrome: it may be secondary to an organic neurological involvement or related to the poor quality of life and the relapsing course of disease. Parenchymal CNS involvement represents a serious morbidity of disease, often leading to disability and to mortality, if not treated early. On the other hand, dural sinus thrombosis is associated to a more favorable outcome than parenchymal involvement.

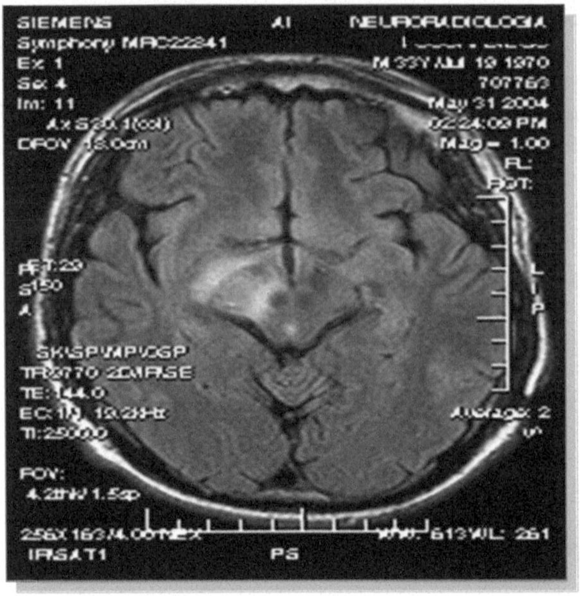

Fig. 20.3 Neurological involvement in BD

Notably, the onset of CNS involvement seems to occur in the first 10 years, with a higher incidence rate in the first 5 years [15]. These data have surely important clinical implications, since the timing of onset of neuro-BD strongly affects the scheduling of the follow-up timing. Indeed, neuro-BD is still related to high rates of morbidity and mortality: early recognition of severe organ involvements could certainly represent an important element to prevent irreversible damages due to the chronic-relapsing course of the disease. As suggested in other systemic autoimmune diseases, a disease-specific set of quality assessment tools should help physicians deliver a high quality of care in neuro-BD patients [16, 18].

20.5 Vascular Involvement

The prevalence of vascular involvement varies from 20 to 35 % of BD patients and it may involve all types of vessels within the arterial and venous system. It is characterized by a clear male preponderance. Vascular involvement in BD represents a serious risk for multiple vessel-related complications, including thromboses, stenoses, occlusions, and aneurysms. While arterial involvement is less common (3–12 %), the most common feature is represented by venous thrombosis, which mainly occurs in the lower limbs; other specific features are represented by vena cava thrombosis, pulmonary artery aneurysm (PAA), Budd-Chiari syndrome, peripheral artery aneurysms, dural sinus thrombosis and abdominal aorta aneurysms [19–22].

The prognosis of BD patients with arterial involvement is poor, with a death rate of 13.5 % in the patients with arterial lesions, above all in presence of the pulmonary artery and thoracic aorta involvement. PAA is a well known cause of morbidity in BD and is associated to the highest mortality rates despite an aggressive therapeutic approach [10]. The typical clinical picture of PAA is represented by hemoptysis, dyspnea, fever, chest pain and cough. Cardiac involvement is a rare manifestation in BD and comprises pericarditis, endocarditis with valvular lesions, myocarditis, intracardiac thrombosis, endomyocardial fibrosis, coronary vasculitis and myocardial aneurysm formation [20–22].

Younger men patients affected by BD are more suitable to have a more severe disease, due to an increased frequency both of morbidity and mortality, secondary to ocular, vascular and neurological involvement. The variable prognosis, associated to different gender or age, may represent an essential and useful element to tailor the management not only to the type and severity of symptoms, but also to the epidemiological profile of BD patients.

20.6 Therapy of BD

Subsequent mucocutaneous and joint involvement may be bothersome and impair the quality of life, but do not cause organ-threatening damage. Their management depends on the type, frequency, and severity of the symptoms. Management of serious organ involvement such as ocular, vascular, gastrointestinal, and neurologic disease is with corticosteroids and immunosuppressives, which should be started early to prevent irreversible damage. So far, more aggressive treatment for serious organ involvement with traditional DMARDs and biologic agents has improved the outcome of BD patients [23–29]. In particular, TNF-alpha antagonists are exceptionally promising for treatment of resistant cases and their efficacy and safety have been proven in recalcitrant ocular, vascular, gastrointestinal and neurologic involvements.

References

1. Seyahi E, Tahir Turanli E, Mangan MS et al (2010) The prevalence of Behçet's syndrome, familial Mediterranean fever, HLAB51 and MEFV gene mutations among ethnic Armenians living in Istanbul, Turkey. Clin Exp Rheumatol 28(Suppl 60):S67–S75
2. Talarico R, Baldini C, Della Rossa A et al (2012) Large- and small-vessel vasculitis: a critical digest of the 2010–2011 literature. Clin Exp Rheumatol 30(Suppl 70):S130–S138. 3
3. Yazici H, Seyahi E, Yurdakul S (2008) Behçet's syndrome is not so rare: why do we need to know? Arthritis Rheum 58:3640–3643
4. Hatemi G, Yazici Y, Yazici H (2013) Behçet's syndrome. Rheum Dis Clin North Am 39(2):245–261
5. International Study Group for Behçet's Disease (1990) Criteria for diagnosis of Behçet's disease. Lancet 335:1078–1080

6. Yazici H (2003) Behçet syndrome: an update. Curr Rheumatol Rep 5:195–219
7. Talarico R, Baldini C, Della Rossa A et al (2012) Large- and small-vessel vasculitis: a critical digest of the 2010–2011 literature. Clin Exp Rheumatol 30(Suppl 70):S130–S138
8. Talarico R, Baldini C, Della Rossa A, Carli L, Tani C, Bombardieri S (2013) Systemic vasculitis: a critical digest of the recent literature. Clin Exp Rheumatol 31(Suppl 75):S84–S88
9. Tursen U, Gurler A, Boyvat A (2003) Evaluation of clinical findings according to sex in 2313 Turkish patients with Behçet's disease. Int J Dermatol 42:346–351
10. Yazici H, Esen F (2008) Mortality in Behçet's syndrome. Clin Exp Rheumatol 26(Suppl 51):S138–S140
11. Yazici Y, Yurdakul S, Yazici H (2010) Behçet's syndrome. Curr Rheumatol Rep 12(6):429–435
12. Tugal-Tutkun I, Opal S, Altan-Yaylacioglu R et al (2004) Uveitis in Behcet disease: an analysis of 880 patients. Am J Ophthalmol 138(3):373–380
13. Cho YJ, Kim WK, Lee JH et al (2008) Visual prognosis and risk factors for Korean patients with Behcet uveitis. Ophthalmologica 222(5):344–350
14. Serdaroglu P, Yazici H, Ozdemir C et al (1989) Neurological involvement in Behçet syndrome – a prospective study. Arch Neurol 46:265–269
15. Talarico R, d'Ascanio A, Figus M, Stagnaro C, Ferrari C, Elefante E, Baldini C, Tani C, Mosca M, Bombardieri S (2012) Behçet's disease: features of neurological involvement in a dedicated centre in Italy. Clin Exp Rheumatol 30(3 Suppl 72):S69–S72
16. Kidd D, Steuer A, Denman AM, Rudge P (1999) Neurological complications in Behçet's syndrome. Brain 122(Pt 11):2183–2194
17. Akman-Demir G, Serdaroglu P, Tasçi B (1999) Clinical patterns of neurological involvement in Behçet's disease: evaluation of 200 patients. The Neuro-Behçet Study Group. Brain 122(Pt 11):2171–2182
18. Mosca M, Tani C, Aringer M et al (2011) Development of quality indicators to evaluate the monitoring of SLE patients in routine clinical practice. Autoimmun Rev 10:383–388
19. Tohme A, Aoun N, El-Rassi B et al (2003) Vascular manifestations of Behçet's disease: eighteen cases among 140 patients. Joint Bone Spine 70(5):384–389
20. Saadoun D, Asli B, Wechsler B et al (2012) Long-term outcome of arterial lesions in Behçet disease: a series of 101 patients. Medicine (Baltimore) 91(1):18–24
21. Seyahi E, Melikoglu M, Akman C et al (2012) Pulmonary artery involvement and associated lung disease in Behçet disease:a series of 47 patients. Medicine (Baltimore) 91(1):35–48
22. Geri G, Wechsler B, Thi Huong du L et al (2012) Spectrum of cardiac lesions in Behçet disease: a series of 52 patients and review of the literature. Medicine (Baltimore) 91(1):25–34
23. Yazici H, Hatemi G (2012) Empiricism in managing Behçet's syndrome. Clin Exp Rheumatol 30(3 Suppl 72):S7–S9
24. Ozguler Y, Hatemi G, Yazici H (2014) Management of Behçet's syndrome. Curr Opin Rheumatol 26(3):285–291
25. Hatemi G, Seyahi E, Fresko I, Talarico R, Hamuryudan V (2014) Behçet's syndrome: a critical digest of the 2013–2014 literature. Clin Exp Rheumatol 32(4 Suppl 84):S112–S122
26. Vaiopoulos AG, Sfikakis PP, Kanakis MA, Vaiopoulos G, Kaklamanis PG (2014) Gastrointestinal manifestations of Behçet's disease: advances in evaluation and management. Clin Exp Rheumatol 32(4 Suppl 84):S140–S148
27. Arida A, Sfikakis PP (2014) Anti-cytokine biologic treatment beyond anti-TNF in Behçet's disease. Clin Exp Rheumatol 32(4 Suppl 84):S149–S155
28. Interlandi E, Leccese P, Olivieri I, Latanza L (2014) Adalimumab for treatment of severe Behçet's uveitis: a retrospective long-term follow-up study. Clin Exp Rheumatol 32(4 Suppl 84):S58–S62
29. Hatemi G, Seyahi E, Fresko I, Hamuryudan V (2013) Behçet's syndrome: a critical digest of the 2012–2013 literature. Clin Exp Rheumatol 31(3 Suppl 77):108–117

Chapter 21
Small Vessel Vasculitis of the Skin

Robert G. Micheletti

Abstract Small vessel vasculitis of the skin typically manifests with palpable purpura on the lower extremities. This clinical presentation should prompt confirmatory biopsies for standard processing, as well as direct immunofluorescence. A thorough history, review of systems, and physical examination, as well as targeted laboratory workup, are essential to identifying patients with potentially severe systemic vasculitis or underlying medical conditions. For those patients with small vessel vasculitis confined to the skin, the prognosis is excellent. Most cases resolve in about 1 month. However, an important subset of patients can develop chronic, recurrent disease, for which effective management is often difficult.

Keywords Small vessel vasculitis of the skin • Cutaneous vasculitis • Leukocytoclastic vasculitis • Hypersensitivity vasculitis

21.1 Introduction

Small vessel vasculitis of the skin commonly presents with palpable purpura on the lower extremities. It is referred to by a variety of terms, often used interchangeably, including "leukocytoclastic vasculitis," "hypersensitivity vasculitis," "cutaneous leukocytoclastic angiitis," and "cutaneous small vessel vasculitis," the term used in the revised 2012 Chapel Hill Consensus Criteria to refer to small vessel vasculitis limited to the skin [1]. Each of these terms carries a slightly different shade of meaning that may differ according to the user. For example "hypersensitivity vasculitis" may suggest the process is drug-induced, and "leukocytoclastic vasculitis" is a histologic pattern seen in other types of vasculitis, not unique to small vessel vasculitis of the skin.

Ultimately, regardless of the terminology used, it is important to recognize that these lesions (e.g. purpuric papules) are best thought of as a symptom rather than a disease entity in and of themselves. In order to diagnose skin-limited vasculitis, one

R.G. Micheletti (✉)
Perelman School of Medicine, University of Pennsylvania, Philadelphia, PA, USA
e-mail: Robert.Micheletti@uphs.upenn.edu

F. Dammacco et al. (eds.), *Systemic Vasculitides: Current Status and Perspectives*, DOI 10.1007/978-3-319-40136-2_21

must first rule out important systemic manifestations (such as renal, joint, and gastrointestinal complications) as well as underlying conditions that affect management and prognosis. Patients with these complications may have precisely the same initial presentation of cutaneous vasculitis as those who do not. Patients may also develop systemic manifestations over time, necessitating close follow-up. For most patients, however, small vessel vasculitis of the skin is an acute and self-limited disease with a favorable prognosis, particularly when internal involvement and underlying conditions are absent.

21.2 Epidemiology

Small vessel vasculitis of the skin affects male and female patients equally and has an annual incidence of 30 cases per million adults. In children, small vessel vasculitis of the skin is less common; IgA vasculitis is more frequently seen in that population. The presence of systemic vasculitis, associated connective tissue disease, or malignancy is much more common in adults than in children [2–4].

21.3 Pathophysiology

Small vessel vasculitis is an immune complex-mediated process [5]. Circulating antigens, such as those related to medication exposure, infection, connective tissue disease, and neoplasia, become bound by antibodies, forming immune complexes. These complexes may become lodged within small vessels of various organs, including the joints, gastrointestinal tract, and renal glomerulus, as well as in the superficial dermis. Once there, the complement cascade is activated, inducing an inflammatory response that leads to vessel destruction, tissue damage, and extravasation of red blood cells.

In the case of palpable purpura, the small vessel involvement accounts for the small size of the lesions, the complement cascade and subsequent inflammation account for lesion palpability and symptomatology (burn, itch), and red blood cell extravasation results in non-blanching purpura [6].

21.4 Etiology

About half of cases are idiopathic [7–9]. Most of the remaining cases are either drug-induced or post-infectious. Among drugs, antibiotics, particularly β-lactams, are common culprits, but almost any drug or additive can cause vasculitis [10]. Upper respiratory infections, group A Strep, and Hepatitis C are common infectious causes, but numerous infectious triggers have been reported [11, 12]. In many cases, determining a specific cause can be difficult, especially in the hospitalized setting, where concomitant infection and antibiotic exposure are common.

In addition to infectious and medication triggers, small vessel vasculitis of the skin can also be a manifestation, even a presenting sign, of autoimmune connective tissue disease. Vasculitis is most commonly seen in systemic lupus and Sjögren syndrome but also dermatomyositis and rheumatoid arthritis, and its presence may signal more significant internal involvement [13]. Lastly, in a small but important subset of patients (<5 %), vasculitis occurs secondary to an underlying hematologic or solid organ malignancy [14, 15].

In general, vasculitis tends to appear 7–10 days following a drug or infectious trigger and 6 months after onset of an underlying medical condition, though timing and onset can be quite varied [16]. A careful history and review of systems is essential for uncovering possible triggers and identifying patients with more significant underlying disease.

21.5 Clinical Features

21.5.1 Physical Exam

Small vessel vasculitis of the skin typically presents with small, round, purpuric papules, usually a few millimeters in diameter (Fig. 21.1). Because the process is immune complex-mediated, lesions have a preference for the lower extremities (particularly the legs below the knees) and other dependent areas (such as the back, if the patient is bed-bound) due to gravity, but lesions may occur anywhere. Lesions are also accentuated in areas of pressure, such as beneath socks or sequential

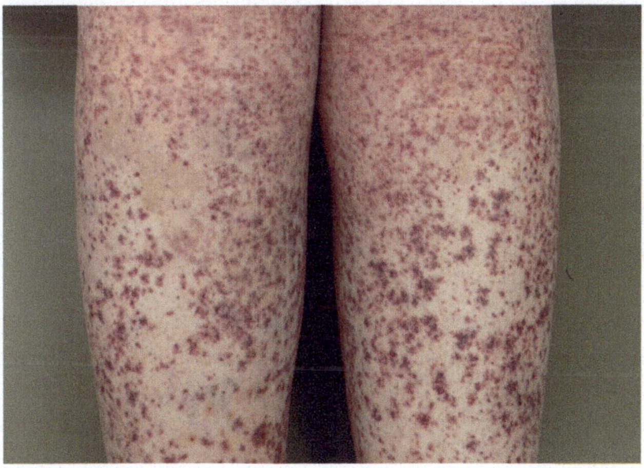

Fig. 21.1 Typical palpable purpura of small vessel vasculitis on the lower legs

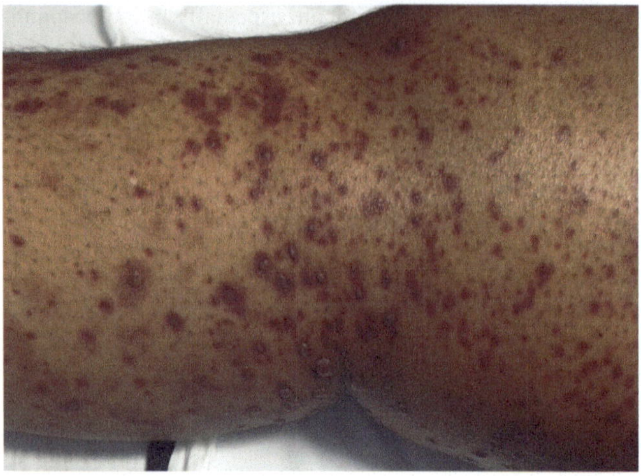

Fig. 21.2 Hemorrhagic vesicles due to exuberant inflammation in small vessel vasculitis

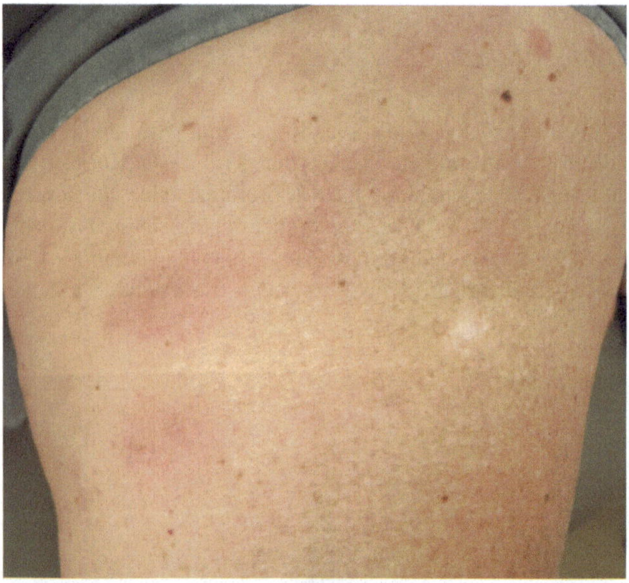

Fig. 21.3 Small vessel vasculitis presenting with persistent, bruise-like urticarial lesions

compression devices in those who are hospitalized. The total number of lesions varies from a few to several hundred. New lesions appear daily during flares and slowly resolve over 2–3 weeks, with residual post-inflammatory hyperpigmentation [17].

In addition to palpable purpura, non-palpable petechiae and purpuric macules are also frequently present. Significant inflammatory reactions can result in vesicles, pustules, and small hemorrhagic bullae (Fig. 21.2). Urticarial papules may also be seen (Fig. 21.3). Patients with hive-like lesions which do not resolve within 24 h, leave behind bruise-like marks, or are associated with fever, arthralgias, and other

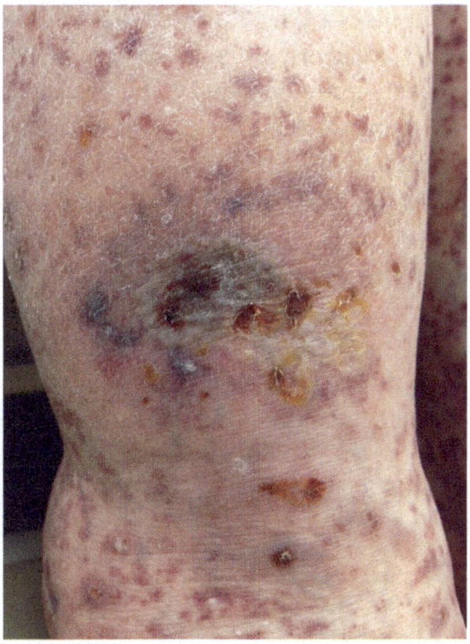

Fig. 21.4 Rounded ulcer due to coalescing lesions of small vessel vasculitis

symptoms, may in fact have a form of small vessel vasculitis called urticarial vasculitis, either with low or normal complement levels [18].

When widespread, lesions can coalesce into larger areas of purpura and can ulcerate. Such ulcers are typically rounded ulcers made up of smaller, round purpuric papules (Fig. 21.4). Absent are manifestations more typical of medium vessel vasculitis, such as livedo reticularis; subcutaneous nodules; retiform or stellate purpura; and more significant cutaneous infarction or ulceration. The larger size of such lesions and their jagged, net-like shape reflects the involvement of medium-sized vessel in the deep dermis or subcutis, which feed a network of downstream smaller vessels. If such lesions are seen, suspicion should be raised for polyarteritis nodosa, a pure medium vessel vasculitis, as well as cryoglobulinemic vasculitis and the ANCA-associated vasculitides, which impact both small and medium-sized vessels [19].

21.5.2 History

Patients with small vessel vasculitis of the skin may experience itching or burning pain as well as swelling and aching of the affected limbs. Or, they may be completely asymptomatic.

In most cases, no significant systemic manifestations are present, but a thorough review of systems should be performed to help identify those patients with evidence of vasculitis involving other organ systems. The presence of constitutional symptoms such as fever and weight loss; joint pains or swelling; myalgias; abdominal pain or melena; hematuria or frothy urine; cough or dyspnea; sinusitis or rhinitis; and paresthesias, weakness, or foot drop are particularly relevant in signifying the presence of systemic vasculitis. In the case of IgA vasculitis, which generally has a cutaneous presentation indistinguishable from other small vessel vasculitis of the skin, nearly two-thirds of patients experience gastrointestinal and joint symptoms, while 40 % develop manifestations of glomerulonephritis [20].

Arthralgias are fairly common during flares, but frank synovitis or arthritis is rare and suggests the presence of systemic disease [9]. If one or more of the above symptoms is present, a targeted laboratory workup should proceed to identify potentially severe extracutaneous manifestations and underlying disease states. Questions about preceding infections, prescribed and non-prescribed medications, and any comorbid medical conditions should also be asked. Finally, those who give a history of cutaneous vasculitis that is not self-limited or that is refractory or recurrent may be more likely to have an underlying disease. Further workup may be required to elucidate the underlying cause in these patients.

21.6 Initial Workup

When a patient presents with lesions suspicious for vasculitis, initial workup should try to answer three basic questions:

1. Are the lesions due to vasculitis?
2. Are organ systems involved?
3. Are there findings which help establish a particular diagnosis?

21.6.1 Histologic Findings

The diagnosis of small vessel vasculitis of the skin can be confirmed by a biopsy showing neutrophilic inflammation with leukocytoclasis as well as fibrinoid necrosis and disruption of small vessels with extravasation of red blood cells into the superficial dermis. Due to the potential for misdiagnosis and the existence of numerous conditions which can mimic vasculitis, skin biopsy should be performed whenever possible. Similarly, pathology reports should always be read carefully and interpreted in the correct clinical context, as certain entities (e.g. arthropod bites, ulcers, neutrophilic dermatoses) can show secondary vasculitic changes histologically. Clinical-pathologic correlation is always essential. A partial list of the differential diagnosis of purpuric macules and papules is shown below:

- Skin-limited small vessel vasculitis
- IgA vasculitis
- Cryoglobulinemic vasculitis
- ANCA-associated vasculitis
- Arthropod bites

- Macular purpura due to trauma, skin fragility, or anticoagulation
- Platelet dysfunction or deficiency
- Pigmented purpuric dermatosis (capillaritis)
- Cholesterol emboli
- Septic emboli
- Livedoid vasculopathy
- Sweet syndrome

A well-established but not old lesion (1–2 days of age) should be biopsied. Lesions are dynamic, so timing and location are critical. Biopsy of a poorly-established early lesion may fail to demonstrate definitive changes of vasculitis, while biopsy of an old lesion may demonstrate nonspecific inflammation and tissue damage. A 4 mm punch biopsy is generally sufficient to sample a lesion of small vessel vasculitis adequately. Deeper subcutaneous nodules or retiform purpura should be biopsied using a deep punch or wedge biopsy technique in order to sample medium-sized vessels in the deep dermis or subcutis.

Beyond merely confirming the diagnosis of vasculitis, some literature suggests the depth and severity of inflammation may predict systemic involvement or even underlying malignancy [21]. The presence of tissue eosinophilia may suggest drug-induced vasculitis [22].

In addition to performing a biopsy for standard processing, biopsy for direct immunofluorescence studies should also be performed [23]. Because immune complexes may be quickly degraded, a fresh lesion (ideally less than 24 h old) should be selected. The presence of IgA suggests a diagnosis of IgA vasculitis (Henoch-Schonlein purpura) and greater incidence of joint, gastrointestinal, and renal vasculitis associated with that disease [1]. The presence of IgM may correlate with renal involvement [24], or cryoglobulinemia [25]. C3 and IgG at the dermal-epidermal junction (positive lupus band test) may suggest hypocomplementemic urticarial vasculitis and underlying systemic lupus.

21.6.2 Laboratory Studies

Once the diagnosis of vasculitis has been confirmed, and after a thorough history, review of systems, and physical exam, a systematic and targeted laboratory workup should be performed to help establish whether other organ systems are involved. Since no standard protocol for this workup exists, it should be guided by clinical signs and symptoms. Most episodes of small vessel vasculitis of the skin are skin-limited and resolve within 3–4 weeks [9]. With this in mind, not every test need be ordered in every patient. Excessive testing should be avoided, as false positive or irrelevant results can be confusing. Nevertheless, serious internal organ dysfunction does rarely occur, so some basic workup is always indicated, along with whatever additional testing is dictated by the exam and review of systems.

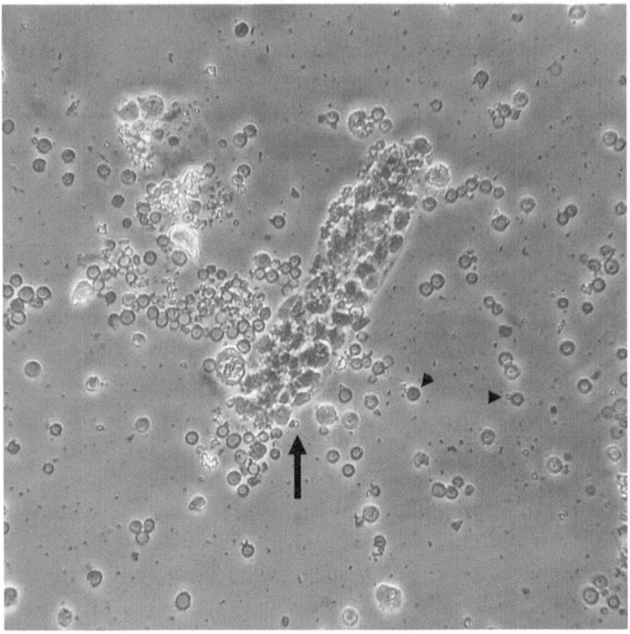

Fig. 21.5 Red blood cell cast (*arrow*) surrounded by dysmorphic red blood cells (*arrowheads*), consistent with glomerulonephritis (Image courtesy of Dr. Amar Bansal)

When the presentation is straightforward and the review of systems negative, nothing more than a complete blood count, complete metabolic panel, and urinalysis with microscopic evaluation may be required. Of these tests, the urinalysis is the most essential, as the presence of glomerulonephritis is most likely to change management. The presence of red blood cells on urinalysis should be followed by direct microscopy of the urine sediment to look for red cell casts and dysmorphic red blood cells (Fig. 21.5), possible indications of active glomerulonephritis. Fecal occult blood testing should be considered in all patients and certainly if abdominal symptoms or evidence of gastrointestinal bleeding is present. A chest x-ray or chest CT should be ordered if the patient complains of cough or dyspnea. Other organ-specific, targeted workup should proceed, if warranted, based on the review of systems and examination.

For those with concerning symptoms or chronic/recurrent lesions without an obvious cause, reasonable workup includes the testing discussed above as well as infectious serologies, including hepatitis B and C virus, HIV, and antistreptolysin-O, and rheumatologic workup, including anti-nuclear antibodies (ANA), rheumatoid factor (which screens for rheumatoid arthritis and is usually highly elevated in the presence of cryoglobulinemic vasculitis), and anti-neutrophilic cytoplasmic antibodies (ANCA) (which are strongly suggestive of ANCA-associated vasculitis). Other testing to consider includes serum protein electrophoresis (SPEP)/immunofixation to look for evidence of a paraprotein; complement 3 and 4 levels, which may be low in the context of urticarial vasculitis, systemic lupus, or cryoglobulinemic vasculitis; and cryoglobulins; as well as any other workup warranted by the presenting signs and symptoms. Acute phase reactants (erythrocyte sedimentation rate and C-reactive protein) may be elevated in the presence of cutaneous small vessel vasculitis but unfortunately are not useful in identifying patients more likely to have systemic involvement.

21.7 Management

21.7.1 Initial Management

Initial therapy (and prognosis) are dictated by the workup. More aggressive systemic therapy is necessary in the case of renal or other organ involvement. Any underlying condition or trigger identified should be addressed and treated, if possible.

If systemic involvement has been excluded, the treatment of skin-limited vasculitis should be symptom-focused. Because most cases are minimally symptomatic and self-limited, aggressive immunosuppression is generally not advisable. General measures such as rest and elevation, compression stockings to decrease stasis-related immune complex deposition, and topical steroids for itch may provide symptomatic relief and decrease the number of lesions. More than half of patients require no systemic treatment [7].

21.7.2 Subsequent Management

Systemic therapy is indicated for skin-limited small vessel vasculitis if it is severe, intractable, or recurrent. Approximately 8–10 % of cases become chronic [9]. Systemic therapy may lessen discomfort, complications such as ulceration, and the psychosocial impact of the disease. For any episode that is not self-limited and lasts longer than a few weeks, systemic therapy is indicated even if the condition is relatively asymptomatic.

Unfortunately, there is a dearth of high-quality data to guide management. Preferred target doses and durations are unknown, though several weeks of therapy may be required for an adequate therapeutic trial. Only one small randomized controlled trial for cutaneous small vessel vasculitis has been performed, for colchicine. The study was limited to 20 patients and failed to show a significant benefit after 1 month of therapy, though some patients flared upon withdrawal of the medication [26]. Colchicine has been reported useful for skin and joint symptoms in open label studies, typically at doses of 0.6 mg 2–3 times daily, if gastrointestinal side effects are not limiting [27].

Dapsone, a medication with anti-neutrophilic effects dosed at 50–200 mg daily, is supported by anecdotal experience and small case series [28]. It is contraindicated in those with glucose-6-phosphate dehydrogenase deficiency and can cause methemoglobinemia and anemia due to hemolysis, necessitating regular laboratory monitoring. It can be combined with colchicine when monotherapy with either medication is unsuccessful.

Other options include hydroxychloroquine (200–400 mg daily), which may be most helpful in urticarial vasculitis associated with systemic lupus erythematosus [29]; pentoxifylline (400–1200 mg daily), which can also be combined with dapsone [30], and non-steroidal anti-inflammatory agents, which may help alleviate symptoms. The use of these agents is supported only by case series and anecdote.

Systemic corticosteroids dosed at 0.5–1.0 mg/kg/day prednisone equivalent are appropriate for those with severe, necrotic lesions or serious systemic manifestations, such as renal or gastrointestinal involvement. The response to therapy is usually rapid, and steroids should be tapered slowly to prevent rebound. Long-term corticosteroid therapy may not be required if the process is self-limited. For chronic cutaneous vasculitis, systemic steroids, with their associated side effects, are not a good long-term option.

When none of these agents is effective or tolerated, or if the vasculitis is chronic and refractory, various immunosuppressive agents can be considered, making sure to balance potential toxicities against the severity of the skin condition and whatever systemic manifestations are present. These include azathioprine (50–200 mg daily) [31], which is contraindicated if thiopurine methyltransferase levels are low; methotrexate (15–25 mg weekly) [32], with folic acid supplementation; mycophenolate mofetil (2–3 g daily) [33], which may be limited by gastrointestinal side effects. Agents such as cyclophosphamide, infliximab, and rituximab have been utilized in some refractory and severe cases but are rarely indicated [34, 35].

21.8 Prognosis

Most episodes of small vessel vasculitis are self-limited, resolving in 3–4 weeks. In general, systemic involvement (if present) is minimal. However, serious internal organ dysfunction does rarely occur. Prognosis ultimately depends on the severity of organ involvement as well as any underlying associated medical disorder. While 10 % of patients develop chronic or recurrent vasculitis, most cases have an overall good prognosis [7, 9].

21.9 Future Directions

Despite this overall good prognosis, small vessel vasculitis can be a significant source of morbidity for those with systemic disease as well as those with chronic and recurrent lesions. Because of the relative lack of data to guide management, preferred treatment can vary widely, and patients frequently try more than one agent before success is achieved. Larger series and well-designed therapeutic trials are needed to help develop evidence-based guidelines for evaluation and treatment of this condition. One such multicenter, randomized trial of colchicine, dapsone, and azathioprine for cutaneous small vessel vasculitis is currently in development [36].

References

1. Jennette JC, Falk RJ, Bacon PA et al (2013) 2012 revised international chapel hill consensus conference nomenclature of vasculitides. Arthritis Rheum 65(1):1–11
2. García-Porrúa C, González-Gay MA (1999) Comparative clinical and epidemiological study of hypersensitivity vasculitis versus Henoch-Schönlein purpura in adults. Semin Arthritis Rheum 28(6):404–412
3. Blanco R, Martínez-Taboada VM, Rodríguez-Valverde V, García-Fuentes M (1998) Cutaneous vasculitis in children and adults. Associated diseases and etiologic factors in 303 patients. Medicine (Baltimore) 77(6):403–418
4. Arora A, Wetter DA, Gonzalez-Santiago TM et al (2014) Incidence of Leukocytoclastic Vasculitis, 1996 to 2010: a population-based study in Olmsted county, Minnesota. Mayo Clin Proc 89(11):1515–1524
5. Mackel SE, Jordon RE (1982) Leukocytoclastic vasculitis. A cutaneous expression of immune complex disease. Arch Dermatol 118(5):296–301
6. Carlson JA (2010) The histological assessment of cutaneous vasculitis. Histopathology 56(1):3
7. Martinez-Taboada VM, Blanco R, Garcia-Fuentes M (1997) Clinical features and outcome of 95 patients with hypersensitivity vasculitis. Am J Med 102:186–191
8. Gyselbrecht L, De Keyser F, Ongenae K (1996) Etiological factors and underlying conditions in patients with leukocytoclastic vasculitis. Clin Exp Rheum 14:665–668
9. Russell JP, Gibson LE (2006) Primary cutaneous small vessel vasculitis: approach to diagnosis and treatment. Int J Dermatol 45(1):3–13
10. Lowry MD, Hudson CF, Callen JP (1994) Leukocytoclastic vasculitis caused by drug additives. J Am Acad Dermatol 30(5 Pt 2):854–855
11. Greco F, Sorge A, Salvo V, Sorge G (2007) Cutaneous vasculitis associated with Mycoplasma pneumoniae infection: case report and literature review. Clin Pediatr (Phila) 46:451
12. Kim HM, Park YB, Maeng HY, Lee SK (2006) Cutaneous leukocytoclastic vasculitis with cervical tuberculous lymphadenitis: a case report and literature review. Rheumatol Int 26:1154
13. Barile-Fabris L, Hernandez-Cabrera MF, Barragan-Garfias JA (2014) Vasculitis in systemic lupus erythematosus. Curr Rheumatol Rep 16(9):440
14. Loricera J, Calvo-Rio V, Ortiz-Sanjuan F et al (2013) The spectrum of paraneoplastic cutaneous vasculitis in a defined population: incidence and clinical features. Medicine (Baltimore) 92(6):331–343
15. Podjasek JO, Wetter DA, Pittelkow MR, Wada DA (2012) Cutaneous small-vessel vasculitis associated with solid organ malignancies: the Mayo Clinic experience, 1996 to 2009. J Am Acad Dermatol 66(2):e55–e65
16. Carlson JA, Cavaliere LF, Grant-Kels JM (2006) Cutaneous vasculitis: diagnosis and management. Clin Dermatol 24(5):414–429
17. Fiorentino DF (2003) Cutaneous vasculitis. J Am Acad Dermatol 48(3):311–340
18. Loricera J, Calvo-Rio V, Mata C et al (2014) Urticarial vasculitis in northern Spain: clinical study of 21 cases. Medicine (Baltimore) 93(1):53–60
19. Xu LY, Esparza EM, Anadkat MJ et al (2009) Cutaneous manifestations of vasculitis. Semin Arthritis Rheum 38(5):348–360

20. Calvo-Río V, Loricera J, Mata C et al (2014) Henoch-Schönlein purpura in northern Spain: clinical spectrum of the disease in 417 patients from a single center. Medicine (Baltimore) 93(2):106–113
21. Podjasek JO, Wetter DA, Wieland CN et al (2014) Histopathologic findings in cutaneous small-vessel vasculitis associated with solid-organ malignancy. Br J Dermatol 171(6):1397–1401
22. Bahrami S, Malone JC, Webb KG, Callen JP (2006) Tissue eosinophilia as an indicator of drug-induced cutaneous small-vessel vasculitis. Arch Dermatol 142:155
23. Minz RW, Chhabra S, Singh S et al (2010) Direct immunofluorescence of skin biopsy: perspective of an immunopathologist. Indian J Dermatol Venereol Leprol 76:150
24. Takeuchi S, Soma Y, Kawakami T (2010) IgM in lesional skin of adults with Henoch-Schönlein purpura is an indication of renal involvement. J Am Acad Dermatol 63(6):1026–1029
25. Gibson LE (2001) Cutaneous vasculitis update. Dermatol Clin 19:603–615
26. Sais G, Vidaller A, Jucqla A (1995) Colchicine in the treatment of cutaneous leukocytoclastic vasculitis. Results of a prospective, randomized controlled trial. Arch Dermatol 131:1399
27. Callen JP (1985) Colchicine is effective in controlling chronic cutaneous leukocytoclastic vasculitis. J Am Acad Dermatol 13(2 Pt 1):193–200
28. Fredenberg MF, Malkinson FD (1987) Sulfone therapy in the treatment of leukocytoclastic vasculitis. Report of three cases. J Am Acad Dermatol 16(4):772–778
29. Lopez LR, Davis KC, Kohler PF, Schocket AL (1984) The hypocomplementemic urticarial-vasculitis syndrome: therapeutic response to hydroxychloroquine. J Allergy Clin Immunol 73:600–603
30. Nurnberg W, Grabbe J, Czarnetzki BM (1995) Urticarial vasculitis syndrome effectively treated with dapsone and pentoxifylline. Acta Derm Venereol 75:54–56
31. Callen JP, Spencer LV, Burruss JB, Holtman J (1991) Azathioprine. An effective, corticosteroid-sparing therapy for patients with recalcitrant cutaneous lupus erythematosus or with recalcitrant cutaneous leukocytoclastic vasculitis. Arch Dermatol 127:515–522
32. Jorizzo JL, White WL, Wise CM et al (1991) Low dose weekly methotrexate for unusual neutrophilic vascular reactions: cutaneous polyarteritis nodosa and Behcet's disease. J Am Acad Dermatol 24:973–978
33. Haeberle MT, Adams WB, Callen JP (2012) Treatment of severe cutaneous small-vessel vasculitis with mycophenolate mofetil. Arch Dermatol 148(8):887–888
34. Mang R, Ruzicka T, Stege H (2004) Therapy for severe necrotizing vasculitis with infliximab. J Am Acad Dermatol 51:321
35. Chung L, Funke AA, Chakravarty EF et al (2006) Successful use of rituximab for cutaneous vasculitis. Arch Dermatol 142(11):1407–1410
36. Taumra RN, Krischer JP, Pagnoux C et al (2016) A small n sequential multiple assignment randomized trial design for use in rare disease research. Contemp Clin Trials 46:48–51

Chapter 22
Adult Primary Central Nervous System Vasculitis

Carlo Salvarani, Robert D. Brown Jr., Caterina Giannini, and Gene G. Hunder

Abstract Primary central nervous system vasculitis is an uncommon disorder of unknown cause that is limited to the brain and spinal cord. The median age of onset is 50 years. The neurological manifestations are diverse, but commonly include headache, altered cognition, focal weakness, or stroke. Serological markers of inflammation are usually normal. Cerebrospinal fluid is abnormal in approximately 80–90 % of the cases. The diagnosis is unlikely in the presence of a normal magnetic resonance imaging (MRI) of the brain. Biopsy of central nervous system tissue showing vasculitis is the only definitive test, however angiography has often been used for diagnosis even though it has only moderate sensitivity and specificity. The size of the vessels involved is varied and influences outcomes and response to treatment. Early recognition is important because treatment with corticosteroids, with or without cytotoxic drugs, may prevent serious outcomes. The differential diagnosis includes reversible cerebral vasoconstriction syndromes and secondary cerebral vasculitis.

Primary central nervous system vasculitis (PCNSV) is an uncommon and poorly understood form of vasculitis that is limited to the brain and spinal cord. PCNSV represents the most frequent vasculitis involving the CNS (Table 1, from Arthritis Rheum 2006; 55:985–9; Table 22.1) [1]. Modern recognition of PCNSV as a separate entity is generally dated to the mid-1950s when Cravioto and Feigin described several cases with a "non-infectious granulomatous angiitis" with a predilection for the nervous system [2]. PCNSV is a rare condition. The only reported incidence rate

C. Salvarani (✉)
Unit of Rheumatology, Arcispedale S.Maria Nuova, V.le Risorgimento N80, 42100 Reggio Emilia, Italy
e-mail: Salvarani.carlo@asmn.re.it

R.D. Brown Jr.
Department of Neurology, Mayo Clinic, Rochester, MN, USA

Division of Anatomic Pathology, Mayo Clinic, Rochester, MN, USA

C. Giannini • G.G. Hunder
Mayo Clinic College of Medicine, Rochester, MN, USA

F. Dammacco et al. (eds.), *Systemic Vasculitides: Current Status and Perspectives*, DOI 10.1007/978-3-319-40136-2_22

Table 22.1 Vasculitides and connective tissue diseases associated with a diagnosis of CNS vasculitis or angiitis at the Mayo Clinic over a 17-year period[a]

Condition	No. patients
Isolated CNS vasculitis	73
ANCA-associated vasculitis	13
Behçet's disease	8
Giant cell arteritis	3
Undefined vasculitis	3
Systemic lupus erythematosus	9
Sjögren's syndrome	2
Rheumatoid arthritis	2
Undefined connective tissue disease	1

Table from Table 1 manuscript: Salvarani et al. [1]

[a]*CNS* central nervous system, *ANCA* anti-neutrophil cytoplasmic antibody

for PCNSV derives from Olmsted County, Minnesota (USA) and is 2.4 cases per 1,000,000 person years. Men and women are similarly affected. The median age at diagnosis is approximately 50 years [3]. Early case reports described PCNSV as a fatal condition scarcely responsive to the immunosuppressive treatment [4]; in contrast, successive cohort studies have reported a more favorable course [3]. Recent studies from Mayo Clinic have recognized that PCNSV is a heterogeneous condition composed of clinical subsets that differ in terms of outcomes and response to treatment [3, 5–10].

Various types of primary systemic vasculitides may affect the central nervous system (CNS) [11, 12]. PCNSV needs to be differentiated from the systemic vasculitides with CNS involvement to avoid therapeutic and prognostic errors.

The aim of this review is to provide an update on the major recent advances on adult primary CNS vasculitis.

22.1 Diagnosis

Diagnostic criteria for PCNSV were proposed by Calabrese and Mallek in 1988 on the basis of clinical experience and literature evidence (Table 22.2) [13]. Angiographic changes indicating a high probability of vasculitis include alternating areas of smooth walled narrowing and dilated cerebral arteries, or occlusions, affecting multiple cerebral vessels in the absence of proximal vessel atherosclerosis or other recognized abnormalities. A single abnormality in multiple arteries or multiple abnormalities in a single vessel are less consistent with PCNSV [3, 11, 12].

Because of the more invasive nature of CNS biopsy, angiography has become preferred for confirming the diagnosis in patients with suggestive clinical findings.

Table 22.2 Diagnostic criteria for primary central nervous system vasculitis proposed by Calabrese and Mallek in 1988

1.	A history or clinical findings of an acquired neurologic deficit, which remain unexplained after a thorough initial basic evaluation
2.	Either classic angiographic or histopathologic features of vasculitis within the central nervous system
3.	No evidence of systemic vasculitis or of any other condition to which the angiographic or pathologic features could be secondary

A diagnosis of primary central nervous system vasculitis is made if all the above criteria are satisfied

However, angiographic changes typical of vasculitis may be seen in non–vasculitic conditions such as vasospasm, CNS infections, lymphomas, cerebral arterial emboli, and also atherosclerosis [4, 14].

Furthermore, among pathologically documented cases, cerebral angiography may be normal reflecting vascular abnormalities in small vessels beyond the resolution of angiography [5].

Overall, the sensitivity of angiography varies between 40 and 90 %, while the specificity has shown to be as low as 30 % [3, 15].

It is important to emphasize that the diagnosis of PCNSV should not be based on positive angiography alone, and that angiography results should always be interpreted in conjunction with clinical, laboratory, and MRI findings.

Magnetic resonance angiography (MRA) is less sensitive than conventional angiography at detecting lesions involving the posterior circulation and distal vessels [3, 16]. MRA is a reasonable initial modality in the investigation of suspected PCNSV, however in cases highly suspected with normal MRA, a cerebral angiography should be performed.

PCNSV is unlikely in the presence of a normal MRI. Several studies have reported abnormalities of MRI close to 100 % [3, 15]. Abnormal findings on MRI are non specific and include cortical and subcortical infarction, parenchimal and leptomeningeal enhancement, intracranial hemorrhage, tumor-like mass lesions, and non-specific areas of increased signal intensity on FLAIR or T2-weighted images.

Wall thickening and intramural contrast enhancement could be specific findings in patients with active cerebral vasculitis affecting large arteries. Occasionally, enhancement may be marked and extend into the adjacent leptomeningeal tissue (perivascular enhancement) [17, 18].

High-resolution 3-tesla contrast-enhanced MRI may be able to differentiate enhancement patterns of intracranial atherosclerotic plaques (eccentric), inflammation (concentric), and other wall pathologies. However, the sensitivity and specificity of this technique remains to be determined [19].

Cerebral and meningeal biopsy remains the gold standard for diagnosis of PCNSV [11, 12, 20]. The procedure in expert hands is safe, being the risk of a persistent problem only about 1 %. Diagnostic histopathological features include transmural vascular inflammation affecting small and medium-sized leptomeningeal and parenchymal arterial vessels [20]. Granulomatous vasculitis is the most common pattern of vasculitis. Beta-4 amyloid deposition is present in almost 50 % of biopsies with this pattern, but rarely seen in other histopathological patterns. Lymphocytic vasculitis is the second most predominant pattern. Necrotizing vasculitis is the least frequently seen pattern, characterized by acute necrotizing vasculitis, similar to the pattern seen in polyarteritis nodosa, with transmural fibrinoid necrosis.

Vasculitis affects vessels in a skipped and segmental pattern, therefore because of sampling error a negative biopsy does not exclude the diagnosis of vasculitis.

Biopsy of a radiographically abnormal area is preferable to random sampling of the nondominant frontal lobe or temporal tip. Miller et al showed that 78 % of the targeted biopsies were diagnostic, whereas none of the blind biopsies demonstrated vasculitis [20].

Stereotactic guidance may be used for deeper lesions but is usually unnecessary for the commonly biopsied, more superficial lesions.

22.2 Clinical Manifestations and Laboratory

Clinical manifestations at diagnosis are listed in Table 22.3 (from Ann Neurol 2007; 62:442–51, Table page 445) [3]. They are non-specific and multiple manifestations are usually present. The onset of the disease may be acute, however it is more frequently insidious and slowly progressive.

Headache, the most common symptoms of PCNSV, may be generalized or localized, often slowly worsening and may spontaneously remit for periods. Cognitive impairment is the second most frequent manifestation. Focal neurological manifestations are present in a large proportion of patients. Other manifestations such as ataxia, seizures, and intracranial hemorrhage are less frequent. Systemic symptoms such as fever and weight loss are uncommon. Symptoms related to spinal cord involvement may occasionally be the presenting manifestation.

Laboratory blood tests in PCNSV are usually normal including acute phase reactants.

CSF analysis is abnormal in 80–90 % of cases [3]. Changes include a mildly increased leukocyte count and total protein concentration. CSF analysis should include appropriate stains, culture, serologic and molecular tests, and flow cytometry studies to exclude infections or malignancy.

22.3 Special Subsets

Several subsets of PCNSV have been identified which may differ in terms of prognosis and optimal management.

Spinal cord involvement has been documented in 5 % of patients, but it is rarely the only manifestation [21]. The thoracic cord is the predominantly affected site. A care-

Table 22.3 Clinical manifestations at presentation

Characteristics	All patients (N = 101), n (%)	Patients diagnosed by biopsy (n = 31), n (%)	Patients diagnosed by angiography (n = 70), n (%)
Headache	64(63)	16(52)	48(69)
Altered cognition	50(50)	22(71)	28(40)
Hemiparesis	44(44)	6(19)	38(54)
Persistent neurological deficit or stroke	40(40)	8(26)	32(46)
Aphasia	28(28)	11(36)	17(24)
Transient ischemic attack	28(28)	5(16)	23(33)
Ataxia	19(19)	5(16)	14(20)
Seizure	16(16)	2(7)	14(20)
Visual symptom (any kind)	42(42)	9(29)	33(47)
Visual field defect	21(21)	5(16)	16(23)
Diplopia (persistent or transient)	16(16)	5(16)	11(16)
Blurred vision or decreased visual acuity	11(11)	0(0)	11(16)
Monocular visual symptoms or amaurosis fugax	1(1)	0(0)	1(1)
Papilledema	5(5)	2(7)	3(4)
Intracranial hemorrhage	8(8)	2(7)	6(9)
Amnestic syndrome	9(9)	4(13)	5(7)
Paraparesis or quadriparesis	7(7)	4(13)	3(4)
Parkinsonism or extrapyramidal sign	1(1)	0(0)	1(1)
Prominent constitutional symptoms	9(9)	4(13)	5(7)
Fever	9(9)	4(13)	5(7)
Nausea or vomiting	25(25)	6(19)	19(27)
Vertigo or dizziness	9(9)	3(10)	6(9)
Dysarthria	15(15)	2(7)	13(19)
Unilateral numbness	13(13)	0(0)	13(19)

ful medical evaluation must be performed to confirm the diagnosis of PCNSV and to exclude other conditions associated with acute or subacute transverse myelitis.

Angiography-negative, biopsy-positive PCNSV is constituted by patients with vasculitis limited to small vessels beyond the resolution of conventional angiography [5]. These patients often have a cognitive dysfunction at presentation, marked elevation in CSF total protein level, meningeal or parenchymal enhancing lesions on MRI, respond favorably to treatment, and have a good outcome.

PCNSV may present with prominent leptomeningeal enhancement on MRI [6]. These patients have an acute clinical onset, a cognitive dysfunction is frequently present, and negative cerebral angiography or MRA. CNS biopsies show a granulomatous vascular inflammation, often associated with cerebral amyloid angiopathy (CAA).

Almost all patients respond to corticosteroid therapy (alone or combined with immunosuppressive agents) with resolution of MRI enhancement and an overall favorable course.

About a quarter of patients with biopsy-positive PCNSV have evidence of cerebral amyloid vascular deposition [7, 22–24]. This condition is defined as amyloid-β related angiitis or ABRA [22]. These patients usually have cognitive dysfunction and/or seizures/spells at presentation, higher concentrations of CSF protein, and enhancing leptomeningeal lesions on MRI. They usually are angiography-negative because the vasculitis is limited to small cortical and leptomeningeal vessels beyond the resolution of conventional angiography. Cerebral biopsy is required for the diagnosis. Brain biopsy samples show a granulomatous histopathological pattern plus vascular deposits of amyloid-β. Patients with ABRA usually respond favorably to treatment and have a good outcome. Early recognition and treatment of ABRA will help avoid serious outcomes. Two pathologic inflammatory reactions to the deposition of amyloid-β in the cerebral vessels have been described: one with a vasculitic transmural, often granulomatous, inflammation (ABRA), and the second with a perivascular non-destructive inflammatory infiltration, so called CAA-related inflammation [22–24]. ABRA and CAA-RI more closely resemble PCNSV than CAA without vascular inflammation and likely are part of the same pathologic spectrum [22].

Rapidly progressive PCNSV represents the worst end of the clinical spectrum of this vasculitis [8]. Patients have a rapidly progressive course and often fatal outcome. They are characterized by bilateral, multiple, large cerebral vessel lesions on angiograms, and multiple bilateral cerebral infarctions. The predominant histopathological pattern is granulomatous and/or necrotizing. These patients respond poorly to the traditional immunosuppressive treatment.

Approximately 4 % of PCNSV patients present with a solitary tumor-like mass lesion [25]. An association with CAA was observed in 29 % of these patients. Excision of the lesion may be curative, however in some patients aggressive immunosuppressive therapy has resulted in a favorable outcome, obviating the need of surgery.

Intracranial hemorrhage (IH) is a presenting manifestation in 11–12.2 % of patients [9]. Intracerebral hemorrhage is the most common, followed by subarachnoid hemorrhage. These patients have less frequently altered cognition, a persistent neurologic deficit, or MRI evidence of cerebral infarctions during the disease course. Necrotizing vasculitis is the predominant histopathologic pattern on biopsy.

22.4 Differential Diagnosis

It is essential to differentiate between PCNSV and both its mimics and secondary causes of CNS vasculitis given the different therapeutic and prognostic implications.

The most common mimics of PCNSV are reversible cerebral vasoconstriction syndromes (RCVS) which group various disorders reported on different appellations (Call-Fleming syndrome, postpartum angiopathy, migrainous vasospasm, drug-induced cerebral vasculopathy) with symptoms caused by vasocostrinction rather than vasculitis [26]. Differentiation is crucial since immunosuppressive therapy beyond a short course of prednisone is not warranted for syndromes caused by vasoconstriction (Table 22.4).

Infectious agents reported in association with CNS vasculitis, include varicella zoster virus, HIV, treponema pallidum, hepatitis C virus, parvovirus B19, cytomegalovirus, mycoplasma pneumoniae, borrelia burgdorferi, mycobacterium tuberculosis, bartonella, and rickettsiae.

Fungal infections (aspergillosis, mucormycosis, coccidioidomycosis, and candidiasis) and subarachnoid cysticercosis have also been implicated in some cases.

CNS vasculitis is most commonly reported in anti-neutrophil cytoplasmic antibody (ANCA)-associated vasculitis, and Behçet's disease. It has been also reported, although rarely, in polyarteritis nodosa, Henoch-Schonlein purpura, Kawasaki disease, giant cell arteritis, and Takayasu arteritis.

CNS vasculitis is an uncommon manifestation of neuro-SLE, rheumatoid arthritis, primary Sjögren syndrome, dermatomyositis, and mixed connective tissue disease.

PCNSV has been associated with lymphoma, particularly Hodgkin disease.

Secondary CNS vasculitis has also been described in patients with neurosarcoidosis, inflammatory bowel disease, graft versus host disease, and with the use of some drugs (cocaine, amphetamine, ephedrine, and phenylpropanolamine).

Table 22.4 Discriminating features of PCNSV and RVCS

Characteristics	PCNSV	RVCS
Precipitating factor	no	Onset in postpartum phase or after exposure to vasoactive substances
Onset	More insidious, progressive course	Acute onset followed by a monophasic course
Headaches	Chronic and progressive	Acute, thunderclap type
CSF findings	Abnormal (leukocytosis and elevated total protein)	Normal to near normal
MRI	Abnormal in almost all patients	Normal in 70 % of patients
Angiography	It may be normal, otherwise diffuse abnormalities often indistinguishable from RCVS. Irregular and asymmetrical arterial stenoses or multiple occlusions are more suggestive of vasculitis. The abnormalities may be irreversible	Always abnormal, strings of beads appearance of cerebral arteries, abnormalities reversible within 6–12 weeks
Cerebral biopsy	Vasculitis	No vasculitic changes
Drug treatment	Prednisone with/without cytotoxic agents	Nimodipine

Abbreviations: *PCNSV* primary central nervous system vasculitis, *RVCS* reversible vasoconstriction syndrome, *CSF* cerebrospinal fluid, *MRI* magnetic resonance imaging

22.5 Treatment and Outcomes

There are no randomized clinical trials of medical management in PCNSV. Therefore, treatment for PCNSV has been derived from therapeutic strategies used in other vasculitides, from anecdotal reports, and from PCNSV cohort studies. Earliest reports suggested a poor prognosis with fatal outcome in the majority of the cases, and transient or doubtful efficacy of glucocorticoids.

Cupps and Fauci in 1983 first found cyclophosphamide (CYC) in combination with corticosteroids to be effective also in PCNSV [27].

The most recent PCNSV cohort studies have described a more favorable course of PCNSV. In a cohort study of 101 patients, a favorable response to glucocorticoids alone or in combination with CYC was achieved in 81 % of the cases [3].

Glucocorticoid therapy should be initiated as soon as the diagnosis of PCNSV is established. We recommend an initial dose of prednisone of 1 mg/Kg/day (or equivalent) as a single or divided dose. If the patient does not respond promptly, CYC should be started. A short course of oral CYC for 3–6 months could be recommended also in PCNSV to induce remission, as in other vasculitides. Intravenous pulses of CYC are probably safer than daily oral therapy, while it is unclear whether the two regimes differ in terms of efficacy in PCNSV. Subsequently, one could consider a lower-risk immunosuppressant such as azathioprine, methotrexate, or mycophenolate mofetil (MMF) for maintenance of remission. A treatment course of 12–18 months is adequate in the majority of cases.

Tumor necrosis factor-α (TNF-α) blockers and anti-CD20 therapy with rituximab have successfully been used to treat patients with PCNSV resistant to glucocorticoids and immunosuppressants [28–30].

There is emerging evidence that all patients with PCNSV do not require the same therapy. Relapses/recurrences were recorded in only 26 % of the Mayo Clinic series [3]. Patients with relapsing disease required longer therapy, but otherwise had outcomes similar to those without relapses.

Therapy appears to be associated with a favorable outcome in most patients. In the Mayo Clinic series most patients with low disability at diagnosis continued to have low disability at last follow-up, while most of the 22 patients with severe disability at diagnosis had less disability at follow-up [3].

Serial MRI and MRA (4–6 weeks after the beginning of treatment, then every 3–4 months during the first year of treatment, or in case of new neurological deficit), as well as repeat careful neurological examinations are useful to monitor the disease course. In patients with stable imaging but worsening clinical symptoms, repeat spinal fluid exam and repeat angiography may be indicated.

22.5.1 Lesson from the Mayo Clinic Cohort of Patients with Adult PCNSV

A recent report described the treatments and outcomes and evaluated the findings at diagnosis predicting the response to treatment and outcome in a series of 163 consecutive patients with PCNSV who were seen at the Mayo Clinic over a 29-year period [31]. 157 patients received glucocorticoids (median starting dose of oral prednisone: 60 mg/day). In 66 patients, intravenous (IV) pulse methylprednisolone therapy preceded oral prednisone. 75 patients were initially treated with prednisone alone, while in 82 patients prednisone was combined with a second drug, mainly CYC (72 patients were treated with oral or intermittent IV pulses). A favorable response was observed in most of the patients treated with prednisone alone or in combination with CYC. Response rates were similar (about 83 %) in both treatment groups with improvement of disability (Rankin scale scores) over time. 72 % of the patients achieved a sustained therapeutic response (no relapses) during follow-up. The median duration of all therapy was around 11 months in both treatment groups. No differences in outcomes (disability and mortality) were observed in the 2 treatment groups; the only difference was the frequency of relapses. Treatment with glucocorticoids alone was associated with more frequent relapses (39 % versus 18 %, $p = 0.006$). Patients with relapses needed longer therapy compared with those without relapses (median duration: 18 months versus 9 months; $p < 0.001$), but relapses were not associated with increased mortality or worse disability (Rankin score) at the last follow-up visit.

This study also evaluated clinical characteristics by diagnosis associated with treatment response, relapses and inability to discontinue treatment at the last follow up. Large-vessel involvement (OR 6.14) and cerebral infarcts on MRI at diagnosis (OR 3.32) were significantly associated with a poor response to treatment, while prominent gadolinium-enhanced cerebral lesions or meninges assessed by MRI (OR 2.28) were significantly associated with longer therapy, which was often being continued at the time of last follow-up. No other findings except the treatment with prednisone alone (OR 2.90) were associated with relapse.

Most of the 15 patients initially treated with an immunosuppressive agent different from CYC (mainly azathioprine or mycophenolate mofetil) had a favorable response, suggesting in some patients the possible use of a less toxic alternative to CYC for the induction of remission [31].

We also evaluated the association of clinical findings at diagnosis with Rankin score outcomes at last follow-up and survival [23]. High disability scores at last follow-up and increased mortality were both significantly associated with increasing age (OR 1.44 and HR 1.39, respectively) and presence of cerebral infarction observed on MRI at presentation (OR 3.74 and HR4.44, respectively), while patients with gadolinium-enhanced meninges or lesions on MRI (OR 0.35 and HR 0.20, respectively) had lower disability and less risk of death. Patients with amyloid angiopathy (0R 0.24) had lower disability at follow-up, while diagnosis by angiography alone (HR 3.28) compared with biopsy and the presence of large vessel involvement on angiograms (HR 4.98) were significantly associated with an increased mortality. These differences were related to the different size of cerebral vessels involved by the inflammatory process. Patients with rapidly progressive PCNSV and often fatal outcome were characterized by the angiographic presence of bilateral, multiple, large vessel lesions and MRI evidence of multiple cerebral infarctions. They represented the worst end of the clinical spectrum of PCNSV [8, 10]. A more benign course was associated with angiography-negative patients but biopsy evident involvement of small cortical and leptomeningeal vessels often presenting with a cognitive disorder and MRI evidence of prominent leptomeningeal enhancement [5, 6]. Patients with Aβ-related angiitis (ABRA) defined by deposition of amyloid-β in the media and adventitia of small cortical and leptomeningeal vessels belong to this clinically less aggressive subset [7, 22]. In view of these findings we proposed a treatment algorithm mainly based on the size of the vessels involved by the inflammation (Fig. 22.1) [23]. In patients with inflammation restricted to small cortical and leptomeningeal vessels who have a more benign disease, prednisone alone was recommended as initial therapy (initial dose of 1 mg/Kg/d), while in patients with more severe large/proximal vessel disease and in those with a rapidly progressive course, high-dose intravenous methylprednisolone (1,000 mg daily for 3–5 days) and CYC can be used to attempt to induce remission immediately after diagnosis. There is insufficient reported experience to suggest replacing CYC with the less

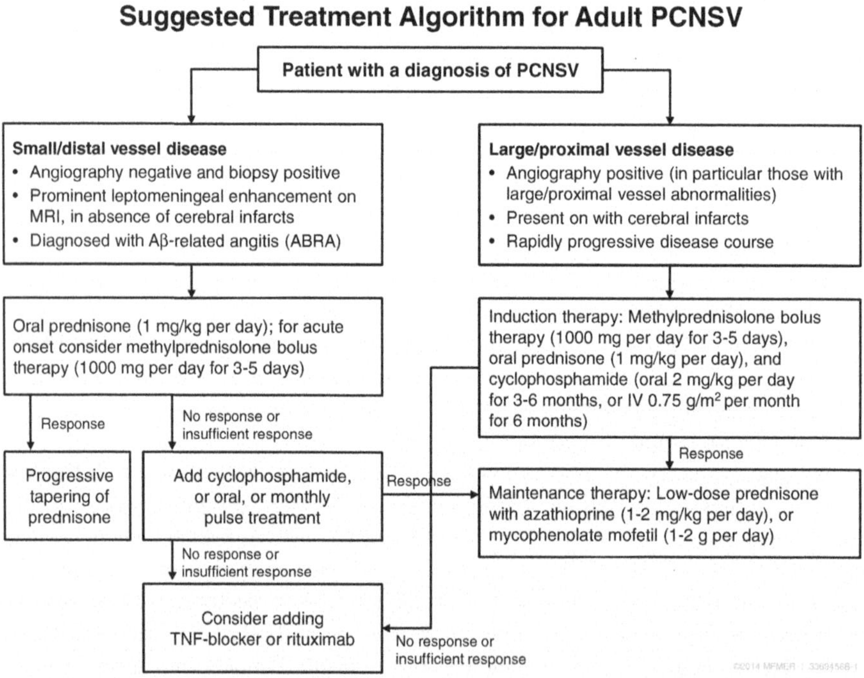

Fig. 22.1 Suggested treatment algorithm for primary central nervous system vasculitis (PCNSV) (Reproduced from Salvarani and colleagues [23], by permission of Wolters Kluwer Health, Inc.)

toxic AZA or MMF for the induction of remission. However, these two immunosuppressors appear to be effective for the maintenance of remission. Methotrexate was seldom used in published studies, and therefore was not included in the algorithm. A small number of case reports have shown the efficacy of tumor necrosis factor-alpha blockers and rituximab in adult PCNSV [28–30], indicating these agents may be helpful in patients who are intolerant or respond poorly to CYC.

A recent manuscript evaluated the efficacy and safety of MMF in PCNSV and compared long-term neurological outcomes in patients treated with this drug and in those receiving other therapies [32]. Most of the 11 patients who were initially treated with MMF or received this drug for a recurrence of vasculitis went into disease remission and did not have flares while on treatment. Four of the 5 patients who received MMF as maintenance treatment continued in remission without flares at last visit and all 4 were able to suspend/reduce GCs. Furthermore, the patients treated with MMF had less severe disability at last follow-up compared to those receiving CYC and prednisone. Only 1 patient had important toxicity (leukopenia) requiring stopping the drug. The overall results indicated that MMF is an effective and safe therapy for adult PCNSV. However, more experience is needed to determine if MMF combined with GCs reliably works as first line induction or maintenance therapy in PCNSV. The efficacy of MMF was also found in children with PCNSV. Hutchinson et al observed that, in children with small vessel childhood PCNSV, MMF was more effective and safer than AZA as maintenance steroid-sparing therapy after induction therapy with GCs and CYC [33]. A different small case series reported that MMF was also effective in 3 children with PCNSV refractory to the combination of GCs and another first-line immunosuppressive agent [34].

Our understanding of PCNSV and the delineation of its spectrum and subsets has advanced but further study is needed to clarify methods of diagnosis and optimal management.

References

1. Salvarani C, Giannini C, Miller DV, Hunder G (2006) Giant cell arteritis: involvement of intracranial arteries. Arthritis Rheum 55:985–989
2. Cravioto H, Feigin I (1959) Noninfectious granulomatous angiitis with a predilection for the nervous system. Neurology 9:599–609
3. Salvarani C, Brown RD Jr, Calamia KT, Christianson TJ, Weigand SD, Miller DV, Giannini C, Meschia JF, Huston J 3rd, Hunder GG (2007) Primary central nervous system vasculitis: analysis of 101 patients. Ann Neurol 62:442–451
4. Moore PM (1989) Diagnosis and management of isolated angiitis of the central nervous system. Neurology 39:167–173
5. Salvarani C, Brown RD Jr, Calamia KT et al (2008) Angiography-negative primary central nervous system vasculitis: a syndrome involving small cerebral vessels. Medicine (Baltimore) 87:264–271
6. Salvarani C, Brown RD Jr, Calamia KT et al (2008) Primary central nervous system vasculitis with prominent leptomeningeal enhancement: a subset with a benign outcome. Arthritis Rheum 58:595–603

7. Salvarani C, Brown RD Jr, Calamia KT et al (2008) Primary central nervous system vasculitis: comparison of patients with and without cerebral amyloid angiopathy. Rheumatology (Oxford) 47:1671–1677
8. Salvarani C, Brown RD Jr, Calamia KT et al (2011) Rapidly progressive primary central nervous system vasculitis. Rheumatology (Oxford) 50:349–358
9. Salvarani C, Brown RD Jr, Calamia KT et al (2011) Primary central nervous system vasculitis presenting with intracranial hemorrhage. Arthritis Rheum 63:3598–3606
10. Salvarani C, Brown RD Jr, Morris JM et al (2014) Catastrophic primary central nervous system vasculitis. Clin Exp Rheumatol 32(Suppl 82):S3–S4
11. Salvarani C, Brown RD Jr, Hunder GG (2012) Adult primary central nervous system vasculitis. Lancet 380(9843):767–777
12. Salvarani C, Brown RD Jr, Hunder GG (2012) Adult primary central nervous system vasculitis: an update. Curr Opin Rheumatol 24:46–52
13. Calabrese LH, Mallek JA (1988) Primary angiitis of the central nervous system. Report of 8 new cases, review of the literature, and proposal for diagnostic criteria. Medicine (Baltimore) 67:20–39
14. Kadkhodayan Y, Alreshaid A, Moran CJ, Cross DT 3rd, Powers WJ, Derdeyn CP (2004) Primary angiitis of the central nervous system at conventional angiography. Radiology 233:878–882
15. Duna GF, Calabrese LH (1995) Limitations of invasive modalities in the diagnosis of primary angiitis of the central nervous system. J Rheumatol 22:662–667
16. Eleftheriou D, Cox T, Saunders D, Klein NJ, Brogan PA, Ganesan V (2010) Investigation of childhood central nervous system vasculitis: magnetic resonance angiography versus catheter cerebral angiography. Dev Med Child Neurol 52(9):863–867
17. Küker W, Gaertner S, Nagele T, Dopfer C, Schoning M, Fiehler J, Rothwell PM, Herrlinger U (2008) Vessel wall contrast enhancement: a diagnostic sign of cerebral vasculitis. Cerebrovasc Dis 26:23–29
18. Salvarani C, Brown RD Jr, Huston J 3rd, Hunder GG (2008) Prominent perivascular enhancement in primary central nervous system vasculitis. Clin Exp Rheumatol 26(3 Suppl 49):S111
19. Swartz RH, Bhuta SS, Farb RI, Agid R, Willinsky RA, Terbrugge KG, Butany J, Wasserman BA, Johnstone DM, Silver FL, Mikulis DJ (2009) Intracranial arterial wall imaging using high-resolution 3-tesla contrast-enhanced MRI. Neurology 72:627–634
20. Miller DV, Salvarani C, Hunder GG, Brown RD, Parisi JE, Christianson TJ, Giannini C (2009) Biopsy findings in primary angiitis of the central nervous system. Am J Surg Pathol 33:35–43
21. Salvarani C, Brown RD Jr, Calamia KT, Christianson TJ, Huston J 3rd, Meschia JF, Giannini C, Miller DV, Hunder GG (2008) Primary CNS vasculitis with spinal cord involvement. Neurology 70:2394–2400
22. Salvarani C, Hunder GG, Morris JM et al (2013) Aβ-related angiitis: comparison with CAA without inflammation and primary CNS vasculitis. Neurology 81:1596–1603
23. Salvarani C, Brown RD Jr, Christianson T et al (2015) An update of the Mayo Clinic cohort of patients with adult primary central nervous system vasculitis: description of 163 patients. Medicine (Baltimore) 94(21):e738. doi:10.1097/MD.0000000000000738
24. Giannini C, Salvarani C, Hunder G, Brown RD (2012) Primary central nervous system vasculitis: pathology and mechanisms. Acta Neuropathol 123:759–772
25. Molloy ES, Singhal AB, Calabrese LH (2008) Tumour-like mass lesion: an under-recognised presentation of primary angiitis of the central nervous system. Ann Rheum Dis 67:1732–1735
26. Ducros A, Bousser MG (2009) Reversible cerebral vasoconstriction syndrome. Pract Neurol 9:256–267
27. Cupps TR, Moore PM, Fauci AS (1983) Isolated angiitis of the central nervous system. Prospective diagnostic and therapeutic experience. Am J Med 74:97–105

28. Salvarani C, Brown RD Jr, Calamia KT, Huston J 3rd, Meschia JF, Giannini C, Miller DV, Hunder GG (2008) Efficacy of tumor necrosis factor alpha blockade in primary central nervous system vasculitis resistant to immunosuppressive treatment. Arthritis Rheum 59:291–296
29. De Boysson H, Arquizan C, Guillevin L, Pagnoux C (2013) Rituximab for primary angiitis of the central nervous system: report of 2 patients from the French COVAC cohort and review of the literature. J Rheumatol 40:2102–2103
30. Salvarani C, Brown RD Jr, Huston J 3rd et al (2014) Treatment of primary central nervous system vasculitis with rituximab: case report. Neurology 82:1287–1288
31. Salvarani C, Brown RD Jr, Christianson TJ et al (2015) Adult primary central nervous system vasculitis treatment and course: analysis of one hundred sixty-three patients. Arthritis Rheumatol 67:1637–1645
32. Salvarani C, Brown RD Jr, Christianson TJ et al (2015) Mycophenolate mofetil in primary central nervous system vasculitis. Semin Arthritis Rheum 45:55–59
33. Hutchinson C, Elbers J, Halliday W et al (2010) Treatment of small vessel primary CNS vasculitis in children: an open-label cohort study. Lancet Neurol 9:1078–1084
34. Sen ES, Leone V, Abinun M et al (2010) Treatment of primary angiitis of the central nervous system in childhood with mycophenolate mofetil. Rheumatology (Oxford) 49:806–811

Chapter 23
Diagnosis and Therapy for Peripheral Vasculitic Neuropathy

Franz Blaes

Abstract Vasculitic neuropathy may occur as a part of systemic vasculitis or an isolated vasculitis of the peripheral nervous system. The typical clinical syndrome is mononeuropathia multiplex, but distal-symmetric neuropathy can also be observed. Neurophysiological examination reveals axonal damage in most cases and nerve biopsy shows inflammatory infiltrates together with vessel wall damage. Treatment includes steroids, cyclophosphamide, azathioprine, rituximab and other immunosuppressants. This chapter provides an overview about clinical, laboratory and histopathological diagnostic criteria and the current treatment options for vasculitic neuropathy.

Keywords Vasculitis • Neuropathy • Immunosuppressive treatment • Rheumatology

23.1 Introduction

Vasculitic neuropathies are a group of inflammatory neuropathies, characterised by inflammation of the vasa nervorum, mainly the epineural arteries of the nerve. They can occur as non-systemic vasculitis (non-systemic vasculitic neuropathy NSVN, exclusively affecting the peripheral nervous system) or part of a systemic vasculitis including the involvement of other organs (systemic vasculitic neuropathy) (Table 23.1). The differential diagnosis may be difficult, if the neuropathy is the first manifestation of vasculitis. The exact incidence of vasculitic neuropathy has never been investigated in bigger studies. However, the annual incidence of systemic vasculitis is 60–140/million, which includes a part of 30 % secondary systemic vasculitis [1, 2]. In nerve biopsy specimen obtained in neuropathy of unknown reason, about 1 % show vasculitic neuropathy [3, 4]. Some systemic vasculitic diseases are rarely associated with neuropathy, whereas in others, such as Churg-Strauss syndrome, it belongs to the diagnostic criteria (Table 23.2) [5].

F. Blaes (✉)
Department of Neurology, Gummersbach Academic Teaching Hospital, Wilhelm-Breckow-Allee 20, 51643 Gummersbach, Germany
e-mail: franz.blaes@klinikum-oberberg.de

F. Dammacco et al. (eds.), *Systemic Vasculitides: Current Status and Perspectives*, DOI 10.1007/978-3-319-40136-2_23

Table 23.1 Classification of vasculitic diseases according to [8, 9]

Primary systemic vasculitis
1. Small vessel vasculitis
Microscopic polyangiitis
Eosinophilic granulomatosis with polyangiitis (Churg-Strauss-Syndrome)
Granulomatosis with polyangiitis (GPA)
Essential mixed cryoglobulinemia (non-Hepatitis C)
Henoch-Schönlein purpura
2. Medium vessel vasculitis
Polyarteritis nodosa
3. Large vessel vasculitis
Giant cell arteritis
Secondary systemic vasculitis
1. Connective tissue diseases
Rheumatoid arthritis
Systemic lupus erythematosus
Sjögren's syndrome
Systemic sclerosis
Dermatomyositis
Mixed connective tissue disease
2. Sarcoidosis
3. Behçet's disease
4. Infections (Hepatitis B and C, HIV, CMV and others)
5. Drugs
6. Malignancy
7. Inflammatory bowel disease
8. Hypocomplementemic urticarial vasculitis syndrome
Non-systemic or localised vasculitis
1. Non-systemic vasculitic neuropathy
2. Diabetic/non-diabetic radiculoplexus neuropathy (DRPLN/RPLN)
3. Localised cutaneous or neuropathic vasculitis

23.2 Classification

In 1990, the American College of Rheumatology presented classification criteria for PAN, CSS, GPA and others. However, this classification was criticized because (1) there was no discrimination between vasculitis from non-vasculitic disease and (2) ANCA testing was not included in the classification. The Chapel Hill consensus conference (CHCC) in 1994 proposed definitions for most vasculitic diseases according to the size and pathology of the involved vessels [6]. In 2007, a new classification incorporating both ACR and CHCC criteria was established and Wegener's

Table 23.2 Frequency of neuropathy in vasculitic diseases (according to [77, 85])

Disease	Frequency	Reference
Primary systemic vasculitis		
Giant cell arteritis	Rare	[101, 102]
Polyarteritis nodosa	65–85 %	[50, 103]
Churg-Strauss syndrome	65–80 %	[22]
Granulomatosis with polyangiitis	5–50 %	[50, 104]
Microscopic polyangitis	6–75 %	[50]
Cryoglobulinemia	30–70 %	[44, 105]
Secondary systemic vasculitis		
1. Connective tissue diseases		
Systemic lupus erythematosus	20–27 %	[106]
Rheumatoid arthritis	15–70 %	[107, 108]
Sjögren syndrome	30–45 %	[109, 110]
Systemic sclerosis (scleroderma)	5–30 %	[44, 111]
2. Others		
Infections (HBV, HCV, HIV)	5–70 %	[44, 50, 73]
Sarcoidosis	5–10 %	[68]
Malignancy	Rare	[112]
Drugs	Rare	[75, 113, 114]

granulomatosis was renamed granulomatosis with polyangiitis [7]. The 2012 second CHCC updated some of the diagnostic criteria for vasculitis. More recently, it has been proposed to divide the vasculitic neuropathies in two groups according to the size of the affected nerve vessels, distinguishing a nerve large arteriole vasculitis from a nerve microvasculitis. The latter involves arterioles <40 μm and endoneurial microvessels and includes the most non-systemic vasculitic neuropathy, Sjögren syndrome and some virus-associated vasculitic neuropathies [8]. Most neurologists also use the classification developed from the Peripheral Nerve Society Task Force. Vasculitic neuropathy is categorized in primary systemic vasculitis, secondary systemic vasculitis or non-systemic/localized vasculitis depending on the disease-associations (Table 23.1) [9].

23.3 Pathogenesis

Blood supply in the peripheral nervous system is secured by the regional vasa nervorum, which feed an extensive network between the epineurial and the endoneurial vessels. This allows a functionality of the peripheral nerve even under anaerobic conditions and makes the nerve quite resistant to ischemic damage [10]. The underlying etiology of the vasculitic disease may be different, and, in some diseases, not completely understood. However, occlusion of vessels by vascular inflammation leading to ischemic nerve damage is the common final path in vasculitic neuropathy.

This ischemic damage occurs diffusely in the whole nerve, but with a maximum effect in the proximal and middle parts of the nerves, which is the most vulnerable zone to ischemia [11]. In cryoglobulinemia, anti-sulfatide antibodies can be found, which may be involved in the pathogenesis [12]. An increased expression of nerve growth factor (NGF) may be involved in the pain development of vasculitic neuropathy [13].

23.4 Clinical Features and Diagnostic Procedures

The typical clinical presentation of vasculitis is mononeuropathia multiplex. However, about 30–60 % of patients with vasculitis have other clinical types of neuropathy including painful sensorimotor axonal neuropathy or pure sensory neuropathy or asymmetric neuropathy. Pain is a regular symptom in most vasculitic neuropathies. In biopsy-proven vasculitic neuropathy, 10–40 % are distal-symmetric pattern [14–16]. The peroneal and tibial nerves on the lower and the ulnar nerve on the upper limb are most frequently involved in vasculitic neuropathy. However, there is no association of a distinct clinical pattern with a special type of vasculitic disease. The majority of vasculitic neuropathies develop subacute within days to weeks and only in a few cases, a chronic, slowly progressing neuropathy has been observed. About 80 % of neuropathies associated with systemic vasculitis, but also 50 % of patients with non-systemic vasculitic neuropathies have general symptoms (weight loss, fever, myalgia, or fatigue).

The neurophysiological examination shows multifocal axonal neuropathy including reduced CMAP amplitudes with only slightly reduced nerve conduction velocities (NCV) in neurography [14, 17, 18]. However, if the CMAP amplitudes are massively reduced, the NCV can also be reduced because of the loss of thick myelinated fibres. A transient nerve conduction block can be observed early after symptom onset and represents an ongoing Wallerian degeneration, which has not been completed distal of the affected nerve; however, within 1–2 weeks, this phenomenon disappears. Electromyography reveals neurogenic pattern including spontaneous muscle fiber activity, and polyphasic, extended, or high-amplitude motor unit action potentials (MUAP). However, if an additional vasculitic myopathy is also present, myopathic and neurogenic changes can be observed at the same time [9, 19].

Whenever systemic vasculitis or another reason for the neuropathy is not known, a variety of laboratory investigations should be performed. As a first step, routine laboratory investigations should be performed in neuropathies with unknown reason. If an inflammatory neuropathy is suspected, a more detailed laboratory investigation should be performed (Table 23.3). Analysis of the cerebrospinal fluid will not increase the sensitivity to detect vasculitic neuropathy, but can be an important investigation in the differential diagnosis of an as yet unclear neuropathy. Other routine technical diagnostic procedures depend on the suspected diagnosis (Table 23.4).

Table 23.3 Laboratory investigations in suspected vasculitic neuropathy (modified according to [85])

Basic neuropathy screening	Vasculitis suspected
Full blood count	Antinuclear antibodies (ANA)
Erythrocyte sedimentation rate	Anti-neutrophil cytoplasmic antibodies (ANCA,
Vitamin B6, B12	including exact determination of the antibody)
C-reactive protein	Extractable nuclear antigens (ENA)
Fasting glucose (2 consec. days)	Rheumatoid factor
Electrolytes	Anti-CCP antibodies
Renal and liver function	Cryoglobulins
Creatinkinase	HIV serology
Serum protein immunofixation	Urine analysis (microalbuminuria ?)
Hepatitis B and C serology	Cerebrospinal fluid analysis
Thyroid function	Angiotensin-converting enzyme
	Soluble interleukin-2 receptor
	Antineuronal antibodies
	Serum complement C3, C4

Table 23.4 Technical investigations in suspected vasculitic neuropathy

Method	Neuropathy
Chest X-Ray	Sarcoidosis, paraneoplastic neuropathy
CT Chest/Abdomen	Paraneoplastic neuropathy
Visceral angiography	Polyarteritis nodosa (PAN)
Salivary gland biopsy	Sjögren syndrome, Mikulicz syndrome

If a patient develops subacute asymmetrical neuropathy or mononeuropathia multiplex without evidence for systemic vasculitis, nerve biopsy should be performed. In most cases, the sural nerve is used with or without muscle biopsy. The combined biopsy of the superficial peroneal nerve together with the peronaeus brevis muscle is an alternative to the sural nerve biopsy [20]. Although controlled studies are lacking, the combined nerve/muscle biopsy has a slightly higher sensitivity to detect vasculitis and the biopsy of the peronaeus brevis muscle may be more effective than the gastrocnemius muscle. One study used a proximal muscle for biopsy (quadriceps) and showed no increased yield for vasculitic neuropathy compared to nerve biopsy alone, indicating that distal muscles are more suitable for biopsy in suspected vasculitis [14].

23.4.1 Pathology

The histopathological diagnosis of vasculitic neuropathy requires different criteria, which have been set up in detail in a guideline of the Peripheral Nerve Society [9]. The definite diagnosis of vasculitis includes both intramural inflammation (Fig. 23.1)

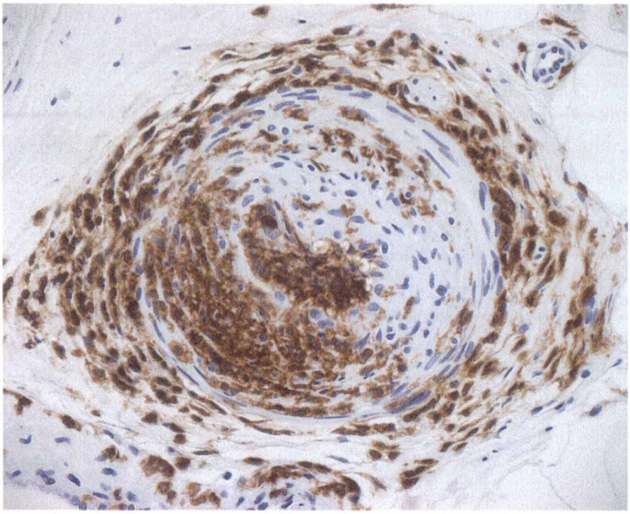

Fig. 23.1 Peripheral nerve vasculitis. Epineurial vessel with lymphocyte infiltration and subtotal stenosis. Lymphocytes stained with anti-LCA (lymphocyte common antigen) antibody (Courtesy of Prof. J. Weis, Aachen)

Table 23.5 Diagnostic criteria for definite vasculitic neuropathy (according to [9])

1. Active lesion: nerve biopsy showing collection of inflammatory cells in vessel wall and one or more signs of acute vascular damage:
(a) Fibrinoid necrosis
(b) Loss/disruption of endothelium
(c) Loss/fragmentation of smooth muscle cells in media
(d) Acute thrombosis
(e) Vascular/perivascular hemorrhage
(f) Leucocytoclasia
2. Chronic lesion with signs of healing/repair: nerve biopsy showing collection of mononuclear inflammatory cells in vessel wall and one or more signs of chronic vascular damage with repair:
(a) Intimal hyperplasia
(b) Fibrosis of media
(c) Adventitial/periadventitial fibrosis
(d) Chronic thrombosis with recanalisation

and additionally vessel damage (Table 23.5). It can be distinguished between findings in active lesions and chronic lesions. However, not all patients meet the criteria for definite vasculitic neuropathy, although vasculitis is clinically highly suspected. For these cases, criteria for probable vasculitic neuropathy have been established (Table 23.6). The histopathological picture shows mainly axonal damage, and the predominant blood vessels affected are epineurial more than peri- or endoneurial. The cellular infiltrates include mainly T-lymphocytes and macrophages, B-cells are rarely seen. In PAN and CSS eosinophils may also be present. Immune complex

Table 23.6 Diagnostic criteria for probable vasculitic neuropathy (according to [9])

1. Pathologic criteria for definite vasculitic neuropathy not fulfilled and
2. Predominantly axonal changes and
3. Perivascular inflammation accompanied by signs of active or chronic vascular damage; or Perivascular/vascular inflammation plus at least one additional class II or III pathologic predictor of definite vasculitic neuropathy:
(a) Vascular deposition of complement, IgM, or fibrinogen by direct immunofluorescence
(b) Hemosiderin deposits
(c) Asymmetric nerve fibre loss or degeneration
(d) Prominent active axonal degeneration
(e) Myofiber necrosis, regeneration or infarcts in peronaeus brevis muscle biopsy

deposits consisting of complement proteins, immunoglobulins and fibrinogen may be present [21, 22].

23.5 Primary Systemic Vasculitides

Neuropathies associated with primary systemic vasculitides have been classified according to the diameter of the affected blood vessels into three groups: large-vessel-, medium-vessel-, and small-vessel vasculitis. Vasculitic neuropathy can be observed predominantly in the small-vessel- and medium-vessel vasculitides, since the vessel diameters in nerves and muscles are mainly in a range between 50 and 300 μm.

23.5.1 Large Vessel Vasculitides

The large vessel vasculitides include Takayasu arteritis and giant cell arteritis. Takayasu arteritis is normally not associated with neuropathies, but shows significant central nervous system involvement, mainly strokes. Giant cell arteritis patients often have central nervous system involvement, in some rare cases, neuropathies have been described [23].

23.5.2 Medium-Sized Vessel Vasculitides

23.5.2.1 Polyarteritis Nodosa (PAN)

The polyarteritis nodosa is a very rare disease, the annual incidence decreased to 0.1–1.6 cases/million in developed countries after establishment of hepatitis B vaccination [24]. The histopathological features have been revised in the CHCC. PAN is a disease of the small and medium-sized arteries, sparing arterioles, capillaries and venules. The vascular inflammation is segmental, often predominantly seen in branching points with mixed inflammatory infiltrates [25]. In later stages vascular remodelling with intima hyperplasia and diffuse fibrotic changes can be found. PAN can develop as idiopathic form, but also associated with viral infections. Hepatitis B was the main virus inducing PAN, but hepatitis C virus, cytomegalovirus, HIV, parvovirus B19 and even streptococci can also trigger the disease. The clinical symptoms include constitutional symptoms, such as fever, weight loss and arthralgia, but also organ involvement of different organ symptoms, of which skin symptoms and nervous system involvement are the most frequent. In contrast to other primary systemic vasculitides, PAN is a typical monophasic disease with a relapse rate of less than 10 %. Heart and central nervous system involvement determines a poor prognosis. The French vasculitis group proposed a five factor score for estimation of the prognosis of PAN. Neuropathies can be found in about 75 % of PAN patients [26]. The typical manifestation is a painful mononeuropathia multiplex, but distal-symmetric sensorimotor neuropathies and even chronic inflammatory demyelinating polyneuropathy (CIDP) can be observed as peripheral nervous system involvement in PAN [5]. There is no typical laboratory abnormality, blood sedimentation rate and C-reactive protein are often elevated and PAN is not associated with ANCA. The diagnosis can be confirmed by the typical histological features in biopsy and/or the detection of microaneurysms in the intestinal angiography.

23.5.3 Small Vessel Vasculitides

23.5.3.1 Churg-Strauss Syndrome (CSS)

This vasculitis type was first described in 1951 by Churg and Strauss. A development over different phases can be observed: (1) a prodromal phase with asthma or rhinitis, which can last for years before the vasculitic symptoms develop, (2) an intermittent phase with eosinophilia and pulmonary eosinophilic infiltrates, and (3) the main vasculitic manifestation including cutaneous, gastrointestinal symptoms, sinusitis, arthralgia and vasculitic neuropathy. PNS involvement is frequent (75–80 % of patients) and neuropathy can be the initial manifestation in a substantial proportion of patients [27]. The main clinical syndrome is mononeuropathia multiplex, some patients may have pure sensory or sensorimotor distal neuropathy [28, 29]. Neurophysiological examination shows typically an axonal damage. Eosinophilia in the peripheral blood can be found regularly and in 40–70 % of patients, pANCA are positive [30]. Eosinophilic infiltrates are present in biopsy specimen of nerve and other tissues [29]. One study suggests two different histopathological types of neuropathy in CSS patients. In this study, ANCA-positive CSS patients have predominantly necrotizing vasculitis in the nerve biopsy, whereas ANCA-negative CSS nerve biopsies show a large number of eosinophilic infiltrates in the epineurium [29].

23.5.4 Granulomatosis with Polyagiitis (GPA)

GPA is a vasculitic disease, regularly associated with c-ANCA, which are directed against proteinase 3. Mostly, patients have granulomatous involvement of the upper and lower respiratory tract and additional rapid-rapidly progressive glomerulonephritis. Subsequent development of vasculitic disease is common and many patients show the full-generalised form including pulmonary and renal involvement. Interestingly, the c-ANCA seems to be directly involved in the physiopathology by activation and degranulation of granulocytes. The cANCA bind to surface-expressed proteinase-3, which has been translocated from inside the cell by the proinflammatory cytokines TNF-α and IL-1. The degranulation of the granulocytes then induces a necrotizing vasculitis with endothelial damage.

Vasculitic neuropathy can be observed in 20–25 % of GPA patients [31, 32]. Most of them show mononeuritis multiplex, in a few patients, cranial nerve involvement has been described [33]. The neuropathy may be the first symptom of GPA in some patients, which underlines the importance of ANA/ANCA diagnostic tests in patients with newly diagnosed asymmetric neuropathy [27].

23.5.5 Microscopic Polyangiitis (MPA)

MPA is a systemic, necrotizing vasculitis of the small vessels, mainly in older patients with a slight male predominance. Patients have lung involvement, glomerulonephritis, skin lesions and abdominal pain. In 50–75 % of patients p-ANCA can be found, some patients can have c-ANCA [34]. About 10–50 % of MPA patients develop neuropathy. The most frequent clinical manifestation is mononeuropathia multiplex, mainly affecting the peroneal, median and ulnar nerves [33, 35, 36].

23.6 IgG4-Related Disease

Immunoglobulin G4-related disease was recently recognized as a common pathophysiology in a heterogeneous group of diseases and can affect a variety of organ systems [37]. One of the first diseases described is the Mikulicz syndrome, an inflammatory disease of the salivary gland, which was already described in the late nineteenth century [38]. The main finding is an infiltration of the affected tissue with IgG4+ plasma cells; the IgG4 level in the serum may be elevated [39]. Single cases of an IgG4-associated neuropathy were reported, but the incidence of IgG4-related neuropathy is unknown yet [40].

23.7 Secondary Systemic Vasculitis

Secondary systemic vasculitis is a heterogeneous disease group. Infectious diseases, connective tissue diseases, malignancies and drugs can cause vasculitis including vasculitic neuropathy.

23.7.1 Systemic Lupus Erythematosus (SLE)

A variety of neurological disturbances can be observed in SLE, including cerebral vasculitis, which often leads to stroke, cerebral venous thrombosis, transverse myelitis, or peripheral neuropathy. The latter can be observed in about 10–20 % of SLE patients [41, 42]. Clinically, the neuropathy in SLE is mainly distal-symmetric sensory or sensorimotor neuropathy and less frequent mononeuropathia multiplex or small fibre [41]. Neurophysiological investigation reveals axonal type in 80–90 % and demyelinating type in 10–20 % of SLE-associated neuropathy. Autonomic disturbances are present in one third of the neuropathy patients, including both sympathetic and parasympathetic nervous system involvement [43].

23.7.2 Systemic Sclerosis (SSc)

About 30 % of systemic sclerosis patients have neuropathy, predominantly sensory or small-fibre type [44]. Additionally, autonomic dysfunction causes mainly gut motility disturbances in these patients [45]. In some cases of systemic sclerosis, neuropathy can be the initial presentation of the disease.

23.7.3 Sjögren's Syndrome

Due to an affection of the exocrine glands, Sjögren's syndrome is clinically defined by the so-called sicca complex (dry eyes/dry mouth). Anti-SSA and –SSB autoantibodies, a subgroup of ENA can be found regularly in these patients [46]. Interestingly, autoantibodies against the muscarinic M3 receptor seem to have an important pathophysiological role in the sialoadenitis and could also be involved in the nervous system manifestations of these patients [47]. Both central and peripheral nervous system involvement including trigeminal nerve affection has been described [48]. Neuropathies have been found in 2–64 % of the patients. These neuropathies can be symmetric or asymmetric or resemble small fibre neuropathy and only a small part of them are vasculitic [49].

23.7.4 Rheumatoid Arthritis (RA)

About 15–50 % of RA patients develop neuropathy. However, peripheral nervous system disease in RA can have a variety of origins, such as drug-induced neuropathy or entrapment mononeuropathy. The main form of polyneuropathy is an axonal, distal-symmetric sensorimotor neuropathy [50].

23.7.5 Paraneoplastic Vasculitic Neuropathy

Until now, every clinical type of neuropathy was found in paraneoplastic (tumor-associated) neuropathy. However, pure sensory neuronopathy (Denny-Brown) is the most classical type of paraneoplastic neuropathy, and this special neuropathy is mainly associated with small cell lung cancer (SCLC) and anti-Hu antibodies can be found (Fig. 23.2). In general, SCLC, breast and ovarian cancer as well as lymphoma are the most frequent tumors found in paraneoplastic neuropathy [51]. Patients with vasculitic paraneoplastic neuropathy have mostly mononeuropathia multiplex or asymmetric neuropathy. Torvik and Berntzen in 1968 and Johnson et al. 1979 described the first patients with vasculitic neuropathy associated with tumors (renal cell carcinoma, SCLC or lymphoma) [52, 53]. Oh described 26 patients with paraneoplastic vasculitic neuropathy, mainly associated with SCLC and lymphoma. Many of them had elevated CSF protein and high ESR [54]. Clinically, mononeuritis multiplex and distal-symmetric sensorimotor neuropathy can be observed. If paraneoplastic vasculitic neuropathy is suspected, antineuronal autoantibodies, as well as antinuclear antibodies (ANA) and ANCA should be examined. In some patients with paraneoplastic neuropathies, both vasculitic and non-vasculitic, only ANA can be found [55]. Non-vasculitic paraneoplastic neuropathies, such as the anti-Hu positive sensory neuronopathy Denny-Brown, do not respond to immunotherapy, whereas vasculitic paraneoplastic neuropathy may respond to immunosuppressants (steroids, cyclophosphamide) in the majority of patients.

23.7.6 Hepatitis C/Cryoglobulinemia

Cryoglobulinemia can occur essential or in the context of chronic infections or lymphoproliferative diseases. The cryoglobulins are mono- or polyclonal immunoglobulins, that precipitate at cool temperatures and these proteins can induce vascular damage by occlusion of microvessels or induction of immune complex deposits in small vessels [56–58]. Fifty per cent of hepatitis C patients have mixed cryoglobulinemia, but only 15 % of them develop the clinical syndrome of mixed cryoglobulinemic vasculitis (MCV). About 80 % of MCV lacking other reasons are caused by

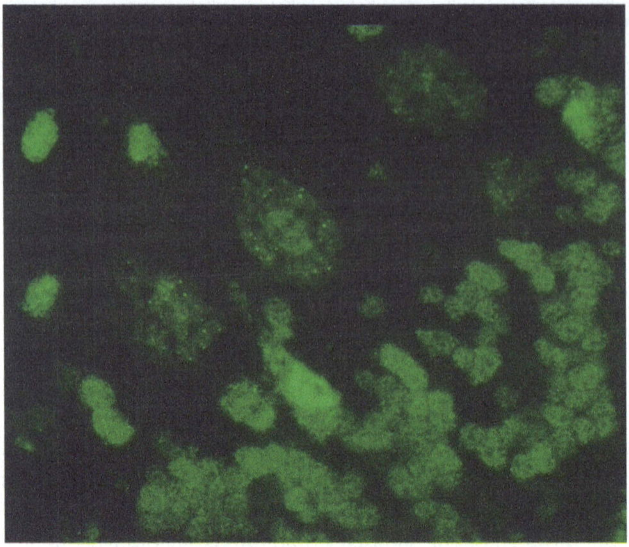

Fig. 23.2 Antineuronal antibody (antu-Hu) from a patient with vasculitic neuropathy and small cell lung cancer. Staining of neuronal nuclei and (to a lesser extent) cytoplasm of cerebellar neurons

chronic hepatitis C. The clinical symptoms consist of skin lesions (purpura), neuropathy, glomerulonephritis, ulcers, arthritis, and sicca syndrome. Rheumatoid factor can be found in more than 80 % and decreased complement factors C3 and C4 occur in 70 % of MCV patients [56]. The pathophysiological role of anti-sulfatide and anti-ganglioside antibodies, which can be detected in a part of MCV-associated neuropathy, is still not elucidated yet [12].

Neuropathy can be asymmetric, mononeuropathia multiplex or distal-symmetric, the latter often showing a slowly progressive course, whereas the others develop acute or subacute [59, 60]. Pure sensory neuropathy has also been described [50]. Pain occurs in 50 % of MCV-associated neuropathy and predominant small-fibre neuropathies have also been observed [59]. Neurophysiological examination reveals axonal neuropathy and even without clinical symptoms, most patients with MCV show at least neurophysiological abnormalities suggestive of neuropathy [61–63]. Nerve biopsy studies show axonal damage, perivascular infiltration with mononuclear cells predominantly in the epineurium and IgM and complement deposits in the affected vessel walls [59, 63, 64].

23.7.7 Other Secondary Systemic Vasculitides

Sarcoidosis can be associated with mononeuropathia multiplex, small fibre neuropathy or typical CIDP. The latter shows a good response to intravenous immunoglobulins (IVIg), the others respond regularly to steroids [65–68].

Behcet's disease and inflammatory bowel disease both are rarely associated with neuropathy and most of them are non-vasculitic. In both diseases, central nervous system involvement is much more frequent [69, 70].

A vasculitic neuropathy in HIV has been described in association with cytomegalovirus (CMV) or with lymphoma. This disease may be the result of immune complex deposition more than direct HIV infection of the nervous system [71–73]. However, there is an increased risk for HIV patients to develop other secondary vasculitides, including PAN or MPA.

23.7.8 Drug-Induced Vasculitic Neuropathy

A variety of drugs can induce vasculitis, both ANCA-positive and –negative. These drugs include many of the new biologicals (etanercept, adalimumab, infliximab and others) as well as other medications, such as carbimazole or levamisole. Some drugs have especially been reported to induce vasculitic neuropathy. However, to prove a causal relationship is sometimes difficult, since some of the drugs are given as a treatment of vasculitic diseases. The antibiotic minocycline can induce a non-systemic vasculitic neuropathy [74, 75]. Other drugs, such as naproxen, penicillin or the antiepileptic phenytoin induced systemic vasculitis including neuropathy, and the discontinuation of the drug resulted in an improvement of the vasculitis [76]. Table 23.7 gives an overview about drugs, which can induce vasculitic neuropathy.

23.8 Non-systemic Vasculitis of the Peripheral Nervous System

If vasculitic neuropathy occurs without detectable systemic involvement, the term non-systemic vasculitic neuropathy (NSVN) is used. About 25 % of all vasculitic neuropathies belong to this group [44]. The course of NSVN is often subacute, but about one third of the patients show progressive disease. The typical clinical picture is an asymmetric, progressive and painful neuropathy with severe paresis [77]. The neurophysiological examination shows axonal sensorimotor or motor neuropathy. Mostly, NSVN remains a localized vasculitis; however, in some cases additional skin manifestations or generalization was observed and the neuropathy tends to relapse, when immunosuppressive treatment is reduced [78, 79]. Nerve biopsy

Table 23.7 Drugs inducing vasculitic neuropathy

Drug	Disease	Reference
Valacyclovir	Mononeuritis multiplex	[115]
Minocycline	Non-systemic vasculitic neuropathy	[75, 113, 114]
Ipililumab	Biopsy-proven vasculitic neuropathy	[116]
Bortezomib	Microvasculitic motor predominant neuropathy	[117]
Rituximab	Mononeuritis multiplex	[118]
Naproxene	Leukocytoclastic vasculitis including neuropathy	[76]
Propylthiouracil	ANCA-positive vasculitis including neuropathy	[119]

should be performed, if NSVN is suspected and histopathology shows nerve microvasculitis. Recently, the main clinical and histopathological features were published in a guideline of the peripheral nerve society, describing criteria for definite, probable and possible vasculitic neuropathy [9].

Treatment of NSVN includes corticosteroids, cyclophosphamide, methotrexate and azathioprine in the first line and is described in detail in the treatment section. In a recent single-center cohort of 60 patients with histologically proven NSVN, initially all patients improved after iv methylprednisolone. However, after 4 years, 48 % of the patients still had immunosuppressive treatment and an age <64 years was associated with a better prognosis [80].

23.9 Diabetic and Non-Diabetic Lumbosacral Radiculoplexus Neuropathy (DLRPN/LRPN)

In patients with diabetes, the lumbosacral plexus including roots and peripheral nerves can be affected in a often painful disease, termed diabetic lumbosacral radiculoplexus neuropathy (DLRPN) [81]. The disease starts often with acute, severe neuropathic pain, followed by asymmetric paresis of the lower limbs, which become disabling during the acute stage. DLRPN is a monophasic disease, but half of the patients have walking difficulties or become wheelchair bound in this stage [8]. Autonomic symptoms can be found in 50 % of DLRPN patients. Spontaneous recovery may occur, but most patients have incomplete recovery including weakness and sensory disturbances. Interestingly, it mainly affects patients with mild diabetes mellitus in a stable situation; a non-diabetic form (LRPN) has also been described. Although upper limb nerves can be involved in DLRPN, a separate upper limb variant exists (DCRPN) [82]. Moreover, an overlap with CIDP has been described in some patients [83]. The histopathological picture is a focal nerve ischemia, caused by microvasculitis [84]. Neurophysiological examination shows axonal involvement of the lumbosacral plexus and very often includes paraspinal denervation. The cerebrospinal analysis reveals elevated total protein and a normal cell count in most patients.

23.10 Treatment

23.10.1 Treatment of Non-Systemic Vasculitic Neuropathy (NSVN)

There are no randomized controlled studies (RCT) for NSVN yet. However, the Peripheral Nerve Society published recommendations for the treatment of NSVN [9]. It is recommended to treat NSVN patients with corticosteroids (initially prednisolone 1 mg/kg/day) with a slow tapering over months. Initial high-dose prednisolone pulses (500–1000 mg prednisolone for 3–5 days) can be used alternatively, followed by daily treatment with 1 mg/kg. Osteoporosis prophylaxis should be given; it is unclear yet, whether steroid treatment increases the risk of peptic ulcers and whether patients should be treated with proton pump inhibitors prophylactically. In case of rapid progressive neuropathy, cyclophosphamide (CYC) pulse treatment should be considered and long-term immunosuppression with methotrexate or azathioprine is necessary [85]. To reduce the risk of hemorrhagic cystitis, mesna should always be used in CYC pulse therapy and for toxicity reason, CYC should only be given for 6–12 months. There are two cohort studies, which implicate a better efficiency of a combination therapy [16, 79]. Other treatment regimens are poorly investigated. Intravenous immunoglobulins have been used successfully in a few otherwise treatment-resistant vasculitis patients [86]. Plasma exchange seems to have little effect, even in combination treatment [87–89]. Since rituximab treatment shows a good effect also on the neuropathic symptoms in generalized vasculitic diseases, it may also be used for NSVN in otherwise treatment-refractory patients [90]. Since vasculitic neuropathies usually are predominantly axonal, no significant improvement may be seen in the first weeks or even months. However, there is no surrogate marker, by which the treatment efficacy during the treatment can be verified. The first symptom, which may improve under sufficient immunosuppression, is the neuropathic pain.

23.10.2 Treatment of Non-Viral Systemic Vasculitic Neuropathy (SVN)

The treatment of vasculitic neuropathy associated with systemic vasculitis (SVN) should be performed according to the guidelines of the underlying systemic disease. As in NSVN, corticosteroids are used as initial treatment of SVN in the same dosages. Improvement of SVN may last weeks or months because of the axonal damage of the nerves. Therefore, erythrocyte sedimentation rate or C-reactive protein can be used to control the efficacy of the treatment. Suppiah and colleagues recently reported a rate of 15 % clinical apparent neuropathy in a cohort of ANCA-associated vasculitis and 40 % improved after treatment [33]. In severe SVN cases, cyclophosphamide (CYC) is used either additionally or subsequently to corticosteroid treatment. Daily oral cyclophosphamide shows serious side effects, therefore pulse therapy (0.6–0.75 g/m^2 every 2–4 weeks) should be preferred and mesna should be given to avoid hemorrhagic cystitis. Cyclophosphamide treatment should be limited to 6–12 months, since there are a variety of long-term immunosuppressive drugs with less toxic side effects. Methotrexate (20–25 mg weekly) or azathioprine (1–2 mg/kg daily) are the classical long-term immunosuppressants to maintain remission. Leflunomide can be used in the long-term treatment of GPA [91].

Another immunosuppressant in the treatment of vasculitis is mycophenolate mofetil (MMF): However, its effectiveness is not completely clear. One open-label pilot trial showed remission maintenance in 13 out of 17 patients with GPA, in another study relapses were more frequent and earlier in comparison to azathioprine [25, 92]. In lupus treatment, MMF is equally effective to azathioprine but has less side effects [93]. However, no data are available regarding its effect on vasculitic neuropathy.

Rituximab, an anti-CD20 monoclonal antibody, targets mainly B-cells and is established as an effective treatment in MPA and GPA. It has recently been licensed for ANCA-associated vasculitis. In the meantime, rituximab is a first-line therapy of ANCA-associated vasculitis and is as effective as cyclophosphamide [90]. It is also effective in cryoglobulinemic vasculitis and is usually given in a weekly dosage of 375 mg/m^2 for four times [94].

23.10.3 Vasculitis Associated with Infections

Neuropathy associated with mixed cryoglobulinemia/HCV infection includes both antiviral and immunosuppressive treatment. Antiviral treatment includes pegylated interferon-α, ribavirin, telaprevir and boceprevir and, more recently, the direct-acting antiviral agents. Interferon- α (IFN- α) alone or in combination with ribavirin may improve neuropathic symptoms in a smaller part of patients [64, 95, 96]. However, IFN- α is also able to induce inflammatory neuropathies and can also exacerbate other symptoms of mixed cryoglobulinemic vasculitis [97, 98]. Therefore, corticosteroids, cyclophosphamide or plasma exchange should be added in patients with severe neuropathy or if neuropathic symptoms do not improve under antiviral treatment. To remove circulating cryoglobulins, plasma exchange is used in MCV, although there are no RCT yet and only a part of MCV patients respond.

Ferri and collegues reported a cohort of MCV patients responding to rituximab, independently of HCV status [99]. In their study, 95 % of the neuropathic symptoms improved and rituximab was considered safe and effective. In HCV-associated MCV, additional rituximab showed a better response than antiviral treatment alone [100].

References

1. Gonzalez-Gay MA, Garcia-Porrua C (1999) Systemic vasculitis in adults in northwestern Spain, 1988–1997. Clinical and epidemiologic aspects. Medicine 78:292–308
2. Watts RA, Carruthers DM, Scott DG (1995) Epidemiology of systemic vasculitis: changing incidence or definition? Semin Arthritis Rheum 25:28–34
3. Davies L (1994) Vasculitic neuropathy. Bailliere's Clin Neurol 3:193–210
4. Kissel JT, Slivka AP, Warmolts JR, Mendell JR (1985) The clinical spectrum of necrotizing angiopathy of the peripheral nervous system. Ann Neurol 18:251–257
5. Basu N, Watts R, Bajema I, Baslund B, Bley T, Boers M et al (2010) EULAR points to consider in the development of classification and diagnostic criteria in systemic vasculitis. Ann Rheum Dis 69:1744–1750
6. Jennette JC, Falk RJ, Andrassy K, Bacon PA, Churg J, Gross WL et al (1994) Nomenclature of systemic vasculitides. Proposal of an international consensus conference. Arthritis Rheum 37:187–192
7. Watts R, Lane S, Hanslik T, Hauser T, Hellmich B, Koldingsnes W et al (2007) Development and validation of a consensus methodology for the classification of the ANCA-associated vasculitides and polyarteritis nodosa for epidemiological studies. Ann Rheum Dis 66:222–227
8. Gwathmey KG, Burns TM, Collins MP, Dyck PJ (2014) Vasculitic neuropathies. Lancet Neurol 13:67–82
9. Collins MP, Dyck PJ, Gronseth GS, Guillevin L, Hadden RD, Heuss D et al (2010) Peripheral Nerve Society Guideline on the classification, diagnosis, investigation, and immunosuppressive therapy of non-systemic vasculitic neuropathy: executive summary. J Peripher Nerv Syst 15:176–184
10. Beggs J, Johnson PC, Olafsen A, Watkins CJ, Cleary C (1991) Transperineurial arterioles in human sural nerve. J Neuropathol Exp Neurol 50:704–718
11. Dyck PJ, Conn DL, Okazaki H (1972) Necrotizing angiopathic neuropathy. Three-dimensional morphology of fiber degeneration related to sites of occluded vessels. Mayo Clin Proc Mayo Clinic 47:461–475
12. Alpa M, Ferrero B, Cavallo R, Naretto C, Menegatti E, Di Simone D et al (2008) Anti-neuronal antibodies in patients with HCV-related mixed cryoglobulinemia. Autoimmun Rev 8:56–58
13. Yamamoto M, Ito Y, Mitsuma N, Hattori N, Sobue G (2003) Pain related differential expression of NGF, GDNF, IL-6, and their receptors in human vasculitic neuropathies. Intern Med 42:1100–1103
14. Bennett DL, Groves M, Blake J, Holton JL, King RH, Orrell RW et al (2008) The use of nerve and muscle biopsy in the diagnosis of vasculitis: a 5 year retrospective study. J Neurol Neurosurg Psychiatry 79:1376–1381
15. Claussen GC, Thomas TD, Goyne C, Vazquez LG, Oh SJ (2000) Diagnostic value of nerve and muscle biopsy in suspected vasculitis cases. J Clin Neuromuscul Dis 1:117–123
16. Davies L, Spies JM, Pollard JD, McLeod JG (1996) Vasculitis confined to peripheral nerves. Brain 119(Pt 5):1441–1448
17. Cruz Martinez A, Barbado FJ, Ferrer MT, Vazquez JJ, Perez Conde MC, Gil AA (1988) Electrophysiological study in systemic necrotizing vasculitis of the polyarteritis nodosa group. Electromyogr Clin Neurophysiol 28:167–173
18. Seo JH, Ryan HF, Claussen GC, Thomas TD, Oh SJ (2004) Sensory neuropathy in vasculitis: a clinical, pathologic, and electrophysiologic study. Neurology 63:874–878
19. Marcaud V, Gauthier C, Maisonobe T, Hogenhuis J, Morizot-Koutlidis R, Degos CF (2002) Vasculitis confined to the peripheral nervous system: atypical clinical presentation. La Revue de medecine interne/fondee par la Societe nationale francaise de medecine interne 23:558–562

20. Agadi JB, Raghav G, Mahadevan A, Shankar SK (2012) Usefulness of superficial peroneal nerve/peroneus brevis muscle biopsy in the diagnosis of vasculitic neuropathy. J Clin Neurosci 19:1392–1396
21. Collins MP, Periquet-Collins I, Sahenk Z, Kissel JT (2010) Direct immunofluoresence in vasculitic neuropathy: specificity of vascular immune deposits. Muscle Nerve 42:62–69
22. Hattori N, Ichimura M, Nagamatsu M, Li M, Yamamoto K, Kumazawa K et al (1999) Clinicopathological features of Churg-Strauss syndrome-associated neuropathy. Brain 122(Pt 3):427–439
23. Caselli RJ, Daube JR, Hunder GG, Whisnant JP (1988) Peripheral neuropathic syndromes in giant cell (temporal) arteritis. Neurology 38:685–689
24. Hernandez-Rodriguez J, Alba MA, Prieto-Gonzalez S, Cid MC (2014) Diagnosis and classification of polyarteritis nodosa. J Autoimmun 48–49:84–89
25. Hiemstra TF, Walsh M, Mahr A, Savage CO, de Groot K, Harper L et al (2010) Mycophenolate mofetil vs azathioprine for remission maintenance in antineutrophil cytoplasmic antibody-associated vasculitis: a randomized controlled trial. JAMA 304:2381–2388
26. Guillevin L, Pagnoux C, Seror R, Mahr A, Mouthon L, Le Toumelin P (2011) The Five-Factor Score revisited: assessment of prognoses of systemic necrotizing vasculitides based on the French Vasculitis Study Group (FVSG) cohort. Medicine 90:19–27
27. Wolf J, Schmitt V, Palm F, Grau AJ, Bergner R (2013) Peripheral neuropathy as initial manifestation of primary systemic vasculitides. J Neurol 260:1061–1070
28. Moosig F, Bremer JP, Hellmich B, Holle JU, Holl-Ulrich K, Laudien M et al (2013) A vasculitis centre based management strategy leads to improved outcome in eosinophilic granulomatosis and polyangiitis (Churg-Strauss, EGPA): monocentric experiences in 150 patients. Ann Rheum Dis 72:1011–1017
29. Oka N, Kawasaki T, Matsui M, Shigematsu K, Unuma T, Sugiyama H (2011) Two subtypes of Churg-Strauss syndrome with neuropathy: the roles of eosinophils and ANCA. Mod Rheumatol 21:290–295
30. Sable-Fourtassou R, Cohen P, Mahr A, Pagnoux C, Mouthon L, Jayne D et al (2005) Antineutrophil cytoplasmic antibodies and the Churg-Strauss syndrome. Ann Intern Med 143:632–638
31. Cattaneo L, Chierici E, Pavone L, Grasselli C, Manganelli P, Buzio C et al (2007) Peripheral neuropathy in Wegener's granulomatosis, Churg-Strauss syndrome and microscopic polyangiitis. J Neurol Neurosurg Psychiatry 78:1119–1123
32. Cisternas M, Soto L, Jacobelli S, Marinovic MA, Vargas A, Sobarzo E et al (2005) Clinical features of Wegener granulomatosis and microscopic polyangiitis in Chilean patients. Rev Med Chil 133:273–278
33. Suppiah R, Hadden RD, Batra R, Arden NK, Collins MP, Guillevin L et al (2011) Peripheral neuropathy in ANCA-associated vasculitis: outcomes from the European Vasculitis Study Group trials. Rheumatology (Oxford) 50:2214–2222
34. Schonermarck U, Lamprecht P, Csernok E, Gross WL (2001) Prevalence and spectrum of rheumatic diseases associated with proteinase 3-antineutrophil cytoplasmic antibodies (ANCA) and myeloperoxidase-ANCA. Rheumatology (Oxford) 40:178–184
35. Agard C, Mouthon L, Mahr A, Guillevin L (2003) Microscopic polyangiitis and polyarteritis nodosa: how and when do they start? Arthritis Rheum 49:709–715
36. Ribi C, Cohen P, Pagnoux C, Mahr A, Arene JP, Puechal X et al (2010) Treatment of polyarteritis nodosa and microscopic polyangiitis without poor-prognosis factors: a prospective randomized study of one hundred twenty-four patients. Arthritis Rheum 62:1186–1197
37. Beyer G, Schwaiger T, Lerch MM, Mayerle J (2014) IgG4-related disease: a new kid on the block or an old aquaintance? United Eur Gastroenterol J 2:165–172
38. Pieringer H, Parzer I, Wohrer A, Reis P, Oppl B, Zwerina J (2014) IgG4- related disease: an orphan disease with many faces. Orphanet J Rare Dis 9:110
39. Perez Alamino R, Martinez C, Espinoza LR (2013) IgG4-associated vasculitis. Curr Rheumatol Rep 15:348

40. Ohyama K, Koike H, Takahashi M, Kawagashira Y, Iijima M, Watanabe H et al (2015) Immunoglobulin G4-related pathologic features in inflammatory neuropathies. Neurology 85:1400–1407
41. Jasmin R, Sockalingam S, Ramanaidu LP, Goh KJ (2015) Clinical and electrophysiological characteristics of symmetric polyneuropathy in a cohort of systemic lupus erythematosus patients. Lupus 24:248–255
42. Oomatia A, Fang H, Petri M, Birnbaum J (2014) Peripheral neuropathies in systemic lupus erythematosus: clinical features, disease associations, and immunologic characteristics evaluated over a twenty-five-year study period. Arthritis Rheumatol 66:1000–1009
43. Tekatas A, Koca SS, Tekatas DD, Aksu F, Dogru Y, Pamuk ON (2014) R-R interval variation and sympathetic skin response in systemic lupus erythematosus. Clin Rheumatol 33:65–70
44. Collins MP, Periquet MI (2008) Isolated vasculitis of the peripheral nervous system. Clin Exp Rheumatol 26:S118–S130
45. Di Ciaula A, Covelli M, Berardino M, Wang DQ, Lapadula G, Palasciano G et al (2008) Gastrointestinal symptoms and motility disorders in patients with systemic scleroderma. BMC Gastroenterol 8:7
46. Kyriakidis NC, Kapsogeorgou EK, Tzioufas AG (2014) A comprehensive review of autoantibodies in primary Sjogren's syndrome: clinical phenotypes and regulatory mechanisms. J Autoimmun 51:67–74
47. Sumida T, Tsuboi H, Iizuka M, Hirota T, Asashima H, Matsumoto I (2014) The role of M3 muscarinic acetylcholine receptor reactive T cells in Sjogren's syndrome: a critical review. J Autoimmun 51:44–50
48. Morgen K, McFarland HF, Pillemer SR (2004) Central nervous system disease in primary Sjogrens syndrome: the role of magnetic resonance imaging. Semin Arthritis Rheum 34:623–630
49. Birnbaum J (2010) Peripheral nervous system manifestations of Sjogren syndrome: clinical patterns, diagnostic paradigms, etiopathogenesis, and therapeutic strategies. Neurologist 16:287–297
50. Schaublin GA, Michet CJ Jr, Dyck PJ, Burns TM (2005) An update on the classification and treatment of vasculitic neuropathy. Lancet Neurol 4:853–865
51. Blaes F (2012) Paraneoplastic neurological syndromes–diagnosis and management. Curr Pharm Des 18:4518–4525
52. Johnson PC, Rolak LA, Hamilton RH, Laguna JF (1979) Paraneoplastic vasculitis of nerve: a remote effect of cancer. Ann Neurol 5:437–444
53. Torvik A, Berntzen AE (1968) Necrotizing vasculitis without visceral involvement. Postmortem examination of three cases with affection of skeletal muscles and peripheral nerves. Acta Med Scand 184:69–77
54. Oh SJ (1997) Paraneoplastic vasculitis of the peripheral nervous system. Neurol Clin 15:849–863
55. Tschernatsch M, Stolz E, Strittmatter M, Kaps M, Blaes F (2005) Antinuclear antibodies define a subgroup of paraneoplastic neuropathies: clinical and immunological data. J Neurol Neurosurg Psychiatry 76:1702–1706
56. De Vita S, Soldano F, Isola M, Monti G, Gabrielli A, Tzioufas A et al (2011) Preliminary classification criteria for the cryoglobulinaemic vasculitis. Ann Rheum Dis 70:1183–1190
57. Ferri C (2008) Mixed cryoglobulinemia. Orphanet J Rare Dis 3:25
58. Ramos-Casals M, Stone JH, Cid MC, Bosch X (2012) The cryoglobulinaemias. Lancet 379:348–360
59. Gemignani F, Brindani F, Alfieri S, Giuberti T, Allegri I, Ferrari C et al (2005) Clinical spectrum of cryoglobulinaemic neuropathy. J Neurol Neurosurg Psychiatry 76:1410–1414
60. Taieb G, Maisonobe T, Musset L, Cacoub P, Leger JM, Bouche P (2010) Cryoglobulinemic peripheral neuropathy in hepatitis C virus infection: clinical and anatomical correlations of 22 cases. Rev Neurol 166:509–514

61. Ammendola A, Sampaolo S, Ambrosone L, Ammendola E, Ciccone G, Migliaresi S et al (2005) Peripheral neuropathy in hepatitis-related mixed cryoglobulinemia: electrophysiologic follow-up study. Muscle Nerve 31:382–385
62. Ciompi ML, Marini D, Siciliano G, Melchiorre D, Bazzichi L, Sartucci F et al (1996) Cryoglobulinemic peripheral neuropathy: neurophysiologic evaluation in twenty-two patients. Biomed Pharmacother 50:329–336
63. Valli G, De Vecchi A, Gaddi L, Nobile-Orazio E, Tarantino A, Barbieri S (1989) Peripheral nervous system involvement in essential cryoglobulinemia and nephropathy. Clin Exp Rheumatol 7:479–483
64. Cacoub P, Maisonobe T, Thibault V, Gatel A, Servan J, Musset L et al (2001) Systemic vasculitis in patients with hepatitis C. J Rheumatol 28:109–118
65. Marques IB, Giovannoni G, Marta M (2014) Mononeuritis multiplex as the first presentation of refractory sarcoidosis responsive to etanercept. BMC Neurol 14:237
66. Saito H, Yamaguchi T, Adachi Y, Yamashita T, Wakai Y, Saito K et al (2015) Neurological symptoms of Sarcoidosis-induced small fiber neuropathy effectively relieved with high-dose steroid pulse therapy. Intern Med 54:1281–1286
67. Singhal NS, Irodenko VS, Margeta M, Layzer RB (2015) Sarcoid polyneuropathy masquerading as chronic inflammatory demyelinating polyneuropathy. Muscle Nerve 52:664–668
68. Vital A, Lagueny A, Ferrer X, Louiset P, Canron MH, Vital C (2008) Sarcoid neuropathy: clinico-pathological study of 4 new cases and review of the literature. Clin Neuropathol 27:96–105
69. Kararizou E, Davaki P, Speggos K, Stamboulis E (2012) Rare association of polyneuropathy and Crohn's disease: a clinicopathological study of 4 cases. Pol J path 63:261–266
70. Noel N, Drier A, Wechsler B, Piette JC, De Paz R, Dormont D et al (2014) Neurological manifestations of Behcet's disease. La Revue de medecine interne/fondee par la Societe nationale francaise de medecine interne 35:112–120
71. Johnson RM, Barbarini G, Barbaro G (2003) Kawasaki-like syndromes and other vasculitic syndromes in HIV-infected patients. AIDS 17(Suppl 1):S77–S82
72. Roullet E, Assuerus V, Gozlan J, Ropert A, Said G, Baudrimont M et al (1994) Cytomegalovirus multifocal neuropathy in AIDS: analysis of 15 consecutive cases. Neurology 44:2174–2182
73. Said G, Lacroix C (2005) Primary and secondary vasculitic neuropathy. J Neurol 252:633–641
74. Baratta JM, Dyck PJ, Brand P, Thaisetthawatkul P, Dyck PJ, Engelstad JK et al (2016) Vasculitic neuropathy following exposure to minocycline. Neurol Neuroimmunol Neuroinflamm 3:e180
75. Thaisetthawatkul P, Sundell R, Robertson CE, Dyck PJ (2011) Vasculitic neuropathy associated with minocycline use. J Clin Neuromuscul Dis 12:231–234
76. Schapira D, Balbir-Gurman A, Nahir AM (2000) Naproxen-induced leukocytoclastic vasculitis. Clin Rheumatol 19:242–244
77. Vrancken AF, Said G (2013) Vasculitic neuropathy. Handb Clin Neurol 115:463–483
78. Cassereau J, Baguenier-Desormeaux C, Letournel F, Lavigne C, Leclech C, Dubas F et al (2012) Necrotizing vasculitis revealed in a case of multiple mononeuropathy after a 14-year course of spontaneous remissions and relapses. Clin Neurol Neurosurg 114:290–293
79. Collins MP, Periquet MI, Mendell JR, Sahenk Z, Nagaraja HN, Kissel JT (2003) Nonsystemic vasculitic neuropathy: insights from a clinical cohort. Neurology 61:623–630
80. Uceyler N, Geng A, Reiners K, Toyka KV, Sommer C (2015) Non-systemic vasculitic neuropathy: single-center follow-up of 60 patients. J Neurol 262:2092–2100
81. Collins MP (2012) The vasculitic neuropathies: an update. Curr Opin Neurol 25:573–585
82. Massie R, Mauermann ML, Staff NP, Amrami KK, Mandrekar JN, Dyck PJ et al (2012) Diabetic cervical radiculoplexus neuropathy: a distinct syndrome expanding the spectrum of diabetic radiculoplexus neuropathies. Brain 135:3074–3088

83. Garces-Sanchez M, Laughlin RS, Dyck PJ, Engelstad JK, Norell JE, Dyck PJ (2011) Painless diabetic motor neuropathy: a variant of diabetic lumbosacral radiculoplexus Neuropathy? Ann Neurol 69:1043–1054
84. Dyck PJ, Norell JE, Dyck PJ (1999) Microvasculitis and ischemia in diabetic lumbosacral radiculoplexus neuropathy. Neurology 53:2113–2121
85. Blaes F (2015) Diagnosis and therapeutic options for peripheral vasculitic neuropathy. Ther Adv Musculoskelet Dis 7:45–55
86. Levy Y, Uziel Y, Zandman GG, Amital H, Sherer Y, Langevitz P et al (2003) Intravenous immunoglobulins in peripheral neuropathy associated with vasculitis. Ann Rheum Dis 62:1221–1223
87. Geltner D (1981) The place of plasmapheresis in immune complex disease. Harefuah 101:77–78
88. Geltner D, Kohn RW, Gorevic P, Franklin EC (1981) The effect of combination therapy (steroids, immunosuppressives, and plasmapheresis) on 5 mixed cryoglobulinemia patients with renal, neurologic, and vascular involvement. Arthritis Rheum 24:1121–1127
89. Monti G, Saccardo F (2007) Emergency in cryoglobulinemic syndrome: what to do? Dig Liver Dis 39(Suppl 1):S112–S115
90. Stone JH, Merkel PA, Spiera R, Seo P, Langford CA, Hoffman GS et al (2010) Rituximab versus cyclophosphamide for ANCA-associated vasculitis. N Engl J Med 363:221–232
91. Metzler C, Miehle N, Manger K, Iking-Konert C, de Groot K, Hellmich B et al (2007) Elevated relapse rate under oral methotrexate versus leflunomide for maintenance of remission in Wegener's granulomatosis. Rheumatology (Oxford) 46:1087–1091
92. Silva F, Specks U, Kalra S, Hogan MC, Leung N, Sethi S et al (2010) Mycophenolate mofetil for induction and maintenance of remission in microscopic polyangiitis with mild to moderate renal involvement–a prospective, open-label pilot trial. Clin J Am Soc Nephrol 5:445–453
93. Maneiro JR, Lopez-Canoa N, Salgado E, Gomez-Reino JJ (2014) Maintenance therapy of lupus nephritis with mycophenolate or azathioprine: systematic review and meta-analysis. Rheumatology (Oxford) 53:834–838
94. De Vita S, Quartuccio L, Isola M, Mazzaro C, Scaini P, Lenzi M et al (2012) A randomized controlled trial of rituximab for the treatment of severe cryoglobulinemic vasculitis. Arthritis Rheum 64:843–853
95. Ferri C, Marzo E, Longombardo G, Lombardini F, La Civita L, Vanacore R et al (1993) Interferon-alpha in mixed cryoglobulinemia patients: a randomized, crossover-controlled trial. Blood 81:1132–1136
96. Rieu V, Cohen P, Andre MH, Mouthon L, Godmer P, Jarrousse B et al (2002) Characteristics and outcome of 49 patients with symptomatic cryoglobulinaemia. Rheumatology (Oxford) 41:290–300
97. Boonyapisit K, Katirji B (2002) Severe exacerbation of hepatitis C-associated vasculitic neuropathy following treatment with interferon alpha: a case report and literature review. Muscle Nerve 25:909–913
98. Stubgen JP (2009) Interferon alpha and neuromuscular disorders. J Neuroimmunol 207:3–17
99. Ferri C, Cacoub P, Mazzaro C, Roccatello D, Scaini P, Sebastiani M et al (2011) Treatment with rituximab in patients with mixed cryoglobulinemia syndrome: results of multicenter cohort study and review of the literature. Autoimmun Rev 11:48–55
100. Dammacco F, Tucci FA, Lauletta G, Gatti P, De Re V, Conteduca V et al (2010) Pegylated interferon-alpha, ribavirin, and rituximab combined therapy of hepatitis C virus-related mixed cryoglobulinemia: a long-term study. Blood 116:343–353
101. Chia L, Fernandez A, Lacroix C, Adams D, Plante V, Said G (1996) Contribution of nerve biopsy findings to the diagnosis of disabling neuropathy in the elderly. A retrospective review of 100 consecutive patients. Brain 119(Pt 4):1091–1098

102. Koike H, Sobue G (2013) Clinicopathological features of neuropathy in anti-neutrophil cytoplasmic antibody-associated vasculitis. Clin Exp Nephrol 17:683–685
103. Ohyama K, Koike H, Masuda M, Sone J, Hashimoto R, Tomita M et al (2013) Autonomic manifestations in acute sensory ataxic neuropathy: a case report. Auton Neurosci 179:155–158
104. de Souza FH, Radu Halpern AS, Valente Barbas CS, Shinjo SK (2010) Wegener's granulomatosis: experience from a Brazilian tertiary center. Clin Rheumatol 29:855–860
105. Cacoub P, Saadoun D (2008) Hepatitis C virus infection induced vasculitis. Clin Rev Allergy Immunol 35:30–39
106. Servioli L, Perez C, Consani S, Suarez A, Sehabiaga G, Collazo C et al (2007) Prevalence and characteristics of immunomediated neuropathies in a group of patients with autoimmune diseases. J Clin Neuromuscul Dis 9:285–290
107. Agarwal V, Singh R, Wiclaf CS, Tahlan A, Ahuja CK et al (2008) A clinical, electrophysiological, and pathological study of neuropathy in rheumatoid arthritis. Clin Rheumatol 27:841–844
108. Voskuyl AE, Hazes JM, Zwinderman AH, Paleolog EM, van der Meer FJ, Daha MR et al (2003) Diagnostic strategy for the assessment of rheumatoid vasculitis. Ann Rheum Dis 62:407–413
109. Delalande S, de Seze J, Fauchais AL, Hachulla E, Stojkovic T, Ferriby D et al (2004) Neurologic manifestations in primary Sjogren syndrome: a study of 82 patients. Medicine 83:280–291
110. Terrier B, Lacroix C, Guillevin L, Hatron PY, Dhote R, Maillot F et al (2007) Diagnostic and prognostic relevance of neuromuscular biopsy in primary Sjogren's syndrome-related neuropathy. Arthritis Rheum 57:1520–1529
111. Dyck PJ, Hunder GG, Dyck PJ (1997) A case-control and nerve biopsy study of CREST multiple mononeuropathy. Neurology 49:1641–1645
112. Oh SJ, Slaughter R, Harrell L (1991) Paraneoplastic vasculitic neuropathy: a treatable neuropathy. Muscle Nerve 14:152–156
113. Kermani TA, Ham EK, Camilleri MJ, Warrington KJ (2012) Polyarteritis nodosa-like vasculitis in association with minocycline use: a single-center case series. Semin Arthritis Rheum 42:213–221
114. Ogawa N, Kawai H, Yamakawa I, Sanada M, Sugimoto T, Maeda K (2010) Case of minocycline-induced vasculitic neuropathy. Rinsho Shinkeigaku 50:301–305
115. Pary LF, Henszel A, Kelkar P (2004) Vasculitic mononeuritis multiplex induced by valacyclovir. Neurology 62:1906–1907
116. Manousakis G, Koch J, Sommerville RB, El-Dokla A, Harms MB, Al-Lozi MT et al (2013) Multifocal radiculoneuropathy during ipilimumab treatment of melanoma. Muscle Nerve 48:440–444
117. Mauermann ML, Blumenreich MS, Dispenzieri A, Staff NP (2012) A case of peripheral nerve microvasculitis associated with multiple myeloma and bortezomib treatment. Muscle Nerve 46:970–977
118. Mauermann ML, Ryan ML, Moon JS, Klein CJ (2007) Case of mononeuritis multiplex onset with rituximab therapy for Waldenstrom's macroglobulinemia. J Neurol Sci 260:240–243
119. Frigui M, Kechaou M, Haddouk S, Masmoudi A, Kaddour N, Masmoudi H et al (2008) Benzylthiouracil induced ANCA-positive vasculitis: study of three cases and review of the literature. Ann Endocrinol (Paris) 69:517–522

Chapter 24
Childhood Uveitis

Alice Brambilla, Rolando Cimaz, and Gabriele Simonini

Abstract Pediatric uveitis embraces a group of inflammatory diseases affecting the vascular layer of the eye. Among the broad spectrum of possible etiologies, juvenile idiopathic arthritis stands for the most common cause of anterior chronic uveitis in Western countries. Despite being considered a rare disease, non-infectious chronic uveitis is a serious and disabling sight-threatening condition accounting for up to 10 % of pathologies leading to blindness. Visual complications arise as a consequence of persistent or recurrent ocular inflammation, but also as result of chronic steroid treatment. Targeted antimicrobial treatment is necessary for infectious uveitis. On the other hand, non-infectious uveitis is managed trough a "step-by-step" approach, in order to control local inflammation, achieve a corticosteroid-sparing effect and reduce the risk of visual complications. Therapeutic options include corticosteroids, conventional immune-modulatory therapy and tumor necrosis factor α [TNF-α] antagonists. Preliminary evidence suggests a possible role of non anti-TNF-α modifier immunosuppressive treatment for refractory cases, accounting for about 25 % of total patients. Given the high cost and the lack of long-term safety data, the experience with these agents is still limited to few cases managed in highly specialized centers.

24.1 Introduction

Uveitis embraces a group of severe and disabling inflammatory diseases affecting the vascular layer of the eye (uvea). Different population-based studies reported an annual incidence of roughly 22.6–52.4/100,000 person-years, with higher rates documented in developed Western countries. Differences by sex have been reported, women being affected more than men at almost any age. Approximately 5–10 % of total cases develop during childhood, with an estimated incidence of 4–7/100,000 children/year and a prevalence of 28/100,000 children/year [1]. Despite being considered a rare disease, it represents a sight-threatening condition accounting for up to 10 % of pathologies leading to blindness.

A. Brambilla • R. Cimaz • G. Simonini (✉)
Rheumatology Unit-NEUROFARBA Department, Anna Meyer Children Hospital, University of Florence, Florence, Italy
e-mail: gabriele.simonini@unifi.it

F. Dammacco et al. (eds.), *Systemic Vasculitides: Current Status and Perspectives*, DOI 10.1007/978-3-319-40136-2_24

According to the anatomical extension of ocular inflammation they can be classified as anterior (iritis and iridocyclitis), intermediate (pars-planitis, hyalitis) and posterior uveitis (choroiditis, chorioretinitis, retinitis, neuroretinitis). The term panuveitis refers to the broad involvement of anterior chamber, vitreous and choroid/retina.

The clinical course may be acute, chronic or recurrent. Acute uveitis is generally symptomatic and tends to a complete resolution within 3 months from its onset. Children commonly show conjunctival hyperemia, photophobia, ocular pain, lacrimation and visual loss [2].

Acute-recurrent uveitis is characterized by the relapse of disease after 3 months of remission, whereas the persistence of disease with prompt (within 3 months) relapses after discontinuation of therapy defines the condition of chronic uveitis. Children affected by chronic uveitis may be asymptomatic and frequently present bilateral involvement.

The most common causes of childhood uveitis are reported in Table 24.1. Among the broad spectrum of possible etiologies, juvenile idiopathic arthritis (JIA) stands for the most common cause of anterior chronic uveitis in childhood. It is responsible for 1.8–47 % of total cases, with the higher incidence recorded in Western countries.

Infectious diseases are obviously more frequent in underdeveloped countries. The main role is played by Toxoplasmosis and Herpes virus infection, but also HIV and CMV infections are becoming more frequent.

Systemic vasculitis and autoimmune conditions may be complicated with ocular inflammation as well. Kawasaki disease, Systemic Lupus Erithematosus, Bechet's disease and Inflammatory Bowel Diseases need periodic ophthalmological evaluation to assess potential ocular involvement.

Masquerade syndrome refers to those conditions presenting with intraocular infiltrating cells not related to immune-mediated mechanisms. Hematologic malig-

Table 24.1 Common causes of uveitis in children

Etiologic group	Disease
Infectious disease	**Bacterial**: Syphilis, Tuberculosis, Lyme Disease, Brucellosis, Cat Scratch Disease, Leprosy
	Viral: Herpes simplex virus 1–2, Cytomegalovirus, Epstein-Barr Virus, Varicella-Zoster Virus, Mumps, Rubella
	Fungal: Aspergillosis, Coccidioidomycosis, Histoplasmosis, Blastomycosis, Candidiasis, Cryptococcosis
	Parasitic: Toxocariasis Toxoplasmosis, Pneumocystosis
Chronic inflammatory disease	Juvenile idiopathic arthritis, Psoriasis, Inflammatory bowel diseases
Autoimmune condition	Systemic lupus erythematosus, Sjögren Disease
Tumor	Leukemia, Lymphoma, Neuroblastoma
Vasculitis	Behçet's disease, Systemic lupus erythematosus, Kawasaki disease, Sarcoidosis, Polyarteritis nodosa, Wegener's granulomatosis
Other	Vogt-Koyanagi Harada Syndrome, Blau Disease, Tubulo-interstitial nephritis and uveitis

nancies (i.e. leukemia, intraocular lymphoma), retinoblastoma, retinal detachment or degeneration and intraocular trauma stand for the most common causes of masquerade syndrome.

The cases without an identifiable origin are addressed as "idiopathic" and represent nearly half of total patients.

A close collaboration between pediatric rheumatologist and pediatric ophthalmologist is fundamental in order to define the proper diagnostic work up and therapeutic pathway.

24.2 JIA-Associated Uveitis

According to International League of Associations for Rheumatology (ILAR) criteria, JIA is classified in seven different subtypes: Systemic, Oligoarthritis, Polyarthritis (Rheumatic Factor-negative), Polyarthritis (RF-positive), Psoriatic, Enthesitis-related arthritis and Undifferentiated arthritis. Among these, Oligoarthritis, Polyarthritis RF-negative and Psoriatic arthritis have the higher risk to develop secondary uveitis, especially in female patients. Enthesitis-related arthritis is complicated with uveitis in up to 20 % of cases, generally presenting with acute uveitis affecting male teen-agers [3].

Regardless of the subtype of arthritis, a younger age at diagnosis is associated with a higher risk of secondary uveitis. This seems especially true for girls under 7 years of age.

Positive ANA titre is considered a risk factor as well, since 65–90 % of patient with JIA-associated arthritis present incremented ANA levels. Conversely, no correlation with Rheumatic Factor has been documented.

Children affected by JIA develop uveitis in up to 50 % within 3 months and in up to 90 % within 4 years from the diagnosis. Only 2–7 % of patients are diagnosed with uveitis before the onset of arthritis. Ocular inflammation may also appear for the first time during adult age.

Patients affected by uveitis may present a severe articular involvement, however the presence of ocular inflammation does not seem to affect the long-term prognosis of JIA. The clinical course of uveitis and arthritis may be completely independent as well.

24.3 Complications

Compared to adults, childhood uveitis is characterized by poor prognosis and higher risk of secondary complications, with considerable socio-economic burden.

Visual complications are reported in up to 80 % of patients after 3 years and in almost 100 % of patients after 20 years of disease. Nowadays, uveitis represents the third leading cause of blindness in developed countries. Ocular complications arise as a consequence of persistent or recurrent ocular inflammation, but also as result of chronic steroid treatment [4].

The most common complications include cataract (19–81 % of patients), glaucoma (8–38 %), band keratopathy (7–10 %), synechiae (8–75 %), cystoid macular edema (8–42 %), ocular hypotony (19 %), retinal detachment, retinal ischemia and optic atrophy. Up to 30 % of patients show reduced visual acuity and up to 10 % develop blindness. From 28 to 70 % of affected children may require surgical therapy.

The final visual outcome is influenced by negative prognostic factors such as the severity of illness, posterior ocular involvement, presence of complications at the diagnosis, long duration of the disease, younger age at onset and delayed assessment in a specialized centre.

24.4 Screening in JIA

The diagnosis of uveitis is based on complete ocular examination with slit-lamp. According to ILAR, an ophthalmologic evaluation should be performed at the time of diagnosis and periodically repeated in all children, regardless of the absence of symptoms. The frequency of ocular examination is defined on the basis of the subtype of arthritis, the age at onset and the presence of ANA (Table 24.2).

Table 24.2 Suggested screening intervals for uveitis in patients with JIA as classified by International League of Associations for Rheumatology (ILAR) criteria

JIA Subtype	ANA	Age at JIA onset (years)	JIA duration (years)	Recommended screening intervals (months)
OA, RF-PA, PsA, AA	Positive	≤6	≤4	3
OA, RF-PA, PsA, AA	Positive	≤6	>4	6
OA, RF-PA, PsA, AA	Positive	≤6	≥7	12
OA, RF-PA, PsA, AA	Positive	>6	≤2	6
OA, RF-PA, PsA, AA	Positive	>6	>2	12
OA, RF-PA, PsA, AA	Negative	≤6	≤4	6
OA, RF-PA, PsA, AA	Negative	≤6	>4	12
OA, RF-PA, PsA, AA	Negative	>6	NA	12
ERA	NA	NA	NA	12
RF + PA, SyA	NA	NA	NA	12

OA Oligoarthritis, *RF-PA* Polyarthritis (RF-negative), *FR + PA* Polyarthritis (RF-positive), *PsA* Psoriatic arthritis, *ERA* Enthesitis-related arthritis, *SyA* Systemic arthritis, *AA* Other arthritis. *NA* not applicable

24.5 Treatment

Targeted antimicrobial treatment is necessary for infectious uveitis.

Treatment for non-infectious uveitis is based on a "step-by-step" approach, in order to control local inflammation, achieve a corticosteroid-sparing effect and reduce the risk of visual complications [5, 6]. Local steroid therapy associated to mydriatics is proposed for mild-to-moderate conditions, especially in case of anterior involvement. Severe ocular inflammation may instead require oral or intravenous systemic steroid treatment. Chronic corticosteroid administration is limited by the possible side effects, such as cataract, glaucoma, hypertension, hyperglycemia, failure to thrive, hypercholesterolemia and gastro-intestinal bleeding.

In corticosteroid-resistant and corticosteroid-dependent cases systemic immunomodulatory agents should be considered. They comprise both immunosuppressive agents (Methotrexate [MTX], Azathioprine, Mycophenolate Mofetil, Cyclosporine) and biologic response modifiers.

For patients intolerant or non-responders to MTX, biologic therapies represent a valid alternative option. Biologics are genetically engineered proteins aimed at inhibition of specific components of the immune system (cytokines) responsible for inducing and maintaining inflammation. The main adverse events include potential allergic reaction and increased risk of reactivation of latent infections.

24.5.1 Methotrexate

Methotrexate is a competitive inhibitor of dihydrofolate reductase, the enzyme implied in the conversion of dihydrofolate to the active tetrahydrofolate. The blockage of this pathway leads to the inhibition of RNA transcription and DNA synthesis, especially in B and T lymphocytes. MTX represents the first choice among systemic long-term therapies for uveitis [7]. The recommended dosage is 10–15 mg/m^2 once a week (orally or subcutaneously) and the therapeutic effect is generally achieved 6–10 weeks later. In a recent meta-analysis MTX induced a reduction of ocular inflammation in up to 73 % of children affected by autoimmune chronic uveitis refractory to steroid therapy [8]. The main collateral events due to MTX administration involve gastrointestinal problems (nausea, vomiting, stomach-ache, diarrhea) and increased transaminase levels. The association of folic acid 24 h after MTX administration and the preference for subcutaneous injections are recommended in order to reduce the incidence of nausea. Moreover, periodical clinical and laboratory follow-up are requested (every 4 weeks initially, then every 3 months).

24.5.2 Tumor Necrosis Factor Alpha (TNF- α) Inhibitors

Successful anti-TNF alpha therapies have been reported for uveitis associated with JIA, Behçet's disease, inflammatory bowel diseases, sarcoidosis as well as idiopatic uveitis.

Infliximab is a chimeric monoclonal antibody targeting TNF-α, administered intravenously at a dosage of 5–10 mg/kg at weeks 0, 2 and 6 (induction phase), then every 4–8 weeks (maintenance phase). Pediatric patients may require higher doses by weight or more frequent infusions compared with adults. It represents an efficient alternative for short-term treatment of uveitis.

Adalimumab is a fully humanized monoclonal antibody against TNF alpha. The route of administration is subcutaneous injection at 24 mg/m^2 once a week. Compared to Infliximab, it demonstrated similar efficacy and tolerance during induction phase, whereas it seemed to be superior for long-term maintenance therapy [9, 10]. Etanercept is not routinely recommended due to unsatisfactory response [11].

24.5.3 Non Anti-TNF Alpha Biologic Activity Modifiers

A subset of patients fails to respond to TNF-alpha blockers or is unable to tolerate these therapies and may therefore benefit from switching to another drug. Overall, about 25 % of children with autoimmune chronic uveitis who receive Adalimumab and Infliximab do not respond to these treatments. Preliminary evidence suggests the promising role of non anti-TNF alpha, such as Abatacept, Rituximab, Tocilizumab, Golimumab, Canakinumab, Gevokizumab and Alemtuzumab, for severe sight-threatening uveitis refractory to previous course of immunosuppressive treatment, DMARDs and anti-TNFα.

Abatacept is a soluble recombinant protein aimed at the inhibition of antigen-presenting cells, preventing the co-stimulation signal necessary for T cell activation. Recommended administration route is intravenous injection at 10 mg/kg at weeks 0, 2 then every 4 weeks. Data concerning the use of Abatacept for the treatment of uveitis are still limited. While first case reports and a small study have shown good efficacy [12–15], a subsequent retrospective study from Tappeiner et al. [16] on 21 patients demonstrated a partial and short-lasting efficacy of abatacept. Another recent case series seems to suggest the potential role of abatacept in inducing a persistent near-complete response in the subgroup of patients with severe chronic idiopathic uveitis when treatment options are limited [17]. A systematic review by Simonini at al [18]. evaluated the evidence regarding the effectiveness and the safety of non-anti-TNF biologic modifier (Abatacept, Rituximab, Tocilizumab, Alemtuzumab, Anakinra) in non-infectious chronic uveitis refractory to previous treatments. Their data suggested that switching to these agents has a favorable effect in the improvement of intraocular inflammation in children and adults affected by autoimmune chronic uveitis.

Given the unavailability of large-scale randomized controlled trials and the high cost of the treatment, the experience with these drugs is still limited to few cases managed in highly specialized centres.

Suggested therapeutic pathway for non-infectious chronic uveitis is reported in Fig. 24.1.

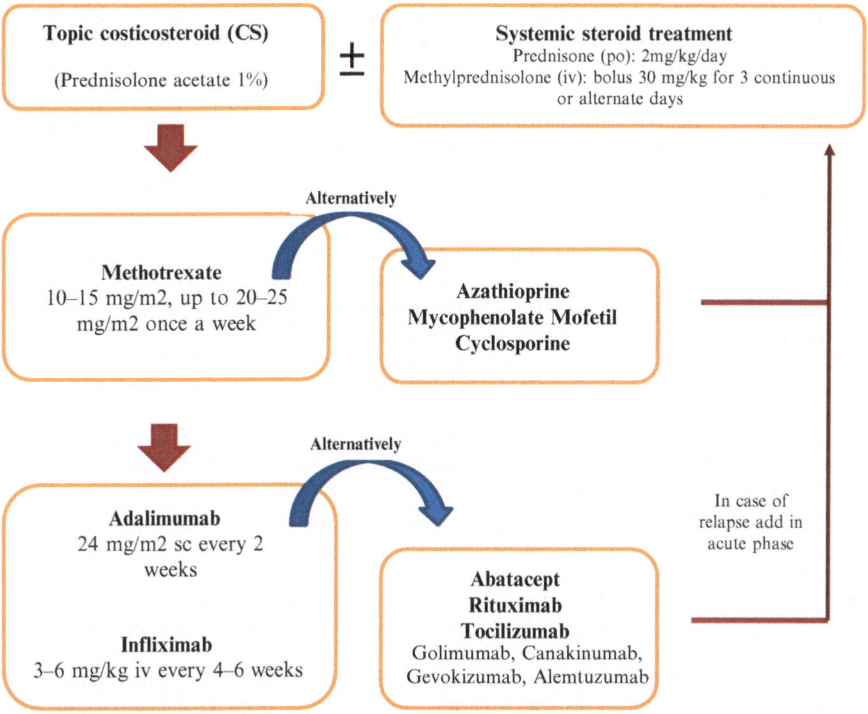

Fig. 24.1 Proposed therapeutic pathway for non infectious uveitis

References

1. Miserocchi E, Fogliato G, Modorati G, Bandello F (2013) Review on the worldwide epidemiology of uveitis. Eur J Ophthalmol 23(5):705–717
2. Petty RE, Rosenbaum JT (2011) Uveitis in juvenile idiopatic arthritis. In: Cassidy J (ed) Textbook of pediatric rheumatology 6th edn. Saunders, Philadelphia, pp 305–314
3. Saurenmann RK, Levin AV, Feldman BM, Rose JB, Laxer RM, Schneider R, Silverman ED (2007) Prevalence, risk factors, and outcome of uveitis in juvenile idiopathic arthritis: a long-term followup study. Arthritis Rheum 56:647–657
4. Rosenberg K, Feuer WJ, Davis JL (2004) Ocular complications of pediatric uveitis. Ophthalmology 111:2299–2306
5. Simonini G, Cantarini L, Bresci C, Lorusso M, Galeazzi M, Cimaz R (2010) Current therapeutic approaches to autoimmune chronic uveitis in children. Autoimmun Rev 9:674–683
6. Cantarini L, Simonini G, Frediani B, Pagnini I, Galeazzi M, Cimaz R (2012) Treatment strategies for childhood noninfectious chronic uveitis: an update. Expert Opin Investig Drugs 21:1–6
7. Rabinovich CE (2011) Treatment of juvenile idiopathic arthritis-associated uveitis: challenges and update. Curr Opin Rheumatol 23:432–436
8. Simonini G, Paudyal P, Jones GT, Cimaz R, Macfarlane GJ (2012) Current evidence of methotrexate efficacy in childhood chronic uveitis: a systematic review and meta-analysis approach. Rheumatology 52:825–831

9. Simonini G, Taddio A, Cattalini M, Caputo R, libero C, Naviglio S, Bresci C, Lorusso M, Lepore L, Cimaz R (2011) Prevention off flare recurrences in childhood-refractory chronic uveitis: an open-label comparative study of adalimumab versus infliximab. Arthritis Care Res 63:612–618
10. Simonini G, Taddio A, Cattalini M, Caputo R, de Libero C, Parentin F, Pagnini I, Lepore L, Cimaz R (2013) Superior efficacy of adalimumab in treating childhood refractory chronic uveitis when used as first biologic modifier drug: adalimumab as starting anti-TNF-α therapy in childhood chronic uveitis. Pediatr Rheumatol Online J 15:11–16
11. Simonini G, Druce K, Cimaz R, Macfarlane GJ, Jones GT (2014) Current evidence of anti-tumor necrosis factor α treatment efficacy in childhood chronic uveitis: a systematic review and meta-analysis approach of individual drugs. Arthritis Care Res 66:1073–1084
12. Angeles-Han S, Flynn T, Lehman T (2008) Abatacept for refractory juvenile idiopathic arthritis-associated uveitis-a case report. J Rheumatol 35:1897–1898
13. Elhai M, Deslandre CJ, Kahan A (2011) Abatacept for refractory juvenile idiopathic arthritis-associated uveitis: two new cases. Comment on the article by Zulian et al. Arthritis Care Res 63:307–308
14. Kenawy N, Cleary G, Mewar D, Beare N, Chandna A, Pearce I (2011) Abatacept: a potential therapy in refractory cases of juvenile idiopathic arthritis-associated uveitis. Graefes Arch Clin Exp Ophthalmol 249:297–300
15. Zulian F, Balzarin M, Falcini F, Martini G, Alessio M, Cimaz R et al (2010) Abatacept for severe anti-tumor necrosis factor alpha refractory juvenile idiopathic arthritis-related uveitis. Arthritis Care Res 62:821–825
16. Tappeiner C, Miserocchi E, Bodaghi B, Kotaniemi K, Mackensen F, Gerloni V et al (2015) Abatacept in the treatment of severe, longstanding, and refractory uveitis associated with juvenile idiopathic arthritis. J Rheumatol 42:706–711
17. Marrani E, Paganelli V, de Libero C, Cimaz R, Simonini G (2015) Long-term efficacy of abatacept in pediatric patients with idiopathic uveitis: a case series. Graefes Arch Clin Exp Ophthalmol 253:1813–1816
18. Simonini G, Cimaz R, Jones GT, Macfarlane GJ (2015) Non-anti-TNF biologic modifier drugs in non-infectious refractory chronic uveitis: the current evidence from a systematic review. Semin Arthritis Rheum 45:238–250

Chapter 25
Cogan's Syndrome

Rosanna Dammacco

Abstract Cogan' syndrome is a multisystemic disorder characterized by the almost unfailing combination of inflammatory ocular disease and sensorineural hearing loss. In addition to constitutional symptoms, clinical features may also include cardiovascular, gastrointestinal, and neurological manifestations with variable frequency. Skin and mucous membranes, kidney and urogenital apparatus are rarely involved. The etiology remains undefined whereas an autoimmune pathogenesis is sustained by experimental and clinical observations. Given its rarity and clinical heterogeneity, the diagnosis is often delayed with possible progression to impairment of visual acuity and auditory dysfunction up to deafness. Therapy is based on the administration of glucocorticoids and immunosuppressive agents. Tumor necrosis factor-alpha blockers, the B-cell depleting monoclonal antibody rituximab, the anti-interleukin-6 receptor monoclonal antibody tocilizumab, and the immunomodulatory leflunomide have also been employed in single patients. Cochlear implantation can be used for hearing rehabilitation. Finally, stent-graft insertion and prosthetic aortic replacement can be extremely useful to correct the consequences of aortitis and large vessel aneurysms.

Keywords Aortitis • Audio-vestibular dysfunction • Cogan's syndrome • Interstitial keratitis • Inflammatory eye diseases • Vasculitis

25.1 Definition and Epidemiology

Cogan syndrome (CS) is a variable vessel vasculitis first described by the American ophthalmologist David Glendenning Cogan in 1945 [1], characterized by recurrent non-syphilitic interstitial keratitis associated to vestibule-auditory dysfunction that leads to progressive deafness. Before its nature of potentially systemic vasculitis was detected, it was also inappropriately termed "eye and ear syndrome".

R. Dammacco (✉)
Department of Basic Medical Sciences, Neurosciences and Sensory Organs, University of Bari Medical School, 70124 Bari, Italy
e-mail: rosanna.dammacco@uniba.it

F. Dammacco et al. (eds.), *Systemic Vasculitides: Current Status and Perspectives*, DOI 10.1007/978-3-319-40136-2_25

The invariable occurrence of progressive hearing loss immediately recalls the pathology that afflicted the eminent German composer Ludwig van Beethoven almost two centuries before Cogan's description [2, 3]. However, the fragmentary information available to the historians of medicine does not allow a definite conclusion as to whether Beethoven was indeed a forerunner of CS.

The rarity of the condition, which is now considered the archetypal inner ear autoimmune disease, is supported by the fact that a total number of 623 papers have been retrieved from PubMed under the voice "Cogan's syndrome" from October 1951 to the time of this writing (February 2016). The disease is not hereditary, is diagnosed less rarely in whites than in blacks, does not appear to have a clear sex prevalence, and affects mainly young adults of Caucasian origin [4], although in a retrospective review of a cohort of 60 CS patients, collected at the Mayo Clinic in a time period of over 60 years, the age at diagnosis ranged from 9 to 70 years (mean age 38 years) [5].

Two clinical variants of CS have been identified, named typical and atypical CS respectively, whose features will be described below.

25.2 Etiopathogenesis

The etiology of CS is still unknown. Because an infection of the upper respiratory tract has been reported in a large proportion of CS patients before or at the outset of the disease, an infectious origin has been suggested, and several agents such as Borrelia and Chlamidiae species, Cytomegalovirus, Treponema pallidum, Parvovirus B19, Herpes virus, and so on, have from time to time been suspected. However, no specific infectious agent has so far been unequivocally identified, the mentioned agents possibly playing a role of triggering factor.

The autoimmune pathogenesis is sustained by several reports that have been summarized in a recent review [6]. Starting from the first demonstration, which dates back to 1984 [7], of autoantibodies against human cornea and human inner ear tissues in a CS patient, antibodies directed to the epithelial structure of the cornea, healthy inner ear antigens, and endothelial autoantigens have been repeatedly described [8]. Autoantibodies have also been revealed against autoantigens such as an immunodominant peptide structurally similar to the antigen SSA/Ro of the Sjögren's syndrome, reovirus III major core protein lambda-1, DEP-1/CD148, connexin 26, and against the cell-density enhanced protein tyrosine phosphatase-1, which is consistently expressed on the sensory epithelia of the inner ear as well as on endothelial cells [9]. The passive injection of these autoantibodies into mice or their active induction by immunization resulted in the appearance of hearing loss and corneal disease, thus mimicking CS [9]. It has therefore been hypothesized that CS is an autoimmune disease, in which undefined pathogens share antigenic determinants of both the eye and the inner ear, and the host immune response results in damage of these and possibly other organs through the phenomenon known as "antigenic mimicry". The infectious or inflammatory process would also be able to trigger an immune response by unmasking hidden epitopes of the target tissues [10–12].

An interesting finding is the demonstration of anti-heat shock protein (Hsp)70 antibodies in 92.9 % of patients with typical CS, in 16.6 % of those with atypical CS, and in 52.7 % of the patients with autoimmune sensorineural hearing loss, compared with a positivity rate of 5.2 % in the control group. These striking results indicate that anti-Hsp70 antibodies should be considered a serological marker of typical CS [3]. Additional non-specific antibodies that can be found in CS patients include c-ANCA and p-ANCA, rheumatoid factor, anti-cardiolipin, and anti-nuclear antibodies [13, 14].

25.3 Typical and Atypical Variants of CS

As already mentioned, two clinical variants of CS can be recognized. Typical CS is characterized by the following features: (a) non-syphilitic interstitial keratitis, of which Fig. 25.1 is a typical example; (b) audio-vestibular involvement that can mimic Ménière's syndrome; (c) a relatively short lapse of time between the occurrence of ocular and audio-vestibular symptoms, that is usually less than 2 years. Atypical CS is instead defined on the basis of the followings characteristics: (a) ocular involvement of variable severity that may or may not include interstitial keratitis; (b) audio-vestibular symptoms largely different from those of Ménière's syndrome; (c) a greater than 2 years' time interval between the onset of ocular and audio-vestibular manifestations; (d) a significantly higher frequency of systemic disorders. Based on a total number of 222 CS patients retrieved from the literature, typical CS was diagnosed in 58.5 % and atypical CS in 41.5 % of the patients [15].

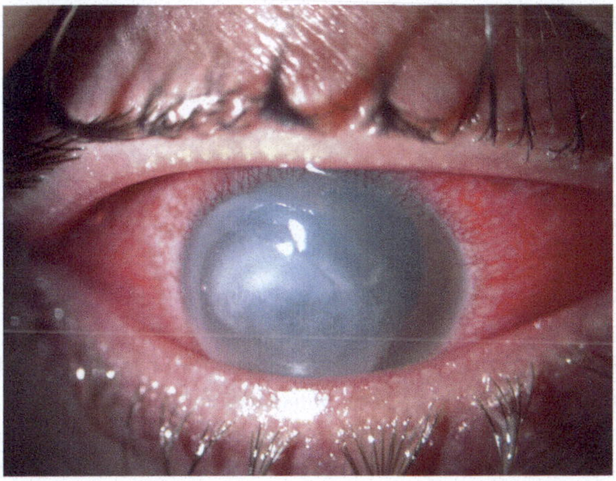

Fig. 25.1 A typical example of severe interstitial keratitis

It should, however, be emphasized that a clear differentiation of typical from atypical cases is often difficult, given the large variability of the clinical features among patients, especially those of the pediatric age [16], and the possible onset of interstitial keratitis as a later rather than the starting feature [6, 15].

25.4 Clinical Features of CS

In the large majority of patients, the first signs of the disease include bilateral conjunctival hyperemia, eye pain, photophobia, blurred vision, lacrimation, diplopia, and foreign body sensation in one or both eyes. Concomitant with the appearance of the ocular manifestations, but more often following them at variable distance, the patient complains of Ménière-like vestibular symptoms such as vertigo, nausea, vomiting, and tinnitus. A sudden sensory hearing loss is not rarely found among the initial symptoms. Over time, roughly 60–80 % of patients develop systemic manifestations, thus accounting for the heterogeneity and the variable severity of CS clinical spectrum [15]. Table 25.1 summarizes the clinical manifestations of CS in a roughly decreasing order of prevalence.

During follow-up, a variable outcome can be observed. Less than one third of the patients achieve clinical remission, whereas death has been found to occur in approximately 10 % of the patients, the most frequent causes being rupture of an aortic aneurysm, bilateral coronary ostial stenosis resulting in myocardial infarction, renal amyloidosis or end-stage renal disease, cerebrovascular accident, subarachnoid hemorrhage, and generalized sepsis enhanced by the immunosuppressive therapy [15]. When aortitis develops, it is usually indistinguishable from Takayasu's disease and it is therefore advisable to submit patients to positron emission tomography/computed tomography (PET/CT) examination, which may detect pathological fluorodeoxyglucose uptake at the level of the aorta ascendens, with or without involvement of additional large vessels [19]. Ocular disease may undergo a single or recurrent flares. However, although an impairment of the visual acuity is observed in a large proportion of patients, permanent loss of vision of remarkable degree is unusual. On the contrary, in addition to sensorineural hearing loss and decreased auditory acuity, unilateral or bilateral permanent deafness is not a rare outcome [20].

25.5 Diagnosis

Given its rarity, the ocular and inner ear involvements that may occur at variable time intervals, the later appearance of a heterogeneous clinical spectrum, and the lack of a properly validated confirmatory test, an early diagnosis of CS, that would result in a much better response to treatment, is frequently missed. On the contrary, a delayed diagnosis may be responsible for a more severe and often permanent damage of the visual and auditory functions, and the occurrence of single organ or more

Table 25.1 Clinical features of Cogan's syndrome

Organ or system involvement	Clinical symptoms	Relative frequency (%)	References
Non-specific constitutional symptoms	Headache, recurrent fever, anorexia, weight loss, fatigue, coughing, arthro-myalgias, lymphadenopathy	30–50	[10, 15]
Eye disease	Interstitial keratitis largely prevalent and mostly bilateral	80–90	[13, 14]
	Uveitis, episcleritis, iridocyclitis, conjunctivitis, optic neuropathy, papilledema, central vein occlusion, papillitis, glaucoma, tenonitis	5–30	
Vestibulo-auditory dysfunction	Uni- or bilateral hearing loss, often of sudden onset	70	[5]
	Vertigo, dizziness, nystagmus	15–20	
Cardio-vascular manifestations	Aortitis of the ascending aorta and the arch with aortic insufficiency, congestive heart failure, coronaritis, pericarditis	15–20	[17]
Large and medium-vessel vasculitis	Systemic necrotizing vasculitis, occurring almost exclusively in atypical CS	5–10	[10, 38]
Gastro-intestinal involvement	Abdominal pain consequent to mesenteric arteritis, melena, diarrhea, splenomegaly, hepato-steatosis, esophagitis	10–15	[15]
Neurological manifestations	Aphasia, ataxia, hemiparesis or hemiplegia, lymphocytic meningitis, encephalitis, peripheral neuropathy	8–10	[18]
Skin and mucous membranes	Skin rash, photosensitivity, oral aphthous ulcers, vitiligo, Raynaud's phenomenon	3–5	[4, 15]
Renal manifestations	Renal artery stenosis, membrano-proliferative glomerulonephritis	2–3	[10, 13]
Urogenital manifestations	Orchitis, bending or arching of the penis	1–3	[15]

extensive vasculitis [10, 11, 14, 21]. It is therefore recommended that complete vestibulo-cochlear and ophthalmologic examinations be performed at the initial suspect of the disease and throughout a long-term follow-up. In addition to the functional tests aimed at detecting dysfunctions in auditory acuity, magnetic resonance imaging and/or computed tomographic scanning may reveal obliteration or stricture of the vestibular labyrinth that often prelude to the onset of permanent deafness.

In terms of differential diagnosis, granulomatosis with polyangiitis (Wegener's disease) can simulate CS given its possible association with auditory and ocular features, although the pulmonary involvement, the typical picture of membrano-proliferative glomerulonephritis, and the high prevalence of proteinase3-specific ANCA (that are of rare occurrence in CS) may help differentiate from each other. Also, CS may sometimes be misdiagnosed as rheumatoid arthritis, in which scleritis, keratitis, and keratoconjunctivitis sicca, as well as sensorineural hearing impairment can be found with variable, though significant frequency [14]. Several additional systemic diseases whose clinical manifestations often include ocular and auditory symptoms should also be considered in the differential diagnosis from CS, such as systemic lupus erythematosus, Still's disease, anti-phospholipid antibody syndrome, eosinophilic granulomatosis with polyangiitis, and Vogt-Koyanagi-Harada syndrome [13].

25.6 Therapy

The main drawback of therapy is its relatively late beginning, given the frequently delayed diagnosis of CS. This may result, as already stated, in a more severe and frequently irreversible deafness, as well as in the appearance of signs and symptoms ascribable to vasculitis of other body districts [22].

A wide array of drugs have been proposed and employed in CS patients, according to the clinical severity and the oligo- or multiple-organ involvement. Table 25.2 summarizes the major clinical manifestations of the disease and the corresponding therapeutic approaches. It should, however, be emphasized that, because of the relative rarity of CS, in the large majority of patients the evidence for the clinical

Table 25.2 Therapeutic approaches to the main clinical features of Cogan's syndrome

Clinical manifestations	Suggested treatment	References
Interstitial keratitis	Topical glucocorticoids (GC)	[23]
	Topical cyclosporine-A	
	Topical atropine to achieve mydriasis	
Episcleritis, uveitis, iridocyclitis, optic neuropathy, tenonitis	Systemic GC ± immunosuppressive drugs (azathioprine, cyclophosphamide, methotrexate, cyclosporine-A)	[22, 32]
Uni- or bilateral hearing loss, Ménière's-like syndrome	Remission-inducing agents: systemic GC, in the most severe cases preceded by pulsed intravenous GC	[6, 10, 11, 13, 23–25, 27, 28, 30–33]
	Treatment of refractory patients: consider the addition of azathioprine, cyclophosphamide, methotrexate, cyclosporine-A, mycophenolate mofetil. TNF-alpha blockers (etanercept, infliximab) as possible stabilizing agents. Rituximab, tocilizumab, and plasmapheresis may be considered.	
	Maintenance in responsive patients: gradual tapering of GC ± methotrexate	
Bilateral, profound deafness	Cochlear implantation	[13, 34–37]
Aortitis and aortic insufficiency	Systemic GC, rituximab, tocilizumab, aortic valve replacement	[31, 32]
Multiple carotid and aortic aneurysms	High-dose GC, methotrexate, stent-graft insertion	[38–41]

efficacy of certain drugs is circumstantial, given that it has been assessed on small cohorts of patients or on single clinical observations. In the absence of a standardized therapy, glucocorticoids (GC) still represent the primary drug to treat the presenting ocular manifestations, but failure to respond, and above all the association of inflammatory ocular disease with sensorineural hearing loss require variable combinations of immunosuppressive drugs.

Based on the experience gained in the ANCA-associated vasculitides as well as in large vessel vasculitides, practically all immunosuppressive drugs have been tested in CS, the dose of each drug and their variable combinations depending on whether remission induction, or treatment of refractory patients to the initial treatment, or response maintenance are to be achieved (Table 25.2). Methotrexate has been successfully employed in certain patients [24, 25], but it has not been shown to be effective in the maintenance of the hearing improvement induced by GC [26]. Even so, methotrexate should be considered a useful steroid-sparing agent. Mycophenolate mofetil has been proposed as effective treatment of steroid-dependent CS in childhood [27]. Tumor necrosis factor (TNF)-alpha blockers, namely infliximab and etanercept, are biological drugs that have also achieved a steroid tapering and sparing effect [28, 29], and have been able to significantly improve auditory and ocular disease, provided they are administered in early stages [11]. However, the more suitable timing for initiating TNF-alpha blockers and to prevent permanent auditory disability has not been established.

The possibility of employing biologic drugs in large vessel vasculitis has emerged more recently, based on a better knowledge of the pathogenetic mechanisms underlying these disorders. The inner ear autoimmune nature of CS has induced to use the B-cell depleting monoclonal antibody rituximab with encouraging results [30]. In addition, since serum interleukin-(IL) 6 has been shown to be increased in the serum and upregulated in the arteries of patients with large vessel vasculitis, the administration of a monoclonal antibody directed to the receptor of IL-6, called tocilizumab, in preliminary studies has been able to induce positive results for the treatment of CS-associated aortitis [31, 32]. In one single patient with juvenile CS the immunomodulator leflunomide was also able to induce clinical remission and appears therefore a promising drug [33]. Obviously, prospective randomized multicenter clinical trials will be necessary to establish the best combination(s) related to the heterogeneity of CS in terms of clinical efficacy and sparing of GC toxicity.

In patients with established deafness, remarkably good hearing rehabilitation can be achieved by a cochlear implantation [34, 35]. It is highly advisable, however, that the patient receives exhaustive counselling before undergoing implantation, so that he/she is properly informed on the actual limits of this procedure and the potential risk of progressive cochlear ossification or obliteration [36, 37].

The aortitis and periaortitis of CS are histologically characterized by a mixed inflammatory pattern [17], and more frequently affect the ascending aorta and the arch [38], although thoraco-abdominal aneurysms can also be diagnosed [39–41]. In addition to the steroid and immunosuppressive treatment during the acute phase of the disease, cardio-surgery procedures may include stent-graft insertion and prosthetic thoraco-abdominal aortic replacement [40].

References

1. Cogan DG (1945) Syndrome of nonsyphilitic interstitial keratitis and vestibuloauditory symptoms. Arch Ophthalmol 33:144–149
2. Larkin E (1971) Beethoven's illness – a likely diagnosis. Proc R Soc Med 64:493–496
3. Bonaguri C, Orsoni J, Russo A, Rubino P, Bacciu S, Lippi G et al (2014) Cogan's syndrome: anti-Hsp70 antibodies are a serological marker in the typical form. Isr Med Assoc J 16:285–288
4. Vollertsen RS, McDonald TJ, Younge BR, Banks PM, Stanson AW, Ilstrup DM (1986) Cogan's syndrome: 18 cases and a review of the literature. Mayo Clin Proc 61:344–361
5. Gluth MB, Baratz KH, Matteson EL, Driscoll CL (2006) Cogan syndrome: a retrospective review of 60 patients throughout a half century. Mayo Clin Proc 81:483–488
6. Kessel A, Vadasz Z, Toubi E (2014) Cogan syndrome –pathogenesis, clinical variants and treatment approaches. Autoimmun Rev 13:351–354
7. Arnold W, Gebbers JO (1984) Serum antibodies against corneal and internal ear tissues in Cogan's syndrome. Laryngol Rhinol Otol (Stuttg) 63:428–432
8. Berti E, Vannucci G, Lunardi C, Bianchi B, Bason C, Puccetti A et al (2013) Identification of autoantibodies against inner ear antigens in a cohort of children with idiopathic sensorineural hearing loss. Autoimmunity 46:525–530
9. Lunardi C, Bason C, Leandri M, Navone R, Lestani M, Millo E et al (2002) Autoantibodies to inner ear and endothelial antigens in Cogan's syndrome. Lancet 360:915–921
10. Haynes BF, Kaiser-Kupfer MI, Mason P, Fauci AS (1980) Cogan syndrome: studies in thirteen patients, long-term follow-up, and a review of the literature. Medicine (Baltimore) 59:426–441
11. Tayer-Shifman OE, Ilan O, Tovi H, Tal Y (2014) Cogan's syndrome - clinical guidelines and novel therapeutic approaches. Clin Rev Allergy Immunol 47:65–72
12. Greco A, Gallo A, Fusconi M, Magliulo G, Turchetta R, Marinelli C, Macri GF, De Virgilio A, de Vincentiis M (2013) Cogan's syndrome: an autoimmune inner ear disease. Autoimmun Rev 12:396–400
13. Zenone T (2013) Cogan syndrome. Presse Med 42:951–960
14. Espinoza GM, Prost A (2015) Cogan's syndrome and other ocular vasculitides. Curr Rheumatol Rep 17:24. doi:10.1007/s11926-015-0499-y
15. Grasland A, Pouchot J, Hachulla E, Blétry O, Papo T, Vinceneux P (2004) Study group for Cogan's syndrome. Typical and atypical Cogan's syndrome: 32 cases and review of the literature. Rheumatology (Oxford) 43:1007–1015
16. Pagnini I, Zannin ME, Vittadello F, Sari M, Simonini G, Cimaz R et al (2012) Clinical features and outcome of Cogan syndrome. J Pediatr 160:303–307
17. Stone JR, Bruneval P, Angelini A, Bartoloni G, Basso C, Batoroeva L et al (2015) Consensus statement on surgical pathology of the aorta from the society for cardiovascular pathology and the association for european cardiovascular pathology: I. Inflammatory diseases. Cardiovasc Pathol 24:267–278
18. Bicknell JM, Holland JV (1978) Neurologic manifestations of Cogan syndrome. Neurology 28:278–281
19. Balink H, Bruyn GA (2007) The role of PET/CT in Cogan's syndrome. Clin Rheumatol 26:2177–2179
20. Cundiff J, Kansal S, Kumar A, Goldstein DA, Tessler HH (2006) Cogan's syndrome: a cause of progressive hearing deafness. Am J Otolaryngol 27:68–70
21. Baumann A, Helbling A, Oertle S, Häusler R, Vibert D (2005) Cogan's syndrome: clinical evolution of deafness and vertigo in three patients. Eur Arch Otorhinolaryngol 262:45–49
22. Jančatová D, Zeleník K, Komínek P, Matoušek P (2015) Atypical Cogan's syndrome: a case report and summary of current treatment options. Int J Pediatr Otorhinolaryngol 79:428–431
23. Shimura M, Yasuda K, Fuse N, Nakazawa M, Tamai M (2000) Effective treatment with topical cyclosporin A of a patient with Cogan syndrome. Ophthalmologica 214:429–432

24. Richardson B (1994) Methotrexate therapy for hearing loss in Cogan's syndrome. Arthritis Rheum 37:1559–1561
25. Salley LH Jr, Grimm M, Sismanis A, Spencer RF, Wise CM (2001) Methotrexate in the management of immune mediated cochleovesitibular disorders: clinical experience with 53 patients. J Rheumatol 28:1037–1040
26. Harris JP, Weisman MH, Derebery JM, Espeland MA, Gantz BJ, Gulya AJ et al (2003) Treatment of corticosteroid-responsive autoimmune inner ear disease with methotrexate: a randomized controlled trial. JAMA 290:1875–1883
27. Hautefort C, Loundon N, Montchilova M, Marlin S, Garabedian EN, Ulinski T (2009) Mycophenolate mofetil as a treatment of steroid dependent Cogan's syndrome in childhood. Int J Pediatr Otorhinolaryngol 73:1477–1479
28. Matteson EL, Choi HK, Poe DS, Wise C, Lowe VJ, McDonald TJ et al (2005) Etanercept therapy for immune-mediated cochleovestibular disorders: a multi-center, open-label, pilot study. Arthritis Rheum 53:337–342
29. Fricker M, Baumann A, Wermelinger F, Villiger PM, Helbling A (2007) A novel therapeutic option in Cogan diseases?TNF-alpha blockers. Rheumatol Int 27:493–495
30. Orsoni JG, Laganà B, Rubino P, Zavota L, Bacciu S, Mora P (2010) Rituximab ameliorated severe hearing loss in Cogan's syndrome: a case report. Orphanet J Rare Dis 5:18
31. Shibuya M, Fujio K, Morita K, Harada H, Kanda H, Yamamoto K (2013) Successful treatment with tocilizumab in a case of Cogan's syndrome complicated with aortitis. Mod Rheumatol 23:577–581
32. Unizony S, Stone JH, Stone JR (2013) New treatment strategies in large-vessel vasculitis. Curr Opin Rheumatol 25:3–9
33. Xie L, Cai Y, Bao X, Cao L (2009) Leflunomide for the successful management of juvenile Cogan's syndrome. Clin Rheumatol 28:1453–1455
34. Kawamura S, Sakamoto T, Kashio A, Kakigi A, Ito K, Suzuki M et al (2010) Cochlear implantation in a patient with atypical Cogan's syndrome complicated with hypertrophic cranial pachymeningitis. Auris Nasus Larynx 37:737–741
35. Wang JR, Yuen HW, Shipp DB, Stewart S, Lin VY, Chen JM et al (2010) Cochlear implantation in patients with autoimmune inner ear disease including Cogan syndrome: a comparison with age- and sex-matched controls. Laryngoscope 120:2478–2483
36. Bovo R, Ciorba A, Trevisi P, Aimoni C, Cappiello L, Castiglione A et al (2011) Cochlear implant in Cogan syndrome. Acta Otolaryngol 131:494–497
37. Bacciu A, Pasanisi E, Di Lella F, Guida M, Bacciu S, Vincenti V (2015) Cochlear implantation in patients with Cogan syndrome: long-term results. Eur Arch Otorhinolaryngol 272:3201–3207
38. Cochrane AD, Tatoulis J (1991) Cogan's syndrome with aortitis, aortic regurgitation, and aortic arch vessel stenoses. Ann Thorac Surg 52:1166–1167
39. Tseng JF, Cambria RP, Aretz HT, Brewster DC (1999) Thoracoabdominal aortic aneurysm in Cogan's syndrome. J Vasc Surg 30:565–568
40. Bisdas TE, Teebken OE, Wilhelmi M, Lotz J, Bredt M, Haverich A et al (2009) Surgical treatment of a thoracoabdominal aneurysm in Cogan's syndrome. Ann Thorac Surg 88:1668–1670
41. Su JW, Low AH, Tay KH, Sebastian MG, Thumboo J, Sin KY (2006) Recurrent aortic aneurysms following thoracic aortic stent-graft repair in a patient with Cogan syndrome. J Endovasc Ther 13:779–782

Chapter 26
Non-infectious Retinal Vasculitis

Shiri Shulman and Zohar Habot-Wilner

Abstract Retinal vasculitis is a common clinical feature in intermediate and posterior uveitis. It is defined as either clinical or angiographic evidence of vascular inflammation accompanied by intraocular inflammation. It may be a primary inflammation of the blood vessel or, more commonly, secondary to inflammation of adjacent structures such as the retina or the choroid. In most series, the most common systemic diseases associated with retinal vasculitis is Behçet's disease. Primary vasculitides such as polyarteritis nodosa or ANCA-associated vasculitis are rarely associated with retinal vasculitis. The most common clinical findings in retinal vasculitis are vitritis and vascular sheathing. The retinal veins are more commonly involved than the arteries. Treatment of retinal vasculitis is aimed at improving vision as well as preventing complications caused either by the intraocular inflammation or from ischemia and involves both systemic immunomodulatory therapy as well as ocular treatment such as intravitreal steroids and anti-VEGF injections. With proper therapy the prognosis is good.

26.1 Introduction

The retinal vasculature is commonly involved in intermediate and posterior uveitic diseases. The clinical term vasculitis, as used in internal medicine, refers to histologically proven vasculitis due to inflammation of the vessel wall [1]. The Standardization of Uveitis Nomenclature (SUN) working group concluded that the term retinal vasculitis referred to a clinical description, which should be used in cases with evidence of ocular inflammation and retinal vascular changes. Perivascular sheathing as seen in ophthalmologic examination and vascular leakage or occlusion on fluorescein angiogram (FA) were depicted as evidence of retinal vascular disease [2]. Retinal vasculitis may cause significant visual acuity impairment and visual field changes; these are related to involvement of posterior retinal blood vessels, severe intraocular inflammation, central or branch vein or artery occlusions, macular ischemia, macular edema, and the presence of ischemic retinal areas that can

S. Shulman (✉) • Z. Habot-Wilner
Ophthalmology division, Tel-Aviv medical center, Tel Aviv, Israel
e-mail: shulmanshiri@gmail.com

F. Dammacco et al. (eds.), *Systemic Vasculitides: Current Status and Perspectives*, DOI 10.1007/978-3-319-40136-2_26

lead to new blood vessel formation and thereafter to vitreous hemorrhage, traction retinal detachment, rubeosis iridis, and neovascular glaucoma [4–7].

This chapter will discuss the characteristics and treatments of non-infectious retinal vasculitis and the associated systemic diseases.

26.2 Classification

Retinal vasculitis is classified into two main subclasses. The first is primary vasculitis in which the vessel is the primary target of the inflammatory process. This may be limited to the eye or related to primary systemic vasculitis which involves the eye (primary systemic vasculitides). The second subclass is secondary vasculitis, in which inflammation of the vessel wall occurs secondarily to an inflammatory process involving adjacent structures. This may be localized to the eye or associated with systemic disease, of infectious or non-infectious etiology [3, 8]. This chapter will focus on non-infectious retinal vasculitis. Table 26.1 shows the classification of non-infectious retinal vasculitis.

26.3 Pathogenesis

Our understanding of retinal vasculitis pathology stems from clinical observations, enucleation and postmortem studies and experimental models. Ocular histopathologic and immunopathologic studies were published for the four principal systemic disease-associated retinal vasculitis: Behçet's disease, sarcoidosis, systemic lupus erythematosus (SLE) and multiple sclerosis (MS) [9–17]. The vascular changes in retinal vasculitis are characterized by perivascular infiltration of lymphocytes of the vessel wall with disruption of tight endothelial junctions (the inner blood retinal barrier) in the retinal capillaries. Cell-mediated immunity, with CD4 + T helper cells within and around the retinal vessels, plays a role in disease pathology. Retinal vasculitis-induced ischemia is suggested to be either thrombotic or obliterative secondary to inflammatory cells infiltration. Thrombotic vascular changes can occur due to local endothelial injury or increased prothrombin activity, as was observed in Behçet's disease. The retinal arteriolar occlusion without vessel inflammation seen in SLE may be due to immune complex deposition within the lumen by filtration or secondary to anti-endothelial cell antibody binding [18]. Vasculitis may involve the retinal arteries, veins, and/or capillaries. The retinal veins were found to be most commonly affected [19, 20]. Retinal vasculitis affecting predominantly the veins has been described in association with Behçet's disease, sarcoidosis, multiple sclerosis, pars planitis and Birdshot chorioretinopathy. Arterial involvement is more commonly seen in SLE, IRVAN, Susac syndrome, Crohn's disease and the primary systemic vasculitides: Giant cell arteritis, Takayasu's arteritis, PAN, GPA, Churg-Strauss syndrome.

Table 26.1 Classification of non-infectious retinal vasculitis

Primary retinal vasculitis		Secondary retinal vasculitis	
Localized to the eye	Primary systemic vasculitides	Localized to the eye	Associated with systemic disease
Idiopathic retinal vasculitis	Giant cell arteritis	Birdshot chorioretinopathy	Behçet's disease
Pars planitis	Takayasu's arteritis	Primary ocular lymphoma	Sarcoidosis
Frosted branch angiitis	Polyarteritis nodosa (PAN)		Systemic lupus erythematosus
Idiopathic retinal vasculitis, aneurysms and neuro-retinitis (IRVAN)	Granulomatosis with Polyangitis (GPA)		Multiple sclerosis
Acute multifocal hemorrhagic retinal vasculitis	Churg-Strauss syndrome		Inflammatory bowel disease
	Essential cryoglobulinemic vasculitis		HLA B27 – associated-seronegative spondyloarthropathy
	Cutaneus leucocytoclastic angiitis		Susac's syndrome
			Relapsing polychondritis
			Rheumatoid arthritis
			Sjögren's syndrome
			Drug induced: intravenous immunoglobulins, post vaccinations
			Secondary to malignancies: cancer associated retinopathy, acute leukemia, CNS lymphoma

26.4 Clinical Manifestations

The Standardization of Uveitis Nomenclature (SUN) Working Group [2] concluded that the term retinal vasculitis referred to a clinical description which should be used in cases with evidence of ocular inflammation and retinal vascular changes. Retinal vasculitis may be either symptomatic or asymptomatic. This depends on the area involved by vasculitis, the severity of the ocular inflammatory process and the presence of ocular complications. Patients may have minimal or no symptoms, in cases of peripheral retinal vascular changes with a mild intraocular inflammation (anterior/intermediate/posterior uveitis). On the other hand, inflammation of the posterior retinal blood vessels and/or significant intraocular inflammation and/or the presence of ocular complications, may cause a decrease in vision. Patients may complain on floaters if there is an accompanying vitritis and on visual field scotomata which are usually related to the areas of ischemia. Other symptoms may include alteration of color vision, metamorphopsia, and rarely pain [19, 21]. As retinal vasculitis may be secondary to a systemic disease, systemic symptoms and signs may be present and suggest an underlying etiology. Therefore, a comprehensive medical history and review of systems is crucial to the evaluation of patients with retinal vasculitis.

26.5 Clinical Findings and Imaging

Perivascular sheathing/cuffing as seen in ophthalmological examination and vascular leakage or occlusion on fluorescein angiogram (FA) were depicted by the Standardization of Uveitis Nomenclature (SUN) Working Group [2] as evidence of retinal vascular disease for the classification of retinal vasculitis. Several studies showed that non- infectious retinal vasculitis is mostly a bilateral disease [21–23]. The inflammatory process may involve blood vessels in a varying distribution including the central and peripheral retina as well as the optic nerve head. Inflammation of macular blood vessels may cause macular edema, or occlusive vasculitis with macular ischemia, both are risk factors for vision loss. Occlusive vasculitis may cause cotton-wool spots and intraretinal hemorrhages, with late changes as telangiectasis, microaneurysms and neovascularization which may progress to vitreous hemorrhage or traction retinal detachment. Ali et al. [20] showed that occlusive vasculitis is more likely to be associated with ocular complications such as epiretinal membrane, cystoid macular edema and neovascularisation.

We published a study that included 82 eyes of 45 patients with non-infectious retinal vasculitis in Israel [19]. Various presentations of vasculitis were found in our study: true vasculitis with clinical and angiographic evidence of vascular occlusion was found among patients with Behçet's disease, SLE, Crohn's disease and Multiple Myeloma. The second vasculitis type characterized by vascular sheathing without clinical or angiographic evidence of occlusion, was found among patients with sarcoidosis and sympathetic ophthalmia. The third type of vasculitis was demonstrated in patients with birdshot chorioretinopathy, in which vasculitis was identified by vascular angiographic leakage without occlusion and with minimal clinical findings of sheathing.

Intraocular inflammation is essential in order to define retinal vasculitis, as per the SUN classification. The inflammation may involve the anterior or posterior segment of the eye. We found vitritis to be the most common finding of ocular inflammation in our study [19], as was also found by Saurabh et al. [25].

FA has an important role in disease diagnosis and may allow better evaluation of retinal vasculitis, than clinical assessment. FA demonstrates inflammatory changes as vascular staining and leakage and ischemic manifestations as vascular occlusion, capillary nonperfusion, retinal neovascularization, and sclerosis of vessels. Another common finding by FA is diffuse capillary leakage. The use of wide-field angiography may offer an advantage in detecting peripheral retinal ischemia and neovascularization as compared to traditional FA imaging as was shown by Leder et al. [24].

26.6 Retinal Vasculitis Associated with Systemic Diseases

Several studies published on characteristics of patients with non-infectious retinal vasculitis [19, 21, 22]. Shulman et al. [19] showed that 53 % of their patients were diagnosed with secondary vasculitis, with Behçet's disease being the most common systemic disease. Other two publications from London, by Stanford et al. [21] and Graham et al. [22] showed similar results. In contrast to these findings, clinical data of retinal vasculitis patients from clinics in North America [11, 20] showed that most of their patients were not found to have a systemic disease related to their retinal vasculitis. Only 2 studies reported to have patients with systemic vasculitides related to retinal vasculitis; Graham and coauthors [22] found 6 out of 150 patients (4 %): three cases of PAN and 3 cases of GPA and Rosenbaum and colleagues [26] found 3 cases out of 207 (1.4 %) cases in a mixed cohort of infectious and non-infectious systemic diseases. The difference between the cohorts may lie in the variability in the ethnic origin of the patients in different continents. In addition, these studies emphasize that diagnostic evaluation should not include extensive laboratory tests as they have a low yield. The most common related systemic disease was found to be Behçet's disease which is a clinical diagnosis and does not require laboratory testing. Laboratory workup should be focused and based on medical and ocular findings.

26.6.1 Systemic Primary Vasculitides

Surprisingly, the systemic vasculitides including polyarteritis nodosa (PAN) and ANCA-associated vasculitides (granulomatosis with polyangiitis (GPA), microscopic polyangiitis (MPA), and eosinophilic granulomatosis with polyangiitis (EGPA); Churg–Strauss syndrome (CSS)) are rarely associated with retinal vasculitis.

Also, looking at a series of patients with GPA with a 16 year follow-up period, 40 out of 140 patients had ocular involvement, the most common being orbital inflammation or scleritis/episcleritis. Only 4 or 2.9 % had retinal vasculitis [27]. A French study showed similar results with ocular involvement in 214 out of 1,286 patients with systemic vasculitis (16.6 %) but only 16 cases of retinal vasculitis in this group (1.3 %). Seven out of these patients had PAN [28].

26.6.2 Behçet's Disease

Ocular involvement in Behçet's disease (BD) occurs in approximately 70 % of the patients and is associated with a high risk of visual loss [29, 30]. Taylor et al. [31] retrospectively reviewed 107 patients with ocular BD and found a 13 % 10-year risk rate of severe visual loss (6/60 or worse). The most common cause of irreversible visual loss was macular ischemia due to branch retinal vein occlusion.

Retinal vasculitis in BD involves both arteries and veins though venular involvement is more common. It manifests as either diffuse vascular leakage on FA or may be accompanied by vascular occlusion [31]. Visual acuity may be effected by macular edema (due to vascular leakage), macular ischemia or hemorrhages.

Neovascularisation of the optic nerve head or the retina is a serious complication. It may be a result of pro-angiogenesis factors secreted as a response to either ischemia or inflammation. It was reported in 4 % of 1,567 BD patients in Turkey [32].

26.6.3 Sarcoidosis

Ocular involvement occurs in up to 60 % of patients with systemic sarcoidosis with retinal vasculitis in up to 37 % of patients with ocular involvement [33]. Inflammation of the retinal veins (periphlebitis) is a common ocular manifestation and was considered by the first International Workshop on Ocular Sarcoidosis as one of seven clinical signs that comprise the diagnosis of ocular sarcoidosis [34]. Typical features of the involved vessels include segmental involvement and perivenous exudates, with a typical appearance of "candle wax drippings".

26.6.4 Multiple Sclerosis

The presence of peripheral periphlebitis is reported in up to 20 % of patients with MS [35]. In an autopsy of 93 eyes from patients with an established diagnosis of MS, seven showed segmental perivenular infiltrates of lymphocytes and plasma cells [35]. A similar finding was noted around central nervous system veins in two patients with MS, suggesting that periphlebitis is an early event preceding demyelinization [17].

In some cases, occlusive vasculitis with secondary retinal neovascularization is seen. Complications include vitreous hemorrhage and neovascular glaucoma [36].

Despite these complications, visual prognosis of MS related uveitis is generally good, probably due to mostly peripheral involvement [17].

26.6.5 Systemic Lupus Erythematosus (SLE)

Retinopathy may affect up to 30 % of patients with SLE [37]. The pathogenesis of vascular occlusion is not clear, possible explanations include immune-complex deposition, complement activation and fibrinoid degeneration of the vascular wall resulting in thrombosis and occlusion [38]. Clinical findings include cotton-wool spots (retinal microinfarctions) as well as larger artery occlusion. On FA retinal ischemia may be evident as well as neovascularization.

26.7 Treatment

The purpose of treatment of retinal vasculitis is to improve visual acuity by reducing ocular inflammation (uveitis) and prevent complications such as macular edema and neovascularization.

26.7.1 Systemic Immunosuppressive Agents

Systemic corticosteroids with or without immunomodulatory (IMN) agents are the mainstay of treatment in noninfectious vasculitis patients. BD with retinal vasculitis, is initially treated with a combination of corticosteroids and IMN agents such as azathioprine or cyclosporine [39, 40].

In ocular sarcoidosis, the presence of retinal vasculitis requires the use of systemic corticosteroids and often the addition of IMN agents, most commonly methotrexate [41]. In SLE, systemic corticosteroids and IMN, such as azathioprine and mycophenolate mofetil, are established treatments that can reduce vasculopathy and resolve cotton wool spots [42], though there is little evidence supporting their role in preventing the progression of retinal vasocclusion.

26.7.2 Biologic Agents

Anti -tumor necrosis factor alpha (TNF- α) drugs such as infliximab and adalimumab have been used successfully in the management of sight threatening uveitis and retinal vasculitis mainly in cases of severe ocular BD as a first line treatment [43] or used in cases refractory to other treatments [31, 44–46]. Anti-TNF- α agents are also used for the treatment of refractive cases of ocular sarcoidosis [47, 48]. Anti-TNF agents are not recommended for the treatment of MS-related retinal vasculitis as it may cause and exacerbate demyelinative manifestations [49].

Interferon alfa (INF-α) has been used in refractory cases of retinal vasculitis associated with Behçet's disease and was shown to induce disease remission in 55 % of patients even after cessation of the drug [50].

26.7.3 Retinal Laser Photocoagulation and Intravitreal Anti-Vascular Endothelial Growth Factor (VEGF)

Scatter laser photocoagulation is recommended in the presence of retinal neovascularization, and is also considered in cases of severe retinal ischemia, as preventive treatment for neovascularization formation. It intends at reduction of the metabolic consumption of the retina and reduction of the production of angiogenic factors. However, it is not effective in all cases [24]. There are no randomized clinical trials assessing the efficacy of laser treatment.

Intravitreal bevacizumab can be used in eyes with persistent neovascularization following or in combination with laser photocoagulation [51, 52].

26.8 Prognosis

Ali et al. [20] examined the correlation between characteristics of retinal vasculitis with ocular complications, or the response to different lines of treatment. They identified several factors that correlate with a worse prognosis: (A) smoking was significantly related to worsening of vision from baseline visit to final visit (B) age at onset of presentation below 40 years was associated with greater use of steroid sparing agents (C) occlusive vasculitis was associated with retinal neovascularization; (D) vasculitis with cotton wool spots and intraretinal hemorrhage was associated with epiretinal membrane formation. We had a long-term follow-up study, with a median follow-up of 46 months [19] and showed that visual acuity significantly improved during follow-up as compared to baseline visual acuity. We identified two prognostic factors for visual acuity improvement: worse visual acuity at baseline visit and Behçet's disease. An explanation for this finding may be that patients with low visual acuity and patients diagnosed with Behçet's disease are more likely to be treated aggressively which results in a better final outcome.

26.9 Conclusions

Retinal vasculitis may be an isolated condition or associated with systemic inflammatory diseases, the most common are Behçet's disease, Sarcoidosis and SLE. Primary vasculitides are rarely associated with retinal vasculitis. When treated properly with systemic immunosuppression, it is associated with good prognosis.

References

1. Hughes EH, Dick AD (2003) The pathology and pathogenesis of retinal vasculitis. Neuropathol Appl Neurobiol 29:325–340
2. Jabs DA, Nussenblatt RB, Rosenbaum JT (2005) Standardization of uveitis nomenclature working group. Standardization of uveitis nomenclature for reporting clinical data. Results of the first international workshop. Am J Ophthalmol 140:509–516
3. Herbort CP, Cimino L, Abu El-Asrar AM (2005) Ocular vasculitis: a multidisciplinary approach. Curr Opin Rheumatol 17:25–33
4. Walton RC, Ashmore ED (2003) Retinal vasculitis. Curr Opin Ophthalmol 14:409–413
5. Abu El-Asrar AM, Herbort CP, Tabbara KF (2005) Retinal vasculitis. Ocul Immunol Inflamm 13:415–433
6. Levy-Clarke GA, Nussenblatt R (2005) Retinal vasculitis. Int Ophthalmol Clin 45:99–113
7. Abu El-Asrar AM, Tabbara KF (2010) A clinical approach to the diagnosis of retinal vasculitis. Int Ophthalmol 30:149–173
8. Gupta A, Gupta V, Herbort CP et al (2009) Uveitis, text and imaging. Jaypee Brothers Medical Publishers, New Delhi
9. Fenton RH, Easom HA (1964) Behcet's syndrome. A histopathologic study of the eye. Arch Ophthalmol 72:71–81
10. Charteris DG, Champ C, Rosenthal AR et al (1992) Behcet's disease: activated T lymphocytes in retinal perivasculitis. Br J Ophthalmol 76:499–501
11. George RK, Chan CC, Whitcup SM et al (1997) Ocular immunopathology of Behcet's disease. Surv Ophthalmol 42:157–162
12. Frank KW, Weiss H (1983) Unusual clinical and histopathological findings in ocular sarcoidosis. Br J Ophthalmol 67:8–16
13. Graham EM, Spalton DJ, Barnard RO et al (1985) Cerebral and retinal vascular changes in systemic lupus erythematosus. Ophthalmology 92:444–448
14. Vine AK, Barr CC (1984) Proliferative lupus retinopathy. Arch Ophthalmol 102:852–854
15. Arnold AC, Pepose JS, Hepler RS et al (1984) Retinal periphlebitis and retinitis in multiple sclerosis. I. Pathologic characteristics. Ophthalmology 91:255–262
16. Lucarelli MJ, Pepose JS, Arnold AC et al (1991) Immunopathologic features of retinal lesions in multiple sclerosis. Ophthalmology 98:1652–1656
17. Engell T, Jensen OA, Klinken L (1985) Periphlebitis retinae in multiple sclerosis. A histopathological study of two cases. Acta Ophthalmol 63:83–88
18. Hughes EH, Dick AD (2003) The pathology and pathogenesis of retinal vasculitis. Neuropathol Appl Neurobiol 29:325–340
19. Shulman S, Kramer M, Amer R et al (2015) Characteristics and long-term outcome of patients with noninfectious retinal vasculitis. Retina. May 27. [Epub ahead of print]
20. Ali A, Ku JH, Suhler EB et al (2014) The course of retinal vasculitis. Br J Ophthalmol 98:785–789
21. Stanford MR, Graham E, Kasp E et al (1988) A longitudinal study of clinical and immunological findings in 52 patients with relapsing retinal vasculitis. Br J Ophthalmol 72:442–444

22. Graham EM, Stanford MR et al (1989) A point prevalence study of 150 patients with idiopathic retinal vasculitis: 1. Diagnostic value of ophthalmological features. Br J Ophthalmol 73:714–721
23. George RK, Walton RC, Whitcup SM, Nussenblatt RB (1996) Primary retinal vasculitis. Systemic associations and diagnostic evaluation. Ophthalmology 103:384–389
24. Leder HA, Campbell JP, Sepah YJ et al (2013) Ultra-wide-field retinal imaging in the management of non-infectious retinal vasculitis. J Ophthalmic Inflamm Infect 3(1):30
25. Saurabh K, Das RR, Biswas J, Kumar A (2011) Profile of retinal vasculitis in a tertiary eye care center in eastern India. Indian J Ophthalmol 59:297–301
26. Rosenbaum JT, Ku J, Ali A et al (2012) Patients with retinal vasculitis rarely suffer from systemic vasculitis. Semin Arthritis Rheum 41:859–865
27. Bullen CL, Liesegang TJ, McDonald TJ, DeRemee RA (1983) Ocular complications of Wegener's granulomatosis. Ophthalmology 90:279–290
28. Rothschild PR, Pagnoux C, Seror R et al (2013) Ophthalmologic manifestations of systemic necrotizing vasculitides at diagnosis: a retrospective study of 1286 patients and review of the literature. Semin Arthritis Rheum 42:507–514
29. Kac¸maz RO, Kempen JH, Newcomb C et al (2008) Ocular inflammation in Behc¸et disease: incidence of ocular complications and of loss of visual acuity. Am J Opthalmol 146:828–836
30. Verity DH, Vaughan GR, Stanford MR (2003) Behc¸et's disease: from hippocrates to the third millennium. Br J Ophthalmol 87:1175–1183
31. Taylor SRJ, Singh J, Menezo V et al (2011) Behet disease: visual prognosis and factors influencing the development of visual loss. Am J Ophthalmol 152:1059–1066
32. Tugal-Tutkun I, Onal S, Altan-Yaycioglu R et al (2004) Uveitis in Behc¸et disease: an analysis of 880 patients. Am J Ophthalmol 138:373–380
33. Khalatbari D, Stinnett S, McCallum RM, Jaffe GJ (2004) Demographic-related variations in posterior segment ocular sarcoidosis. Ophthalmology 111:357–362
34. Herbort CP, Rao NA, Mochizuki M (2009) International criteria for the diagnosis of ocular sarcoidosis: results of the first international workshop on ocular sarcoidosis (IWOS). Ocul Immunol Inflamm 17:160–169
35. Graham EM, Francis DA, Sanders MD, Rudge P (1989) Ocular inflammatory changes in established multiple sclerosis. J Neurol Neurosurg Psychiatry 52:1360–1363
36. Towler HM, Lightman S (2000) Symptomatic intraocular inflammation in multiple sclerosis. Clin Exp Ophthalmol 28:97–102
37. Klinkhoff AV, Beattie CW, Chalmers A (1986) Retinopathy in systemic lupus erythematosus: relationship to disease activity. Arthritis Rheum 29:1152–1156
38. Belmont HM, Abramson SB, Lie JT (1996) Pathology and pathogenesis of vascular injury in systemic lupus erythematosus interactions of inflammatory cells and activated endothelium. Arthritis Rheum 39:9–22
39. Masuda K, Nakajima A, Urayama A et al (1989) Double-masked trial of cyclosporin versus colchicine and long-term open study of cyclosporin in Behcet's disease. Lancet 1:1093–1096
40. Saadoun B, Wechsler B, Terrada C et al (2010) Azathioprine in severe uveitis of Behc¸et's disease. Arthritis Care Res 62:1733–1738
41. Baughman RP, Lower EE, Ingledue R, Kaufman AH (2012) Management of ocular sarcoidosis. Sarcoidosis Vasc Diffuse Lung Dis 29:26–33
42. Palejwala NV, Walia HS, Yeh S (2012) Ocular manifestations of systemic lupus erythematosus: a review of the literature. Autoimmune Dis 2012:29089
43. Levy-Clarke G, Jabs DA, Read RW et al (2013) Expert panel recommendations for the use of anti-tumor necrosis factor biologic agents in patients with ocular inflammatory disorders. Ophthalmology 121:785.e3–796.e3
44. Al Rashidi SA, Al Fawaz A, Kangave D, Abu El-Asrar AM (2013) Long-term clinical outcomes in patients with refractory uveitis associated with Behc¸et disease treated with infliximab. Ocul Immunol Inflamm 21:468–474

45. Bawazeer A, Raffa LH, Nizamuddin N (2010) Clinical experience with adalimumab in the treatment of ocular Behc¸et disease. Ocul Immunol Inflamm 18:226–232
46. Petropoulos IK, Vaudaux JD, Guex-Crosier Y (2008) AntiTNF-α therapy in patients with chronic non-infectious uveitis: the experience of Jules Gonin Eye Hospital. Klin Monatsbl 225:457–461
47. Erckens RJ, Mostard RLM, Wijnen AHM et al (2012) Adalimumab successful in sarcoidosis patients with refractory chronic non-infectious uveitis. Graefes Arch Clin Exp Ophthalmol 250:713–720
48. Cruz BA, Reis DD, Araujo CAA (2007) Refractory retinal vasculitis due to sarcoidosis successfully treated with infliximab. Rheumatol Int 27:1181–1183
49. Mohan M, Edwards ET, Cupps TR et al (2001) Demyelination occurring during anti-tumor necrosis factor alpha therapy for inflammatory arthritides. Arthritis Rheum 44:2862–2869
50. Kotter I, Eckstein AK, St¨ ubiger N, Zierhut M (1998) Treatment of ocular symptoms of Behc¸et's disease with interferon alpha 2a: a pilot study. Br J Ophthalmol 82:488–494
51. Kurup S, Lew J, Byrnes G et al (2009) Therapeutic efficacy of intravitreal bevacizumab on posterior uveitis complicated by neovascularization. Acta Ophthalmol 87:349–352
52. Lee WJ, Cho HY, Lee YJ et al (2011) Intravitreal bevacizumab for severe vaso-occlusive retinopathy in systemic lupus erythematosus. Rheumatol Int 33:247–251

Chapter 27
IgG4-Related Disease

Emanuel Della Torre

Abstract IgG4-related disease (IgG4-RD) is a fibro-inflammatory condition characterized by fibrous swelling of affected organs, elevated serum IgG4 concentration in the majority of patients, and a prompt response to corticosteroid therapy. The concept of IgG4-RD as a novel clinical entity emerged at the beginning of the twenty-first century, when Hamano and others noted that, by analogy with sarcoidosis, several seemingly unrelated disorders encompassing a wide range of organs shared common and unique histopathological features. Indeed, many different conditions once regarded as singular entities are now considered part of the IgG4-RD spectrum, such as autoimmune pancreatitis, hypertrophic pachymeningitis, idiopathic retroperitoneal fibrosis, orbital pseudotumor, and Mikulicz's disease, among others. Vascular involvement has been reported in a proportion of cases of periaortitis and inflammatory aneurysms of the thoracic and abdominal aorta. Given the recent birth of IgG4-RD, little is known with certainty about its immunopathogenesis. A putative pathogenic contribution of the humoral immune response is supported by the rapid clinical responses obtained in these patients with the anti-CD20 monoclonal antibody, rituximab. Novel insights into the pathophysiology of IgG4-RD will likely derive from the identification of possible microbial or self-antigens.

27.1 Introduction

IgG4-RD is a rare relapsing-remitting fibro-inflammatory condition characterized by expansive lesions and serum IgG4 elevation in the majority of cases [1]. IgG4-RD was originally described in the context of the disorder now termed type 1 autoimmune pancreatitis (AIP) [1, 2], and subsequently recognized in nearly every organ system [3]. The diagnosis of IgG4-RD might be challenging because clinical manifestations are largely non-specific and overlap with other inflammatory and neoplastic conditions. Thus, the current gold standard for the diagnosis of IgG4-RD is the identification of characteristic histological features that are strikingly similar

E. Della Torre (✉)
Unit of Medicine and Clinical Immunology, Università Vita-Salute San Raffaele – IRCCS-San Raffaele Scientific Institute, Milan, Italy
e-mail: dellatorre.emanuel@hsr.it

F. Dammacco et al. (eds.), *Systemic Vasculitides: Current Status and Perspectives*, DOI 10.1007/978-3-319-40136-2_27

across the full range of organs affected. Glucocorticoids (GCs) represent the first-line treatment for induction of remission, and typically lead to dramatic clinical responses in the majority of cases [4]. However, since IgG4-RD has a marked tendency to relapse, a variety of GC-sparing agents have been employed in order to maintain disease remission [4]. Recently, rituximab (RTX), an anti-CD20 monoclonal antibody, has proven to induce rapid clinical responses, supporting a crucial role for B lymphocytes in the pathogenesis of this fibrotic condition [5–7].

27.1.1 Epidemiology and Predisposing Factors

IgG4-RD still represents an overlooked clinical entity and its prevalence remains largely underestimated. As a consequence, predisposing and risk factors are difficult to identify because of limited study populations. To date, all available epidemiological studies are based on IgG4-related AIP, the most frequently encountered IgG4-RD manifestation. According to these statistical analyses, IgG4-RD has a predilection for middle-aged to elderly men (male to female ratio 2.8:1), and seems to associate with HLA-DRB1 aplotypes [1]. Researchers have also hypothesized an allergic trigger in the pathogenesis of IgG4-RD, because a proportion of patients have symptoms that overlap with allergic conditions or long-standing histories of allergy (rhinitis, nasal polyps, asthma). In addition, mild to moderate peripheral eosinophilia, and elevations in serum IgE concentration are frequently observed. However, recent studies have questioned this theory, and failed to identify other environmental risk factors for IgG4-RD [8, 9].

27.2 Clinical Manifestations and Organ Involvement

IgG4-RD typically affects middle-aged to elderly men, with sporadic reports of pediatric cases [1]. The clinical presentation is usually indolent, with signs and symptoms becoming evident over months or even years. High, spiking fevers and other manifestations of systemic inflammation that mimic infections are classically absent, but weight loss can occur during the subclinical period. A long-standing history of allergies is present in 30–40 % of patients at diagnosis, but symptoms that overlap with allergic conditions are also reported in some IgG4-RD patients without histories of atopy. These include bronchial asthma, chronic rhinitis and eczema [9]. IgG4-RD is characterized by pseudotumour-like lesions involving single or multiple organs. Different organs might be affected at the same time or one after the other. Clinical manifestations are largely non-specific and vary according to the spectrum of organs involved. Indeed, IgG4-RD might be asymptomatic or present with signs and symptoms related to the mechanical compression exerted by the fibrotic masses on local structures. IgG4-RD has been described in virtually every anatomical region (Table 27.1) [3], but the most common manifestations include

Table 27.1 Spectrum of organs affected by IgG4-related disease

Conditions commonly attributable to IgG4-RD	Conditions occasionally associated with IgG4-RD
Hypertrophic pachymeningitis	Interstitial pneumonia
Orbital pseudotumor	Pulmonary fibrosis
Mikulicz's disease	Sclerosing mastitis
Angiocentric eosinophilic fibrosis	Constrictive pericarditis
Kuttner's tumor	Sclerosing mesenteritis
Riedel's thyroiditis	Tubulointerstitial nephritis
Inflammatory pseudotumor of the lung	Membranous nephritis
Autoimmune pancreatitis	Pseudotumor of the skin
Sclerosing cholangitis	Midline destructive lesions
Inflammatory pseudotumor of the kidney	
Chronic periaortitis	
Inflammatory aortic aneurysm	
Retroperitoneal fibrosis	

type I AIP, chronic periaortitis, retroperitoneal fibrosis (Ormond's disease) and salivary or lacrimal gland swelling (Mikulicz's disease), conditions regarded as single independent entities for decades.

IgG4-related AIP, the most frequently recognized manifestation of IgG4-RD, can be asymptomatic or present with obstructive jaundice, weight loss or abdominal pain (Fig. 27.1a) [3]. Many patients are misdiagnosed initially as having adenocarcinoma of the pancreas, and, sometimes, a proportion of them undergo modified pancreatectomy for diagnostic purposes. Several cases of pancreatic cancer have been reported in patients with previous IgG4-related AIP, but a clear relationship between the two conditions still needs to be fully verified [3]. Secondary diabetes mellitus and malabsorption might complicate long-standing pancreatic disease.

In 25 % of cases, AIP is associated with gallbladder and bile duct involvement. Gallbladder disease, known as 'IgG4-related lymphoplasmacytic cholecystitis', is generally asymptomatic and not associated with gallstones. Conversely, IgG4-related sclerosing cholangitis is likely to present with jaundice and is often difficult to differentiate from cholangiocarcinoma.

IgG4-related retroperitoneal fibrosis classically affects the connective tissue around the abdominal aorta and its branches, therefore being often referred as to "IgG4-related chronic periaortitis" (Fig. 27.1b, c) [1, 3]. IgG4-related periaortitis represents the unique vascular involvement of IgG4-RD and accounts for 10–50 % of cases of idiopathic inflammatory aortitis [10–12]. Depending on disease localization and on the involvement of periureteral areas, it might be asymptomatic or present with back pain, flank pain, dysuria, hematuria, leg or scrotal edema. Aneurysmal dilatation and hydronephrosis represent the most feared complications of IgG4-RD periaortitis, and might require surgical approaches to prevent aortic dissection or

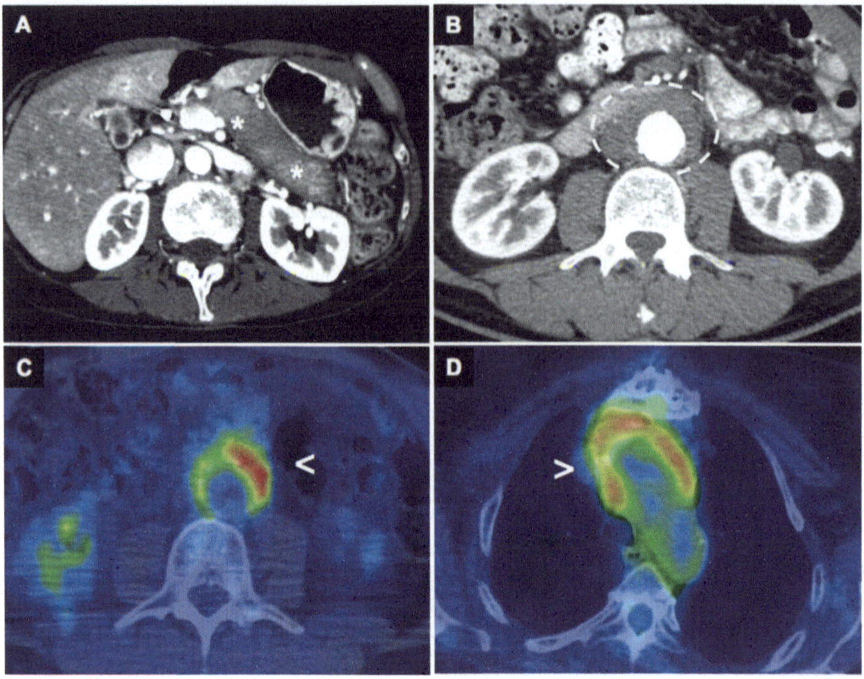

Fig. 27.1 Radiological features of IgG4-related disease. (**a**) "Sausage-like" shape of the pancreas in autoimmune pancreatitis (*asterisks*). (**b–d**) IgG4-related periaortitis of the abdominal (**b** – *circle*; **c** – *arrowhead*) and thoracic aorta (**d** – *arrowhead*)

renal failure, respectively. Chronic periaortitis may also involve the thoracic aorta, occasionally leading to aortic dissection (Fig. 27.1d).

Salivary and lacrimal glands involvement generally causes facial and orbital swelling. Sicca symptoms are often less severe in IgG4-RD than in Sjögren's syndrome and, when present, typically respond well to immunosuppressive therapy [1, 3]. Orbital pseudotumors affect the lacrimal gland most often but can also occur in other orbital regions. Extra-ocular muscle involvement (frequently termed 'orbital myositis' in the days before the recognition of IgG4-RD) can present with exophthalmos and the restriction of ocular movements [1, 3].

Atypical presentations of IgG4-RD that lack pseudotumor-like lesions should be borne in mind, including tubulointerstitial nephritis, glomerulonephritis, midline destructive lesions, interstitial lung disease, pleural and pericardial effusion [1, 3]. In particular, renal involvement might present with variable degrees of proteinuria, hematuria, renal failure or hypocomplementemia [1, 3]. Finally, life-threatening presentations such as rupture of inflammatory aneurysms, coronary syndromes, pachymeningitis and acute neurological events have also been described [1, 3].

27.3 Pathophysiology

The pathogenesis of IgG4-RD remains largely elusive. The presence of a dense fibrotic tissue and of abundant IgG4-positive plasma cells supports a possible underlying "modified Th2 immune-response", which is classically associated with both Th2 (e.g. IL-4 and IL-13) and T regulatory cytokines (e.g. TGF beta and IL-10) production by activated CD4+ T cells. Indeed, Th2 and T regulatory cytokines have been identified by molecular and immunohistochemical analyses of IgG4-RD lesions from different sites of organ involvement [3, 13]. In particular, IL-13 and TGF beta are thought to drive the deposition of extracellular matrix by activated fibroblasts, while IL-4 and IL-10 are considered the major inducers of IgG4 class switch in naïve B lymphocytes [3, 13]. The hypothesis of a "modified Th2 immune-response" is further supported by the presence of allergic symptoms, peripheral eosinophilia, and serum IgE elevation together with serum IgG4 increase in many patients at diagnosis. However, the analysis of circulating T cells for Th2/Tregulatory polarization has led to conflicting results, and showed an expansion of Th2 memory CD4+ T cells only in IgG4-RD patients with a concomitant history of atopy [8, 9]. Therefore, a direct evidence for a role of Th2 cells in IgG4-RD pathogenesis is still lacking.

The pathogenic role of IgG4 itself remains unclear. IgG4 comprises only 1–4 % of the total IgG population in healthy individuals, and, in contrast to other IgG subclasses, IgG4 has limited ability to activate immune responses because of its low affinity for both Fc receptors and the C1 complement molecule [1, 3, 13]. IgG4 is also unique in its ability to undergo "half-antibody exchange", a process whereby the IgG4 heavy chains dissociate from each other and re-associate randomly with other non-symmetrical antibody halves. As a consequence of the half-antibody exchange, IgG4 molecules become functionally "bi-specific" (capable of binding two antigens, neither tightly) and rarely associate with each other to form large immune complexes (Fig. 27.2). In view of this "non-inflammatory" nature of IgG4, its elevated concentrations in IgG4-RD probably do not represent the primary disease driver, but rather signify a secondary process designed to dampen ongoing chronic immune activation or, ultimately, an epiphenomenon. Recently, IgG4-RD activity was shown to correlate with the oligoclonal expansion of circulating plasmablasts, the precursors of plasma cells [14], thus supporting an underlying antigen driven immune-response. Interestingly, B cell depletion with rituximab (a monoclonal anti-CD20 antibody) was shown to induce a swift clinical improvement, a prompt serum IgG4 reduction, and a substantial diminution of circulating plasmablasts, thus further suggesting an important contribution of the B cell lineage to the pathophysiology of this fibrotic disorder [7, 14–16].

All in all, IgG4-RD is likely due to an antigen (whether microbial or autoantigen) driven immune response that induces both the B cell commitment towards IgG4 production and the secretion of pro-fibrotic cytokines by activated T lymphocytes.

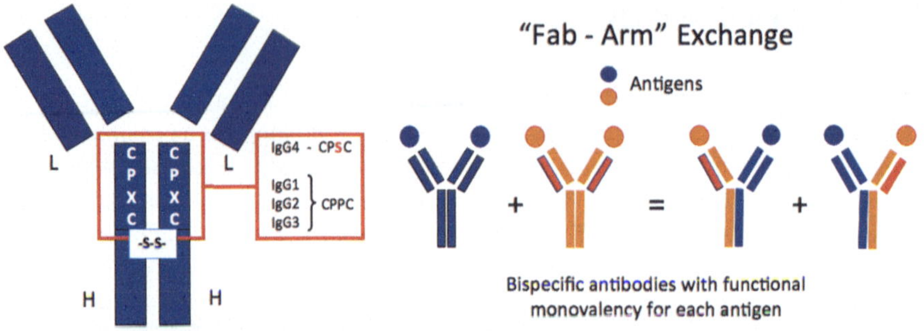

	IgG1	IgG2	IgG3	IgG4
Biological target	Protein	Carbohydrate	Protein	Protein
Functional form	Monomeric, bivalent	Dimeric, bivalent or tetravalent	Monomeric, bivalent	Bispecific, monovalent for each antigen
Serum levels (mg/dl)	5 - 11	1.5 - 6	0.2 - 1	0.08 - 1.4
Proportion of total IgG (%)	43 - 75	16 - 48	1.7 - 7	0.8 - 11.7
Complement fixation	+++	+	+++	-
Binding to FC gamma receptors				
FC gamma RI	++	+	+++	+
FC gamma RII	++	+	+++	-
FC gamma RIII	++	+	++	-

Fig. 27.2 Immunological properties of IgG4 antibodies

27.4 Diagnosis

Actual serological and radiological features of IgG4-RD lack adequate sensitivity and specificity for diagnostic purposes. Therefore, histological confirmation is mandatory in order to achieve a definite diagnosis of IgG4-RD, and to exclude mimicking neoplastic or inflammatory conditions.

27.4.1 Serological Findings

Acute phase reactants such as the erythrocyte sedimentation rate and C-reactive protein are usually (but not always) elevated to a moderate degree. Peripheral blood eosinophilia and increased serum IgE occur in almost 30 % of patients [1, 2, 8, 9]. Some patients might also have positive low-titre ANA, but the presence of anti-Ro/SSA, anti-La/SSB, and ANCA antibodies typically implicates other immune mediated conditions. Other IgG subclasses, particularly IgG1, are frequently elevated in IgG4-RD, though generally not to the extent of IgG4. Complement C3 and C4 levels might decrease in case of renal involvement or concomitant serum IgG1, IgG2 or IgG3 elevation. Elevated serum IgG4 levels are present in almost 60 % of patients and their increase correlates with multi-organ involvement. However, dosage of serum IgG4 concentrations alone has demonstrated a poor specificity and a weak positive predictive value, thus being an unreliable tool for diagnostic and follow-up purposes [17]. Recently, measurement of circulating plasmablasts, the precursors of mature plasma cells, has been proposed as a potential biomarker for IgG4-RD, but larger population studies are required in order to validate this test [14, 15].

27.4.2 Radiological Findings

Radiological findings are largely unspecific and are not sufficient to distinguish IgG4-RD from neoplastic conditions presenting with mass forming lesions. IgG4-related AIP is the sole exception, because computed tomography and magnetic resonance imaging classically show a "sausage-like" shaped pancreas with, often, a surrounding rim of edematous tissue (Fig. 27.1a – asterisks). Positron emission tomography (PET) has an important diagnostic value because active IgG4-RD lesions show an increased ^{18}FDG uptake, unless in case of a long-standing fibrotic scar (Fig. 27.1c, d). In particular, ^{18}FDG PET scan is essential for detecting vascular involvement, might be of help for staging purposes, and represents a useful tool for assessing disease response to treatment.

27.4.3 Histopathological Findings

Histopathological hallmarks of IgG4-RD are (i) a dense storiform fibrosis, (ii) obliterative phebitis, (iii) a lymphoplasmacytic infiltrate rich in IgG4-positive plasma cells and (iv) a mild to moderate eosinophilic infiltrate [1, 3, 18].

The inflammatory infiltrate is mainly composed of oligoclonal B and T lymphocytes. B lymphocytes tend to organize in germinal centers, while T lymphocytes are spread throughout the fibrotic tissue. At a careful immunohistochemical evaluation, the majority of plasma cells stain positive for IgG4 antibodies (Fig. 27.3a – arrows), while other subclasses are less represented. An overall number of IgG4-positive plasma cells > 10/high power field and an IgG4+/IgG+ plasma cells ratio > 40 % are considered diagnostic for IgG4-RD [1, 18], although more detailed organ specific immunohistochemical criteria have been developed in the "Consensus Statement on the Pathology of IgG4-RD" [18]. Interestingly, IgG4-positive plasma cells might be found in other inflammatory, neoplastic and infectious conditions, and thus, a correct diagnosis should rely also on the following histological features.

In particular, "storiform fibrosis" is a hallmark of IgG4-RD (Fig. 27.3b – asterisks). Storiform fibrosis is thought to derive from collagen secretion by activated myofibroblasts in response to different pro-fibrotic stimuli. The term 'storiform' refers to an irregularly whorled organization of the collagen bundles that can be observed in the majority of organs affected by IgG4-RD. For instance, Fig. 27.3c shows the lymphoplasmacytic infiltrate spread within the fibrotic background in the adventitial wall of an IgG4-related aortitis.

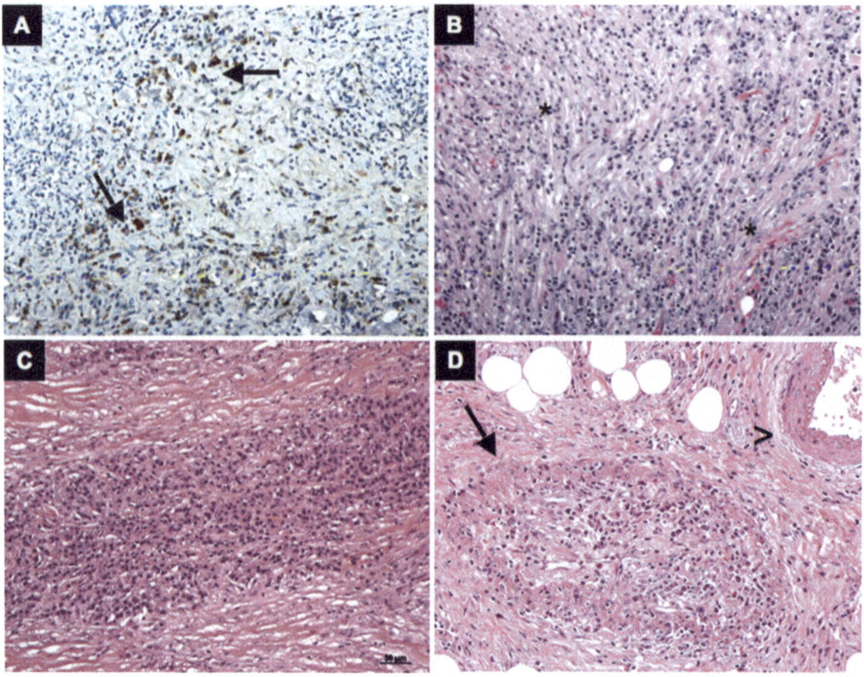

Fig. 27.3 Histopathological features of IgG4-related disease. (**a**) Immunohistochemistry reveals IgG4-positive plasma cells (*arrows*). (**b**) Storiform fibrosis (*asterisks*). (**c**) Lymphoplasmacytic infiltrate spread throughout the storiform fibrosis in the adventitial wall of a thoracic aorta. (**d**) "Obliterative phlebitis": an obliterated vein (*arrow*) close to the accompanying unaffected artery (*arrowhead*)

Histological examination should also carefully assess vascular involvement because "obliterative phlebitis" represents another characteristic aspect of IgG4-RD. "Obliterative phlebitis" is caused by a dense lymphoplasmacytic infiltrate seen both within the wall of the venous channel and within the lumen (Fig. 27.3d – arrow). Accompanying arteries are less likely to be affected by the inflammatory process, and can therefore serve as a guidepost to detecting obliterated venous structures (Fig. 27.3d – arrowhead). The presence of arteritis is a rare finding in IgG4-RD, and, although occasionally observed in cases of autoimmune pancreatitis and in lung lesions, it should suggest alternative small-to-medium sized vessels vasculitis. Interestingly, the arteritis of IgG4-RD is characterized by a non-necrotizing lymphoplasmacytic infiltrate with or without obliteration of the lumen.

The presence of these findings might vary according to disease progression. For instance, epithelial or endothelial damage is not observed until end-stage disease, when fibrosis subverts the parenchymal structure and impairs organ function. At this point, the inflammatory infiltrate is almost completely substituted by a dense fibrous "acellular" tissue ("fibrotic scar"), and IgG4-positive plasma cells might be absent. Neutrophils, necrosis and granulomas are classically absent, and their presence should prompt the exclusion of other potential differential diagnoses.

27.5 Therapy

Treatment is not always necessary in patients with IgG4-RD, and watchful waiting is prudent in some asymptomatic cases. Conversely, when vital organs are involved or patients become symptomatic, aggressive treatment is needed because IgG4-RD can lead to serious organ dysfunction and failure.

Glucocorticoids represent the first line therapy and lead to dramatic clinical responses in the majority of cases with both pancreatic and extra-pancreatic disease. One typical approach is to treat with prednisolone (0.6 mg/kg per day for 2–4 weeks), followed by a gradual taper over a period of 3–6 months once remission is achieved [4, 5].

IgG4-RD has, however, a marked tendency to relapse during or after GC tapers, especially in cases of multiorgan and extrapancreatic involvement [4, 5]. In patients with autoimmune pancreatitis, the risk of disease flare increases substantially with time. In one study, the percentage of patients experiencing relapses was 32 % at 6 months, 56 % at 1 year, 70 % at 2 years, and 92 % at 3 years, despite the maintenance of some GC dose in most patients [4, 5]. Disease flares have been described over the course of GC tapering, after GC withdrawal, and during maintenance therapy with low-dose GCs [4, 5]. Because the durability of treatment responses to GCs is variable and poorly characterized, a variety of GC-sparing agents have been employed as remission-maintenance drugs (e.g. azathioprine, mycophenolate mofetil, methotrexate, cyclophosphamide, and bortezomib) with alternate results [4, 5].

More recently, two doses of rituximab (1gr 15 days apart), an anti-CD20 monoclonal antibody, has proven to induce swift clinical responses in patients with recurrent or refractory disease, suggesting an important contribution of the B cell lineage to the pathophysiology of this fibrotic disorder [5–7, 15, 16]. Indeed, researchers have hypothesized that mass-forming IgG4-RD lesions are more likely to shrink in the presence of a prominent lymphoplasmacytic infiltrate ("active fibroinflammation"), rather than in the presence of tightly organized collagen bundles in which both inflammatory cells and myofibroblasts are rare ("acellular end-stage fibrosis" or "fibrotic scar") [7].

Further studies are needed to provide reliable and standardized guidelines for the long-term management of IgG4-RD.

References

1. Stone JH, Zen Y, Deshpande V (2012) IgG4-related disease. N Engl J Med 366:539–551
2. Hamano H, Kawa S, Horiuchi A, Unno H, Furuya N, Akamatsu T et al (2001) High serum IgG4 concentrations in patients with sclerosing pancreatitis. N Engl J Med 344:732–738
3. Della-Torre E, Lanzillotta M, Doglioni C (2015) Immunology of IgG4-related disease. Clin Exp Immunol 181:191–206. doi:10.1111/cei.12641

4. Khosroshahi A, Stone JH (2011) Treatment approaches to IgG4-related systemic disease. Curr Opin Rheumatol 23:67–71
5. Khosroshahi A, Carruthers M, Deshpande V, Unizony S, Bloch DB, Stone JH (2012) Rituximab for the treatment of IgG4-related disease: lessons from ten consecutive patients. Medicine 91:57–66
6. Khosroshahi A, Bloch DB, Deshpande V, Stone JH (2010) Rituximab therapy leads to rapid decline of serum IgG4 levels and prompt clinical improvement in IgG4-related systemic disease. Arthritis Rheum 62:1755–1762
7. Della-Torre E, Feeney E, Deshpande V, Mattoo H, Mahajan V, Kulikova M et al (2014) B-cell depletion attenuates serological biomarkers of fibrosis and myofibroblast activation in IgG4-related disease. Ann Rheum Dis 74:2236– 2243. doi:10.1136/annrheumdis-2014-205799
8. Mattoo H, Della-Torre E, Mahajan VS, Stone JH, Pillai S (2014) Circulating Th2 memory cells in IgG4-related disease are restricted to a defined subset of subjects with atopy. Allergy 69:399–402
9. Della Torre E, Mattoo H, Mahajan VS, Carruthers M, Pillai S, Stone JH (2014) Prevalence of atopy, eosinophilia, and IgE elevation in IgG4-related disease. Allergy 69:269–272
10. Stone JH, Khosroshahi A, Deshpande V, Stone JR (2010) IgG4-related systemic disease accounts for a significant proportion of thoracic lymphoplasmacytic aortitis cases. Arthritis Care Res (Hoboken) 62:316–322
11. Kasashima S, Zen Y, Kawashima A, Endo M, Matsumoto Y, Kasashima F (2008) A new clinicopathological entity of IgG4-related inflammatory abdominal aortic aneurysm. J Vasc Surg 42:1261–1271
12. Vaglio A, Pipitone N, Salvarani C (2011) Chronic periaortitis: a large-vessel vasculitis? Curr Opin Rheumatol 23:1–6
13. Mahajan VS, Mattoo H, Deshpande V, Pillai SS, Stone JH (2014) IgG4-related disease. Annu Rev Pathol 9:315–347
14. Mattoo H, Mahajan VS, Della-Torre E, Sekigami Y, Carruthers M, Wallace ZS et al (2014) De novo oligoclonal expansions of circulating plasmablasts in active and relapsing IgG4-related disease. J Allergy Clin Immunol 134:679. doi:10.1016/j.jaci.2014.03.034
15. Wallace ZS, Mattoo H, Carruthers M, Mahajan VS, Della Torre E, Lee H et al (2014) Plasmablasts as a biomarker for IgG4-related disease, independent of serum IgG4 concentrations. Ann Rheum Dis 74:190–195. doi:10.1136/annrheumdis-2014–205233
16. Carruthers MN, Topazian MD, Khosroshahi A, Witzig TE, Wallace ZS, Hart PA et al (2015) Rituximab for IgG4-related disease: a prospective, open-label trial. Ann Rheum Dis 74:1171–1177. doi:10.1136/annrheumdis-2014–206605
17. Carruthers MN, Khosroshahi A, Augustin T, Deshpande V, Stone JH (2014) The diagnostic utility of serum IgG4 concentrations in IgG4-related disease. Ann Rheum Dis 74:14–18. doi:10.1136/annrheumdis-2013–204907
18. Deshpande V, Zen Y, Chan JK, Yi EE, Sato Y, Yoshino T et al (2012) Consensus statement on the pathology of IgG4-related disease. Mod Pathol 25:1181–1192

Chapter 28
Urticarial Vasculitis. A Review of the Literature

Giulia De Feo, Roberta Parente, Chiara Cardamone, and Massimo Triggiani

Abstract Urticarial vasculitis (UV) is a clinical entity characterized by the aspects of leukocytoclastic vasculitis (LCV). Since its clinical features often overlap with those of common urticaria, a lesional skin biopsy is often necessary for the diagnosis. UV may be idiopathic or caused by endogenous (tumors) or exogenous (infections) factors. UV is frequently associated to consumption of serum complement factors. In this case, UV is called hypocomplementemic urticarial vasculitis (HUV). Hypocomplementemic urticarial vasculitis syndrome (HUVS) is a more severe form characterized by hypocomplementemia and systemic symptoms. Some patients may have UV with normal complement levels (normocomplementemic UV: NUV) and they usually have minimal or no systemic involvement. Response of UV to treatment is variable.

Keywords Urticarial vasculitis • Complement

Abbreviations

UV	urticarial vasculitis
LCV	leukocytoclastic vasculitis
HUVS	hypocomplementemic urticarial vasculitis syndrome
SLE	systemic lupus erythematosus
NUV/HUV	normocomplementemic/hypocomplementemic urticarial vasculitis
CU	chronic urticaria
COPD	chronic obstructive pulmonary disease
ESR	erythrocytes sedimentation rate
ANA	antinuclear antibodies
CAPS	cryopyrin-associated periodic syndromes
FCAS	familial cold autoinflammatory syndrome

G. De Feo • R. Parente • C. Cardamone • M. Triggiani (✉)
Department of Internal Medicine, Division of Allergy and Clinical Immunology, University of Salerno, Salerno, Italy
e-mail: mtriggiani@unisa.it

F. Dammacco et al. (eds.), *Systemic Vasculitides: Current Status and Perspectives*, DOI 10.1007/978-3-319-40136-2_28

MWS	Muckle-Wells syndrome
CINCA	chronic infantile neurological cutaneous and articular syndromes
NLRP3	NOD-like receptor protein 3

28.1 Introduction

Urticarial vasculitis (UV) is a rare disorder characterized by recurrent episodes of urticaria (each one lasting more than 24 h) and histopathological features of leukocytoclastic vasculitis (LCV) [1, 2]. Hypocomplementemic urticarial vasculitis syndrome (HUVS) represents a more severe form of UV and has characteristic features, other than hypocomplementemia, which resemble systemic lupus erythematosus (SLE) [2, 3]. The diagnostic criteria of HUVS have been defined by Schwartz et al. (1982), as described below [4]. However, some patients do not show significant clinical or laboratory features other than complement consumption and, thus, they are probably affected by hypocomplementemic urticarial vasculitis (HUV), rather than HUVS [2]. Over the years, different names have been used to refer to the clinical and laboratory manifestations of UV. They probably are all variants of the same disease with different appearances, going from urticaria with cutaneous vasculitis without complement consumption (non-hypocomplementemic urticarial vasculitis: NUV), to mild forms with hypocomplementemia as unique manifestation other than cutaneous lesions (HUV), or severe forms with systemic involvement and hypocomplementemia, sometimes associated with minimal urticaria (HUVS) [2].

28.2 Epidemiology

UV is a rare condition and its prevalence and incidence depend on the histological definition of vasculitis, which could differ among the authors. Prevalence ranges from 3 % to 20 % in patients with chronic urticaria (CU) [1]. Women represent from 60 % to 80 % of the affected population [2]. There is a peak of incidence in the fourth decade of life, as it occurs for other immunological diseases. It is rare in children, but it should be considered if urticaria, glomerulonephritis, arthralgia or arthritis, and lung involvement coexist [1, 2]. Renal involvement seems to be more severe in pediatric cases [5, 6].

Table 28.1 Histopathological criteria for the diagnosis of urticarial vasculitis (UV) and differences with histological features of chronic urticaria (CU)

UV	CU
Vessel wall and perivascular infiltrate (mostly neutrophils)	Absent or minimal perivascular infiltrate (various cells, mostly T-lymphocytes)
Injury and swelling of endothelial cells (usually of the venules)	Minimal endothelial swelling (inconstant)
Fibrinoid degeneration of endothelial cells with fibrin deposits	No fibrin deposits
Leukocytoclasis	No leukocytoclasis

28.3 Histopathology and Immunopathology

The most frequent finding in skin biopsy of patients with UV is leukocytoclasis, defined as the fragmentation of leukocytes, particularly neutrophils, and variable numbers of lymphocytes and eosinophils, with generation of nuclear fragments and fibrinoid degeneration with fibrin deposits within and around the venules. Other histological modifications are relatively frequent and they are useful to differentiate UV from CU (Table 28.1) [7]. However, some patients show no major histopathologic change other than a perivascular inflammatory infiltrate, and they could be considered as affected by a less severe form of UV. These observations suggest the existence of a *continuum* in the histopathologic findings which can reflect the severity of clinical manifestations [1, 2, 8].

28.4 Pathophysiology

Pathogenesis of UV seems to be due to a type III hypersensitivity reaction mediated by immune complexes, as it occurs for other leukocytoclastic vasculitides. Immune complexes precipitate in the vessel wall and activate the classical complement pathway with production of C3a and C5a [1, 2]. The antigen of the immune complexes is only rarely identified [2]. Immune complexes are also implicated in lung, cardiac and renal damage [2, 9–14]. The anaphylatoxins C3a and C5a induce mast cell degranulation and production of mediators including histamine, chemokines and cytokines, which increase vessel permeability and chemotaxis of inflammatory cells, particularly neutrophils. As neutrophils migrate to the site of inflammation, they acquire phagocytic properties and release proteolytic enzymes (e.g., protease, collagenase, elastase) causing further tissue damage [1, 2]. The coagulation system seems to be involved in the pathogenesis of UV, probably even more than it is in common urticaria. In both diseases, tissue factor, released mainly by eosinophils infiltrating skin lesions, activates the coagulation pathway leading to generation of thrombin with increased vascular permeability and edema [15].

28.5 Clinical Features

Clinical manifestations of UV can be both cutaneous and systemic (Table 28.2). Cutaneous lesions show different patterns going from the more common wheals with angioedema, to the less common urticarial lesions with residual hyperpigmentation or purpura (Fig. 28.1). Wheals in UV are usually long-lasting (>24 h) and may tend to confluency, reaching the consistency of plaques. In some cases wheals may acquire the typical aspect of "target" lesions with a central area of resolution (Fig. 28.2). Other patterns seem to be rare [2, 16]. Inflammation could also cause epidermal ulceration or subepidermal vesicle formation [9]. These complex skin lesions never occur in common urticaria. Many patients with UV and systemic involvement have arthralgias or arthritis, so that they configure the so-called "AHA (Arthralgias/Arthritis, Hives, Angioedema) Syndrome". In particular, arthralgia occurs in a half of the patients [2].

Arthritis is mainly localized to small joints of the hands, elbows, feet, ankles and knees [2]. Jaccoud's arthropathy (a slowly progressive arthritis causing deformities

Table 28.2 Clinical features of UV

Skin	Common: erythematous urticarial papules, angioedema, dermographism, annular erythema
	Less common: urticarial lesions with residual hyperpigmentation or purpura
	Rare: erythema multiforme-like lesions, *livedo reticularis*, Raynaud's phenomenon
Joint	Common: arthralgia, arthritis
	Less common: Jaccoud's syndrome
Respiratory	Common: cough, dyspnea, COPD, asthma, pleural effusion
	Less common: laryngeal edema, pleuritis, emphysema, hemoptysis
Central Nervous System	Common: *pseudotumor cerebri*, aseptic meningitis
	Less common: transverse myelitis
Peripheral Nervous System	Common: cranial nerve palsies
	Less common: other peripheral neuropathies
Heart and cardiovascular	Common: pericarditis, pericardial effusion, cardiac tamponade
	Less common: cardiac valve disease
Gastrointestinal	Common: substernal and/or abdominal pain, nausea, vomiting, diarrhea
	Less common: hepatomegaly
Renal	Common: hematuria, proteinuria, glomerulonephritis
	Less common: renal failure
Ophthalmologic	Common: conjunctivitis, episcleritis, uveitis
	Less common: geographic serpiginous choroidopathy, visual loss, optic atrophy
Other	Fever, fatigue, splenomegaly, lymphadenopathy, cold sensitivity

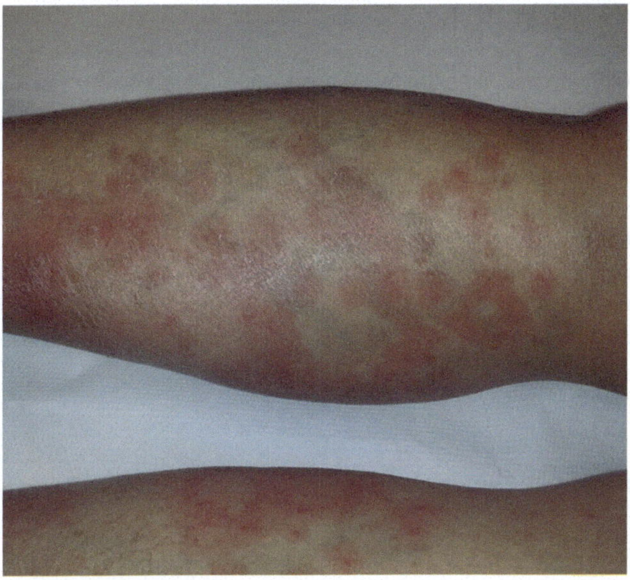

Fig. 28.1 Skin lesions in a patient affected by UV

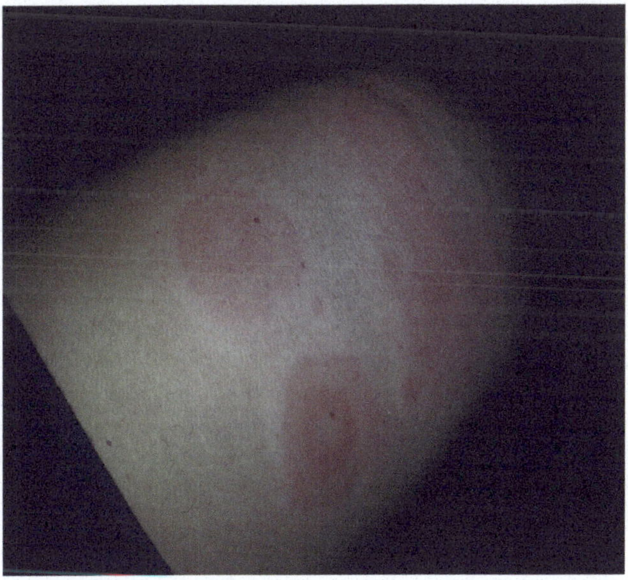

Fig. 28.2 Urticaria with "target lesions" on leg skin

of the fingers and toes) can also occur, and these patients seem to be more frequently interested by cardiac involvement (valvulopathy) [8–10]. Renal disease is less common, particularly in patients with NUV, and usually occurs with moderate-to-severe proteinuria and microhematuria rarely evolving to mild renal failure [5, 6]. Severe respiratory complications are rare (e.g., hemoptysis, emphysema, pleural effusion) whereas asthma and COPD are more frequent [4, 8, 11–14]. Gastrointestinal symptoms may include abdominal pain, nausea, vomiting, diarrhea or gastrointestinal distress, but not bleeding. Ophthalmologic and central nervous system involvement is rare and may include episcleritis, uveitis or conjunctivitis, *pseudotumor cerebri*, aseptic meningitis and cranial nerve palsies [17]. Other systemic symptoms or clinical findings, such as fever, fatigue, hepato-, splenomegaly and lymphadenopathy are considered rare [2].

28.6 Diagnosis, Causes and Associations

Skin biopsy demonstrating the histological findings of urticarial vasculitis remains the gold standard for diagnosis [1]. When the diagnosis has been established, a complete blood count and serum levels of creatinine, blood urea nitrogen, electrolytes, erythrocyte sedimentation rate (ESR), complement (C3, C4, C1q and CH50), circulating immune complexes (present from 30 % to 75 % of patients with UV), total serum immunoglobulins and cryoglobulins in association with liver function tests and urinalysis should be obtained in all patients [2]. An increased ESR, reduced C3, C4 and sometimes C1q and a positive antinuclear antibody (ANA) or anti-dsDNA, are common. Anti-C1q antibodies IgG can be detected (also known as C1q precipitins). An elevated polyclonal immunoglobulin level, cryoglobulinemia and anemia are often present [2]. Anti-Ro/SSa, anti-La/SSb, anti-Sm, antiphospholipid and anti-endothelial cell antibodies may also be present in HUV associated with connective tissue disease and in HUVS. Hypocomplementemia is considered as a marker of systemic involvement [2]. Patients with respiratory symptoms should be evaluated with chest X-ray and pulmonary function tests, whereas those who refer arthralgias/arthritis should undergo joint and skeletal X-ray examination. Suspected causes and association with other diseases should be investigated if physical examination and history are suggestive: connective tissue diseases (e.g., Sjögren's syndrome), infections (e.g., hepatitis B or C, EBV), some drugs (e.g., chemotherapy) [18], cryoglobulinemia and other immunoglobulin abnormalities (IgG/IgD/IgA gammopathy), hematologic disorders (leukemia, lymphoma, polycythemia vera, idiopathic thrombocytopenic purpura, essential thrombocytemia) [19–21], malignancy (e.g., metastatic adenocarcinoma of the colon) and other less common conditions (e.g., inflammatory bowel disease, amyloidosis). Schnitzler syndrome is an association of monoclonal IgM gammopathy with increased markers of systemic inflammation (e.g., elevated C-reactive protein levels, fever, weight loss, arthralgias and bone pain) and the chronic appearance of wheals, which often show signs of urticarial vasculitis [2, 22]. Diagnosis of HUVS (also known as McDuffie

Table 28.3 Diagnostic criteria of HUVS by Schwartz et al. (1982)

Major criteria	Minor criteria
1. Urticaria for more than 6 months	1. Dermal venulitis on biopsy
	2. Arthralgia or arthritis
2. Hypocomplementemia	3. Uveitis or episcleritis
	4. Mild glomerulonephritis
	5. Recurrent abdominal pain
	6. Positive C1q precipitin test by immunodiffusion with reduced C1q level

syndrome) is difficult given its similarities with SLE. Schwartz et al. (1982) defined two major and six minor criteria; both major criteria and at least two of the minor criteria are required for the diagnosis (Table 28.3). Exclusion criteria are: a significant cryoglobulinemia (cryocrit >1 %), an elevated anti-DNA antibody titer, high titers of ANA, hepatitis B antigenemia, decreased C1 esterase inhibitor level and an inherited complement deficiency [2].

UV also needs to be differentiated from other entities that may occur with urticarial skin lesions such as Cryopyrin-Associated Periodic Syndromes (CAPS). Three distinct syndromes are included in CAPS: Familial Cold Autoinflammatory Syndrome (FCAS), Muckle-Wells Syndrome (MWS) and several disorders known as CINCA (Chronic Infantile Neurological Cutaneous and Articular syndromes). These autosomal dominant disorders are due to different mutations of a single gene called NOD-like receptor protein 3 (NLRP3). This gene codifies for a protein which is a cryopyrin. This inflammasome protein is responsible for augmenting serum levels of Interleukin-1β, leading to an autoinflammatory state. Although differential diagnosis may be often difficult, there are some clinical differences between CU/UV and CAPS. For example, skin lesions in CAPS are more symmetrical than CU/UV, they consist of pink macules or plaques rather than typical wheals and usually last less than 24 h. In addition, skin lesions in CAPS are accompanied by pain or burning sensation rather than itching [15].

28.7 Treatment

Treatment of UV depends on the severity of disease. Anti-histamines are usually not effective. During exacerbations most patients require short courses of glucocorticoids to control cutaneous and systemic manifestations. Continuous glucocorticoid treatment is used in patients with more severe forms. The usual dosage is 1 mg prednisone equivalent/Kg of body weight daily *per os* until clinical remission. Alternatively, other immunosuppressive drugs, such as cyclosporin, azathioprine, mycophenolate mofetil and hydroxychloroquine may be used. There is some evidence of effectiveness of dapsone and cyclophosphamide, but these drugs may cause major side effects. Plasmapheresis may be considered in cases not responsive to immunosuppressive treatments. The efficacy of i.v. immunoglobulins has not been fully evaluated. There is evidence that tocilizumab, an anti-IL6 antibody, could be useful in patients refractory to other treatments [23, 24]. Moreover, some patients affected by inflammatory bowel disease and associated UV may benefit from treatment with rituximab [25, 26]. Omalizumab, a monoclonal anti-IgE antibody is effective in anti-histamine-resistant chronic urticaria [27–29]. The mechanism of action of omalizumab in chronic urticaria is complex and goes beyond the binding of free IgE and blockade of IgE binding to their high affinity receptors (FcεRI). It is likely that reduction of IgE binding to FcεRI induces the down-regulation of these receptors on mast cells, basophils and other cells with subsequent functional inhibition of the secretory activities of these cells [23]. Omalizumab also seems to be useful for the treatment of UV, but it is not clear whether it is effective against both normocomplementemic or hypocomplementemic UV [30]. Recently, canakinumab, a monoclonal antibody anti-IL1β used for treatment of CAPS [15], has been proposed for patients with UV. However, since only case reports have been published, further studies are needed to confirm this treatment option [31].

References

1. Venzor J, Lee WL, Huston DP (2002) Urticarial vasculitis. Clin Rev Allergy Immunol 23:201–216
2. Davis MDP, Brewer JD (2004) Urticarial vasculitis and hypocomplementemic urticarial vasculitis syndrome. Immunol Allergy Clin North Am 24:183–213
3. Kolkhir P, Pogorelov D, Olisova O et al (2016) Comorbidity and pathogenic links of chronic spontaneous urticaria and systemic lupus erythematosus – a systematic review. Clin Exp Allergy 46:275–287
4. Schwartz HR, McDuffie FC, Black LF et al (1982) Hypocomplementemic urticarial vasculitis: association with chronic obstructive pulmonary disease. Mayo Clin Proc 57:231–238
5. Renard M, Wounters C, Proesmans W (1998) Rapidly progressive glomerulonephritis in a boy with hypocomplementemic urticarial vasculitis. Eur J Pediatr 157:243–245
6. Balsam L, Karim M, Miller F et al (2008) Crescentic glomerulonephritis associated with hypocomplementemic urticarial vasculitis syndrome. Am J Kidney Dis 52:1168–1173
7. Black AK (1999) Urticarial vasculitis. Clin Dermatol 17:565–569
8. Chen HJL, Bloch KJ (2001) Hypocomplementemic urticarial vasculitis, Jaccoud's arthropathy, valvular heart disease and reversible tracheal stenosis: a surfeit of syndromes. J Rheumatol 28:383–386
9. Babajanians A, Chung-Park M, Wisnieski JJ (1991) Recurrent pericarditis and cardiac tamponade in a patient with hypocomplementemic urticarial vasculitis syndrome. J Rheumatol 18:752–755
10. Kervarrec T, Binois R, Bléchet C et al (2015) Hypocomplementaemic urticarial vasculitis with bullous lesions and pericardial involvement. Ann Dermatol Venereol 19:1–6
11. Hunt DPJ, Weil R, Nicholson AG et al (2006) Pulmonary capillaritis and its relationship to development of emphysema in hypocomplementaemic urticarial vasculitis syndrome. Sarcoidosis Vasc Diffuse Lung Dis 23:70–72
12. Pujara AC, Mohammed TL (2012) Hypocomplementemic urticarial vasculitis syndrome: a rare cause of basilar panacinar emphysema. J Thorac Imaging 27:W50–W51

13. Jamison SC, Brierre S, Sweet J et al (2008) A case of precocious emphysema and lung cancer in a woman with a history of hypocomplementemic urticarial vasculitis. Chest 133:787–789
14. Sjöwall C, Hallbeck M, Sandström P (2015) Clinically suspected recurrence of gastric carcinoid proved to be hypocomplementaemic urticarial vasculitis syndrome with pulmonary involvement. Scand J Rheumatol 44:337–339
15. Marzano AV, Tavecchio S, Venturini M et al (2015) Urticarial vasculitis and urticarial autoinflammatory syndromes. G Ital Dermatol Venereol 150:41–50
16. Lee JSS, Loh TH, Seow SC et al (2007) Prolonged urticaria with purpura: the sprectrum of clinical and histopathologic features in a prospective series of 22 patients exhibiting the clinical features of urticarial vasculitis. J Am Accad Dermatol 56:994–1005
17. Filosto M, Cavallaro T, Pasolini G et al (2009) Idiopathic hypocomplementemic urticarial vasculitis-linked neuropathy. J Neurol Sci 284:179–181
18. Bock VL, Friedlander M, Waring D et al (2014) Cutaneous adverse effects of hormonal adjuvant therapy for breast cancer: a case of localised urticarial vasculitis following anastrozole therapy and a review of the literature. Australas J Dermatol 55:282–285
19. Moulin C, Debarbieux S, Ducastelle-Lepretre S et al (2010) Urticarial vasculitis and asymptomatic acquired C1 esterase inhibitor deficiency revealing an angioimmunoblastic T cell lymphoma. Eur J Dermatol 20:515–516
20. Takao M, Hamada T, Kaji T et al (2016) Hypocomplementemic urticarial vasculitis arising in a patient with immunoglobulin G4-related disease. Int J Dermatol 55:430–433
21. Koudoukpo C, Jachiet M, Zini JM et al (2014) Urticarial vasculitis associated with essential thrombocythaemia progressing to myelofibrosis. Ann Dermatol Venereol 141:773–776
22. Zuberbier T, Maurer M (2014) Urticarial vasculitis and Schnitzler syndrome. Immunol Allergy Clin North Am 34:141–147
23. Makol A, Gibson LE, Michet CJ (2012) Successful use of interleukin 6 antagonist tocilizumab in a patient with refractory cutaneous lupus and urticarial vasculitis. J Clin Rheumatol 18:92–95
24. Greenberger PA (2014) Chronic urticaria: new management options. World Allergy Organ J 7:31–36
25. Swaminath A, Magro CM, Dwyer E (2011) Refractory urticarial vasculitis as a complication of ulcerative colitis successfully treated with rituximab. J Clin Rheumatol 17:281–283
26. Saigal K, Valencia IC, Cohen J et al (2003) Hypocomplementemic urticarial vasculitis with angioedema, a rare presentation of systemic lupus erythematosus: rapid response to rituximab. J Am Acad Dermatol 49:S283–S285
27. Maurer M, Rosen K, Hsie H-J et al (2013) Omalizumab for the treatment of chronic idiopathic or spontaneous urticaria. N Engl J Med 368:924–935
28. Sussman G, Hébert J, Barron C et al (2014) Real-life experiences with omalizumab for the treatment of chronic urticaria. Ann Allergy Asthma Immunol 112:170–174
29. Kai AC, Flohr C, Grattan CE (2014) Improvement in quality of life impairment followed by relapse with 6-monthly periodic administration of omalizumab for severe treatment-refractory chronic urticaria and urticarial vasculitis. Clin Exp Dermatol 39:651–652
30. Ghazanfar MN, Thomsen SF (2015) Omalizumab for urticarial vasculitis: case report and review of the literature. Case Rep Dermatol Med 2015:1. doi:10.1155/2015/576893
31. Krause K, Mahamed A, Weller K et al (2015) Efficacy and safety of canakinumab in urticarial vasculitis: an open-label study. J Allergy Clin Immunol 132:751–754

Part III
Secondary Vasculitides

Chapter 29
HCV-Related Cryoglobulinemic Vasculitis: An Overview

Franco Dammacco, Sabino Russi, and Domenico Sansonno

Abstract Cryoglobulinemic vasculitis (CV) is a small and medium-size vasculitis characterized by the occurrence in the serum of reversibly precipitating proteins, named cryoglobulins and immunochemically formed by an IgM component (monoclonal or polyclonal) with rheumatoid factor activity and a polyclonal IgG component (mixed cryoglobulins). CV is almost invariably associated to chronic HCV infection. In addition to the typical purpura/asthenia/arthralgia syndrome, the pleomorphic clinical picture often includes membrano-proliferative glomerulonephritis and motor-sensory axonopathy. Hemorrhagic alveolitis, gastrointestinal vasculitis, heart failure and hyperviscosity syndrome are less frequently observed. The amount of cryoprecipitate, named cryocrit, is not strictly related to the clinical severity of CV and to the viral load. Rheumatoid factor activity and low levels of the complement C4 (sometimes also of C3 and CH50) are unfailing serological abnormalities. The pathogenetic mechanism of CV is still incompletely defined, but it can be essentially ascribed to the formation of HCV particles/IgG/IgM macromolecular complexes that are good acceptors of C1q and can therefore bind to the C1q receptors on the endothelial cells. This would eventually trigger the onset of a leukocytoclastic vasculitis. Although with wide geographic variations, CV can progress to non-Hodgkin lymphoma (NHL), possibly through an impaired regulatory control of B-cell growth. In Italy, approximately 5 % of B-cell NHLs seem to be HCV-related. Therapy of CV should be adapted to each patient's condition. Low daily doses of corticosteroids can mitigate arthralgias and possibly prevent flares, but their long-term administration should be avoided for their inevitable side effects. Pulsed intravenous infusions of corticosteroids can prevent organ damage in the course of severe vasculitis flares. High rates of sustained virologic responses can be achieved with the use of the new interferon-free, all-oral direct acting antiviral agents. In patients with refractory/relapsing or with severely active CV, characterized by

F. Dammacco (✉)
Department of Biomedical Sciences and Human Oncology, Section of Internal Medicine, University of Bari Medical School, 70124 Bari, Italy
e-mail: francesco.dammacco@uniba.it

S. Russi • D. Sansonno
Department of Internal Medicine and Human Oncology, University of Bari Medical School, 70124 Bari, Italy

F. Dammacco et al. (eds.), *Systemic Vasculitides: Current Status and Perspectives*, DOI 10.1007/978-3-319-40136-2_29

B-cell clonal expansion, a B-cell depleting therapy with rituximab often results in a satisfactory control of clinical features, although the viral load may sometimes transiently increase. Low-grade and indolent HCV-related NHLs have been found to undergo complete or partial remission following anti-HCV therapy. Finally, double filtration plasmapheresis can contribute to the improvement and possibly healing of chronic ulcers on the legs.

Keywords Cryoglobulinemic vasculitis • Hepatitis C virus • Mixed cryoglobulinemia • Rheumatoid factor • Rituximab

29.1 Introduction

Mixed cryoglobulinemia (MC) is a small and midium vessel vasculitis characterized by the presence in the serum of cold-precipitable proteins named "cryoglobulins" [1]. Although single (type I) monoclonal immunoglobulins (IgG or IgM, more rarely IgA) can behave as cryoglobulins in patients with lymphoproliferative disorders, the large majority of mixed cryoglobulins are formed by IgG/IgM immune-complexes in which the IgG fraction is always polyclonal, whereas the IgM fraction can be monoclonal (type II MC) or polyclonal (type III MC). HCV infection is typically associated with type II MC in which the IgMk monoclonal component is consistently endowed with rheumatoid factor (RF) activity and shows anti-idiotypic activity [2].

In early studies and for several years thereafter, given the ignorance of its etiology, MC has been termed "essential" [3]. At the beginning of the 1990s, following the availability of reagents and methods to detect the serum occurrence of anti-HCV antibodies and shortly after of HCV RNA, the large majority of MC patients were indeed found to be HCV-infected [4–7], thus restricting the percentage of "essential" MC to less than 10 %.

Based on this striking association, MC is now considered the most typical and unquestionable extra-hepatic manifestation of HCV infection. In terms of prevalence, although variable amounts of cryoglobulins can be detected in 25–30 % of HCV-positive patients, most of them remain asymptomatic whereas a cryoglobulin-related illness, commonly defined cryoglobulinemic vasculitis (CV), appears only in a minority (10–15 %) of patients.

29.2 Clinical Features

CV includes a wide spectrum of symptoms, extensively described in recent reviews [8, 9], that range in severity from sporadic petechial eruptions to life-threatening manifestations. The triad purpura/asthenia/arthralgia has long been recognized to be almost invariably associated with MC [3]. Patients usually complain of recurrent

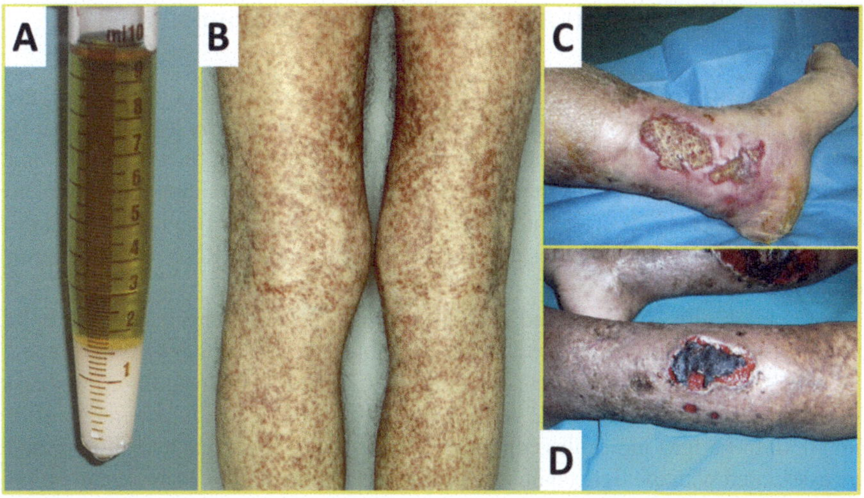

Fig. 29.1 Features of HCV-related cryoglobulinemic vasculitis. (**a**) Whitish cryoprecipitate after keeping a Wintrobe tube at 4 °C for 5 days and centrifuging at 1,400 rpm for 10 min at 4 °C. (**b**) Purpuric eruptions extensively involving the legs. Torpid perimalleolar (**c**) and necrotic pre-tibial (**d**) ulcers

episodes of palpable purpura, confined or largely prevalent to the lower limbs and less often to the buttocks. Over the years, the deposition of hemosiderin in the sites previously affected with crops of purpura results in the local appearance of brownish, irreversible dyschromias which can by themselves raise the suspicion of CV. Weakness and arthralgia, on the other hand, are recorded in approximately 80 % and 70 % respectively, whereas myalgia/fibromyalgia and non-erosive oligoarthritis are observed much less frequently. Raynaud's phenomenon and livedo reticularis are also commonly seen, but unhealing leg and/or malleolar ulcers and digital gangrene are of much less frequent occurrence (Fig. 29.1).

According to our experience in treating 246 patients with HCV-positive CV [10], kidney involvement (with its obvious prognostic implications) is diagnosed in approximately half of the CV patients, one third of whom develop membranoproliferative glomerulonephritis and arterial hypertension, and an additional 15 % a nephrotic syndrome. Membranous glomerulonephritis and focal glomerular sclerosis can more rarely be seen. The renal damage is revealed by the detection of microscopic hematuria and proteinuria, whose amount can range from a little higher than the upper limit of normal up to the nephrotic range.

Another relatively frequent feature of CV is motor-sensory axonopathy, that may result in burning sensation of the legs and diffuse paresthesias, and is exacerbated in patients given interferon-alfa therapy. Neurocognitive impairment and mononeuritis multiplex are unusual complications [11, 12]. Finally, rare symptoms include hemorrhagic alveolitis and interstitial lung fibrosis, gastrointestinal vasculitis, heart failure and dilated cardiomyopathy, hyperviscosity syndrome (that is much more frequent in type I cryoglobulinemia) and painful osteosclerosis [13]. The pathogenesis of these rare manifestations remains undefined but is possibly related to diffuse vessel disease.

29.3 Serological Findings and Laboratory Parameters

The amount of cold-induced serum precipitate, expressed as percentage of the whole serum and named cryocrit by analogy with hematocrit, is usually higher in type I compared with types II and III cryoglobulins. Cryocrit, however, does not seem to be strictly related to the clinical severity of CV [8], although a complete response to therapy is associated to the disappearance of the cryoprecipitate. Obviously, anti-HCV antibodies and variable levels of HCV RNA are detected in virtually all CV patients.

Compared with HCV-infected patients without CV, a striking feature of HCV-positive patients with active CV is the remarkable decrease in their serum of complement C4 and, to a lesser extent, of CH50 and C3 levels [14]. In addition, as stated above, an unfailing mark of MC is the RF activity of the IgM component. Most, if not all, of these IgM molecules exhibit the Wa (V_H1-69) cross-reactive idiotype and are associated with the light chain cross-idiotype 17–109 and the heavy chain cross-idiotype G6. IgM-Wa cross-reactive idiotypes have been shown to be expressed within the cryoprecipitates of patients with type II, but not type III, MC [15]. Further immunological abnormalities of CV patients stem from the peripheral lymphocytogram, that usually shows a higher percentage of CD4-positive and above all CD19/CD20-positive cells.

The prevalence of autoantibodies may range from 9 % of anti-mitochondrial antibodies to 18 % of anti-smooth muscle antibodies and up to 30 % of anti-nuclear antibodies [16].

29.4 Histopathology, Immunofluorescence and Immunohistochemistry

The histological vessel involvement of CV encompasses a broad spectrum of cutaneous lesions of variable severity. Skin biopsies of CV patients include leukocytoclastic vasculitis, urticarial vasculitis, and lymphocytic or granulomatous microvasculitis, and endoarteritis obliterans that may sometimes progress to necrotizing vasculitis, fibrinoid necrosis, and thrombotic occlusion of the vessel lumens [17] (Fig. 29.2a).

Monoclonal IgM with RF activity, complement C3 and C4 components, IgG and less frequently IgA are often detected in the vessel wall by direct immunofluorescence. An important finding to sustain the pathogenetic role of HCV in the onset of cutaneous vasculitis is the demonstration by immunohistochemistry of HCV-related proteins (though not HCV RNA genomic sequences) within the lumen of skin papillary vessels, in vessel walls and in perivascular spaces. Specific HCV immunode-

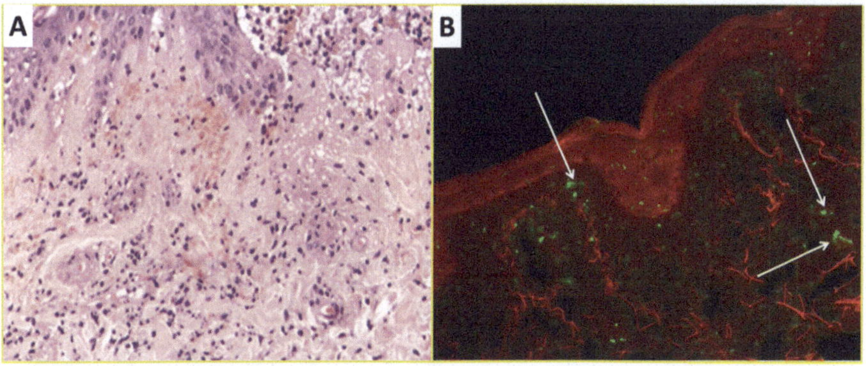

Fig. 29.2 In (**a**), a representative hematoxylin-eosin histological picture of a skin sample from HCV-related CV; in (**b**), specific immunodeposits of HCV core protein can be seen in skin papillary blood vessels (*arrows*)

posits of both core and E2 proteins can also be detected within endothelial cells, the basal lamina, keratinocytes, and sometimes even in areas of normal skin [18, 19] (Fig. 29.2b).

An immune complex mechanism underlies the renal involvement in CV patients and is clinically diagnosed as type I membrano-proliferative glomerulonephritis. Renal biopsies show marked lobular endocapillary proliferation, variable expansion of the mesangial matrix, and glomerular capillary wall thickening. More severe cases may progress to intravascular hyaline thrombi and necrotizing granulomatous vasculitis. Immunofluorescence usually shows a granular staining of the glomerular basement membrane for IgM, IgG, C3 and in a lower percentage for C1q [20]. Tissue deposition of HCV-containing immune complexes is obviously suspected to account for the renal damage in CV patients. However, similarly to the situation of cutaneous vasculitis, HCV RNA genomic sequences have not been unequivocally demonstrated by immunohistochemistry [21].

Peripheral neuropathy in patients with HCV-related CV may also be ascribed to vasculitis. Lymphocytic microvasculitis of small-size arteries is relatively frequent and usually associated with perivascular infiltrates, marked depletion of myelinated fibers and/or axonal degeneration. In a lower number of cases, necrotizing vasculitis can be diagnosed, based on transmural fibrinoid necrosis of the vessel walls, occlusive microangiopathy of epineurial arteries and a rich polymorphonuclear infiltration [22]. Electron microscopy often reveals abnormal deposits of proteinaceous material and widening of myelin lamellae in nerve biopsy samples [23]. HCV-induced chronic sensory neuropathy is likely related to vasculitis of the vasa nervorum consequent to the deposition of HCV-related proteins along the vessel walls.

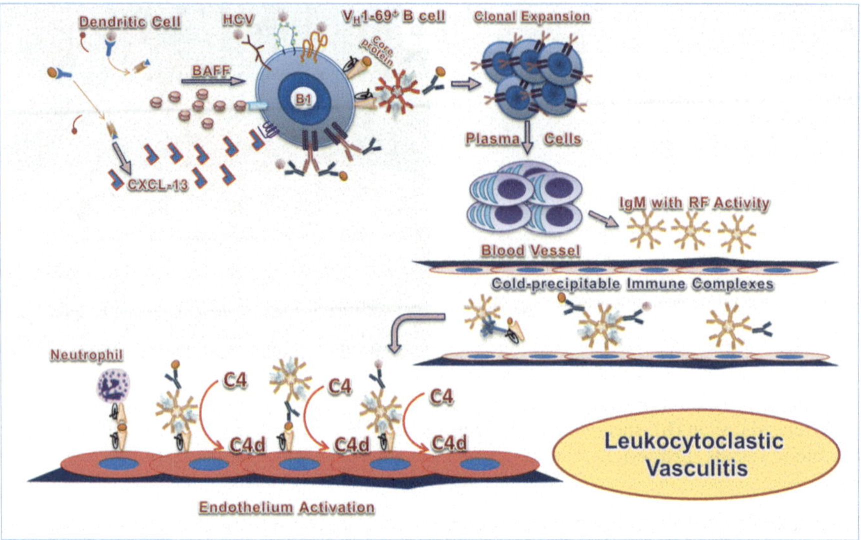

Fig. 29.3 Hypothetical pathogenetic model of HCV-related cryoglobulinemic vasculitic

29.5 Pathogenesis of HCV-Related CV

The pathogenetic mechanism(s) underlying the onset of CV is poorly defined. It can be hypothesized that, through their Toll-like receptors, dendritic cells (DC) are able to capture HCV particles and core protein, and release variable amounts of the cytokine B-cell activating factor (BAFF) that belongs to the tumor necrosis factor-α family. $V_H1\text{-}69^+$ B1 cells and marginal zone B-cells are also able to capture HCV particles and core protein through the action of glycosaminoglycans, scavenger receptors, CD81 molecules as well as C1qR globular domain and B-cell receptors. As a consequence of their responsiveness to BAFF, clonally expanded B-cells synthesize IgM molecules with RF activity, which bind HCV particles, thus resulting in the formation of cryoprecipitable multi-molecular immune complexes that, being good acceptors of C1q protein, can bind to the surface of C1qR-bearing endothelial cells [24]. The specific binding of immune complexes to the globular domain of the C1qR may enhance the recruitment of neutrophils and other inflammatory cells, and this would eventually trigger a leukocytoclastic vasculitis (Fig.29.3).

29.6 HCV-Related B-Cell Clonal Expansion and Non-Hodgkin's Lymphoma

An overwhelming literature consistently points to a significant association between HCV infection and HCV-related CV on one side and the occurrence of B-cell non-Hodgkin's lymphoma (B-NHL) on the other. It has been calculated that in Italy 1 of

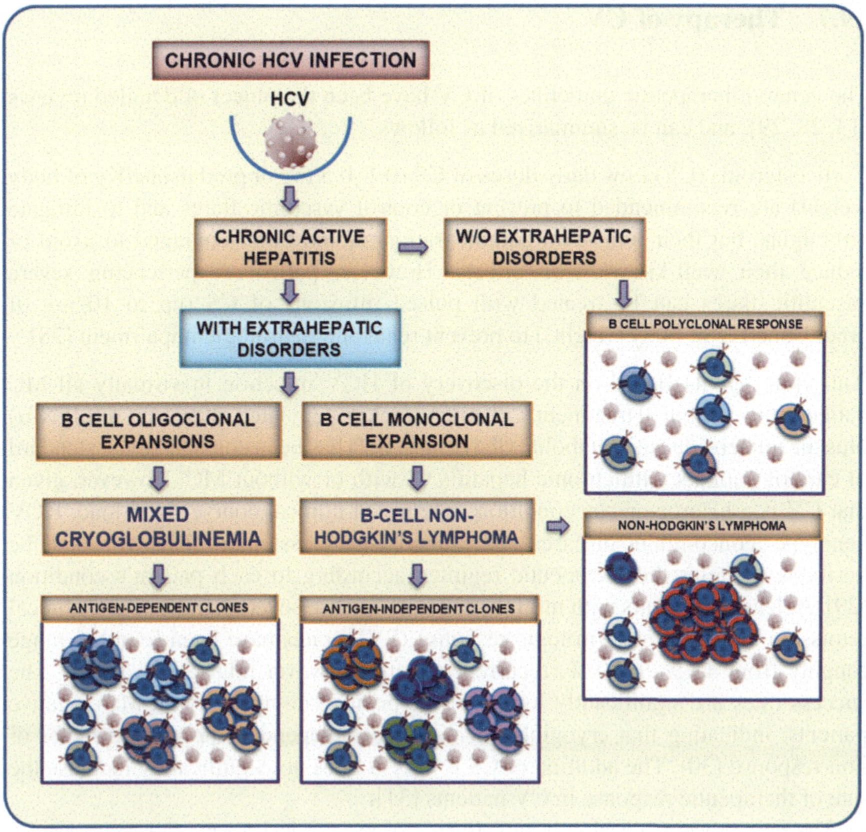

Fig. 29.4 Schematic model of B-cell expansion in chronic HCV infection

20 instances of B-NHL may be ascribed to HCV infection [25]. At 10 years from the initial diagnosis, the prevalence of B-NHL in chronically HCV-infected patients was significantly higher in patients with than in those without MC (13 % vs.6.2 %; p = 0.003) [10]. As discussed by Zignego et al. [26], the same association has been confirmed in studies from other countries, though with wide geographical variations.

In subtype-specific analyses, marginal zone lymphoma, diffuse large B-cell lymphoma, and lymphoplasmacytic lymphoma, but not follicular lymphoma, appear to be more frequently associated to HCV infection [27].

A hypothetical pathogenetic mechanism underlying the onset of HCV-related B-NHLs is a multi-step process depicted in Fig. 29.4. E2 and NS3 proteins are able to stimulate the occurrence of a benign lympho-proliferative disorder, potentially susceptible of progression to a frank lymphoid malignancy through an impaired regulatory control of B-cell growth. Sequential genetic translocations and activation of proto-oncogenes are likely to be involved.

29.7 Therapy of CV

The general therapeutic guidelines of CV have been the object of detailed reviews [13, 28, 29], and can be summarized as follows.

Corticosteroids (CS) Low daily doses of CS (0.1–0.5 mg of prednisone/Kg of body weight) are recommended to prevent or control vasculitic flares and to mitigate arthralgias, but their long-term administration should be discouraged to avoid or reduce their well-known side effects. However, patients experiencing severe vasculitic flares can be treated with pulsed infusions of CS (up to 10 mg of prednisone/Kg of body weight)) to prevent renal and neurologic impairment [28].

Anti-viral Agents Based on the discovery of HCV infection in virtually all MC patients, the combined treatment with once-weekly, pegylated interferon-α (pIFN-α) plus the nucleoside antimetabolite ribavirin (RBV) has been considered the standard of care of patients with chronic hepatitis C, with or without MC. However, given that CV is a heterogeneous condition in terms of clinical course, viral load, HCV genotype, concomitant diseases and potential progression to NHL, it would be desirable to tailor the therapeutic regimen according to each patient's condition [29]. At least in patients with mild to moderate disease severity and activity, clinical remission and sustained virologic response (SVR) rates have been found to range roughly from 45 % to 70 %, recurrences being however relatively frequent. The success rates are significantly lower in MC-positive compared with MC-negative patients, indicating that cryoglobulinemia is an independent prognostic factor of non-response [30]. The addition of CS to IFN-α does not significantly increase the rate of therapeutic response in CV patients [31].

The more recent introduction of direct-acting antiviral agents has dramatically increased the rates of viral eradication in MC-negative patients with chronic hepatitis C, thus making the previous standard of care totally outdated. These all-oral, first-line treatment regimens, with the important limitation of their extremely high cost, have been able to induce SVR rates of approximately 90–95 % even in patients with difficult-to-treat genotypes and in those with decompensated liver cirrhosis [32–34]. The extension of these remarkably successful therapeutic advancements to patients with HCV-related CV is hopefully expected to result in roughly comparable therapeutic successes.

B-Cell-Depleting Therapy Clonal expansion of IgM RF-producing B-cells in HCV-positive patients with CV has been repeatedly shown in several studies [35–37], suggesting that the deletion of B-cell clonalities might be a targeted therapy. Two consecutive reports [38, 39] indeed showed that rituximab (RTX), a chimeric monoclonal antibody specifically directed against CD20 antigen, was able to induce a complete response of clinical and laboratory features in approximately 80 % of refractory and relapsing CV patients. These results have been confirmed in additional studies, summarized in [13]. In one of them, patients were given for 1 year a triple-drug regimen (pIFN-α plus RBV plus RTX) and were followed-up for 36-months. This regimen resulted in a higher rate of complete responses compared with the results achieved in patients receiving pIFN-α plus RBV without RTX (54.5 % vs. 33.3 %; $p < 0.05$), a dramatic depletion of B-cell clones and no increase in the viral load [40].

Therapy of CV Progression to B-NHL A strong, although indirect support to the pathogenetic relationship between chronic HCV infection and a subset of NHLs came from the observation that 7 out of 9 (78 %) HCV-infected patients with splenic lymphoma with villous lymphocytes achieved SVR and complete tumor regression after treatment with IFN-α alone or associated to RBV [41]. In a multicenter Italian study Arcaini et al. [42], studying a cohort of 704 HCV-positive indolent B-NHLs, showed that 44 of the 100 patients treated with first-line antiviral treatment achieved a complete remission and 33 a partial response of their NHLs. In parallel, HCV-RNA clearance was achieved in 80 patients and was related to lymphoma response. Based on these and additional studies [43–45], it seems worth emphasizing that patients with HCV-related, indolent NHL are susceptible to achieve a high rate of tumor remission following HCV RNA clearance induced by direct-acting antiviral agents.

In HCV-positive patients with rapidly progressing, high-grade NHL histotypes, given the unlikeliness of hematologic response in step with eradication of HCV infection, the priority aim should be to restrain the tumor with the use of suitable chemotherapeutic regimens, with or without RTX, and assuring a close follow-up of these patients given the hepatic toxicity and the risk of hepatitis flares consequent to their administration [46–48]. The IFN-free, all-oral antiviral treatment would then be given as a second step, hopefully after remission of NHL.

Plasmapheresis Therapeutic apheresis is a symptomatic procedure that, by removing circulating cryoglobulins and viral particles from plasma by means of extra-corporeal circuits, may help control clinical symptoms, although no strictly controlled clinical studies have so far been published to unequivocally demonstrate its effectiveness. In the variant procedure called double filtration plasmapheresis, plasma is initially separated from the cellular blood components, and is then processed by a filtration system that allows harmful molecules to be selectively removed. The membrane permeable components are then returned to the patient. In our experience, double filtration plasmapheresis can significantly contribute to the improvement and often the healing of chronic, indolent ulcers of the legs in CV patients [49].

Acknowledgments This work was supported by grants from the University of Bari and the Italian Association for Cancer Research (AIRC), Milano.

Potential Conflict of Interest Nothing to report.

References

1. Dammacco F, Sansonno D, Piccoli C, Tucci FA, Racanelli V (2001) The cryoglobulins: an overview. Eur J Clin Invest 31:628–638
2. Agnello V (1997) The etiology and pathophysiology of mixed cryoglobulinemia secondary to hepatitis C virus infection. Springer Semin Immunopathol 19:111–129
3. Meltzer M, Franklin EC, Elias K, McCluskey RT, Cooper N (1966) Cryoglobulinemia – a clinical and laboratory study. II. Cryoglobulins with rheumatoid factor activity. Am J Med 40:837–856
4. Ferri C, Greco F, Longombardo G, Palla P, Moretti A, Marzo E et al (1991) Antibodies against hepatitis C virus in mixed cryoglobulinemia patients. Infection 19:417–420
5. Agnello V, Chung RT, Kaplan LM (1992) A role for hepatitis C virus infection in type II cryoglobulinemia. N Engl J Med 327:1490–1495
6. Dammacco F, Sansonno D (1992) Antibodies to hepatitis C virus in essential mixed cryoglobulinaemia. Clin Exp Immunol 87:352–356
7. Misiani R, Bellavita P, Fenili D, Borelli G, Marchesi D, Massazza M et al (1992) Hepatitis C virus infection in patients with essential mixed cryoglobulinemia. Ann Intern Med 117:573–577
8. Ferri C, Sebastiani M, Giuggioli D, Colaci M, Fallahi P, Piluso A et al (2015) Hepatitis C virus syndrome: a constellation of organ- and non-organ specific autoimmune disorders, B-cell non-Hodgkin's lymphoma, and cancer. World J Hepatol 7:327–343
9. Cacoub P, Comarmond C, Domont F, Savey L, Saadoun D (2015) Cryoglobulinemia vasculitis. Am J Med 128:950–955
10. Lauletta G, Russi S, Conteduca V, Sansonno L, Dammacco F, Sansonno D (2013) Impact of cryoglobulinemic syndrome on the outcome of chronic hepatitis C virus infection: a 15-year prospective study. Medicine (Baltimore), 26 August 2013 [Epub ahead of print]
11. Monaco S, Ferrari S, Gajofatto A, Zanusso G, Mariotto S (2012) HCV-related nervous system disorders. Clin Dev Immunol 2012;2012:236148
12. Monaco S, Mariotto S, Ferrari S, Calabrese M, Zanusso G, Gajofatto A et al (2015) Hepatitis C virus-associated neurocognitive and neuropsychiatric disorders: advances in 2015. World J Gastroenterol 21:11974–11983
13. Dammacco F, Sansonno D (2013) Therapy for hepatitis C virus-related cryoglobulinemic vasculitis. N Engl J Med 369:1035–1045
14. Gorevic PD (2012) Rheumatoid factor, complement, and mixed cryoglobulinemia. Clin Dev Immunol 2012;2012:439018
15. Gorevic PD (2012) Mixed cryoglobulinemia (MC) cross-reactive idiotypes (CRI): structural and clinical significance. In: Dammacco F (ed) HCV infection and cryoglobulinemia. Springer, Milan, pp 99–105
16. Ferri C, Antonelli A, Sebastiani M, Colaci M, Zignego AL (2012) The expanding spectrum of clinical features in HCV-related mixed cryoglobulinemia. In: Dammacco F (ed) HCV infection and cryoglobulinemia. Springer, Milan, pp 155–162
17. Carlson JA, Cavaliere LF, Grant-Kels JM (2006) Cutaneous vasculitis: diagnosis and management. Clin Dermatol 24:414–429
18. Sansonno D, Cornacchiulo V, Iacobelli AR, Di Stefano R, Lospalluti M, Dammacco F (1995) Localization of hepatitis C virus antigens in liver and skin tissues of chronic hepatitis C virus-infected patients with mixed cryoglobulinemia. Hepatology 21:305–312
19. Agnello V, Abel G (1997) Localization of hepatitis C virus in cutaneous vasculitic lesions in patients with type II cryoglobulinemia. Arthritis Rheum 40:2007–2015
20. Fabrizi F, Plaisier E, Saadoun D, Martin P, Messa P, Cacoub P (2013) Hepatitis C virus infection, mixed cryoglobulinemia, and kidney disease. Am J Kidney Dis 61:623–637
21. Sansonno D, Gesualdo L, Manno C, Schena FP, Dammacco F (1997) Hepatitis C virus-related proteins in kidney tissue from hepatitis C virus-infected patients with cryoglobulinemic membranoproliferative glomerulonephritis. Hepatology 25:1237–1244

22. Mariotto S, Ferrari S, Monaco S (2014) HCV-related central and peripheral nervous system demyelinating disorders. Inflamm Allergy Drug Targets 13:299–304
23. Vallat JM, Magy L, Richard L, Piaser M, Sindou P, Calvo J et al (2008) Intranervous immunoglobulin deposits: an underestimated mechanism of neuropathy. Muscle Nerve 38:904–911
24. Sansonno D, Dammacco F (2005) Hepatitis C virus, cryoglobulinaemia, and vasculitis: immune complex relations. Lancet Infect Dis 5:227–236
25. Mele A, Pulsoni A, Bianco E, Musto P, Szklo A, Sanpaolo MG et al (2003) Hepatitis C virus and B-cell non-hodgkin lymphomas: an italian multicenter case–control study. Blood 102:996–999
26. Zignego AL, Gragnani L, Piluso A, Sebastiani M, Giuggioli D, Fallahi P et al (2015) Virus-driven autoimmunity and lymphoproliferation: the example of HCV infection. Expert Rev Clin Immunol 11:15–31
27. de Sanjose S, Benavente Y, Vajdic CM, Engels EA, Morton LM, Bracci PM et al (2008) Hepatitis C and non-hodgkin lymphoma among 4784 cases and 6269 controls from the international lymphoma epidemiology consortium. Clin Gastroenterol Hepatol 6:451–458
28. Pietrogrande M, De Vita S, Zignego AL, Pioltelli P, Sansonno D, Sollima S et al (2011) Recommendations for the management of mixed cryoglobulinemia syndrome in hepatitis C virus-infected patients. Autoimmun Rev 10:444–454
29. Cacoub P, Terrier B, Saadoun D (2014) Hepatitis C virus-induced vasculitis: therapeutic options. Ann Rheum Dis 73:24–30
30. Gragnani L, Fognani E, Piluso A, Boldrini B, Urraro T, Fabbrizzi A et al (2015) MaSVE study group. Long-term effect of HCV eradication in patients with mixed cryoglobulinemia: a prospective, controlled, open-label, cohort study. Hepatology 61:1145–1153
31. Dammacco F, Sansonno D, Han JH, Shyamala V, Cornacchiulo V, Iacobelli AR et al (1994) Natural interferon-alpha versus its combination with 6-methyl-prednisolone in the therapy of type II mixed cryoglobulinemia: a long-term, randomized, controlled study. Blood 84:3336–3343
32. Feld JJ, Jacobson IM, Hézode C, Asselah T, Ruane PJ, Gruener N et al (2015) Sofosbuvir and velpatasvir for HCV genotypes 1, 2, 4, 5, and 6 infection. N Engl J Med 373:2599. doi:10.1056/NEJMoa1512610
33. Foster GR, Afdhal N, Roberts SK, Bräu N, Gane EJ, Pianko S et al (2015) Sofosbuvir and velpatasvir for HCV genotype 2 and 3 infection. N Engl J Med 373:2608. doi:10.1056/NEJMoa1512612
34. Curry MP, O'Leary JG, Bzowej N, Muir AJ, Korenblat KM, Fenkel JM et al (2015) Sofosbuvir and velpatasvir for HCV in patients with decompensated cirrhosis. N Engl J Med 373:2618. doi:10.1056/NEJMoa1512614
35. Racanelli V, Sansonno D, Piccoli C, D'Amore FP, Tucci FA, Dammacco F (2001) Molecular characterization of B cell clonal expansions in the liver of chronically hepatitis C virus-infected patients. J Immunol 167:21–29
36. Vallat L, Benhamou Y, Gutierrez M, Ghillani P, Hercher C, Thibault V et al (2004) Clonal B cell populations in the blood and liver of patients with chronic hepatitis C virus infection. Arthritis Rheum 50:3668–3678
37. Charles ED, Brunetti C, Marukian S, Ritola KD, Talal AH, Marks K et al (2011) Clonal B cells in patients with hepatitis C virus-associated mixed cryoglobulinemia contain an expanded anergic CD21low B-cell subset. Blood 117:5425–5437
38. Sansonno D, De Re V, Lauletta G, Tucci FA, Boiocchi M, Dammacco F (2003) Monoclonal antibody treatment of mixed cryoglobulinemia resistant to interferon alpha with an anti-CD20. Blood 101:3818–3826
39. Zaja F, De Vita S, Mazzaro C, Sacco S, Damiani D, De Marchi G et al (2003) Efficacy and safety of rituximab in type II mixed cryoglobulinemia. Blood 101:3827–3834
40. Dammacco F, Tucci FA, Lauletta G, Gatti P, De Re V, Conteduca V et al (2010) Pegylated interferon-alpha, ribavirin, and rituximab combined therapy of hepatitis C virus-related mixed cryoglobulinemia: a long-term study. Blood 116:343–353

41. Hermine O, Lefrère F, Bronowicki JP, Mariette X, Jondeau K, Eclache-Saudreau V et al (2002) Regression of splenic lymphoma with villous lymphocytes after treatment of hepatitis C virus infection. N Engl J Med 347:89–94
42. Arcaini L, Vallisa D, Rattotti S, Ferretti VV, Ferreri AJ, Bernuzzi P et al (2014) Antiviral treatment in patients with indolent B-cell lymphomas associated with HCV infection: a study of the Fondazione Italiana Linfomi. Ann Oncol 25:1404–1410
43. Carrier P, Jaccard A, Jacques J, Tabouret T, Debette-Gratien M, Abraham J, Mesturoux L, Marquet P, Alain S, Sautereau D, Essig M, Loustaud-Ratti V (2015) HCV-associated B-cell non-Hodgkin lymphomas and new direct antiviral agents. Liver Int 35:2222–2227
44. Michot JM, Canioni D, Driss H, Alric L, Cacoub P, Suarez F, Sibon D, Thieblemont C, Dupuis J, Terrier B, Feray C, Tilly H, Pol S, Leblond V, Settegrana C,Rabiega P, Barthe Y, Hendel-Chavez H, Nguyen-Khac F, Merle-Béral H, Berger F, Molina T, Charlotte F, Carrat F, Davi F, Hermine O, Besson C; ANRS HC-13 Lympho-C Study Group (2015) Antiviral therapy is associated with a better survival in patients with hepatitis C virus and B-cell non-Hodgkin lymphomas, ANRS HC-13 lympho-C study. Am J Hematol 90: 197–203
45. Tasleem S, Sood GK (2015) Hepatitis C associated B-cell Non-hodgkin lymphoma: clinical features and the role of antiviral therapy. J Clin Transl Hepatol 3:134–139
46. Ennishi D, Maeda Y, Niitsu N, Kojima M, Izutsu K, Takizawa J et al (2010) Hepatic toxicity and prognosis in hepatitis C virus-infected patients with diffuse large B-cell lymphoma treated with rituximab-containing chemotherapy regimens: a Japanese multicenter analysis. Blood 116:5119–5125
47. Marignani M, Mangone M, Cox MC, Angeletti S, Veggia B, Ferrari A et al (2011) HCV-positive status and hepatitis flares in patients with B-cell non-Hodgkin's lymphoma treated with rituximab-containing regimens. Dig Liver Dis 43:139–142
48. Paydas S (2015) Hepatitis C, virus and lymphoma. Crit Rev Oncol Hematol 93:246–256
49. Ramunni A, Lauletta G, Brescia P, Saliani MT, Montrone M, Chironna M, Sansonno D, Dammacco F, Coratelli P (2008) Double-filtration plasmapheresis in the treatment of leg ulcers in cryoglobulinemia. J Clin Apher 23:118–122

Chapter 30
Vasculitis in Connective Tissue Diseases

Patrizia Leone, Sebastiano Cicco, Angelo Vacca, Franco Dammacco, and Vito Racanelli

Abstract Vasculitis is a recurrent complication of connective tissue diseases. It is often an under-recognized manifestation that can lead to significant morbidity and mortality due to vital organ damage. Cutaneous lesions, representing small-vessel vasculitis, dominate clinical presentation, although widespread necrotizing visceral medium- and large-vessel involvement, mimicking primary vasculitic syndromes, may also occur. The pathogenesis of vascular inflammation is not completely understood, but immune complexes are believed to play a crucial role. The diagnosis is usually based on the combination of clinical findings, laboratory testing, tissue biopsy and imaging of the involved blood vessels. Timely and aggressive pulse treatment with high dose of glucocorticosteroids and immunosuppressive agents, followed by gradual tapering, can significantly improve patients' survival.

30.1 Introduction

Connective tissue diseases including systemic lupus erythematosus (SLE), rheumatoid arthritis (RA), systemic sclerosis, polymyositis, dermatomyositis, Sjögren's syndrome and mixed connective tissue disease can be associated to vasculitis, the most common type being small-vessel venulitis involving the skin. Conventionally, vasculitides associated to connective tissue diseases are identified with the word 'vasculitis' preceded by a prefix that indicates the specific disease (e.g., rheumatoid vasculitis). Given their higher frequency, this chapter will be restricted to lupus vasculitis and rheumatoid vasculitis, that can complicate the clinical picture and dramatically influence the course of the corresponding diseases.

P. Leone • S. Cicco • A. Vacca • F. Dammacco • V. Racanelli (✉)
Department of Biomedical Sciences and Human Oncology, Section of Internal Medicine, University of Bari Medical School, 70124 Bari, Italy
e-mail: vito.racanelli1@uniba.it

F. Dammacco et al. (eds.), *Systemic Vasculitides: Current Status and Perspectives*, DOI 10.1007/978-3-319-40136-2_30

30.2 Lupus Vasculitis (LV)

LV is an inflammatory disorder of small- and medium-sized vessels that affects patients with SLE. Only a few studies retrievable from the literature are focused on LV, given that it is difficult to isolate its clinical manifestations from those of the general disease in terms of pathogenesis and response to treatment. In particular, SLE vascular lesions (commonly known as lupus vasculopathy) may be induced not only by immune complex deposition on the vessel wall, but also by thrombotic microangiopathy related to antiphospholipid syndrome and by enhanced atherosclerosis resulting from SLE-associated pro-inflammatory cytokines and steroid treatment.

It is now recognized that the pathogenesis of SLE depends on genetic and environmental factors that promote abnormal innate and adaptive immune responses. These responses result in the generation of autoreactive T and B cells [1–3], enhanced ability of T cells to help B cells, production of pathogenic autoantibodies, and formation of immune complexes that deposit in tissues, activate complement, cause inflammation, and over time lead to irreversible organ damage [4]. Although the large majority of autoantibodies bind nuclear antigens (DNA, RNA, and nuclear proteins), additional antibodies bind cell membrane antigens, plasma proteins, and extra-cellular matrix antigens. Among these, anti-endothelial cell antibodies (AECA) seem to play a direct role in the onset of vasculitis by causing endothelial cell activation, production of tissue factors and cytotoxicity. AECA have been detected in up to 80 % of SLE patients and in several pathological conditions having in common their association with vasculitis [5].

Vasculitis may occur in approximately 50 % of SLE patients [6, 7] with a wide variety of clinical manifestations, dependending on the size of the affected vessels (capillaries, arteries, veins) and the sites involved (skin or internal organs). The prognosis may range from mild to life-threatening. The skin is the most commonly affected site, being involved in over 80 % of SLE vasculitis lesions [6, 7]. Kidney, gastrointestinal tract, nervous system (central and peripheral), lung and heart can be involved with lower frequency [7]. Drenkard et al. [6] observed that, in a cohort of 540 patients with SLE, 194 had vasculitis (35.9 %) which was cutaneous in 160 patients (82.5 %), visceral in 24 (12.4 %) and combined cutaneous plus visceral in 10 (5.1 %). Ramos-Casals et al. [7], studying a cohort of 670 SLE patients, were able to diagnose vasculitis in 76 of them (11.3 %). Cutaneous lesions were largely prevalent, occurring in 68 patients (89.4 %), whereas isolated visceral vasculitis was found in the remaining eight patients (10.5 %), mainly involving peripheral nerves and kidneys.

30.2.1 Skin Vasculitis

Clinical features of skin small vessel vasculitis include palpable purpura, urticaria, petechiae, papulo-nodular lesions, erythematous plaques or macules, panniculitis, livedo reticularis, cutaneous infarction, and superficial ulcerations [6, 7] (Fig. 30.1a,

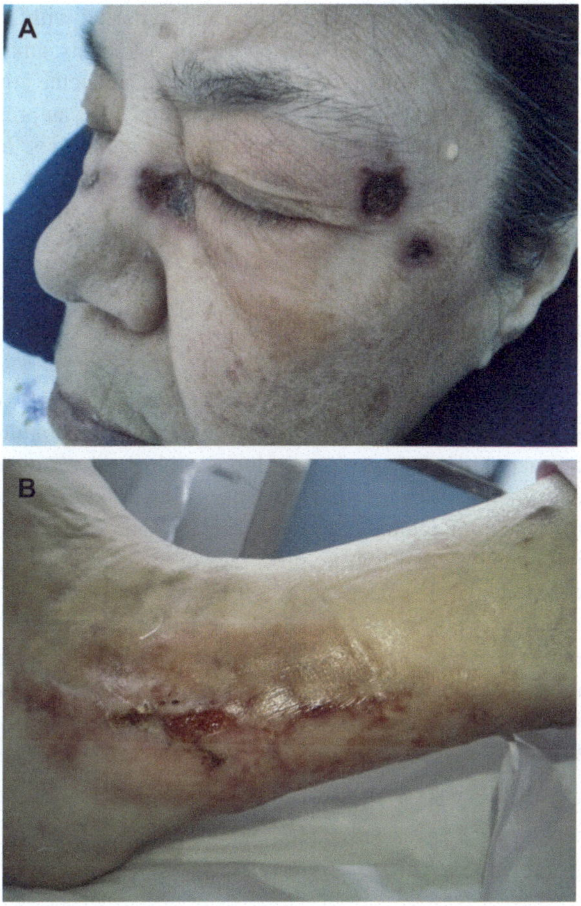

Fig. 30.1 Vasculitic skin lesions in systemic lupus erythematosus. Poorly healing ulcerations of the face (**a**) and low extremity (**b**) of a 62-year-old female patient

b). Medium vessels vasculitis is less frequent and appears as subcutaneous nodules or ischemic ulcers. Lesions of different types may occur simultaneously or in the course of subsequent SLE flares.

30.2.2 Nervous System Vasculitis

Nervous system vasculitis usually affects small or medium-sized vessels, causing nerve infarction and stroke [8]. Large vessel involvement is rare but can be catastrophic and associated with high mortality [9]. The most common clinical manifestation is mononeuritis multiplex. In Drenkard's SLE patient cohort, mononeuritis multiplex was diagnosed in 19 out of 24 patients (79.1 %) with visceral vasculitis [6]. It results from necrotizing or occlusive vasculitis of the vasa nervorum, causing axonal degeneration. The clinical features include painful, asymmetrical, progressive, asynchronous sensory and motor peripheral neuropathy involving at least two separate nerve areas [10].

Neuropsychiatric manifestations are reported with a prevalence widely ranging from 14 to 75 % and include depression, psychosis, anxiety, mood disorders, cognitive dysfunction, memory deficit, delirium, and acute confusional states. These syndromes can occur singly or in combination, and can present at any point during the course of the disease [11]. Anti-neuronal antibodies, anti-ribosomal P antibodies and anti-DNA antibodies that cross-react with the N-methyl-D aspartate receptor have been frequently detected [12].

30.2.3 Kidney Vasculitis

The hallmark of renal involvement is a small vessel vasculitis affecting the renal glomeruli (glomerulonephritis) [13]. Vascular and tubulo-interstitial lesions are less frequent [14]. The serology of lupus nephritis is characterized by the production of anti-double stranded (ds)DNA antibodies in addition to immune complex deposition and complement activation. The anti-dsDNA antibodies have a direct impact on inflammatory and fibrotic processes in resident renal cells and various glomerular components [15]. Depending on the site of immunoglobulin accumulation, their antigen specificity, their capacity to activate complement and to trigger an inflammatory response, it is possible to distinguish three different glomerular patterns of injury: mesangial, endothelial and epithelial [16, 17]. In several patients these different patterns can coexist, leading to a more complex clinical expression of kidney vasculitis.

30.2.4 Pulmonary Vasculitis

Pulmonary vasculitis usually involves small or medium-sized vessels. Pulmonary capillaritis with diffuse alveolar hemorrhage is the most common manifestation and it often coexists with lupus nephritis [18–20]. Clinical symptoms include fever, dyspnea, cough, pulmonary infiltrates and hypoxemia. Over time, most SLE patients progress to pulmonary arterial hypertension associated with Raynaud's syndrome, serositis, digital vasculitis and anti-phospholipid syndrome [21].

30.2.5 Gastrointestinal Vasculitis

The most common lupus manifestation in the gastrointestinal tract is mesenteric vasculitis [22]. It causes acute abdominal pain, nausea, vomiting, diarrhea, melena, hematemesis and bloating [23]. Severe, occlusive damage often leads to ischemia that may result in ulceration and perforation, with mortality rates up to 50 % [22, 24]. Abdominal computed tomography (CT) is the most useful diagnostic tool, given that it allows to visualize both the bowel wall and the abdominal vasculature. It usually shows dilated bowel, focal or diffuse bowel wall thickening and enhancement (target sign), mesenteric edema, stenosis or engorgement of mesenteric vessels (comb sign), and ascites.

Pancreatic vasculitis is rare and often fatal. Vascular damage consists of severe intimal proliferation in the pancreatic vessels, necrotizing vasculitis and occlusion of arteries and arterioles by thrombi. It usually results in SLE-associated acute pancreatitis whose most frequent symptom is severe abdominal pain [25].

30.2.6 Cardiac Involvement and Coronary Vasculitis

Vasculitis of the coronary arteries is a rare condition associated with a high prevalence of myocardial infarction [26]. It has been calculated that the overall risk of myocardial infarction in SLE patients is ten-fold higher than in the general population, because of an 'accelerated atherosclerosis' due to complex interactions between dysfunctional immune regulation, inflammation, traditional cardiac risk factors and treatment of the underlying autoimmune disease [27]. Angiographically, it manifests as a coronary aneurysm or arteritis, sometimes occurring in the absence of clinical SLE flare and with minimal serologic evidence of disease activity [28].

30.2.7 Laboratory Findings

Enzyme-linked immunosorbent assay (ELISA) and immunofluorescence are the main laboratory tests used to detect circulating anti-nuclear antibodies (ANA). ANA and anti dsDNA antibodies have been associated with the formation of immune complexes and vasculitis. Among non-nuclear autoantibodies, the following main associations can be detected with variable frequency: (a) AECA with vasculitis and lupus nephritis; (b) rheumatoid factor with cutaneous vasculitis; (c) anti-neutrophil cytoplasmic autoantibodies (ANCA), usually perinuclear, with pulmonary and renal vasculitis; (d) anti-neuronal antibodies, anti-ribosomal P antibodies and anti-dsDNA antibodies that cross-react with the N-methyl-D aspartate receptor with nervous system vasculitis.

30.2.8 Histological Findings

Biopsy showing disruption or destruction of the vessel wall with inflammatory cell infiltrates is the gold standard for the diagnosis of vasculitis. The conventional assessment of formalin-fixed paraffin-embedded tissues is often accompanied by immunofluorescence staining which reveals the vessel deposition of immunoglobulins and complement.

30.2.9 In Vivo Imaging

Imaging studies may be useful in the diagnosis of LV. Pertinent examples are abdominal ultrasonography and CT in patients with mesenteric vasculitis; x-rays and CT scans when pulmonary vasculitis is suspected; cardiac CT and coronary angiography to demonstrate SLE coronary arteritis; magnetic resonance imaging (MRI) and angiography to support the diagnosis of nervous system vasculitis.

30.2.10 Treatment

Before undergoing therapy, each patient must be carefully studied in order to establish whether the final diagnosis is vasculopathy, where there is with no evidence of underlying inflammation, or true vasculitis because the former requires anticlotting therapy rather than immunosuppression.

Treatment of LV is obviously dependent on the degree and severity of the vasculitis process. While patients with acute manifestations of skin vasculitis are given oral prednisone (0.5–1 mg/kg/body weight) with subsequent tapering, those with chronic cutaneous involvement are usually treated with antimalarial agents (hydroxychloroquine 200 mg twice daily) and/or with azathioprine (100 mg/day), and/or mycophenolate mofetil (0.5–1 g twice daily) [29]. When the kidney or other parenchymal organs are affected, treatment usually includes pulsed intravenous infusions of methylprednisolone (1 g/m^2/day for 3 days) followed by gradual tapering, monthly intravenous or daily oral cyclophosphamide for 6 months, and long-term immunosuppression with oral mycophenolate mofetil (up to 1.5 g twice daily) [30]. High-dose prednisone combined with intravenous pulse or oral daily cyclophosphamide are also used for the treatment of nervous system vasculitis [31, 32].

30.3 Rheumatoid Vasculitis (RV)

RA, a chronic multisystem disease characterized by a symmetric, peripheral polyarthritis, can cause inflammation of vessels of different size. Involvement of small vessels is usually associated with active arthritis, whereas involvement of medium-sized vessels may occur even when joint disease is quiescent.

Genetic predisposition along with environmental factors may trigger the development of RA, with subsequent synovial T cell activation. Although the pathogenetic mechanisms remain unknown, immune complexes are believed to play a key role in RV.

Similarly to LV, most of the data concerning RV are somehow extrapolated from more general studies on RA. The majority of patients with RV have high titers of rheumatoid factor (RF) [33, 34] as well as anti-citrullinated peptide antibodies (ACPA) [35], AECA [36], anti-C1q antibodies [37], and anti-glucose-6-phosphate isomerase antibodies [38]. A case-control study, carried out in 69 RV patients and 138 control RA patients without RV, has highlighted the association between RV and the presence of increased RF levels [39].

Genetic factors that predispose to the onset of RA include the HLA-DRB1 locus and, more specifically, those alleles with a conserved aminoacid sequence in the third hypervariable region of the molecule, termed the shared epitopes [40]. A meta-analysis revealed a close relationship between RV and three specific genotypes of the HLA-DRB1 shared epitope *0401/*0401, *0401/*0404, and *0101/*0401 [40]. In 2006, a study from the Mayo Clinic also described a positive association of HLA-C3 with RV that was independent of linkage disequilibrium with HLA-DRB1 [41]. Additional risk predictors of RV include smoking, long-standing seropositive nodular erosive disease and male sex [39].

RV has a wide range of clinical manifestations and may involve virtually any organ of the body. The most common sites of involvement are skin and peripheral nerves (approximately 90 % and 40 % of patients, respectively [42, 43]), followed by eye and heart. Less frequently, RV may affect central nervous system, kidney, lung and gastrointestinal system [42–44].

30.3.1 Skin

Digital, often periungueal lesions, petechiae, purpura, ulcers and gangrene are the most common cutaneous manifestations [42, 45]. Isolated digital lesions include micro-infarctions (<1 mm), occurring or not in association with other clinical manifestations of systemic vasculitis. They tipically follow a benign course and do not require further treatment in addition to that for RA itself [45, 46]. Petechiae and purpura mostly develop on the lower extremities [47]. Leg ulcers (Fig. 30.2) are painful and frequently follow trivial trauma; they tend to be found on the lower extremities or dorsum of the feet, ankles and the upper calves [48–50]. Ulcers of the digital tips of the fingers and toes often evolve to gangrene [48].

30.3.2 Peripheral Nerves

Mononeuritis multiplex and distal symmetric sensory or sensory-motor neuropathy are typical peripheral nervous system manifestations [42, 45, 51]. Mononeuritis multiplex generally causes foot or wrist drop and, when identified in a patient with long-standing RA, it can be considered diagnostic of RV. Vasculitis-mediated axonal degeneration and demyelinization may also result in milder, symmetric sensory-motor neuropathy of the lower extremities characterized by paresthesias, numbness, burning pain and occasionally mild motor deficits.

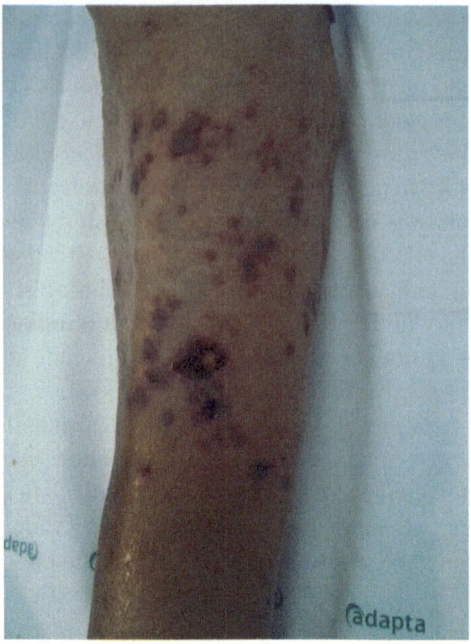

Fig. 30.2 Rheumatoid vasculitis presenting as cutaneous infarctions of the lower limb in a 66-year-old female patient

30.3.3 Central Nervous System

Seizures [52], neuropsychiatric symptoms [53], confusional state [54], hemiparesis [55, 56], dementia and blindness [57] are the relevant clinical manifestations of the rare involvement of the central nervous system in RV patients. This involvement is associated with high morbidity and, in some cases, can be life-threatening.

30.3.4 Heart

Coronary arteritis causing myocardial infarction and aneurysms, pericarditis and aortitis are usually diagnosed at post-mortem in RA patients [58, 59]. Most of these vasculitic lesions are silent during the patient's life and are thus seldom diagnosed. Moreover, RA patients are predisposed to 'accelerated atherosclerosis' and it is very difficult to distinguish between atherosclerotic cardiovascular disease and systemic vasculitis as the actual cause of cardiac manifestations.

30.3.5 Kidneys

Renal involvement includes chronic renal impairment, proteinuria, microscopic hematuria and increased serum creatinine levels, and is reported in approximately one quarter of patients with systemic RV [44]. A focal segmental necrotizing glomerulonephritis, that is typical of vasculitic glomerulonephritis, develops rarely [60].

30.3.6 Eyes

Scleritis and keratitis affect approximately 16 % of RV patients [44]. The most severe ocular manifestation is peripheral ulcerative keratitis or 'corneal melt', characterized by inflammation of the limbal part of the cornea and adjacent sclera, collagen destruction, cellular infiltration and limbal vascular changes that may lead to corneal perforation and blindness. Severe pain, foreign body sensation and photophobia are the main symptoms [42, 61, 62]. Given that peripheral ulcerative keratitis frequently precedes systemic vasculitis, early and aggressive treatment is required to prevent visual loss and a mortality rate up to 30 %.

30.3.7 Lungs

Diffuse alveolar hemorrhage presenting with dyspnea, cough and hemoptysis characterize the pulmonary involvement in RV patients. Lung biopsy shows necrotizing pulmonary capillaritis with acute and chronic alveolar hemorrhage and immune complex deposition [63].

30.3.8 Gastrointestinal System

Small- and large-bowel infarction [64], appendicitis [65], intra-hepatic hemorrhage [66], intra-abdominal hemorrhage from ruptured mesenteric aneurysms [67] and pancreatitis [68] have been reported in approximately 10 % of RV patients [44].

30.3.9 Systemic Manifestations

Systemic manifestations of RV including weight loss, hepatomegaly, splenomegaly and less commonly fever have been recorded in about 80 % of patients [42, 44]. Because these symptoms are non-specific and can resemble those of other autoimmune diseases, the diagnosis of RV must be supported by biopsy-confirmed acute necrotizing arteritis [42, 44].

30.3.10 Laboratory Findings

Laboratory tests in RV are of relatively low diagnostic value, although they are often useful to rule out other conditions such as ANCA-associated vasculitis, HCV-related cryoglobulinemia, HIV-associated vasculitis and Sjögren syndrome. Nevertheless, abnormalities of several serological markers, including anemia, C-reactive protein, erythrocyte sedimentation rate and hypoalbuminemia have been reported in RV patients [42]. Midium-to-high titers of RF and ACPA are strongly associated with the presence of RV [39, 41]. However, in a study carried out in a cohort of 81 patients with suspected RV it was found that only the combination of increased IgA RF levels and decreased C3 level was useful to support histologically-proven RV [69].

30.3.11 Histological Findings

The histological examination of a biopsy specimen from affected organ(s), characterized by the presence of an inflammatory infiltrate associated with destruction of the vessel wall, is essential to support the diagnosis of RV. Occasionally, the so-called 'blind biopsies' from clinically uninvolved sites such as muscle [70], rectum [71] or labial salivary gland samples [72] may be a potential, useful tool for the diagnosis. Perivascular infiltration of ≥ three cell layers in a muscle biopsy specimen is a highly specific and sensitive test to differentiate RV from RA without vasculitis [70].

30.3.12 Electromyography and In Vivo Imaging

Electromyography is used to assess the type and extent of neuropathic involvement. An electrophysiological evaluation of peripheral nerves can help direct appropriate nerve and/or muscle biopsy. Angiography may be helpful to evaluate mesenteric vasculitis or medium vessel vasculitis affecting the extremities.

30.3.13 Treatment

Patients with mild to moderate RV involving the skin and/or the peripheral nerves can be treated with prednisone (0.5–1.5 mg/day) and methotrexate (10–25 mg/week) or azathioprine (50–150 mg/day). Patients with severe RV and multi-organ involvement require a more aggressive therapy with high doses of steroids and cyclophosphamide [73, 74]. Because of the marked toxicity of cyclophosphamide, there is a need for alternative therapies. Clinical evidence based on case reports and case series supports the use of biological agents such as anti-tumor necrosis factor (TNF) alpha antibodies [75–79] and the anti-B-cell monoclonal antibody rituximab [80–83]. However, the role of anti-TNF therapy in RV is doubtful, given that the development of vasculitis has been reported following the administration of anti-TNF antibodies [84–88]. On the contrary, rituximab is emerging as an effective treatment to achieve remission in RV, resulting in a significant decrease in the daily dose of prednisone and an acceptable toxicity profile [80–83], although maintenance therapy is possibly required.

Definite conclusions as to the choice of the best treatment for RV cannot be drawn given the small sample sizes consequent to its relative rarity, the heterogeneity of the clinical manifestations and the highly variable clinical response. Our approach to the management of RV, based on its severity and the occurrence of extra-articular manifestations, is synthesized in a flow-chart (Fig. 30.3).

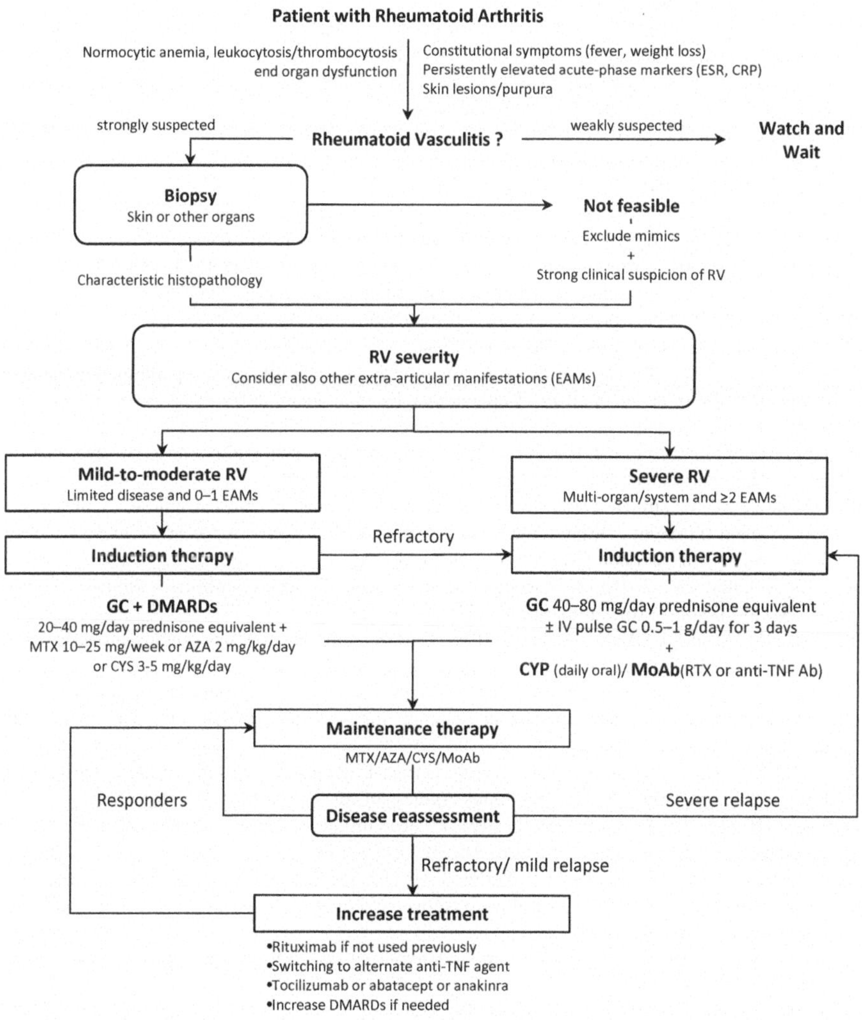

Fig. 30.3 Management of rheumatoid vasculitis. *AZA* azathioprine, *EAMs* extra-articular manifestations, *CYP* cyclophosphamide, *CYS* cyclosporine, *DMARDs* disease-modifying antirheumatic drugs, *GC* glucocorticoids, *IV* intravenous, *MoAb* monoclonal antibody, *MTX* methotrexate, *RA* rheumatoid arthritis, *RTX* rituximab, *RV* rheumatoid vasculitis

References

1. Liossis SN, Kovacs B, Dennis G, Kammer GM, Tsokos GC (1996) B cells from patients with systemic lupus erythematosus display abnormal antigen receptor-mediated early signal transduction events. J Clin Invest 98(11):2549–2557
2. Perl A (2010) Systems biology of lupus: mapping the impact of genomic and environmental factors on gene expression signatures, cellular signaling, metabolic pathways, hormonal and cytokine imbalance, and selecting targets for treatment. Autoimmunity 43(1):32–47
3. Tsokos GC, Kovacs B, Liossis SN (1997) Lymphocytes, cytokines, inflammation, and immune trafficking. Curr Opin Rheumatol 9(5):380–386
4. Inghirami G, Simon J, Balow JE, Tsokos GC (1988) Activated T lymphocytes in the peripheral blood of patients with systemic lupus erythematosus induce B cells to produce immunoglobulin. Clin Exp Rheumatol 6(3):269–276
5. Renaudineau Y, Dugue C, Dueymes M, Youinou P (2002) Antiendothelial cell antibodies in systemic lupus erythematosus. Autoimmun Rev 1(6):365–372
6. Drenkard C, Villa AR, Reyes E, Abello M, Arcon-Segovia D (1997) Vasculitis in systemic lupus erythematosus. Lupus 6(3):235–242
7. Ramos-Casals M, Nardi N, Lagrutta M, Brito-Zeron P, Bove A, Delgado G et al (2006) Vasculitis in systemic lupus erythematosus: prevalence and clinical characteristics in 670 patients. Medicine (Baltimore) 85(2):95–104
8. Younger DS (2004) Vasculitis of the nervous system. Curr Opin Neurol 17(3):317–336
9. Weiner DK, Allen NB (1991) Large vessel vasculitis of the central nervous system in systemic lupus erythematosus: report and review of the literature. J Rheumatol 18(5):748–751
10. Gorson KC (2007) Vasculitic neuropathies: an update. Neurologist 13(1):12–19
11. Hanly JG (2001) Neuropsychiatric lupus. Curr Rheumatol Rep 3(3):205–212
12. Fragoso-Loyo H, Cabiedes J, Orozco-Narvaez A, Dávila-Maldonado L, Atisha-Fregoso Y, Diamond B et al (2008) Serum and cerebrospinal fluid autoantibodies in patients with neuropsychiatric lupus, erythematosus. Implications for diagnosis and pathogenesis. PLoS One 3(10):e3347
13. Schwartz MM, Korbet SM, Lewis EJ (2008) The prognosis and pathogenesis of severe lupus glomerulonephritis. Nephrol Dial Transplant 23(4):1298–1306
14. Omokawa A, Wakui H, Okuyama S, Togashi M, Ohtani H, Komatsuda A et al (2008) Predominant tubulointerstitial nephritis in a patient with systemic lupus erythematosus: phenotype of infiltrating cells. Clin Nephrol 69(6):436–444
15. Yung S, Chan TM (2015) Mechanisms of kidney injury in lupus nephritis – the role of anti-dsDNA antibodies. Front Immunol 6:475
16. Fries JW, Mendrick DL, Rennke HG (1988) Determinants of immune complex-mediated glomerulonephritis. Kidney Int 34(3):333–345
17. Weening JJ, D'Agati VD, Schwartz MM, Seshan SV, Alpers CE, Appel GB et al (2004) The classification of glomerulonephritis in systemic lupus erythematosus revisited. Kidney Int 65(2):521–530
18. Barile LA, Jara LJ, Medina-Rodriguez F, Garcia-Figueroa JL, Miranda-Limon JM (1997) Pulmonary hemorrhage in systemic lupus erythematosus. Lupus 6(5):445–448
19. Orens JB, Martinez FJ, Lynch JP III (1994) Pleuropulmonary manifestations of systemic lupus erythematosus. Rheum Dis Clin North Am 20(1):159–193
20. Zamora MR, Warner ML, Tuder R, Schwarz MI (1997) Diffuse alveolar hemorrhage and systemic lupus erythematosus. Clinical presentation, histology, survival, and outcome. Medicine (Baltimore) 76(3):192–202
21. Asherson RA, Oakley CM (1986) Pulmonary hypertension and systemic lupus erythematosus. J Rheumatol 13(1):1–5
22. Ju JH, Min JK, Jung CK, Oh SN, Kwok SK, Kang KY et al (2009) Lupus mesenteric vasculitis can cause acute abdominal pain in patients with SLE. Nat Rev Rheumatol 5(5):273–281

23. Lee CK, Ahn MS, Lee EY, Shin JH, Cho YS, Ha HK et al (2002) Acute abdominal pain in systemic lupus erythematosus: focus on lupus enteritis (gastrointestinal vasculitis). Ann Rheum Dis 61(6):547–550
24. Janssens P, Arnaud L, Galicier L, Mathian A, Hie M, Sene D et al (2013) Lupus enteritis: from clinical findings to therapeutic management. Orphanet J Rare Dis 8:67
25. Nesher G, Breuer GS, Temprano K, Moore TL, Dahan D, Baer A et al (2006) Lupus-associated pancreatitis. Semin Arthritis Rheum 35(4):260–267
26. Manzi S, Meilahn EN, Rairie JE, Conte CG, Medsger TA Jr, Jansen-McWilliams L et al (1997) Age-specific incidence rates of myocardial infarction and angina in women with systemic lupus erythematosus: comparison with the Framingham Study. Am J Epidemiol 145(5):408–415
27. Skaggs BJ, Hahn BH, McMahon M (2012) Accelerated atherosclerosis in patients with SLE--mechanisms and management. Nat Rev Rheumatol 8(4):214–223
28. Wilson VE, Eck SL, Bates ER (1992) Evaluation and treatment of acute myocardial infarction complicating systemic lupus erythematosus. Chest 101(2):420–424
29. Francis L, Perl A (2009) Pharmacotherapy of systemic lupus erythematosus. Expert Opin Pharmacother 10(9):1481–1494
30. Ginzler EM, Dooley MA, Aranow C, Kim MY, Buyon J, Merrill JT et al (2005) Mycophenolate mofetil or intravenous cyclophosphamide for lupus nephritis. N Engl J Med 353(21):2219–2228
31. Bertsias GK, Ioannidis JP, Aringer M, Bollen E, Bombardieri S, Bruce IN et al (2010) EULAR recommendations for the management of systemic lupus erythematosus with neuropsychiatric manifestations: report of a task force of the EULAR standing committee for clinical affairs. Ann Rheum Dis 69(12):2074–2082
32. Salvarani C, Pipitone N, Hunder GG (2016) Management of primary and secondary central nervous system vasculitis. Curr Opin Rheumatol 28(1):21–28
33. Scott DG, Bacon PA, Allen C, Elson CJ, Wallington T (1981) IgG rheumatoid factor, complement and immune complexes in rheumatoid synovitis and vasculitis: comparative and serial studies during cytotoxic therapy. Clin Exp Immunol 43(1):54–63
34. Westedt ML, Herbrink P, Molenaar JL, de Vries E, Verlaan P, Stijnen T et al (1985) Rheumatoid factors in rheumatoid arthritis and vasculitis. Rheumatol Int 5(5):209–214
35. Laskari K, Ahmadi-Simab K, Lamken M, Csernok E, Gross WL, Hellmich B (2010) Are anti-cyclic citrullinated peptide autoantibodies seromarkers for rheumatoid vasculitis in a cohort of patients with systemic vasculitis? Ann Rheum Dis 69(2):469–471
36. van der Zee JM, Heurkens AH, van der Voort EA, Daha MR, Breedveld FC (1991) Characterization of anti-endothelial antibodies in patients with rheumatoid arthritis complicated by vasculitis. Clin Exp Rheumatol 9(6):589–594
37. Siegert CE, Daha MR, van der Voort EA, Breedveld FC (1990) IgG and IgA antibodies to the collagen-like region of C1q in rheumatoid vasculitis. Arthritis Rheum 33(11):1646–1654
38. van Gaalen FA, Toes RE, Ditzel HJ, Schaller M, Breedveld FC, Verweij CL et al (2004) Association of autoantibodies to glucose-6-phosphate isomerase with extraarticular complications in rheumatoid arthritis. Arthritis Rheum 50(2):395–399
39. Voskuyl AE, Zwinderman AH, Westedt ML, Vandenbroucke JP, Breedveld FC, Hazes JM (1996) Factors associated with the development of vasculitis in rheumatoid arthritis: results of a case-control study. Ann Rheum Dis 55(3):190–192
40. Gorman JD, David-Vaudey E, Pai M, Lum RF, Criswell LA (2004) Particular HLA-DRB1 shared epitope genotypes are strongly associated with rheumatoid vasculitis. Arthritis Rheum 50(11):3476–3484
41. Turesson C, Schaid DJ, Weyand CM, Jacobsson LT, Goronzy JJ, Petersson IF et al (2006) Association of HLA-C3 and smoking with vasculitis in patients with rheumatoid arthritis. Arthritis Rheum 54(9):2776–2783
42. Genta MS, Genta RM, Gabay C (2006) Systemic rheumatoid vasculitis: a review. Semin Arthritis Rheum 36(2):88–98

43. Turesson C, Matteson EL (2009) Vasculitis in rheumatoid arthritis. Curr Opin Rheumatol 21(1):35–40
44. Scott DG, Bacon PA, Tribe CR (1981) Systemic rheumatoid vasculitis: a clinical and laboratory study of 50 cases. Medicine (Baltimore) 60(4):288–297
45. Vollertsen RS, Conn DL (1990) Vasculitis associated with rheumatoid arthritis. Rheum Dis Clin North Am 16(2):445–461
46. Watts RA, Carruthers DM, Scott DG (1995) Isolated nail fold vasculitis in rheumatoid arthritis. Ann Rheum Dis 54(11):927–929
47. Magro CM, Crowson AN (2003) The spectrum of cutaneous lesions in rheumatoid arthritis: a clinical and pathological study of 43 patients. J Cutan Pathol 30(1):1–10
48. Hasegawa M, Nagai Y, Sogabe Y, Hattori T, Inoue C, Okada E et al (2013) Clinical analysis of leg ulcers and gangrene in rheumatoid arthritis. J Dermatol 40(12):949–954
49. McRorie ER, Jobanputra P, Ruckley CV, Nuki G (1994) Leg ulceration in rheumatoid arthritis. Br J Rheumatol 33(11):1078–1084
50. Valencia IC, Falabella A, Kirsner RS, Eaglstein WH (2001) Chronic venous insufficiency and venous leg ulceration. J Am Acad Dermatol 44(3):401–421
51. Puechal X, Said G, Hilliquin P, Coste J, Job-Deslandre C, Lacroix C et al (1995) Peripheral neuropathy with necrotizing vasculitis in rheumatoid arthritis. A clinicopathologic and prognostic study of thirty-two patients. Arthritis Rheum 38(11):1618–1629
52. Neamtu L, Belmont M, Miller DC, Leroux P, Weinberg H, Zagzag D (2001) Rheumatoid disease of the CNS with meningeal vasculitis presenting with a seizure. Neurology 56(6):814–815
53. Gupta VP, Ehrlich GE (1976) Organic brain syndrome in rheumatoid arthritis following corticosteroid withdrawal. Arthritis Rheum 19(6):1333–1338
54. Kim RC (1980) Rheumatoid disease with encephalopathy. Ann Neurol 7(1):86–91
55. Ohno T, Matsuda I, Furukawa H, Kanoh T (1994) Recovery from rheumatoid cerebral vasculitis by low-dose methotrexate. Intern Med 33(10):615–620
56. Pedersen RC, Person DA (1998) Cerebral vasculitis in an adolescent with juvenile rheumatoid arthritis. Pediatr Neurol 19(1):69–73
57. Ramos M, Mandybur TI (1975) Cerebral vasculitis in rheumatoid arthritis. Arch Neurol 32(4):271–275
58. Gravallese EM, Corson JM, Coblyn JS, Pinkus GS, Weinblatt ME (1989) Rheumatoid aortitis: a rarely recognized but clinically significant entity. Medicine (Baltimore) 68(2):95–106
59. van Albada-Kuipers GA, Bruijn JA, Westedt ML, Breedveld FC, Eulderink F (1986) Coronary arteritis complicating rheumatoid arthritis. Ann Rheum Dis 45(11):963–965
60. Harper L, Cockwell P, Howie AJ, Michael J, Richards NT, Savage CO et al (1997) Focal segmental necrotizing glomerulonephritis in rheumatoid arthritis. QJM 90(2):125–132
61. Karampatakis V, Konidaris V, Michailidou M, Gerofotis A, Daniilidis M (2013) Peripheral corneal ulceration associated with rheumatoid arthritis. Am J Case Rep 14:318–321
62. Sule A, Balakrishnan C, Gaitonde S, Mittal G, Pathan E, Gokhale NS et al (2002) Rheumatoid corneal melt. Rheumatology (Oxford) 41(6):705–706
63. Schwarz MI, Zamora MR, Hodges TN, Chan ED, Bowler RP, Tuder RM (1998) Isolated pulmonary capillaritis and diffuse alveolar hemorrhage in rheumatoid arthritis and mixed connective tissue disease. Chest 113(6):1609–1615
64. Babian M, Nasef S, Soloway G (1998) Gastrointestinal infarction as a manifestation of rheumatoid vasculitis. Am J Gastroenterol 93(1):119–120
65. van Laar JM, Smit VT, de Beus WM, Collee G, Jansen EH, Breedveld FC (1998) Rheumatoid vasculitis presenting as appendicitis. Clin Exp Rheumatol 16(6):736–738
66. Mizuno K, Ikeda K, Saida Y, Takenaka R, Shibata M, Takeuchi T (1996) Hepatic hemorrhage in malignant rheumatoid arthritis. Am J Gastroenterol 91(12):2624–2625
67. Achkar AA, Stanson AW, Johnson CM, Srivatsa SS, Dale LC, Weyand CM (1995) Rheumatoid vasculitis manifesting as intra-abdominal hemorrhage. Mayo Clin Proc 70(6):565–569

68. Geboes K, Dalle I (2002) Vasculitis and the gastrointestinal tract. Acta Gastroenterol Belg 65(4):204–212
69. Voskuyl AE, Hazes JM, Zwinderman AH, Paleolog EM, van der Meer FJ, Daha MR et al (2003) Diagnostic strategy for the assessment of rheumatoid vasculitis. Ann Rheum Dis 62(5):407–413
70. Voskuyl AE, van Duinen SG, Zwinderman AH, Breedveld FC, Hazes JM (1998) The diagnostic value of perivascular infiltrates in muscle biopsy specimens for the assessment of rheumatoid vasculitis. Ann Rheum Dis 57(2):114–117
71. Tribe CR, Scott DG, Bacon PA (1981) Rectal biopsy in the diagnosis of systemic vasculitis. J Clin Pathol 34(8):843–850
72. Flipo RM, Janin A, Hachulla E, Houvenagel E, Foulet A, Cardon T et al (1994) Labial salivary gland biopsy assessment in rheumatoid vasculitis. Ann Rheum Dis 53(10):648–652
73. Bartels CM, Bridges AJ (2010) Rheumatoid vasculitis: vanishing menace or target for new treatments? Curr Rheumatol Rep 12(6):414–419
74. Makol A, Matteson EL, Warrington KJ (2015) Rheumatoid vasculitis: an update. Curr Opin Rheumatol 27(1):63–70
75. Armstrong DJ, McCarron MT, Wright GD (2005) Successful treatment of rheumatoid vasculitis-associated foot-drop with infliximab. J Rheumatol 32(4):759–760
76. Garcia-Porrua C, Gonzalez-Gay MA, Quevedo V (2006) Should anti-tumor necrosis factor-alpha be the first therapy for rheumatoid vasculitis? J Rheumatol 33(2):433–434
77. Puechal X, Miceli-Richard C, Mejjad O, Lafforgue P, Marcelli C, Solau-Gervais E et al (2008) Anti-tumour necrosis factor treatment in patients with refractory systemic vasculitis associated with rheumatoid arthritis. Ann Rheum Dis 67(6):880–884
78. Unger L, Kayser M, Nusslein HG (2003) Successful treatment of severe rheumatoid vasculitis by infliximab. Ann Rheum Dis 62(6):587–588
79. van der Bijl AE, Allaart CF, Van VJ, Van DS, Breedveld FC (2005) Rheumatoid vasculitis treated with infliximab. J Rheumatol 32(8):1607–1609
80. Assmann G, Pfreundschuh M, Voswinkel J (2010) Rituximab in patients with rheumatoid arthritis and vasculitis-associated cutaneous ulcers. Clin Exp Rheumatol 28(1 Suppl 57):81–83
81. Hellmann M, Jung N, Owczarczyk K, Hallek M, Rubbert A (2008) Successful treatment of rheumatoid vasculitis-associated cutaneous ulcers using rituximab in two patients with rheumatoid arthritis. Rheumatology (Oxford) 47(6):929–930
82. Maher LV, Wilson JG (2006) Successful treatment of rheumatoid vasculitis-associated foot drop with rituximab. Rheumatology (Oxford) 45(11):1450 1451
83. Puechal X, Gottenberg JE, Berthelot JM, Gossec L, Meyer O, Morel J et al (2012) Rituximab therapy for systemic vasculitis associated with rheumatoid arthritis: results from the AutoImmunity and Rituximab Registry. Arthritis Care Res (Hoboken) 64(3):331–339
84. Jarrett SJ, Cunnane G, Conaghan PG, Bingham SJ, Buch MH, Quinn MA et al (2003) Anti-tumor necrosis factor-alpha therapy-induced vasculitis: case series. J Rheumatol 30(10):2287–2291
85. Mohan N, Edwards ET, Cupps TR, Slifman N, Lee JH, Siegel JN et al (2004) Leukocytoclastic vasculitis associated with tumor necrosis factor-alpha blocking agents. J Rheumatol 31(10):1955–1958
86. Ramos-Casals M, Roberto PA, Diaz-Lagares C, Cuadrado MJ, Khamashta MA (2010) Autoimmune diseases induced by biological agents: a double-edged sword? Autoimmun Rev 9(3):188–193
87. Saint MB, De Bandt M (2006) Vasculitides induced by TNFalpha antagonists: a study in 39 patients in France. Joint Bone Spine 73(6):710–713
88. Sokumbi O, Wetter DA, Makol A, Warrington KJ (2012) Vasculitis associated with tumor necrosis factor-alpha inhibitors. Mayo Clin Proc 87(8):739–745

Chapter 31
Buerger's Disease (Thromboangiitis Obliterans)

Masayuki Sugimoto and Kimihio Komori

Abstract Buerger's disease (also known as thromboangiitis obliterans) is a non-atherosclerotic, inflammatory, segmental peripheral vascular occlusive disease that typically affects young male smokers. The patients typically present with ischemic symptoms caused by stenosis or occlusion of the distal small arteries. Superficial thrombophlebitis often occurs, which is migratory and frequently correlates with disease activity. The disease escalates usually at the age of 30–40 years and thereafter symptoms diminish. This disease had been more frequently reported in the Mediterranean, Middle East, and Far East Asia, but the number of patients is decreasing.

There are two characteristic histopathological findings. One is that inflammatory infiltrating cells are well recognized, predominantly in the thrombi and the intima. The other is that the internal elastic lamina and all layers of the vessel wall structures are well preserved. Buerger's disease is distinguishable from atherosclerosis and other vasculitides by this characteristic preserved elastic lamina. There are several clinical diagnostic criteria for Buerger's disease, which mostly require a compatible history, supportive physical findings, diagnostic vascular abnormalities on imaging studies, and current or past smoking.

The etiology of Buerger's disease still remains to be elucidated and no therapeutic guidelines exist. However, smoking clearly associates with its exacerbation and remission. Absolute smoking cessation is the one and only definitive therapy for Buerger's disease.

The fate of ischemic limb in Buerger's disease is not so poor compared to that in atherosclerotic disease, provided that the patients maintain smoking cessation. As for the life expectancy, long-term survival is generally considered not to be affected by Buerger's disease, due to the rare involvement of cerebral, coronary, and visceral arteries.

M. Sugimoto • K. Komori (✉)
Division of Vascular Surgery, Department of Surgery, Nagoya University Graduates School of Medicine, 65, Tsurumaicho, Showa-ku, Nagoya 466-8550, Japan
e-mail: komori@med.nagoya-u.ac.jp

F. Dammacco et al. (eds.), *Systemic Vasculitides: Current Status and Perspectives*, DOI 10.1007/978-3-319-40136-2_31

31.1 History

The first case report of Buerger's disease was published in 1878 by Winiwarter at the University Clinic of Vienna. He reported a case of a 57-year-old man who underwent a below-the-knee amputation as a consequence of right foot gangrene [52].

In Tokyo, Haga reported clinical features of 10 cases and pathological findings of 13 specimens in 1889, and 9 years later, he contributed a paper with additional cases of spontaneous gangrene [10, 11].

Thereafter, in 1908, Leo Buerger, at Mount Sinai Hospital in New York, described a noteworthy disease of young individuals under the age of 40 years, in whom there were symptoms and signs of progressive and recurrent vascular insufficiency. Buerger named this disease "*thromboangiitis obliterans*" and distinguished it from atherosclerosis [2]. Currently, Buerger's disease is recognized as being a specific vascular disease based on a typical clinical picture, natural history, and histopathology.

31.2 Epidemiology

31.2.1 Patient's Demographics

According to the clinical literature, Buerger's disease is more frequently reported in the Mediterranean, Middle East, and Far East Asia than in Western Europe and North America [27]. Its incidence is highest among natives of India, Japan, Korea, Ceylon, and Ashkenazi Jews [39]. However, the possibility that this difference results from the heterogeneity of diagnostic criteria cannot be excluded. Furthermore, the high prevalence in some areas has been attributed to the use of specific tobacco types, including home-grown Kawung cigarettes in Indonesia and raw tobacco "beedi" cigarettes in Ceylon and Bangladesh.

Since 1950s, the numbers of new Buerger's disease patients have decreased in Western countries [51]. Whether this decline is true or only due to the application of more strict diagnostic criteria with the resultant exclusion of patients with premature atherosclerosis is unclear. However, the prevalence of Buerger's disease is decreasing even in Japan, formerly known as a high prevalence country [30].

On the other hand, the prevalence of the disease is increasing among women. This is likely due to the increase of smoking among women [34, 53]. Whilst Buerger's disease is becoming a relatively rare peripheral arterial disease, recent emerging cases of so-called "cannabis-associated arteritis" are categorized as a subtype of Buerger's disease by some researchers, prompting further investigation of Buerger's disease [5, 8].

31.2.2 HLA-Type

A correlation between several human lymphocyte antigen (HLA) subtypes (A9, B5, A1, B8, and DR4) and a higher incidence of Buerger's disease has been reported by several researchers [9, 32, 48]. Conversely, it has also been suggested that HLA-B12 antigen may be associated with disease resistance [4]. However, these findings are still controversial [33].

31.3 Clinical Presentation

Patients with Buerger's disease typically present with ischemic symptoms caused by stenosis or occlusion of the distal small arteries and veins. Intermittent claudication of the feet, legs, hands, and arms are typical symptoms. Critical ischemic symptoms, such as rest pain, ulceration, and gangrene, are also frequently observed at the initial presentation. Involvement of both the upper and lower extremities and the size and location of affected vessels help distinguish Buerger's disease from atherosclerosis. In most reports, isolated lower extremity involvement is present in approximately 50 % of patients, 30–40 % exhibit both lower and upper extremity ischemia, and in 10 % the signs and symptoms are confined to the upper extremity.

Although Buerger's disease most commonly affects the vessels in the upper and lower extremities, there are several sporadic reports of involvement of cerebral arteries, coronary arteries, visceral arteries, and pulmonary arteries. Notwithstanding that symptoms may begin in the peripheral portion of a single limb, Buerger's disease frequently involves multiple extremities. In the Shionoya's series, all patients with Buerger's disease had more than one limb involved. Two limbs were affected in 16 % of the patients, three limbs in 41 %, and four limbs in 43 %. Even in the patients in whom only one limb is clinically symptomatic, it is often the case that abnormal angiographic findings are already present in other limbs [46].

Superficial thrombophlebitis, which is migratory and frequently correlates with disease activity, occurs in approximately 40 % of the patients. Sometimes, this thrombophlebitis may appear in advance of ischemic symptoms caused by arterial occlusion, and is relatively specific to Buerger's disease. Nevertheless, it can be observed also in Behcet's disease. Raynaud's phenomenon is present in approximately 40 % of the patients and is possibly asymmetrical.

31.4 Etiology

In spite of the research carried out by our predecessors, the etiology of the disease remains unknown. However, there are important specific features of Buerger's disease that distinguish it from atherosclerotic obliterans (ASO) and other types of vasculitides.

31.4.1 Hypothetical Etiology of Buerger's Disease

Despite significant evidence that inflammation and autoimmunity play a central role, the precise pathogenic mechanism underlying Buerger's disease has not been determined yet. The inciting antigen still remains unknown.

Smoking, infection, nutritional deficits, or general autoimmunity may be responsible for the pathogenesis of Buerger's disease [6, 20, 41].

Some researches suggest that Buerger's disease could be a form of endarteritis from the viewpoint of the endothelial function. In patients with Buerger's disease, endothelium-dependent vaso-relaxation in the peripheral vasculature is impaired. On the other hand, non-endothelium-dependent mechanisms are intact [26].

Endothelial cells may play an important role in the initiation and maintenance of the inflammatory response. Increased expression of adhesive molecules, such as VCAM-1, ICAM-1 and selectin, on the surface of endothelial cells from patients with Buerger's disease has been described [12, 19, 24].

Recently, oral bacteria, such as periodontal bacteria, were reported to play an important role in the development of various vascular diseases [3, 18]. Especially in Buerger's disease, the relation between periodontal bacteria and platelet aggregation was discussed. Iwai and colleagues reported that almost two thirds of their patients with Buerger's disease had severe periodontitis, and that the DNA fragments of oral microorganisms were detected in the vascular lesions of Buerger's disease patients [16]. Conclusively, they have hypothesized that oral bacterial infections are one of the etiologies for Buerger's disease, suggesting that smoking cessation might be essential for the remission of Buerger's disease through the improvement of oral conditions [13, 17].

31.4.2 Histopathological Features

Regarding the histopathological features, Buerger described the lesions as being acute and chronic, segmental, with intense inflammatory infiltrate. He believed that the original pathologic process was thromboarteritis or a thrombophlebitic process, rather than a proliferative or obliterating process derived from the intima of the arteries and veins.

31.4.3 Phases of the Disease

Histologically, the disease is subdivided into 3 stages: acute, subacute, and chronic stages and the histopathological findings fluctuate according to the phases of the disease [46] (Fig. 31.1).

In a typical acute-stage lesion, the lumen is occluded by fresh thrombi and intimal thickness with remarkable leukocyte, including neutrophil, infiltration.

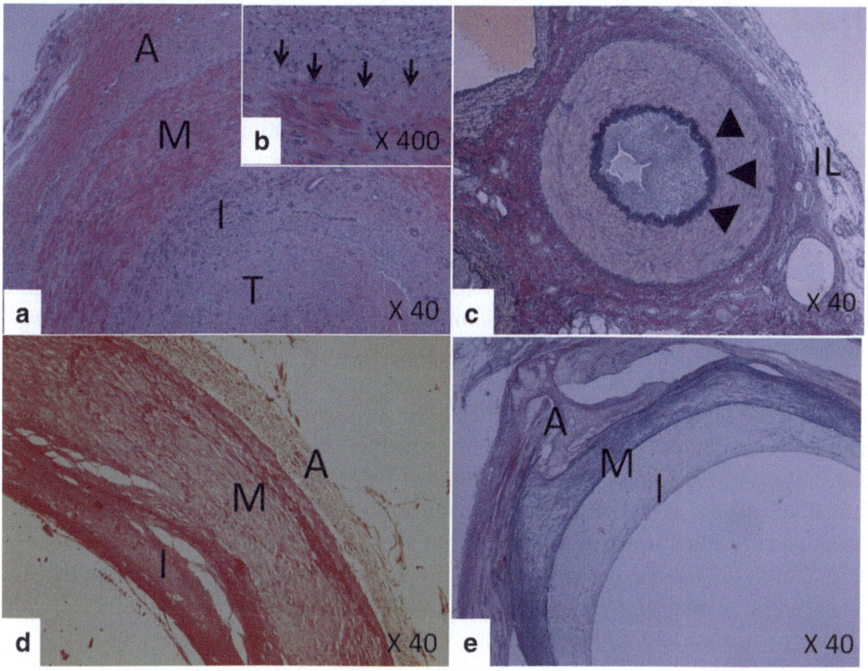

Fig. 31.1 Histological findings of acute (**a**, **b**: popliteal artery) and chronic stage (**c**: anterior tibial artery) lesions. In the acute-stage lesion, fresh and organized thrombi are seen. Remarkable inflammatory cell infiltration is observed in the thrombus and intima. In the chronic stage, the lumen is occluded by organized thrombi with recanalized vessels. Inflammatory cell infiltration can scarcely be seen. Note that in the acute stage, the internal elastic lamina (*arrows* and *arrowheads*) is almost intact. Whereas, in cases of atherosclerosis (**d**, **e**: the superficial femoral artery), fibro-intimal proliferation and hyaline degeneration are seen and cell infiltration is barely observed in any of the 3 layers. The internal elastic lamina and elastic fibers are severely disrupted and fragmented (**a**, **b**, **d**: H&E stain, C, E: elastica-van Gieson stain). *A* adventitia, *M* media, *I* intima, *T* thrombus (Kobayashi et al. 2014)

Multinucleated giant cells are seen in the thrombi, but necrotizing inflammation or granulomatous lesions are not observed. There have been investigators who studied the acute phase. They observed panvasculitis within small and medium-sized arteries and veins [25, 40]. The intense inflammatory infiltration and cellular proliferation are specific of the acute-stage lesion. Decreased fibrinolytic activity of the intima may also contribute to thrombus formation [50].

In the subacute stage, the lumen is occluded by fresh and organized thrombus with partial recanalization. Multinucleated giant cells are diminished in number in thrombi or vessel walls. Inflammatory infiltrating cells such as CD3+ pan-T cells, CD4+ helper-inducer cells, and CD20+ pan-B cells respond to the elastic lamina of the affected vessels in the subacute phase [22, 24].

In the chronic stage, the occlusive thrombi are organized and recanalized extensively, different from the acute or subacute stage lesions. Mild cell infiltration is seen in the intima, media, and adventitia. Prominent vascularization of the media and perivascular fibrosis characterize the chronic phase.

31.4.4 Pathologic Findings

As Leo Buerger stated in his report, there are 2 consistent pathologic findings regardless of its stage. One is that inflammatory infiltrating cells are well recognized, predominantly in the thrombi and the intima. The other is that the internal elastic lamina and all 3 layers of the vessel wall are well preserved. Buerger's disease is distinguishable from atherosclerosis and other vasculitides by this characteristically preserved elastic lamina. No calcification or hyaline degeneration is found in Buerger's disease. In atherosclerosis, in contrast, the general architecture and elastic lamina are destroyed, and inflammatory infiltrating cells are scarcely recognized in any of the 3 layers. These histological features of Buerger's disease are not common among other vasculitides.

We hypothesize that these specific findings might be related to the degree of extracellular matrix (ECM) degradation, which is caused by the plasminogen activator system and matrix metalloproteinases (MMPs). Restriction of the immune reaction (cellular as well as humoral) to the arterial intima defines Buerger's disease as an endarteritis. Furthermore, in Buerger's disease, expression of PAI-1 in media may inhibit the function of uPA and MMP- 3, which disrupt and degenerate ECMs. This can be related to preservation of the wall architecture in vessels affected with Buerger's disease [20, 21] (Fig. 31.2).

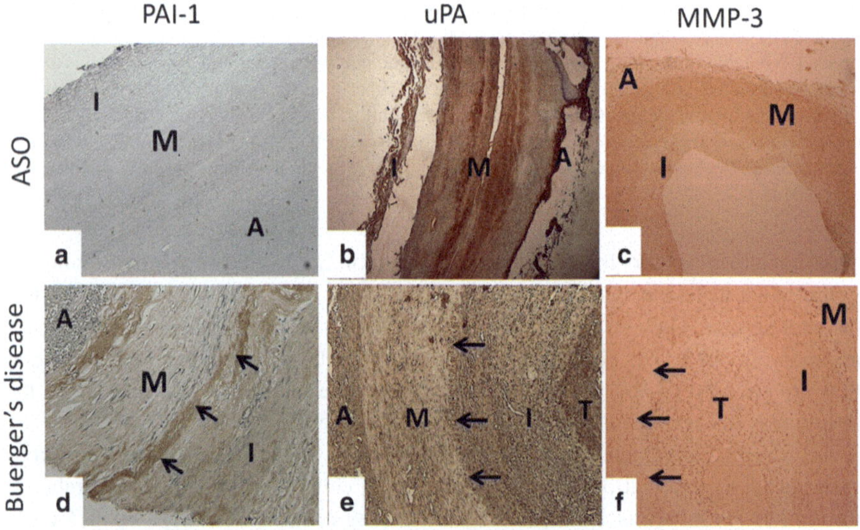

Fig. 31.2 Immunostaining for plasminogen activator inhibitor-1 (PAI-1), urokinase-type plasminogen activator (uPA), and matrix metalloproteinase (MMP-3) in atherosclerotic obliterans (ASO) (**a-c**) and Buerger's disease (**d-f**). In Buerger's disease, the expression of PAI-1 was very well recognized in media, particularly around internal elastic lamina (**d**), however, MMP-3 and uPA were scarce in intima (**e, f**). *A* adventitia, *M* media, *I* intima. (Kobayashi et al. 2014)

31.5 Diagnosis

31.5.1 Diagnostic Criteria

The diagnosis of Buerger's disease remains controversial and has not been settled because of various criteria used across the world. In general, Buerger's disease is a clinical diagnosis that requires a compatible history, supportive physical findings, and diagnostic vascular abnormalities on imaging studies. In most criteria, current or past smoking is a prerequisite for the diagnosis.

Several criteria have been proposed for the diagnosis of Buerger's disease. At our institute, the clinical criteria of Shionoya are used for the diagnosis: (1) smoking history, (2) onset before the age of 50 years, (3) infrapopliteal arterial occlusion, (4) either upper limb involvement or phlebitis migrans, and (5) absence of atherosclerotic risk factors other than smoking. Definitive presentation of Buerger's disease is considered to occur when all 5 criteria are met. Patients meeting all but the 4th criterion were diagnosed as probably suffering from Buerger's disease [46].

Mills et al. have proposed major and minor diagnostic criteria (Table 31.1). The major criteria are essential and the minor criteria support the diagnosis. These criteria may exclude a subset of patients with possible Buerger's disease, but their application defines a group of patients with nearly incontrovertible Buerger's disease [35].

Papa et al. have developed a scoring system based on the "negative" and "positive" criteria (Table 31.2). If the patient has 6 or more points, the diagnosis is definite; if 4–5, probable; if 2–3, suspected; and if 1 or less, diagnosis is excluded [42].

Olin et al. have proposed diagnostic criteria that is are similar to ours; (1) age younger than 45 years; (2) present or recent history of tobacco use; (3) presence of distal extremity ischemia (claudication, pain at rest, ischemic ulcers or gangrene), documented by non-invasive vascular testing; (4) exclusion of autoimmune diseases, hypercoagulable states and diabetes mellitus; (5) exclusion of a proximal source of embolization by echocardiography and arteriography; (6) consistent arteriographic findings [38].

31.5.2 Laboratory Tests

There is no specific laboratory test for the diagnosis of Buerger's disease. Laboratory tests are mainly used to exclude alternative diagnoses. Initial laboratory studies should include a complete blood count, metabolic panel, liver function tests, blood glucose levels, inflammatory markers such as erythrocyte sedimentation rate and C-reactive protein, cold agglutinins, and cryoglobulins. Additionally, serological markers of autoimmune disease, including anti-nuclear antibodies, anti-centromere antibodies, and anti-scl-70 antibodies, should be obtained. Results of serologic tests for the immunologic markers and the autoantibodies are usually normal or negative,

Table 31.1 Diagnostic criteria by Mills et al

Major criteria	
	Onset of distal extremity ischemic symptoms before 45 years old
	Tobacco abuse
	Exclusion of:
	Proximal embolic source
	Trauma and local lesions
	Autoimmune disease
	Hypercoagulable state
	Atherosclerosis
	Diabetes
	Hyperlipidemia
	Hypertension
	Renal failure
	Undiseased arteries proximal to the popliteal or distal brachial level
	Objective documentation of distal occlusive disease by means of:
	Segmental arterial Doppler studies and 4-limb plethysmography
	Arteriography
	Histopathology
Minor criteria	
	Migratory superficial phlebitis
	Raynaud's syndrome
	Upper limb involvement
	Instep claudication

even though an immune reaction has been demonstrated in the arterial intima [20]. Lupus anticoagulant and anti-cardiolipin antibodies are possibly detected in some cases, but these may also indicate an isolated thrombophilia.

31.5.3 Imaging Studies

The typical angiographic features are as follows; (1) involvement of the small and medium-sized vessels, such as the palmar, plantar, tibial, peroneal, radial, and ulnar arteries and the digital arteries of the fingers and toes, (2) segmental occlusive lesions, (3) more severe disease distally, and normal proximal arteries with no

Table 31.2 Diagnostic criteria by Papa et al

Positive points		
Age at onset	<30/30–40 years old	+2/+1
Foot claudication	Present/By history	+2/+1
Upper extremity	Symptomatic/Asymptomatic	+2/+1
Phlebitis migrans	Present/By history only	+2/+1
Raynaud's syndrome	Present/By history only	+2/+1
Angiography; biopsy	If typical, both/either	+2/+1
Negative points		
Age at onset	45–50/>50 years old	−1/-2
Sex/Smoking	Female/Nonsmoker	−1/-2
Location	Single limb/No leg involvement	−1/-2
Absent pulses	Brachial/Femoral	−1/-2
Atherosclerosis, diabetes, hypertension, hyperlipidemia	Discovered after diagnosis 5–10 years/2–5 years	−1/-2
Probability of diagnosis	**Number of points**	
Diagnosis excluded	0–1	
Low likelihood	2–3	
Probable, medium likelihood	4–5	
Definite, high likelihood	6 or higher	

evidence of atherosclerosis, (4) abundant collaterals around areas of occlusion ("corkscrew", "spider legs", or "tree roots" collaterals) (Figs. 31.3 – 31.7).

In general, arterial calcification and atheromatous changes are scarce. There are frequent occlusions of the radial or ulnar artery, or both, at or above the wrist, with marked tortuosity of re-canalized segments. The palmar arches and the digital arteries are often narrowed and diminished. Proximal lower extremity arteriography is generally normal from the supra-inguinal to the popliteal level. Tibial and pedal artery disease is generally segmental with abrupt transitions between normal vessel and sudden occlusions. Abundant collaterals are to be observed in chronic cases. These angiographic findings of Buerger's disease are, however, sometimes indistinguishable from those of other collagen diseases, such as scleroderma, systemic lupus erythematosus, mixed connective tissue diseases, anti-phospholipid antibody syndrome, and so on. Clinicians must remember that these findings are suggestive, but not pathognomonic of Buerger's disease.

31.5.4 Biopsy

Histopathological findings are not always required for the diagnosis of Buerger's disease. Occasionally, biopsy of superficial thrombophlebitis during the acute phase can be pathognomonic. For the diagnosis of Buerger's disease affecting atypical

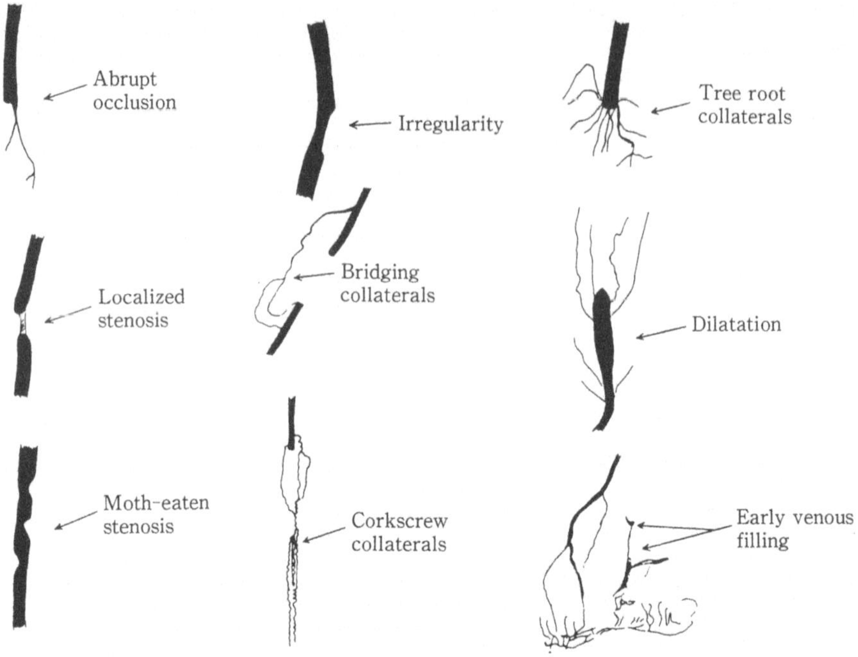

Fig. 31.3 Illustrations of angiographic features of Buerger's disease. [46]

Fig. 31.4 Abrupt occlusion accompanied by tree root collaterals is seen in the superficial femoral artery

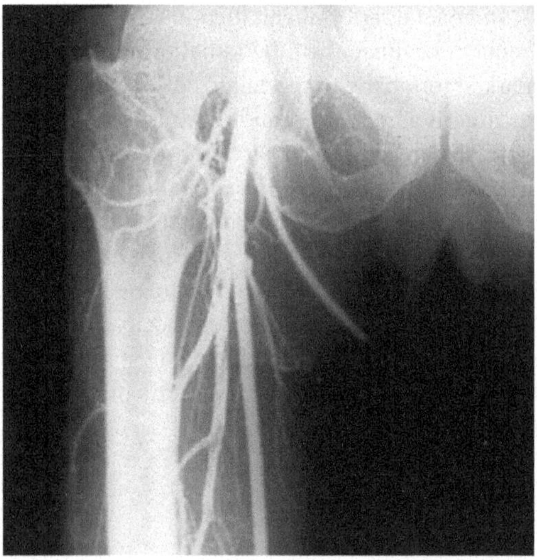

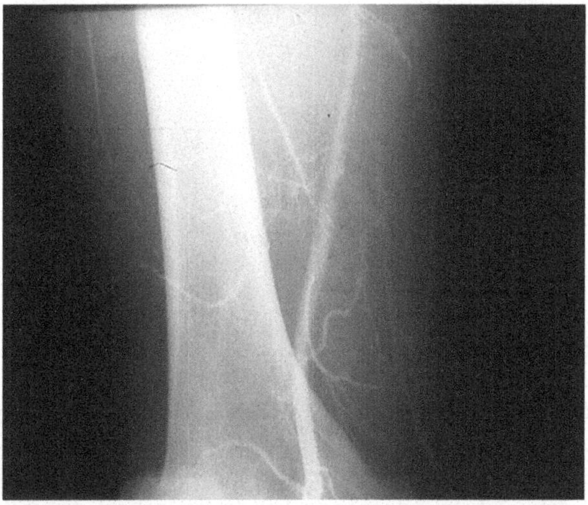

Fig. 31.5 The second portion of the popliteal artery is occluded. The corkscrew collaterals are heading to the lower calf

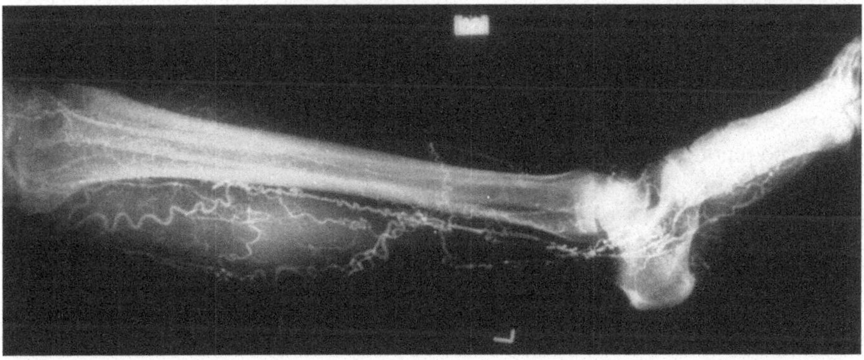

Fig. 31.6 Left femoral arteriogram of a 48-year-old man with Buerger's disease. The popliteal and crural arteries are extensively occluded, and collateral vessels show a corkscrew appearance

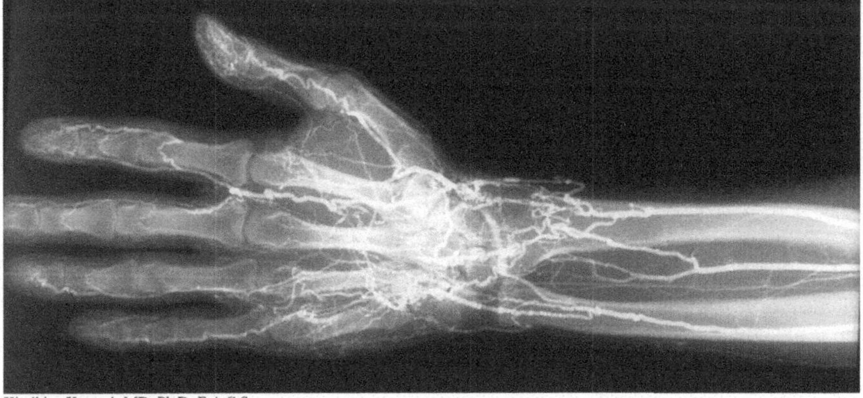

Kimihiro Komori, MD, Ph.D, F.A.C.S

Fig. 31.7 Right brachial arteriogram of a 46-year-old man. Despite multiple occlusions of the arteries in his forearm and hand, he had no symptom in his right upper extremity

locations such as visceral arteries, only the histopathological findings of acute-phase lesions can be definitive.

31.6 Treatments

31.6.1 Smoking Cessation

As mentioned above, the etiology of Buerger's disease still remains to be elucidated and no therapeutic guidelines exist. However, smoking clearly associates with its exacerbation and remission. Absolute smoking cessation is the one and only definitive therapy for Buerger's disease. Any therapeutic approaches without smoking cessation are unsuccessful in treating arterial insufficiency [36, 45].

The measurement of urinary nicotine and cotinine (a metabolic by-product of nicotine) help determine whether the patients is still smoking, using nicotine-replacement products, or being exposed to large amounts of environmental tobacco smoke. We have observed that symptomatic ischemia almost never recurred in patients with persistently low urinary cotinine levels [29].

Nicotine replacement therapy should also be avoided because it may contribute to disease activity even for the purpose of smoking cessation. On the other hand, there is no report that varenicline enhances the activity of Buerger's disease.

31.6.2 Medical Therapy

Various medications, including anti-platelets, anti-coagulants, and vasodilators, have been used for the treatment of Buerger's disease in the clinical practice. However, the clinical evidence to support the use of these drugs is scarce.

In a randomized controlled trial by Fiessinger and Schafer, 133 patients with the disease received intravenous iloprost (stable analogue of prostacyclin) or low-dose aspirin during a 28-day observation period. After 21–28 days, iloprost group achieved higher rates of clinical improvements, such as ulcer healing and pain relief [7].

31.6.3 Surgical Revascularization

The indication of bypass surgery is limited in patients with Buerger's disease, because of infra-popliteal arterial involvement, loss of target vessels, and poor quality of veins due to phlebitis. However, bypass surgery may be considered in patients with severe ischemia if suitable distal target vessels are identified. Successful bypass surgery can improve ischemic symptoms.

The reported outcomes of bypass surgery in patients with Buerger's disease are often suboptimal. In Ohta's series, the primary patency rates were 41 %, 32 %, and 30 % and secondary patency rates were 54 %, 47 %, and 39 % at 1, 5, and 10 years, respectively [37]. At the same time, satisfactory long-term patency is reported, provided that patients do not resume smoking. In Sasajima's series of 71 infra-inguinal bypasses in 61 patients with claudication or CLI due to Buerger's disease, the 5-year patency rate was 66.8 % in ex-smokers compared to 34.7 % in those who continued to smoke [44]. Our series of 23 infra-genicular bypasses in critical ischemic limbs also achieved a comparable primary patency rate of 67 % at 5 years. In this series, 12 patients with 14 affected limbs stopped smoking and no major amputation occurred. On the other hand, 10 patients with 10 limbs affected continued smoking, resulting in two major amputations [49]. Additionally, it was proved that revascularization allows ulcerations to heal in 90 % of cases even if the long-term graft patency is limited [23].

Another revascularization technique is the use of omental transfer for critical ischemia. Several reports from India profess the usefulness of this technique [31, 47].

31.6.4 Sympathectomy

Sympathectomy, particularly lumbar, had been used in patients who could not be candidates for bypass surgery. However, its effectiveness remains unclear. Sporadically, sympathectomy may induce the healing of superficial ischemic ulceration.

31.6.5 Therapeutic Angiogenesis

In a small study of patients with Buerger's disease, Isner et al. reported their results of intramuscular administration of vascular endothelial growth factor (VEGF) genes [15].

Several studies have evaluated autologous bone marrow mononuclear cell implantation for patients with critical limb ischemia resulting from Buerger's disease, and have successfully achieved reduction of pain and healing of ulceration [28, 43]. Others attempted the same procedure with autologous peripheral blood mononuclear cells, achieving the improvement of clinical symptoms [14].

31.7 Prognosis

In comparison with ASO, Buerger's disease results in relatively favorable outcomes for affected limbs. Critical limb ischemia (CLI) in patients with ASO is deemed to predict poor outcomes in terms of both survival and limb salvage. However, this is not always true in Buerger's disease. In contrast to CLI in ASO patients, medical treatment and smoking cessation are sometimes enough for the remission of ischemic symptoms in Buerger's disease.

In Olin and colleagues' series, 94 % of patients who quit smoking remained free of amputation, while 43 % of those who continued smoking required at least one amputation. No patient without gangrenous lesions at the time of smoking cessation required amputation [38,39]. In Ohta's series of 110 patients with Buerger's disease, 43 % of patients underwent 108 amputation procedures. Among those who continued smoking, 19 % required a major amputation. None of those who stopped smoking underwent amputation [37].

The occurrence of periods of escalation and remission is another characteristic feature of Buerger's disease. One study reports that the mean number of escalation was 5.4 times during follow-up [1]. The disease escalates usually at the age of 30–40 years and then symptoms diminish. In persons 60 years of age or older, recurrence is seldom observed.

Still, in most cases, recurrence of the disease is the consequence of returning to smoking. On the other hand, the retrospective analysis of our series of CLI in Buerger's disease implied that smoking had no effect on the recurrence in senior patients once complete remission has been achieved [49].

As for the life expectancy, long-term survival is generally considered not to be affected by Buerger's disease due to the rare involvement of cerebral, coronary, and visceral arteries. The life expectancy of patients with Buerger's disease approaches that of the normal population and has been estimated to be 97 % at 5 years and 94 % at 10 years [34].

References

1. Borner C, Heidrich H (1998) Long-term follow-up of thromboangiitis obliterans. VASA 27:80–86
2. Buerger L (1908) Thromboangiitis obliterans: a study of the vascular lesions leading to presenile spontaneous gangrene. Am J Med Sci 136:567–580
3. Chen YW, Umeda M, Nagasawa T, Takeuchi Y, Huang Y, Inoue Y et al (2008) Periodontitis may increase the risk of peripheral arterial disease. Eur J Endovasc Surg 35:153–158
4. de Moerloose P, Jeannet M, Mirimanoff P, Bouvier CA (1979) Evidence for an HLA-linked resistance gene in Buerger's disease. Tissue Antigens 14:169–173
5. Desbois AC, Cacoub P (2013) Cannabis-associated arterial disease. Ann Vasc Surg 27:996–1005
6. Eichhorn J, Sima D, Linschau C, Turowski A, Schmidt H, Schneider W et al (1998) Antiendothelial cell antibodies in thromboangiitis obliterans. Am J Med Sci 315:17–23

7. Fiessinger JN, Schafer M (1990) Trial of iloprost versus aspirin treatment for critical limb ischaemia of thromboangiitis obliterans. The TAO Study. Lancet 335:555–557
8. Grotenhermen F (2010) Cannabis-associated arteritis. VASA 39:43–53
9. Gulati SM, Singh KS, Thusoo TK, Saha K (1979) Immunological studies in thromboangitis obliterans (Buerger's disease). J Surg Res 27:287–293
10. Haga E (1889) Spontaneous gangrene in Japan. J Tokyo Med Assoc 3:419–422, 560–565, 630–3, 1217–21, 1346–53 (in Japanese)
11. Haga E (1898) Über spontane Gangräne. Arch Pathol Anat 152:26–60 (in German)
12. Halacheva K, Gulubova MV, Manolova I, Petkov D (2002) Expression of ICAM-1, VCAM-1, E-selectin and TNF-alpha on the endothelium of femoral and iliac arteries in thromboangiitis obliterans. Acta Histochem 104:177–184
13. Igari K, Inoue Y, Iwai T (2016) The epidemiologic and clinical findings of patients with buerger disease. Ann Vasc Surg 30:263–269
14. Inaba S, Egashira K, Komori K (2002) Peripheral-blood or bone-marrow mononuclear cells for therapeutic angiogenesis? Lancet 360:2083
15. Isner JM, Baumgartner I, Rauh G, Schainfeld R, Blair R, Manor O et al (1998) Treatment of thromboangiitis obliterans (Buerger's disease) by intramuscular gene transfer of vascular endothelial growth factor: preliminary clinical results. J Vasc Surg 28:964–973
16. Iwai T, Inoue Y, Umeda M, Huang Y, Kurihara N, Koike M et al (2005) Oral bacteria in the occluded arteries of patients with Buerger disease. J Vasc Surg 42:107–115
17. Iwai T, Sato, Kume H, Inoue Y, Umeda M, Kagayama T et al (2012) Clinical study of Phlebitis migrans and incompetence of the leg's superficial vein in Buerger disease. Ann Vasc Dis 5:45–51
18. Iwai T (2009) Periodontal bacteremia and various vascular diseases. J Periodontal Res 44:689–694
19. Joras M, Poredos P, Fras Z (2006) Endothelial dysfunction in Buerger's disease and its relation to markers of inflammation. Eur J Clin Investig 36:376–382
20. Kobayashi M, Ito M, Nakagawa A, Nishikimi N, Nimura Y (1999) Immunohistochemical analysis of arterial wall cellular infiltration in Buerger's disease (endarteritis obliterans). J Vasc Surg 29:451–458
21. Kobayashi M, Nishikimi N, Komori K (2006) Current pathological and clinical aspects of Buerger's disease in Japan. Ann Vasc Surg 20:148–156
22. Kurata A, Machinami R, Schulz A, Fukayama M, Franke FE (2003) Different immunophenotypes in Buerger's disease. Pathol Int 53:608–615
23. Lazarides MK, Georgiadis GS, Papas TT, Nikolopoulos ES (2006) Diagnostic criteria and treatment of Buerger's disease: a review. Int J Low Extrem Wounds 5:89–95
24. Lee T, Seo JW, Sumpio BE, Kim SJ (2003) Immunobiologic analysis of arterial tissue in Buerger's disease. Eur J Vasc Endovasc Surg 25:451–457
25. Lie JT (1988) Thromboangiitis obliterans (Buerger's disease) revisited. Pathol Annu 23:257–291
26. Makita S, Nakamura M, Murakami H, Komoda K, Kawazoe K, Hiramori K (1996) Impaired endothelium-dependent vasorelaxation in peripheral vasculature of patients with thromboangiitis obliterans (Buerger's disease). Circulation 94:II211–II215
27. Małecki R, Zdrojowy K, Adamiec R (2009) Thromboangiitis obliterans in the 21st century–a new face of disease. Atherosclerosis 206:328–334
28. Matoba S, Tatsumi T, Murohara T, Imaizumi T, Katsuda Y, Ito M et al (2008) Long-term clinical outcome after intramuscular implantation of bone marrow mononuclear cells (Therapeutic Angiogenesis by Cell Transplantation [TACT] trial) in patients with chronic limb ischemia. Am Heart J 156:1010–1018
29. Matsushita M, Shionoya S, Matsumoto T (1991) Urinary cotinine measurement in patients with Buerger's disease--effects of active and passive smoking on the disease process. J Vasc Surg 14:53–58

30. Matsushita M, Nishikimi N, Sakurai T, Nimura Y (1998) Decrease in prevalence of Buerger's disease in japan. Surgery 124:498–502
31. Maurya SD, Singhal S, Gupta HC, Elhence IP, Sharma BD (1985) Pedicled omental grafts in the revascularization of ischemic lower limbs in Buerger's disease. Int Surg 70:253–255
32. McLoughlin GA, Helsby CR, Evans CC, Chapman DM (1976) Association of HLA-A9 and HLA-B5 with Buerger's disease. Br Med J 2:1165–1166
33. Mills JL, Taylor LM Jr, Porter JM (1987) Buerger's disease in the modern era. Am J Surg 154:123–129
34. Mills JL, Porter JM (1991) Buerger's disease (thromboangiitis obliterans). Ann Vasc Surg 5:570–572
35. Mills JL, Porter JM (1993) Buerger's disease: a review and update. Semin Vasc Surg 6:14–23
36. Nishikimi N (2000) Fate of limbs with failed vascular reconstruction in Buerger's disease patients. Int J Cardiol 75(Suppl 1):S183–S185
37. Ohta T, Ishioashi H, Hosaka M, Sugimoto I (2004) Clinical and social consequences of Buerger disease. J Vasc Surg 39:176–180
38. Olin JW, Young JR, Graor RA, Ruschhaupt WF, Bartholomew JR (1990) The changing clinical spectrum of thromboangiitis obliterans (Buerger's disease). Circulation 82:IV3–IV8
39. Olin JW (2000) Thromboangiitis obliterans (Buerger's disease). N Engl J Med 343:864–869
40. Olin JW, Shih A (2006) Thromboangiitis obliterans (Buerger's disease). Curr Opin Rheumatol 18:18–24
41. Papa M, Bass A, Adar R, Halperin Z, Schneiderman J, Becker CG et al (1992) Autoimmune mechanisms in thromboangiitis obliterans (Buerger's disease): the role of tobacco antigen and the major histocompatibility complex. Surgery 111:527–531
42. Papa MZ, Rabi I, Adar R (1996) A point scoring system for the clinical diagnosis of Buerger's disease. Eur J Vasc Endovasc Surg 11:335–339
43. Saito S, Nishikawa K, Obata H, Goto F (2007) Autologous bone marrow transplantation and hyperbaric oxygen therapy for patients with thromboangiitis obliterans. Angiology 58:429–434
44. Sasajima T, Kubo Y, Inaba M, Goh K, Azuma N (1997) Role of infrainguinal bypass in Buerger's disease: an eighteen-year experience. Eur J Vasc Endovasc Surg 13:186–192
45. Shigematsu H, Shigematsu K (1999) Factors affecting the long-term outcome of Buerger's disease (thromboangiitis obliterans). Int Angiol 18:58–64
46. Shionoya S (1990) Pathology. In: Shionoya S (ed) Buerger's disease. The University of Nagoya Press, Nagoya, pp 57–79
47. Singh I, Ramteke VK (1996) The role of omental transfer in Buerger's disease: New Delhi's experience. Aust N Z J Surg 66:372–376
48. Smolen JS, Youngchaiyud U, Weidinger P, Kojer M, Endler AT, Mayr WR et al (1978) Autoimmunological aspects of thromboangiitis obliterans (Buerger's disease). Clin Immunol Immunopathol 11:168–177
49. Sugimoto M, Miyachi H, Morimae H, Kodama A, Narita H, Banno H et al (2015) Fate of ischemic limbs in patients with Buerger's disease based on our 30-year experience: does smoking have a definitive impact on the late loss of limbs?. Surg Today 45:466–470
50. Tanaka K (1998) Pathology and pathogenesis of Buerger's disease. Int J Cardiol 66:S237–S242
51. Wessler S, Ming SC, Gurewich V, Freiman DG (1960) A critical evaluation of thromboangitis obliterans. The case against Buerger's disease. N Engl J Med 262:1149–1162
52. Winiwarter F (1878) Über eine eigentümliche Form von Endarteritis und Endophlebitis mit Gangrän des Fusses. Arch kiln Chir 23:202–26 (in German)
53. Yörükoğlu Y, Ilgit E, Zengin M, Nazliel K, Salman E, Yucel E (1993) Thromboangiitis obliterans (Buerger's disease) in women (a reevaluation). Angiology 44:527–532

Chapter 32
Cannabis-Associated Vasculitis

Anne Claire Desbois and Patrice Cacoub

Abstract Cannabis is the most illicit drug consumed in France, especially by young adults and teenagers from all social groups. This drug is implicated in various complications such as psychotic disorders, anxiety, depressive symptoms, but also somatic disorders like cardiovascular disorders.

Cannabis use is associated with stroke, myocardial infarction and lower limb arteritis. Arterial disease affects especially young men. There is a very strong temporal link between arterial complications and cannabis use for strokes and myocardial infarctions. Patient outcome is closely correlated with cannabis withdrawal and relapses associated with cannabis re-challenge.

Cannabis use is associated with particular characteristics of arterial disease. The increased risk of myocardial infarction onset occurs within 1 h of smoking marijuana compared to periods of non-use. Strokes occurs mainly in the posterior cerebral circulation. Compared to cohorts of thromboangiitis obliterans patients, patients with cannabis-associated limb arteritis are younger, more often male and have more frequent unilateral involvement of the lower limbs at clinical presentation.

In conclusion, cannabis use is associated with arterial disease such as stroke, myocardial infarction and limbs arteritis. It appears essential to investigate cannabis use in young patients presenting with such arterial manifestations, as outcome is closely correlated with cannabis withdrawal.

A.C. Desbois
UMR CNRS 7211, INSERM U959, Paris 6, Paris, France

Laboratory I3 "Immunology, Immunopathology, Immunotherapy, UMR CNRS 7211, INSERM U959, Groupe Hospitalier Pitié-Salpêtrière, Université Pierre et Marie Curie, Paris 6, Paris, France

P. Cacoub (✉)
UMR CNRS 7211, INSERM U959, Paris 6, Paris, France

Laboratory I3 "Immunology, Immunopathology, Immunotherapy, UMR CNRS 7211, INSERM U959, Groupe Hospitalier Pitié-Salpêtrière, Université Pierre et Marie Curie, Paris 6, Paris, France

Department of Internal Medicine and Clinical Immunology, AP-HP, Hôpital La Pitié-Salpêtrière, F-75013 Paris, France
e-mail: patrice.cacoub@aphp.fr

F. Dammacco et al. (eds.), *Systemic Vasculitides: Current Status and Perspectives*, DOI 10.1007/978-3-319-40136-2_32

Keywords Cannabis • Arterial disease • Stroke • Myocardial infarction • Lower limb arteritis • Vasculitis

32.1 Introduction

Cannabis is the most illicit drug consumed in France, especially in young adults and teenagers from all social groups [1]. The use of cannabis has strongly increased in recent years in France [1]. A French study reported that 31 % of people aged 14–64 years had used it at least once [2]. Delta-9-trans-tetrahydrocannabinol (delta-9-THC) is considered the main active agent in cannabis. This substance is implicated in various complications such as psychotic disorders, anxiety, depressive symptoms, but also somatic disorders like cardiovascular disorders [3] and neoplasia [1]. The mechanisms of action of delta-9-THC have proven to be complex. Animal studies have shown a peripheral vasoconstrictor effect of delta-9-THC [4], and it seems to have prothrombotic effects [5]. In humans, it causes tachycardia with increased cardiac output and cardiac workload [6–8] and may cause orthostatic hypotension and syncope [8]. As consequences, cannabis use may be associated with many arterial complications, such as stroke, myocardial infarction and limb arteritis, with increased risk of death. Recently, 35 remarkable cardiovascular complications following cannabis use were reported by French Addictovigilance Network [22 cardiac complications (acute coronary syndrome, n = 20), ten peripheral arterial disease and three cerebral complications], with a 25 % death rate [9]. However, to assert the specific cardiovascular effects of cannabis is difficult, because it is often used in combination with other drugs such as alcohol or cocaine and most cannabis users also smoke tobacco.

32.2 Neurological Complications: Ischemic Strokes

Literature reports and studies provided convincing data that demonstrate an association of cannabis use and ischemic strokes. In a recent case-control study of young patients presenting with stroke, the logistic regression analysis adjusted for age, gender, and ethnicity, showed that cannabis use was associated with an increased risk of ischemic stroke/transient ischemic attack [odds ratio (OR), 2.30; 95 % confidence interval (95CI), 1.08–5.08]. However, after adjusting for tobacco use, an association independent of tobacco was not confirmed (OR 1.59; 95CI 0.71–3.70) [10]. The strong temporal relationship between on one hand symptoms onset/relapses and cannabis use and, on the other hand, the improvement of neurological disease with cannabis withdrawal, strongly suggests that cannabis is a trigger of such arterial manifestations. In our literature review, all patients had used cannabis the day before the stroke, i.e. during or 30 min before the symptoms for most of them, and 24 h preceding the symptoms for the remaining patients [11].

Some patients significantly increased their cannabis use during the days preceding the neurological complication. In previous reports, patients who stopped their cannabis use had a favorable outcome and patients who continued smoking cannabis relapsed and did not show reversibility [11]. In a French cohort of patients with a reversible cerebral vasoconstriction syndrome (RCVS), none displayed relapse of neurological deficit or thunderclap headache after cannabis discontinuation [12]. In the literature, most patients are heavy chronic cannabis smokers. Very few patients did present neurological symptoms after acute use of cannabis. Only few patients reported use of other vasoconstrictive drugs (about 10 % for cocaine and/or ecstasy) or had additional cardiovascular risk factors.

The prevalence of cannabis-induced cerebral strokes is difficult to estimate. In a prospective study of patients younger than 45 years admitted to a stroke unit, 13/48 (27 %) patients were cannabis consumers with positive urine tests and no other drug consumption [13]. In a more recent study, Wolf et al. reported a 19 % frequency of cannabis users in a cohort of 159 young patients hospitalized for stroke. The prevalence of cannabis consumers was higher in stroke patients with intracranial stenosis than stroke patients with other etiologies (37 % vs 11 %). [14] In a retrospective cohort of 831 patients who were admitted to a stroke unit between 2004 and 2007, 17 (2 %) patients had a neurologic deficit associated with cannabis use [15]. A retrospective study of 67 patients presenting with a RCVS, 20 (32 %) patients were cannabis consumers including nine patients that consumed only cannabis, whereas the remaining patients used alcohol, nasal decongestants, cocaine or selective serotonin reuptake inhibitors [12].

In a literature review of 71 cases of cannabis-associated strokes, patients were mainly young men (86 % males, mean age of 35 years) [11]. Main symptoms included: unilateral weakness, hypoesthesia/numbness, aphasia, dysarthria, blurred vision, headache and cerebellar syndrome. Some patients presented with a CRVS. Of note, few patients had concomitant cannabis-induced myocardial and cerebral infarctions. Vascular imaging often showed multifocal (50 %) and less frequently monofocal intra-cerebral artery stenosis, but it was normal in 20 %. A prospective study showed that cannabis use was significantly associated with multifocal ischemic strokes [OR 113; 95CI 9–5047; $p < 0.001$] [13]. These findings were confirmed in a more recent study in which all patients with drug abuse-associated strokes with intracranial stenosis had multi-arterial stenoses. [14] Interestingly, more than 50 % of cannabis-associated strokes reported in the literature occurred in the posterior circulation compared to 28 % of 326 reported patients aged 15–50 years presenting with ischemic strokes and 23 % of 8,057 patients reported in another retrospective cohort [16].

Underlying mechanisms of cannabis-associated strokes are not completely understood. Previous studies have shown impaired cerebral blood flow regulation after smoking marijuana, with increased cerebral blood volume in the right frontal region and the left temporal and cerebellar regions [17–19]. Other studies have reported an increase of systolic velocity and pulsatility index in the middle cerebral artery after cannabis smoking [20] and after 3 and 28 days of cannabis abstinence in heavy smokers [21], suggesting a vasoconstrictive phenomenon.

32.3 Cardiac Complications

Myocardial infarction (MI) has also been associated with cannabis use. In a prospective study [22], 12.5 % of patients younger than 50 years who were hospitalized for MI smoked marijuana. Using a crossover analysis, it was found a major increase in the risk of MI within 1 h after smoking marijuana compared to periods of non-use (OR 4.8; 95CI 2.9–9.5; $p < 0.001$). This risk decreased rapidly in the following hours after cannabis smoking cessation. Another prospective cohort study of 65,171 people (aged 15–49 years) who participated in multiphasic health checkups [3], did not find an increased risk of MI (OR 1.1, 95CI 0.7; 1.7), or stroke (OR 1.0; 95CI 0.8; 1.3) in current smoker men/cannabis consumers. Consistently with cannabis-associated neurological complications, none of the patients who discontinued smoking did relapse while recurrences of MI have been observed in patients who did not achieve cannabis withdrawal [11].

In a literature review of 147 patients with cannabis-associated MI, most of them were young men (male 93 %, mean age 42-years). Forty percent of patients had symptoms onset within 24 h following cannabis use. All patients were current or heavy smokers. Most patients also had other cardiovascular risk factors [tobacco use (70 %), hypertension (28 %), obesity (38 %) and diabetes (6 %)] [11]. In another study, compared with non-users, cannabis smokers with MI were more likely to be men ($p < 0.001$), current cigarette smokers ($p < 0.001$) and obese ($p = 0.008$). They were less likely to have a history of angina ($p < 0.001$) or hypertension ($p = 0.002$) [22].

Cannabis use is associated with a poor prognosis in patients who develop cardiovascular manifestations. A prospective study found that marijuana was associated with three-fold higher risk of mortality after MI, both after age and sex adjustment and in more fully adjusted models (cocaine, alcohol and tobacco use) [23]. There was a gradual increase in the risk of mortality with more frequent marijuana use. Age- and sex-adjusted hazard ratios associated with any marijuana use were 1.9 (95CI, 0.6–6.3) for cardiovascular mortality and 4.9 (95CI, 1.6–14.7) for non-cardiovascular mortality (i.e. motor vehicle accident, AIDS and lung cancer).

Cannabis use has been associated with numerous deleterious cardiovascular effects. Aronow [24] showed that in ten patients with chronic stable angina, the time to exercise-induced angina was decreased by 48 % after smoking a single marijuana cigarette compared with 8.6 % after smoking a placebo and 23 % after smoking a high nicotine tobacco cigarette ($p < 0.001$). Cannabis use increases heart rate and blood pressure, particularly when supine. Cannabis use increases oxygen demand and blood carboxyhemoglobin levels, resulting in a decreased oxygen supply for the heart [22]. Another proposed mechanism for cardiovascular toxicity is the disruption of a vulnerable atherosclerotic plaque in response to hemodynamic effects associated with cannabis use [25]. The increased risk of MI within the hours following cannabis use support this hypothesis. Coronarography was normal in 40 % of patients while it showed thrombosis in others. These findings suggest a reversible vasospasm in patients with normal coronary angiography.

32.4 Limb Arteritis

Sterne and Ducastaing first described cannabis-associated arteritis in 1960, in a series of Moroccan males who were heavy cannabis smokers (10–15 pipefuls per day) [26]. These patients presented with lower limb arteritis that showed characteristics similar to those found in thromboangiitis obliterans. In a literature review [11], 80 patients with cannabis-associated arteritis were reported, mostly young men (as observed in neurological and cardiovascular complications), with Raynaud's phenomenon and venous thromboses (both superficial and deep venous thromboses). Almost all patients had lower limb involvement while some of them also had upper limbs involvement. At diagnosis, most patients presented, after several months of claudication, with painful distal necrosis of the lower or upper limbs (stage IV of the Leriche classification). Rarely, arterial lesions were asymptomatic. Vascular imaging of the lower limbs revealed in almost all patients distal arterial occlusions characteristic of thromboangiitis obliterans. Proximal arterial lesions may be observed and lesions are usually bilateral (90 %).

It may be difficult to confirm the impact of cannabis on the onset of arterial complications, notably in patients who are heavy tobacco users. However, vascular flares were associated with cannabis consumption, i.e. relapses were always related to cannabis use, and remissions were associated with cannabis withdrawal (even without tobacco withdrawal). Cannabis users with limb arteritis compared to patients with typical thromboangiitis obliterans were younger, more often males, had more frequent unilateral involvement of the lower limbs and had less frequently thrombophlebitis and Raynaud's phenomenon [27]. These characteristics strongly suggest that cannabis also plays a key role in the pathogenesis of such arteritis in addition to tobacco. Cannabis associated-arteritis (in addition to tobacco) may represent a particular and severe form of thromboangiitis obliterans. The prognosis of cannabis-associated limbs arteritis appears to be very poor, as far as up to 58 % of patients underwent amputations.

Conflict of Interest Statement The authors do not have any conflict of interest to declare.

References

1. Costes J-M, Beck F, Legleye S, Palle C (2005) [Epidemiology of cannabis use]. Rev Prat 55(1):17–22
2. Cottencin O, Karila L, Lambert M et al (2010) Cannabis arteritis: review of the literature. J Addict Med 4(4):191–196
3. Sidney S (2002) Cardiovascular consequences of marijuana use. J Clin Pharmacol 42(11 Suppl):64S–70S
4. Adams MD, Earnhardt JT, Dewey WL, Harris LS (1976) Vasoconstrictor actions of delta8- and delta9-tetrahydrocannabinol in the rat. J Pharmacol Exp Ther 196(3):649–656
5. Deusch E, Kress HG, Kraft B, Kozek-Langenecker SA (2004) The procoagulatory effects of delta-9-tetrahydrocannabinol in human platelets. Anesth Analg 99(4):1127–1130, table of contents
6. Beaconsfield P, Ginsburg J, Rainsbury R (1972) Marihuana smoking. Cardiovascular effects in man and possible mechanisms. N Engl J Med 287(5):209–212
7. Renault PF, Schuster CR, Heinrich R, Freeman DX (1971) Marihuana: standardized smoke administration and dose effect curves on heart rate in humans. Science 174(4009):589–591
8. Weiss JL, Watanabe AM, Lemberger L, Tamarkin NR, Cardon PV (1972) Cardiovascular effects of delta-9-tetrahydrocannabinol in man. Clin Pharmacol Ther 13(5):671–684
9. Jouanjus E, Lapeyre-Mestre M, Micallef J (2014) French Association of the Regional Abuse and Dependence Monitoring Centres (CEIP-A) Working Group on Cannabis Complications*. Cannabis use: signal of increasing risk of serious cardiovascular disorders. J Am Heart Assoc 3(2):e000638
10. Barber PA, Pridmore HM, Krishnamurthy V et al (2013) Cannabis, ischemic stroke, and transient ischemic attack: a case-control study. Stroke J Cereb Circ 44(8):2327–2329
11. Desbois AC, Cacoub P (2013) Cannabis-associated arterial disease. Ann Vasc Surg 27(7):996–1005
12. Ducros A, Boukobza M, Porcher R et al (2007) The clinical and radiological spectrum of reversible cerebral vasoconstriction syndrome. A prospective series of 67 patients. Brain J Neurol 130(Pt 12):3091–3101
13. Wolff V, Lauer V, Rouyer O et al (2011) Cannabis use, ischemic stroke, and multifocal intracranial vasoconstriction: a prospective study in 48 consecutive young patients. Stroke J Cereb Circ 42(6):1778–1780
14. Wolff V, Armspach J-P, Beaujeux R et al (2014) High frequency of intracranial arterial stenosis and cannabis use in ischaemic stroke in the young. Cerebrovasc Dis Basel Switz 37(6):438–443
15. Singh NN, Pan Y, Muengtaweeponsa S, Geller TJ, Cruz-Flores S (2012) Cannabis-related stroke: case series and review of literature. J Stroke Cerebrovasc Dis 21(7):555–560
16. Phillips MCL, Leyden JM, Chong WK et al (2011) Ischaemic stroke among young people aged 15 to 50 years in Adelaide South Australia. Med J Aust 195(10):610–614
17. Volkow ND, Gillespie H, Mullani N et al (1996) Brain glucose metabolism in chronic marijuana users at baseline and during marijuana intoxication. Psychiatry Res 67(1):29–38
18. O'Leary DS, Block RI, Flaum M et al (2000) Acute marijuana effects on rCBF and cognition: a PET study. Neuroreport 11(17):3835–3841
19. Mathew RJ, Wilson WH, Coleman RE, Turkington TG, DeGrado TR (1997) Marijuana intoxication and brain activation in marijuana smokers. Life Sci 60(23):2075–2089
20. Mathew RJ, Wilson WH, Humphreys DF, Lowe JV, Wiethe KE (1992) Changes in middle cerebral artery velocity after marijuana. Biol Psychiatry 32(2):164–169
21. Herning RI, Better WE, Tate K, Cadet JL (2005) Cerebrovascular perfusion in marijuana users during a month of monitored abstinence. Neurology 64(3):488–493
22. Mittleman MA, Lewis RA, Maclure M, Sherwood JB, Muller JE (2001) Triggering myocardial infarction by marijuana. Circulation 103(23):2805–2809

23. Mukamal KJ, Maclure M, Muller JE, Mittleman MA (2008) An exploratory prospective study of marijuana use and mortality following acute myocardial infarction. Am Heart J 155(3):465–470
24. Aronow WS, Cassidy J (1974) Effect of marihuana and placebo-marihuana smoking on angina pectoris. N Engl J Med 291(2):65–67
25. Muller JE, Abela GS, Nesto RW, Tofler GH (1994) Triggers, acute risk factors and vulnerable plaques: the lexicon of a new frontier. J Am Coll Cardiol 23(3):809–813
26. Sterne J, Ducastaing C (1960) [Arteritis caused by Cannabis indica]. Arch Mal Coeur Vaiss 53:143–147
27. Martin-Blondel G, Koskas F, Cacoub P, Sène D (2011) Is thromboangiitis obliterans presentation influenced by cannabis addiction? Ann Vasc Surg 25(4):469–473

Part IV
Diagnostic and Therapeutic Advances of Systemic Vasculitides

Chapter 33
Imaging in Systemic Vasculitis

Mazen Abusamaan, Patrick Norton, Klaus Hagspiel, and Aditya Sharma

Abstract Vasculitis is a heterogeneous and complex group of diseases associated with inflammation of the vessel wall. It can affect both arteries and veins; however seen mostly in arteries. Vasculitis is further divided into small, medium and large vessel vasculitis based on the arteries involved. In this chapter, we provide a detailed discussion on the use of various diagnostic modalities often employed in the diagnosis and treatment of acute as well as chronic vasculitis and their sequelae.

Keywords Vasculitis • Vasculitides • Imaging • Arteritis • Aortitis

33.1 Introduction

Vasculitides are a heterogeneous group of diseases characterized by inflammation of the blood vessel wall. These usually present with certain characteristic constitutional and systemic symptoms in addition to vascular inflammation [35, 46, 73, 83]. Classification is usually based on the size of the predominately involved vessels [35, 46, 73, 83]. Imaging can play a major role in identifying some of the vasculitides especially large and medium vessel vasculitis (Table 33.1) [77]. The great strides made in developing superior non-invasive imaging modalities have played an important role in the current diagnosis and management of vasculitis. Combination of clinical findings along with superior imaging modalities has made it easier to accurately diagnose many vasculitides without the need of invasive procedures such as skin or other tissue biopsy often required in the past. Imaging modalities frequently used for diagnosis are color duplex ultrasound (CDUS), computerized tomography (CT), magnetic resonance imaging (MRI) and magnetic resonance angiography (MRA), and positron emission tomography (PET). The cardinal

M. Abusamaan • A. Sharma (✉)
Division of Cardiovascular Medicine, University of Virginia, Charlottesville, VA, USA
e-mail: MSA6A@hscmail.mcc.virginia.edu; asharma@virginia.edu

P. Norton • K. Hagspiel
Department of Radiology, University of Virginia, Charlottesville, VA, USA
e-mail: PTN7Y@hscmail.mcc.virginia.edu; KDH2N@hscmail.mcc.virginia.edu

F. Dammacco et al. (eds.), *Systemic Vasculitides: Current Status and Perspectives*, DOI 10.1007/978-3-319-40136-2_33

Table 33.1 Vasculitidies in which imaging of the vasculature can play a major role in identification and or treatment

Large vessel Vasculitis:
Giant Cell arteritis
Takayasu's arteritis
Aortitis
Medium vessel Vasculitis:
Polyarteritis nodosa
Kawasaki disease
Medium and Small vessel vasculitis
Primary angiitis of the central nervous system
Buerger's disease
Vasculitis involving arteries and veins of various sizes
Behçet's disease

imaging signs of vasculitis are vessel wall thickening, irregular contours, perivascular inflammation, aneurysm formation and stenosis and occlusion as a result of vascular remodeling secondary to inflammation [77, 83]. These signs are detectable in large and medium vessel vasculitides where imaging is the cornerstone in diagnosis due to inaccessibility of the primary involved vessel to obtain histopathologic sampling. In this chapter, we discuss the role of the different imaging modalities for the diagnosis of different types of vasculitides, as well as their relative advantages and disadvantages.

33.2 Ultrasonography

CDUS is a widely available non-invasive imaging modality which is relatively inexpensive, safe, lacks radiation and does not need contrast. This imaging modality is useful particularly in diagnosis and management of large and medium-vessel vasculitis such as giant cell arteritis (GCA), inflammatory aortitis and Takayasu's arteritis.

Vasculitis in acute stages will present with significant vessel wall inflammation and thickening, which could be identified on CDUS particularly in large and medium-vessel vasculitis (Fig. 33.1). The use of CDUS in the diagnosis of GCA was first reported by Schmidt and colleagues in 1997, who discovered that the inflamed temporary artery wall demonstrated hypoechoic edematous wall swelling called halo sign (Fig. 33.2) [71]. Presence of this sign has good sensitivity and specificity for the diagnosis of GCA [6, 39]. The absence of the halo sign has been reported to have a very high (92–96 %) negative predictive value [42, 71]. This imaging characteristic typically disappears after a mean of 16 days with corticosteroid treatment [42, 64, 71]. Interestingly, LeSar CJ et al. reported greater sensitivity of clinical diagnosis supported by US or MR findings compared to histological exam in untreated patients or patients who received corticosteroid therapy for <2 days [42].

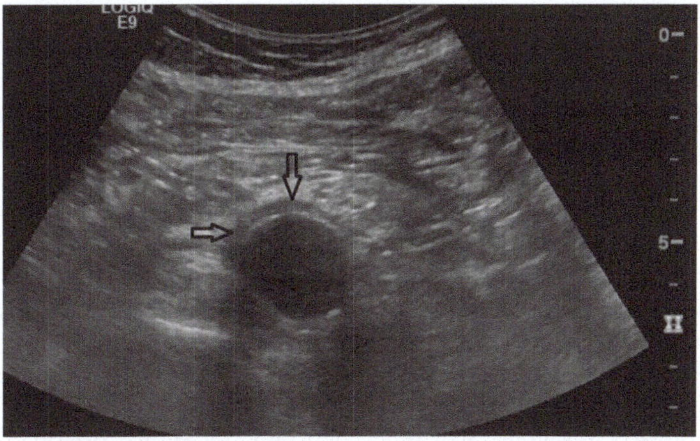

Fig. 33.1 Duplex ultrasound of a 44-year-old man with isolated idiopathic aortitis showing vessel wall inflammation in the form of thickening (see *arrows*) in the abdominal aorta

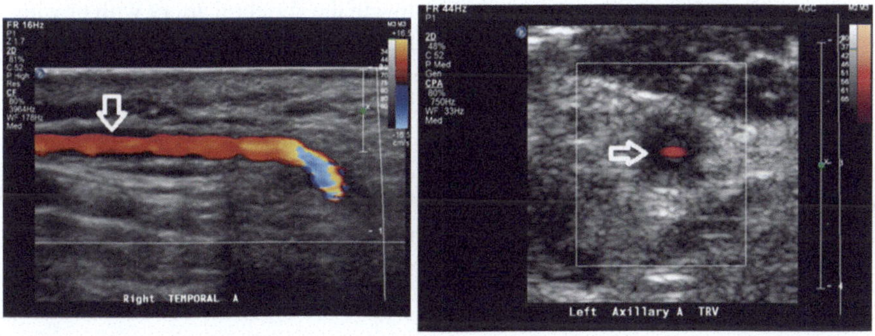

Fig. 33.2 Duplex ultrasound of an elderly man with Giant cell arteritis showing vessel wall inflammation in the form of thickening (see *arrows*) in the right temporal artery and the left axillary artery (Halo sign). Acknowledgement: Image provided by Heather Gornik, MD (Cleveland Clinic)

Although temporal artery biopsy remains the gold standard for the diagnosis of GCA by showing mononuclear cell infiltration of the arterial wall on histopathologic exam but because the inflamed segment in temporal arteritis is not always homogeneously distributed along the temporal artery, temporal artery biopsy can be falsely negative [23, 43, 44, 82]. Due to this limitation and the improvement in resolution in ultrasonic diagnostic modalities, there has been active investigation in using ultrasound more frequently for the diagnosis of GCA.

Other CDUS findings of large and medium-vessel vasculitis include stenosis, as evidenced by narrowing of the vessel lumen on gray scale imaging, increased Doppler velocities and turbulence due to the stenosis. Occlusion of the temporal artery can be demonstrated by absence of color signals in the vessel. During examination, the sonographer can communicate with the patient and correlate areas of tenderness with ultrasound findings [21, 56, 68, 70]. CDUS can assess longer arterial segments and examine neighboring arteries for inflammation, which is not feasible by temporal artery biopsy (Fig. 33.2). Often CDUS is used to identify and mark inflamed segments of superficial temporal artery to guide temporal artery biopsy in order to improve accuracy of histopathologic results [38]. Several studies have shown inflammatory changes in the axillary artery in patients with large vessel vasculitis [12, 69, 72]. In a meta-analysis of 23 studies, CDUS of the temporal artery had a sensitivity and specificity of 87 and 96 % compared with the clinical diagnosis alone [39]. In some experienced centers, the diagnosis is established by clinical assessment and ultrasound without the need for invasive biopsy [1, 12, 39, 52, 56, 68, 69, 72].

In Takayasu's Arteritis (TA), CDUS demonstrates heterogeneous intimal thickening of the affected large vessel with irregular luminal contour and bright appearance known as "Macaroni" sign (Fig. 33.3) [26]. Takayasu's Arteritis does not involve the temporal artery. Homogenous wall thickening of the common carotid or left subclavian artery of more than 1.0 mm before hemodynamic significant stenosis occurs, is suspicious for TA [69]. In later stages, stenosis or occlusion develops and leads to collateral formation and retrograde flow in the vertebral artery with or without subclavian steal syndrome [62]. The limitations of ultrasonography are its inability to detect peri-adventitial small vessel vasculitis or vasculitis of the vasa vasorum, which can only be demonstrated on histopathologic examination, the inability to depict vessels localized behind osseous structures and air such as the thoracic aorta unless performed by transesophageal echocardiography and its dependence on the sonographer's skills [52]. Lastly, on the horizon there are now clinical studies evaluating the use of microbubble contrast-enhanced ultrasonography in the GCA and TA (clinical trials.gov NCT01795456).

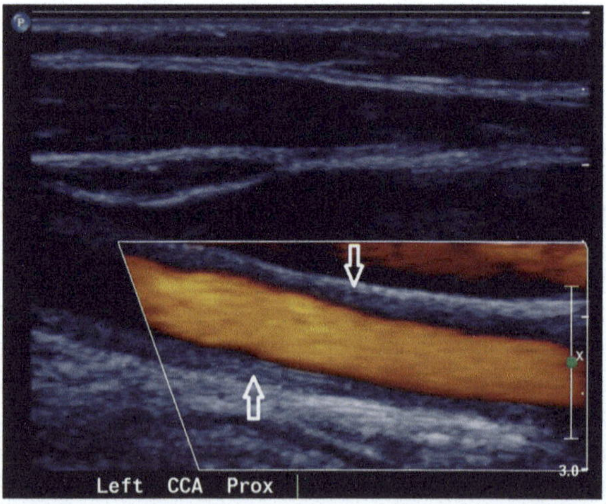

Fig. 33.3 Duplex ultrasound of a young woman with Takayasu's arteritis showing vessel wall inflammation in the form of thickening in the left common carotid artery (see *arrows*). Acknowledgement: Image provided by Esther Kim, MD (Cleveland Clinic)

33.3 Computed Tomography Angiography

CT and CT Angiography (CTA) are both useful to detect inflammatory changes in large and medium sized arteries because of good spatial resolution and fast scanning times. CT/CTA accurately measures vessel diameters, and demonstrates arterial wall thickening, calcifications, thrombosis, stenosis and occlusion [27, 86].

In a series of 40 GCA patients, large-vessel involvement/wall thickening was present in 27; in these 27 cases, the aorta was affected in 65 %, the brachiocephalic trunk in 48 %, the carotid arteries in 35 %, and the subclavian arteries in 43 % [57]. Another study by Park et al. involving 47 patients with TA showed that the left subclavian artery was involved in 55 %, abdominal aorta in 53 %, right renal artery in 45 %, right subclavian and left renal arteries in 38 %, descending thoracic aorta in 32 %, left common carotid artery in 30 %, and coronary arteries in15% [55]. Findings of long segment aortic wall thickening with smooth distal tapering, aneurysm formation, dissection and dilatation are more prevalent in GCA patients compared to the normal population [40, 53, 61]. Thoracic aortic aneurysms more readily develop in an inflamed aorta, which explains their increased incidence compared to healthy subjects of the same age [40, 53, 61]. Aortic involvement most commonly presents as annuloaortic ectasia. Aortic aneurysms, dissections can occur throughout the aorta. Some experts recommend annual chest X-rays and abdominal duplex for aneurysm screening in GCA patients [47]. Coronary and mesenteric artery involvement in GCA has been described [34].

CTA is well suited for the diagnosis of TA. The inflammatory changes in the vessel wall in the acute phase lead to the classic "double ring" finding on CT (Fig. 33.4). The poorly enhancing inner ring usually reflects intimal hyperplasia, whereas

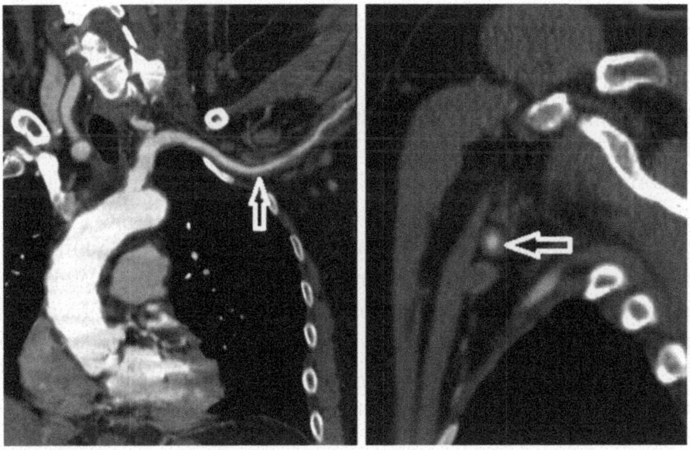

Fig. 33.4 CT angiogram of a 64 year-old man with Giant cell arteritis showing wall thickening (*arrows*) of the left subclavian and axillary arteries (long axis, *left image*; short axis, *right image*) representing inflammatory changes in the acute phase. Normally, the arterial wall is not perceptible on CT angiography

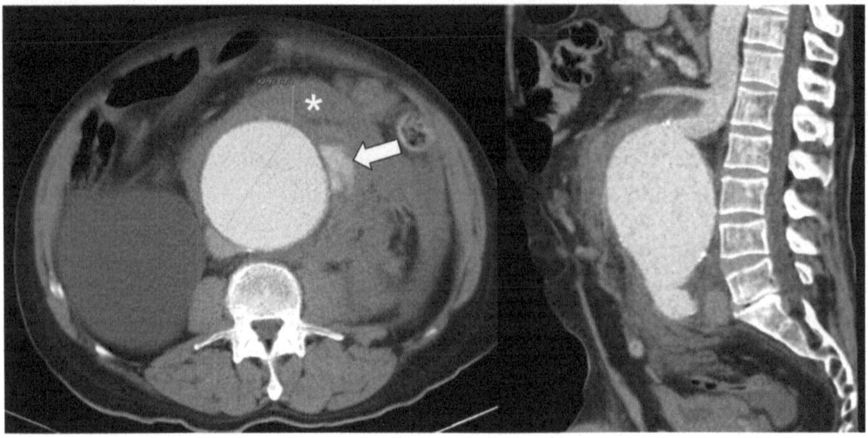

Fig. 33.5 CT Angiogram of a 72-year-old man with 9 cm ruptured abdominal aortic aneurysm with active extravasation (*arrow*) and retroperitoneal hematoma (*asterisk*) in axial (*left*) and sagittal (*right*) planes

the brightly enhancing outer ring indicates active inflammation in the medial and adventitial layers of the artery [30]. After 5 years or later, the chronic stages are characterized by arterial wall calcifications, which are typically linear without involvement of the ascending aorta [30, 60]. Compared to conventional angiography, CT is superior in demonstrating arterial wall thickening, perivascular inflammation, calcifications and mural thrombi [85]. Yamada I et al. had 25 patients with TA undergoing CTA and conventional angiogram. They reported that CTA was able to assess stenotic lesions in TA in all brachiocephalic trunks, in 37 of 40 common carotid arteries, and in 33 of 40 subclavian arteries, with a sensitivity and specificity of 93 % and 98 % respectively [85]. CTA has been shown to detect silent coronary artery involvement in patients affected with TA. In a recent study that included 111 patients with TA, using a 128-section dual-source CTA, coronary involvement was disclosed in more than half of the patients [37]. CTA provides an accurate measurement of aortic aneurysm to identify those with high risk of rupture (>5 cm in ascending aorta, >6 cm in thoracic aorta and >5.5 cm in abdominal aorta or >0.5 expansion at any site in 6 months or more) (Fig. 33.5) [3, 32].

One of the main limitations of CT and CTA is exposure to ionized radiation, particularly as these patients frequently need follow-up imaging. Lately, newer low dose techniques and novel iterative reconstruction algorithms for CTA allow significant reduction in radiation exposure [2]. The spatial resolution of CT is in the order of 0.5 mm, which is limiting accurate assessment of involvement of smaller vessels [79]. The need to use iodinated contrast agents is a further limitation of CTA.

33.4 Magnetic Resonance

Similar to CT, MRI is well-suited to evaluate the aorta and other large and medium sized vessels. It is generally considered superior to CTA in its ability to assess the vessel wall. MRI has a considerable significance in detecting early signs of vasculitis, mainly by detection of subtle changes in the aortic wall. Fat-suppressed T2 black blood sequences can demonstrate increased wall thickness and wall edema. T1-weighted sequences post contrast show mural enhancement (Fig. 33.6) [15, 16]. Post-contrast T1 images are superior to T2 or fat-suppressed images in detecting large-vessel inflammation and are required for detection of more subtle signs [81].

MRI can be used for periodic assessment of TA by providing luminal and vessel wall assessment without radiation or iodinated contrast [81]. The MRI findings of high signal on T2 weighted sequences and mural contrast enhancement is more prominent in active disease stages and tends to diminish after initiation of immunosuppressive therapy. However, their role in assessment of disease activity remains controversial and requires more evidence [81].

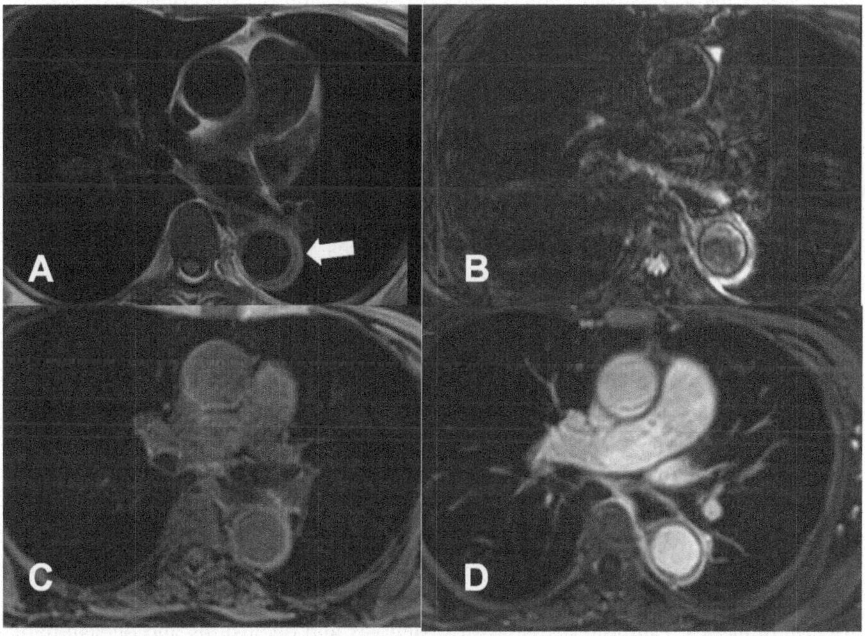

Fig. 33.6 MR angiogram performed with ECG-gating of a 70-year-old man with acute aortitis. (**a**) T1-weighted imaging shows morphologic changes of circumferential thickening of the descending aortic wall (*arrow*). (**b**) T2-weighted image demonstrates high (bright) signal due to edema in the aortic wall. (**c**) The inflamed aortic wall is of higher signal than the blood within the aortic lumen on T1-weighted imaging with fat saturation before the administration of gadolinium contrast media. (**d**) After the administration of contrast media, the marked enhancement of the inflamed aortic media and adventitia with relatively poor enhancement of the swollen intima can be seen, resulting in the "double ring" finding

High resolution MRI has been shown to have the same sensitivity and specificity as CDUS for the detection of temporal artery involvement in GCA, although it has not been used routinely for this reason [10, 41]. Bley TA et al. compared MRI with CDUS in 59 patients with suspected GCA [10]. 36 of the 59 patients had GCA. Sensitivity of MRI and CDUS was 69 % and 67 % respectively. Specificity was 91 % in both groups (38). Another study by the same groups demonstrated that high-resolution 1.5-T MR of the TA has been shown to be 81 % sensitive and 97 % specific for GCA diagnosis [11]. Interestingly, this study showed histology results alone had a sensitivity of 77.8 % and specificity of 100 %. The degree of inflammatory enhancement by high resolution MRI has shown to be affected by steroid therapy [8, 9].

Drawbacks of MRI are its relatively higher cost and lack of universal availability, particularly for high-resolution wall imaging. Another disadvantage of MRI is the limited use in patients with claustrophobia and metal implants such as cardiac pacemaker and defibrillators. In patients with renal insufficiency, the use of gadolinium-based contrast agents can be limited due to the risk of nephrogenic systemic fibrosis. MRI has suboptimal visualization of vessel calcifications. Moreover, similar to CTA, MRA has a limited role for small vessels evaluation, its spatial resolution generally being below that of CTA.

33.5 Digital Subtraction Angiography

Digital subtraction angiography (DSA), once the primary diagnostic modality, has lost its role to other noninvasive imaging techniques, mainly due to its invasiveness and inability to demonstrate changes of the vessel wall [56]. It remains however an important tool for assessing medium and small sized arteries due to its superior spatial resolution and to guide endovascular treatment.

In GCA, DSA can demonstrate aortic aneurysm and irregular wall contours [78]. Frequently, in cases with upper extremity involvement, lesions present as bilateral stenosis or occlusions with a smooth, tapered appearance in the subclavian, axillary and proximal brachial arteries. The presence of affected segments followed by normal segments (skip lesions) is typical for GCA. Occasionally patients present with aneurysmal lesions. GCA generally spares the origins of the great (supra-aortic) vessels [80].

Unlike GCA, TA tends to affect the aorta and the origins of the supraaortic and visceral branches. Luminal irregularities, stenosis and occlusions and occasionally aneurysms can be found. The characteristic angiographic appearance of these lesions was described as proximal, long segment, smooth contours, abrupt and flame-shaped terminations. TA is classified angiographically into six types, and this can be readily achieved with CTA or MR. DSA is not required for this purpose (Table 33.2) [49]. Where DSA still has a diagnostic role is in the realm of medium sized vasculitides, where it detects stenosis, occlusions and aneurysms more accurately than CTA or MRA.

Thromboangiitis obliterans (TAO) or Buerger's disease is a form of vasculitis where DSA still plays a major role in diagnosis. TAO is a non-necrotizing vasculitis affecting medium and small arteries of the upper and lower extremities and rarely the coronary, carotid and visceral arteries. Patients are generally younger than 45 years and heavy smokers. The hallmark angiographic findings of TAO are segmental occlusions of the arteries with distinctive collateral vessels, so-called corkscrew collaterals which are thought to be enlarged vasa vasorum (Fig. 33.7) [4, 54, 76].

Table 33.2 Angiographic classification of Takayasu's Arteritis

Type	Site of involvement
I	Branches of aortic arch
IIa	Ascending aorta, aortic arch and its branches
IIb	Ascending aorta, aortic arch and its branches and thoracic descending aorta
III	Thoracic descending aorta, abdominal aorta and/or renal arteries
IV	Abdominal aorta and/or renal arteries
V	Combination of Types IIb and IV

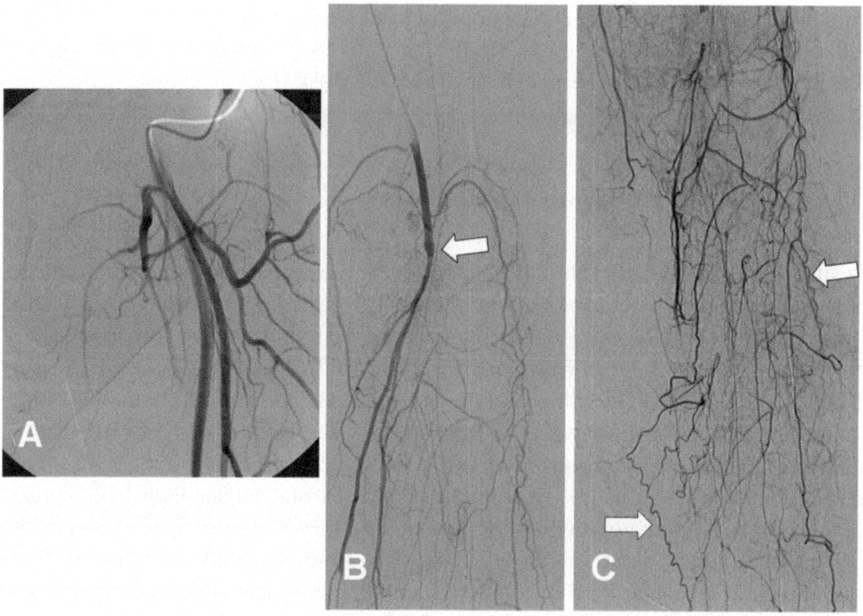

Fig. 33.7 Digital subtraction angiogram of a 41 year-old man with Buerger's disease. (**a**) Contrast media injection at the left common femoral artery reveals normal morphology (appropriate size and tapering) of the arteries of the proximal lower extremity. (**b**) Injection at the level of the above-knee popliteal artery shows occlusion of the popliteal artery with flow into collateral vessels (*arrow*). (**c**) Occlusion of all primary below knee runoff arteries with development of multiple "cork screw" collateral arteries (*arrows*)

These small vessels are best seen by DSA due to its high spatial resolution, but have also been observed on CTA and MRA.

Another disease where DSA is a useful diagnostic tool is Polyarteritis nodosa (PAN). PAN is a systemic necrotizing vasculitis affecting medium-sized arteries. Involved organs are the skin, peripheral nerves, GI tract, kidneys, heart and central nervous system. The most characteristic angiographic appearance for PAN is multiple saccular or fusiform microaneurysms measuring between 1 and 5 mm. They typically occur at the branching point of the affected artery [28]. Occlusions, irregular vessels and stenosis can also be seen. The majority of patients have more than one aneurysm and they commonly involve multiple organs [78]. The small aneurysms encountered in PAN are usually not well seen by CTA [28]. In case of vascular occlusion, renal or hepatic infarcts can be seen on CT or MRI. The hepatic, renal and superior mesenteric arteries are most commonly affected. DSA has been reported to have a sensitivity of 89 % and specificity of 90 % in PAN with 55 % PPV and 98 % NPV [31]. However, it is important to remember that PAN cannot be ruled out by negative angiography and conversely microneurysms can also occur in a number of structural vasculopathies.

The main role of DSA today is to guide endovascular procedures aimed at revascularization of branch vessels causing claudication, mesenteric ischemia, renal failure or refractory hypertension and claudication or to treat aneurysm with embolization of covered stents. The drawbacks of DSA include exposure to ionizing radiation, its invasiveness, cost and the need to use iodinated contrast agents. Cerebral angiography carries also a small risk of stroke [84].

33.6 Positron Emission Tomography

Positron emission tomography (PET) is a nuclear medicine technique that is able to evaluate the degree of uptake of an intravenously injected radiolabeled glucose analogue (fluorodeoxyglucose [FDG]) by activated cells in infections, malignancies, and inflammatory processes [56]. The first use of PET FDG for GCA and TA was in 1999 [13, 29]. PET FDG allows whole body visualization of glucose-consuming inflamed vessels non-invasively. In large vessels, the intensity of FDG uptake is usually classified on a semi-quantitative 4-point scale: none (grade 0), less than liver uptake (grade 1), similar to liver uptake (grade 2), and higher than liver uptake (grade 3) [48]. Uptake exceeding the hepatic one (grades 2 and 3) is necessary to consider the diagnosis of vasculitis [7, 48]. The limited spatial resolution of PET can be improved when PET is combined with CT. A recent study on large vessel vasculitis evaluated the diagnostic role of PET (Fig. 33.8) [24]. PET had an overall sensitivity of 73 %, a specificity of 84 %, a positive predictive value of 82 %, and a negative predictive value of 77 %. The addition of PET to the diagnostic algorithm increased the diagnostic accuracy from 54 to 71 %. On the other hand, the diagnostic accuracy of PET dropped dramatically after initiation of steroid and/or immunosuppressive therapy by nearly 50 % [24]. PET is considered the most sensitive test for the detection of early vessel wall inflammation. Anatomical localization is improved when performed as PET-CT or PET-MRI. PET is limited in the diagnosis

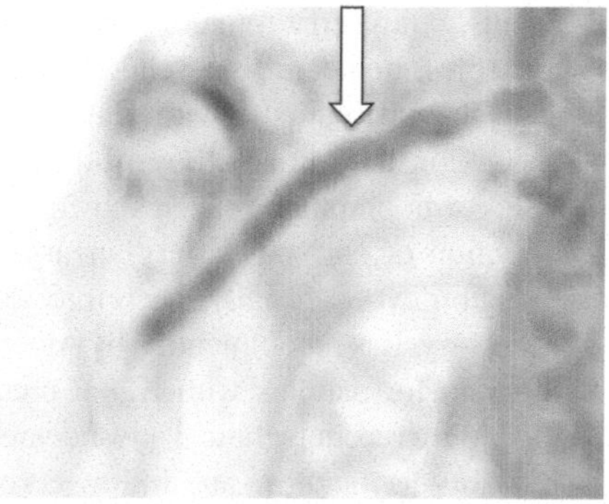

Fig. 33.8 Eighty year old man with Giant cell arteritis. *Arrow* shows increased FDG update in the right subclavian artery

of small vessel vasculitis (vessel diameter <4 mm). In addition, the temporal artery cannot be visualized by this technique due to the superficial location and the proximity to the glucose consuming brain. Other limitations include limited availability, high cost, lack of standardized PET interpretation, and radiation [56].

33.7 Role of Imaging in Other Rare Vasculitis

1. Goodpasture disease is a small vessel vasculitis characterized by the presence of circulating anti–glomerular basement membrane. It predominantly involves renal and pulmonary systems causing glomerulonephritis and pulmonary hemorrhage. Chest radiographs can demonstrate pulmonary involvement which occurs in 50–80 % of the cases. Typical are bilateral nodular or reticular opacities [58]. CT typically shows predominantly sub-pleural or diffuse nodules and/or masses with cavitations [45]. In the head, sinus mucosal edema and thickening, bony destruction, and erosions of surrounding structures can be demonstrated on MR and CT, and then indicate nasal and cranial sinus involvement [25].
2. Churg-Strauss syndrome (CSS) is an ANCA-positive vasculitis affecting small- to medium-sized vessels. It is characterized by early constitutional symptoms with fever, allergic rhinitis and asthma. At a later stage, systemic vasculitis affects various organs including the nervous system causing peripheral neuropathy, GI causing eosinophilic gastroenteritis and ischemic bowel, and the kidney causing segmental necrotizing glomerulonephritis. Eosinophilia in the blood and affected tissues is a hallmark finding. In CSS, Chest X-rays may reveal recurrent lower lobe pulmonary infiltrates and, less frequently, reticulo-nodular opacities, and multiple nodules [17]. Asthma-like findings with hyperinflation and bronchial wall thickening have also been demonstrated. High-resolution chest CT typically shows areas of consolidation or bilateral patchy ground-glass opacities with predominant sub-pleural and lower lobe distribution [17].
3. Kawasaki disease is a medium vessel vasculitis typically seen in young children. It primarily affects the cardiovascular system particularly the coronary arteries. At early stage, children often present with prolonged fever, conjunctivitis, anterior

uveitis, strawberry tongue, adenopathy (especially cervical), erythema of palms and soles, myocarditis or endocarditis. Later, coronary aneurysms develop. In a study of 100 patients with Kawasaki disease, 44 patients had coronary artery lesions on the initial echocardiogram (31 with ectasia, 13 with aneurysm) [5]. Echocardiography has been shown to have high sensitivity and specificity for the detection of coronary artery aneurysms (95 % and 99 % respectively) without radiation exposure [50]. Therefore, conventional coronary angiography is rarely necessary. It is recommended by American Heart Association to consider echocardiogram for patients with KD at presentation, at 2 weeks, and after 6–8 weeks [50]. With recent technical advancement, coronary CTA have become an excellent modality to diagnose and follow coronary artery aneurysms and their complication [36].

4. Behçet's disease is a chronic inflammatory disorder that can affect all vessel types. It is more commonly diagnosed in people of Mediterranean, East Asian, and Middle Eastern ethnicities. Common manifestations include recurrent oro-genital ulcerations, uveitis and retinal vasculitis. Vascular ultrasound can detect superficial and deep venous thrombosis which is the most common vascular manifestation in Behçet's disease [14, 22]. Arterial thrombosis, stenosis and aneurysm have also been observed with pulmonary artery aneurysm demonstrated by CT being the most fearful vascular complication [14, 22].
5. Adult primary central nervous system vasculitis (PCNSV) is a rare small and medium-vessel vasculitis restricted to brain and spinal cord. It is most commonly seen in young to middle-aged men. Headache is often a presenting symptoms, but transient ischemic attacks, cerebral infarction, paralysis, seizures, and visual field defects are among the other manifestations. Biopsy is the only definitive test, but angiography is also used for diagnosis. Characteristic findings are involvement of multiple cerebral arteries with alternating areas of smooth wall narrowing and dilatation (rarely aneurysms) or arterial occlusions in the absence of proximal vascular atherosclerosis [66]. DSA was shown to have a sensitivity of 40–90 % and specificity of 30 % for the diagnosis of PCNSV [67].

   MRI is the screening test of choice for PCNSV with 100 % sensitivity in detecting the secondary vasculitis effect on the brain parenchyma [67]. Non-specific MRI findings include multiple areas of ischemic infarct, meningeal thickening or abnormal white matter areas. The MR vascular findings of PCNSV must be supported by DSA as MRA is 20 % less sensitive than DSA for detailing vascular wall abnormalities in PCNSV [66]. If angiographic findings are negative in a symptomatic patient with high clinical suspicion, brain biopsy should be obtained to confirm diagnosis. This subtype of PCNSV carries a better prognosis [65]. PET has a limited role in PCVNS due to FDG consumption by the brain parenchyma and lack of spatial resolution of small neuro-vascular vessels. A new tracer known as [11C] PK11195, which selectively binds to the peripheral benzodiazepine binding site of activated macrophages has been investigated to detect vascular inflammation using PET for evaluation of patient with PCVNS [59].

6. Idiopathic inflammatory aortic aneurysms and periaortic retroperitoneal fibrosis are disorders characterized by the enhanced fibro-inflammatory thickening of adventitial tissues caused by infiltration of immuno-inflammatory cells and deposition of thickened fibrous tissues [63]. Patients often have constitutional symptoms along with chest, back, or abdominal pain. When associated with retroperitoneal fibrosis it can cause ureteral obstruction and subsequent renal failure.

   Idiopathic inflammatory aortic aneurysms appear as fusiform dilatation with thick aortic wall and a significant amount of perianeurysmal soft tissue (hypoattenuating mass) sparing the posterior wall (Fig. 33.1). Contrast enhanced CT has a sensitivity and specificity of 83 and 100 % respectively when making the diagnosis [33]. Pre-operative evaluation with PET and MR imaging is important to assess the extent of inflammation and adhesions of the aneurysm with the surrounding tissue to avoid injury of the surrounding structure during surgery [20]. In chronic peri-aortitis, CTA will reveal irregular aortic wall thickening with dense adhesions and soft tissue mass surrounding the aorta causing stenosis [20]. The most common CTA findings of periaortic retroperitoneal fibrosis is dense homogeneous soft tissue plaque surrounding the abdominal aorta without displacement of adjacent structures [19].
7. Lastly, radiation induced arteritis is a rare complication of high-dose radiation therapy. It may manifest 5 years or more after radiation exposure. Affected vessel confined to the radiation field may develop stenosis, thrombosis, pseudoaneurysm, rupture or calcification (Fig. 33.9) [18].

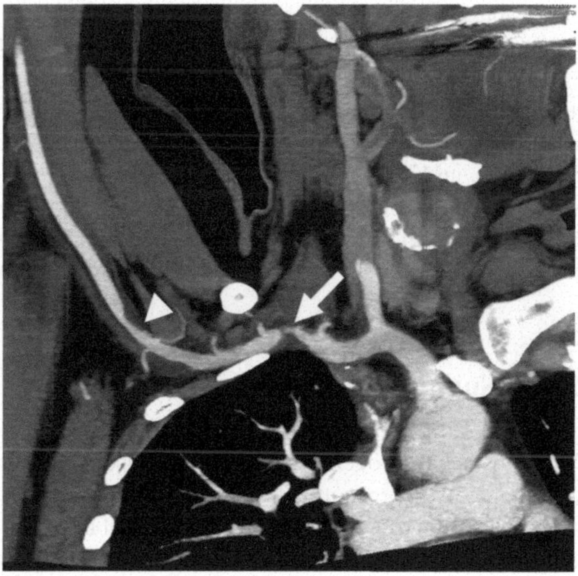

Fig. 33.9 CT Angiogram of the right upper extremity in a 46 year-old female after radiation therapy for breast cancer. Stenosis of the right subclavian (*arrow*) as well as smooth narrowing of the axillary artery (*arrowhead*) typical of radiation arteritis. Notice the lack of atherosclerotic calcification in the vessels

Table 33.3 Mimics of vasculitis on imaging

1.	Connective tissue disorders such as Marfan's syndrome, Ehlers-Danlos syndrome, Loeys-Dietz syndrome
2.	Atherosclerosis
3.	Anti-phospholipid antibody syndrome
4.	Cholesterol embolization syndrome
5.	Vasospastic disease such as Raynaud's disease
6.	Moyamoya
7.	Fibromuscular dysplasia
8.	Standing waves

33.8 Vasculitis Mimics

Accurate knowledge of imaging studies is vital to identify mimics of vasculitis (Table 33.3). Accurate diagnosis often requires imaging interpretation along with clinical findings and laboratory results. CT findings will point to vasculitis if the affected vessels wall appears symmetric, long with smooth stenosis and marked increase in vessel wall thickening which is also concentric in nature. On the other hand, the vessel wall appearance in atherosclerosis is typically asymmetric, with patchy lesions, associated with calcifications and if wall thickening is present, it appears eccentric rather than concentric [56]. In addition, an increased FDG uptake by PET scan favors vasculitis over atherosclerosis, although this has to be considered in the clinical context as often atherosclerosis will have underlying inflammation, thereby resulting in FDG uptake.

Fibromuscular dysplasia can also be confused with vasculitis. The most common form of FMD i.e. multifocal type will have multiple narrowings along a segment of artery appearing like an irregular "string of beads" with no evidence of vessel wall inflammation in the form of thickening (Fig. 33.10) [75]. Standing waves, which is concentric narrowing of the vessels likely from vasospasm, is typically seen after angiogram and can be differentiated by its lack of other vasculitic features and lack of reproducibility [51, 74]. Lastly, connective tissue disorders such as Ehlers-Danlos Syndrome, and Marfan's syndrome should be properly differentiated. They often will have dissections, aneurysms and tortuosity in the arteries similar to vasculitis; however they will lack vessel wall thickening and inflammation. Often these patients have very characteristic clinical findings which should be taken into account [76].

33.9 Summary

Vasculitis is a heterogeneous group of diseases characterized by non-specific clinical symptoms, often non-contributory physical exam and frequently non-specific shared laboratory findings, all of which contribute to a complex diagnostic process. A number of imaging techniques are now available to the clinician, and selection of

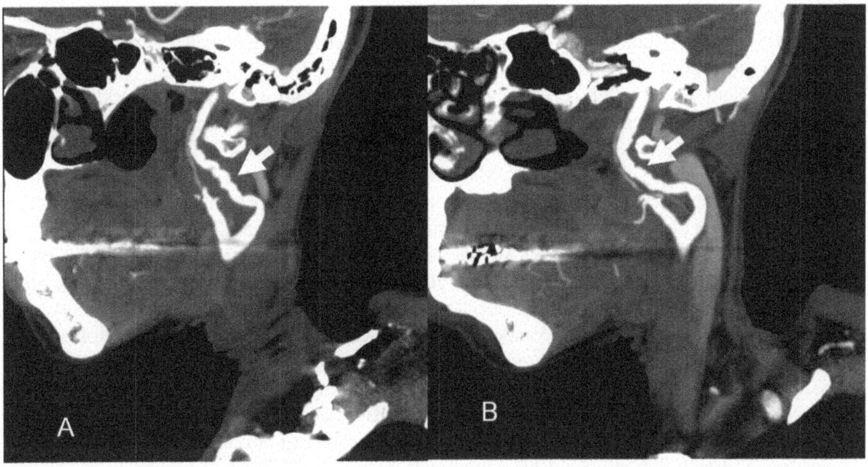

Fig. 33.10 Fort-five-year-old female with multifocal type of right (image **a**) and left (image **b**) internal carotid artery fibromuscular dysplasia (see *arrows*)

the most appropriate imaging study is paramount to aid in proper and early diagnosis. Imaging, particularly the cross-sectional imaging techniques such as CTA and MRI, also play a role in monitoring disease extent and activity and to assess for complications of vasculitis. Pending further studies, imaging can be potentially utilized to monitor therapy and predict treatment outcomes.

References

1. Alberts MS, Mosen DM (2007) Diagnosing temporal arteritis: duplex vs. biopsy. QJM 100:785–789
2. Alkadhi H, Schindera ST (2011) State of the art low-dose CT angiography of the body. Eur J Radiol 80:36–40
3. Anderson JL, Halperin JL, Albert NM, Bozkurt B, Brindis RG, Curtis LH, Demets D, Guyton RA, Hochman JS, Kovacs RJ, Ohman EM, Pressler SJ, Sellke FW, Shen WK (2013) Management of patients with peripheral artery disease (compilation of 2005 and 2011 ACCF/AHA guideline recommendations): a report of the American College of Cardiology Foundation/American Heart Association Task Force on Practice Guidelines. Circulation 127:1425–1443
4. Aqel MB, Olin JW (1997) Thromboangiitis obliterans (Buerger's disease). Vasc Med 2:61–66
5. Baer AZ, Rubin LG, Shapiro CA, Sood SK, Rajan S, Shapir Y, Romano A, Bierman FZ (2006) Prevalence of coronary artery lesions on the initial echocardiogram in Kawasaki syndrome. Arch Pediatr Adolesc Med 160:686–690
6. Ball EL, Walsh SR, Tang TY, Gohil R, Clarke JM (2010) Role of ultrasonography in the diagnosis of temporal arteritis. Br J Surg 97:1765–1771
7. Belhocine T, Blockmans D, Hustinx R, Vandevivere J, Mortelmans L (2003) Imaging of large vessel vasculitis with (18)FDG PET: illusion or reality? A critical review of the literature data. Eur J Nucl Med Mol Imaging 30:1305–1313

8. Bley TA, Markl M, Schelp M, Uhl M, Frydrychowicz A, Vaith P, Peter HH, Langer M, Warnatz K (2008) Mural inflammatory hyperenhancement in MRI of giant cell (temporal) arteritis resolves under corticosteroid treatment. Rheumatology (Oxford) 47:65–67
9. Bley TA, Ness T, Warnatz K, Frydrychowicz A, Uhl M, Hennig J, Langer M, Markl M (2007) Influence of corticosteroid treatment on MRI findings in giant cell arteritis. Clin Rheumatol 26:1541–1543
10. Bley TA, Reinhard M, Hauenstein C, Markl M, Warnatz K, Hetzel A, Uhl M, Vaith P, Langer M (2008) Comparison of duplex sonography and high-resolution magnetic resonance imaging in the diagnosis of giant cell (temporal) arteritis. Arthritis Rheum 58:2574–2578
11. Bley TA, Uhl M, Carew J, Markl M, Schmidt D, Peter HH, Langer M, Wieben O (2007) Diagnostic value of high-resolution MR imaging in giant cell arteritis. AJNR Am J Neuroradiol 28:1722–1727
12. Blockmans D, DE Ceuninck L, Vanderschueren S, Knockaert D, Mortelmans L, Bobbaers H (2006) Repetitive 18F-fluorodeoxyglucose positron emission tomography in giant cell arteritis: a prospective study of 35 patients. Arthritis Rheum 55:131–137
13. Blockmans D, Maes A, Stroobants S, Nuyts J, Bormans G, Knockaert D, Bobbaers H, Mortelmans L (1999) New arguments for a vasculitic nature of polymyalgia rheumatica using positron emission tomography. Rheumatology (Oxford) 38:444–447
14. Calamia KT, Schirmer M, Melikoglu M (2011) Major vessel involvement in Behcet's disease: an update. Curr Opin Rheumatol 23:24–31
15. Choe YH, Kim DK, Koh EM, Do YS, Lee WR (1999) Takayasu arteritis: diagnosis with MR imaging and MR angiography in acute and chronic active stages. J Magn Reson Imaging 10:751–757
16. Choe YH, Lee WR (1998) Magnetic resonance imaging diagnosis of Takayasu arteritis. Int J Cardiol, 66 Suppl 1, S175–S179; discussion S181
17. Choi YH, Im JG, Han BK, Kim JH, Lee KY, Myoung NH (2000) Thoracic manifestation of Churg-Strauss syndrome: radiologic and clinical findings. Chest 117:117–124
18. Chun YS, Cherry KJ (2003) Radiation arteritis. J Am Coll Surg 196:482
19. Cronin CG, Lohan DG, Blake MA, Roche C, Mccarthy P, Murphy JM (2008) Retroperitoneal fibrosis: a review of clinical features and imaging findings. AJR Am J Roentgenol 191:423–431
20. Cullenward MJ, Scanlan KA, Pozniak MA, Acher CA (1986) Inflammatory aortic aneurysm (periaortic fibrosis): radiologic imaging. Radiology 159:75–82
21. Delle SEDIE A, Riente L, Filippucci E, Iagnocco A, Meenagh G, Epis O, Grassi W, Valesini G, Montecucco C, Bombardieri S (2008) Ultrasound imaging for the rheumatologist. XV. Ultrasound imaging in vasculitis. Clin Exp Rheumato 26:391–394
22. Duzgun N, Ates A, Aydintug OT, Demir O, Olmez U (2006) Characteristics of vascular involvement in Behcet's disease. Scand J Rheumatol 35:65–68
23. Espigol-Frigole G, Prieto-Gonzalez S, Alba MA, Tavera-Bahillo I, Garcia-Martinez A, Gilabert R, Hernandez-Rodriguez J, Cid MC (2015) Advances in the diagnosis of large vessel vasculitis. Rheum Dis Clin North Am 41:125–140, ix
24. Fuchs M, Briel M, Daikeler T, Walker UA, Rasch H, Berg S, Ng QK, Raatz H, Jayne D, Kotter I, Blockmans D, Cid MC, Prieto-Gonzalez S, Lamprecht P, Salvarani C, Karageorgaki Z, Watts R, Luqmani R, Muller-Brand J, Tyndall A, Walter MA (2012) The impact of 18F-FDG PET on the management of patients with suspected large vessel vasculitis. Eur J Nucl Med Mol Imaging 39:344–353
25. Ghinoi A, Pipitone N, Cavazza A, Boiardi L, Salvarani C (2008) Wegener granulomatosis with spleen infarction: case report and review of the literature. Semin Arthritis Rheum 37:328–333
26. Gornik HL, Creager MA (2008) Aortitis. Circulation 117:3039–3051
27. Gotway MB, Araoz PA, Macedo TA, Stanson AW, Higgins CB, Ring EJ, Dawn SK, Webb WR, Leung JW, Reddy GP (2005) Imaging findings in Takayasu's arteritis. AJR Am J Roentgenol 184:1945–1950
28. Hagspiel KD, Angle JF, Spinosa DJ, Matsumoto AH (1999) Diagnosis please. Case 13: polyarteritis nodosa–systemic necrotizing vasculitis with involvement of hepatic and superior mesenteric arteries. Radiology 212:359–364

29. Hara M, Goodman PC, Leder RA (1999) FDG-PET finding in early-phase Takayasu arteritis. J Comput Assist Tomogr 23:16–18
30. Hartlage GR, Palios J, Barron BJ, Stillman AE, Bossone E, Clements SD, Lerakis S (2014) Multimodality imaging of aortitis. JACC Cardiovasc Imaging 7:605–619
31. Hekali P, Kajander H, Pajari R, Stenman S, Somer T (1991) Diagnostic significance of angiographically observed visceral aneurysms with regard to polyarteritis nodosa. Acta Radiol 32:143–148
32. Hiratzka LF, Bakris GL, Beckman JA, Bersin RM, Carr VF, Casey DE Jr, Eagle KA, Hermann LK, Isselbacher EM, Kazerooni EA, Kouchoukos NT, Lytle BW,Milewicz DM, Reich DL, Sen S, Shinn JA, Svensson LG, Williams DM. 2010. 2010 ACCF/AHA/AATS/ACR/ASA/SCA/SCAI/SIR/STS/SVM guidelines for the diagnosis and management of patients with Thoracic Aortic Disease: a report of the American College of Cardiology Foundation/American Heart Association Task Force on Practice Guidelines, American Association for Thoracic Surgery, American College of Radiology, American Stroke Association, Society of Cardiovascular Anesthesiologists, Society for Cardiovascular Angiography and Interventions, Society of Interventional Radiology, Society of Thoracic Surgeons, and Society for Vascular Medicine. Circulation 121:e266–e369
33. Iino M, Kuribayashi S, Imakita S, Takamiya M, Matsuo H, Ookita Y, Ando M, Ueda H (2002) Sensitivity and specificity of CT in the diagnosis of inflammatory abdominal aortic aneurysms. J Comput Assist Tomogr 26:1006–1012
34. Jang JJ, Gorevic PD, Olin JW (2007) Images in vascular medicine giant cell arteritis presenting with acute myocardial infarction. Vasc Med 12:379
35. Jayne D (2009) The diagnosis of vasculitis. Best Pract Res Clin Rheumatol 23:445–453
36. Johnson PT, Fishman EK (2010) CT angiography of coronary artery aneurysms: detection, definition, causes, and treatment. AJR Am J Roentgenol 195:928–934
37. Kang EJ, Kim SM, Choe YH, Lee GY, Lee KN, Kim DK (2014) Takayasu arteritis: assessment of coronary arterial abnormalities with 128-section dual-source CT angiography of the coronary arteries and aorta. Radiology 270:74–81
38. Karahaliou M, Vaiopoulos G, Papaspyrou S, Kanakis MA, Revenas K, Sfikakis PP (2006) Colour duplex sonography of temporal arteries before decision for biopsy: a prospective study in 55 patients with suspected giant cell arteritis. Arthritis Res Ther 8:R116
39. Karassa FB, Matsagas MI, Schmidt WA, Ioannidis JP (2005) Meta-analysis: test performance of ultrasonography for giant-cell arteritis. Ann Intern Med 142:359–369
40. Kermani TA, Warrington KJ, Crowson CS, Ytterberg SR, Hunder GG, Gabriel SE, Matteson EL (2013) Large-vessel involvement in giant cell arteritis: a population-based cohort study of the incidence-trends and prognosis. Ann Rheum Dis 72:1989–1994
41. Klink T, Geiger J, Both M, Ness T, Heinzelmann S, Reinhard M, Holl-Ulrich K, Duwendag D, Vaith P, Bley TA (2014) Giant cell arteritis: diagnostic accuracy of MR imaging of superficial cranial arteries in initial diagnosis-results from a multicenter trial. Radiology 273:844–852
42. Lesar CJ, Meier GH, Demasi RJ, Sood J, Nelms CR, Carter KA, Gayle RG, Parent FN, Marcinczyk MJ (2002) The utility of color duplex ultrasonography in the diagnosis of temporal arteritis. J Vasc Surg 36:1154–1160
43. Lie JT (1987) The classification and diagnosis of vasculitis in large and medium-sized blood vessels. Pathol Annu 22(Pt 1):125–162
44. Lie JT (1990) Illustrated histopathologic classification criteria for selected vasculitis syndromes. American College of Rheumatology Subcommittee on Classification of Vasculitis. Arthritis Rheum 33:1074–1087
45. Lohrmann C, Uhl M, Kotter E, Burger D, Ghanem N, Langer M (2005) Pulmonary manifestations of wegener granulomatosis: CT findings in 57 patients and a review of the literature. Eur J Radiol 53:471–477
46. Luqmani RA, Pathare S, Kwok-Fai TL (2005) How to diagnose and treat secondary forms of vasculitis. Best Pract Res Clin Rheumatol 19:321–336
47. Marie I, Proux A, Duhaut P, Primard E, Lahaxe L, Girszyn N, Louvel JP, Levesque H (2009) Long-term follow-up of aortic involvement in giant cell arteritis: a series of 48 patients. Medicine (Baltimore) 88:182–192

48. Meller J, Strutz F, Siefker U, Scheel A, Sahlmann CO, Lehmann K, Conrad M, Vosshenrich R (2003) Early diagnosis and follow-up of aortitis with [(18)F]FDG PET and MRI. Eur J Nucl Med Mol Imaging 30:730–736
49. Moriwaki R, Noda M, Yajima M, Sharma BK, Numano F (1997) Clinical manifestations of Takayasu arteritis in India and Japan--new classification of angiographic findings. Angiology 48:369–379
50. Newburger JW, Takahashi M, Gerber MA, GEWITZ MH, Tani LY, Burns JC, Shulman ST, Bolger AF, Ferrieri P, Baltimore RS, Wilson WR, Baddour LM, Levison ME, Pallasch TJ, Falace DA, Taubert KA (2004) Diagnosis, treatment, and long-term management of Kawasaki disease: a statement for health professionals from the Committee on Rheumatic Fever, Endocarditis and Kawasaki Disease, Council on Cardiovascular Disease in the Young, American Heart Association. Circulation 110:2747–2771
51. Norton PT, Hagspiel KD (2005) Stationary arterial waves in magnetic resonance angiography. J Vasc Interv Radiol 16:423–424
52. Novaro GM, Erim T, Pinski SL (2006) Spondyloarthropathy-associated aortitis and massive thickening of the aortic-mitral curtain: diagnosis by echocardiography. Cardiology 106:98–101
53. Nuenninghoff DM, Hunder GG, Christianson TJ, Mcclelland RL, Matteson EL (2003) Incidence and predictors of large-artery complication (aortic aneurysm, aortic dissection, and/or large-artery stenosis) in patients with giant cell arteritis: a population-based study over 50 years. Arthritis Rheum 48:3522–3531
54. Olin JW, Young JR, Graor RA, Ruschhaupt WF, Bartholomew JR (1990) The changing clinical spectrum of thromboangiitis obliterans (Buerger's disease). Circulation 82:IV3–IV8
55. Park JH, Han MC, Kim SH, Oh BH, Park YB, Seo JD (1989) Takayasu arteritis: angiographic findings and results of angioplasty. AJR Am J Roentgenol 153:1069–1074
56. Pipitone N, Versari A, Salvarani C (2008) Role of imaging studies in the diagnosis and follow-up of large-vessel vasculitis: an update. Rheumatology (Oxford) 47:403–408
57. Prieto-Gonzalez S, Arguis P, Garcia-Martinez A, Espigol-Frigole G, Tavera-Bahillo I, Butjosa M, Sanchez M, Hernandez-Rodriguez J, Grau JM, Cid MC (2012) Large vessel involvement in biopsy-proven giant cell arteritis: prospective study in 40 newly diagnosed patients using CT angiography. Ann Rheum Dis 71:1170–1176
58. Puechal X (2007) Antineutrophil cytoplasmic antibody-associated vasculitides. Joint Bone Spine 74:427–435
59. Pugliese F, Gaemperli O, Kinderlerer AR, Lamare F, Shalhoub J, Davies AH, Rimoldi OE, Mason JC, Camici PG (2010) Imaging of vascular inflammation with [11C]-PK11195 and positron emission tomography/computed tomography angiography. J Am Coll Cardiol 56:653–661
60. Restrepo CS, Ocazionez D, Suri R, Vargas D (2011) Aortitis: imaging spectrum of the infectious and inflammatory conditions of the aorta. Radiographics 31:435–451
61. Robson JC, Kiran A, Maskell J, Hutchings A, Arden N, Dasgupta B, Hamilton W, Emin A, Culliford D, Luqmani RA (2015) The relative risk of aortic aneurysm in patients with giant cell arteritis compared with the general population of the UK. Ann Rheum Dis 74:129–135
62. Roldan-Valadez E, Hernandez-Martinez P, Osorio-Peralta S, Elizalde-Acosta I, Espinoza-Cruz V, Casian-Castellanos G (2003) Imaging diagnosis of subclavian steal syndrome secondary to Takayasu arteritis affecting a left-side subclavian artery. Arch Med Res 34:433–438
63. Sakamoto A, Nagai R, Saito K, Imai Y, Takahashi M, Hosoya Y, Takeda N, Hirano K, Koike K, Enomoto Y, Kume H, Homma Y, Maeda D, Yamada H, Fukayama M, Hirata Y, Ishizaka N (2012) Idiopathic retroperitoneal fibrosis, inflammatory aortic aneurysm, and inflammatory pericarditis--retrospective analysis of 11 case histories. J Cardiol 59:139–146
64. Salvarani C (2003) Large vessel vasculitis. Clin Exp Rheumatol 21:S133–S134
65. Salvarani C, Brown RD Jr, Calamia KT, Christianson TJ, Huston J 3rd, Meschia JF, Giannini C, Miller DV, Hunder GG (2008) Angiography-negative primary central nervous system vasculitis: a syndrome involving small cerebral vessels. Medicine (Baltimore) 87:264–271

66. Salvarani C, Brown RD Jr, Hunder GG (2012) Adult primary central nervous system vasculitis. Lancet 380:767–777
67. Salvarani C, Brown RD Jr, Hunder GG (2012) Adult primary central nervous system vasculitis: an update. Curr Opin Rheumatol 24:46–52
68. Schmidt WA (2007) Technology insight: the role of color and power Doppler ultrasonography in rheumatology. Nat Clin Pract Rheumatol 3:35–42; quiz 59
69. Schmidt WA (2014) Ultrasound in vasculitis. Clin Exp Rheumatol 32:S71–S77
70. Schmidt WA, Blockmans D (2005) Use of ultrasonography and positron emission tomography in the diagnosis and assessment of large-vessel vasculitis. Curr Opin Rheumatol 17:9–15
71. Schmidt WA, Kraft HE, Vorpahl K, Volker L, Gromnica-Ihle EJ (1997) Color duplex ultrasonography in the diagnosis of temporal arteritis. N Engl J Med 337:1336–1342
72. Schmidt WA, Seifert A, Gromnica-Ihle E, Krause A, Natusch A (2008) Ultrasound of proximal upper extremity arteries to increase the diagnostic yield in large-vessel giant cell arteritis. Rheumatology (Oxford) 47:96–101
73. Seo P, Stone JH (2004) Large-vessel vasculitis. Arthritis Rheum 51:128–139
74. Sharma AM, Gornik HL (2012) Standing arterial waves is NOT fibromuscular dysplasia. Circ Cardiovasc Interv 5:e9–e11
75. Sharma AM, Kline B (2014) The United States registry for fibromuscular dysplasia: new findings and breaking myths. Tech Vasc Interv Radiol 17:258–263
76. Sharma AM, Norton PT, Zhu D (2014) Conditions presenting with symptoms of peripheral arterial disease. Semin Intervent Radiol 31:281–291
77. Sharma AM, Singh S, Lewis JE (2014) Diagnostic approach in patients with suspected vasculitis. Tech Vasc Interv Radiol 17:226–233
78. Stanson AW (1990) Roentgenographic findings in major vasculitic syndromes. Rheum Dis Clin North Am 16:293–308
79. Stepansky F, Hecht EM, Rivera R, Hirsh LE, Taouli B, Kaur M, Lee VS (2008) Dynamic MR angiography of upper extremity vascular disease: pictorial review. Radiographics 28, e28
80. Tato F, Hoffmann U (2008) Giant cell arteritis: a systemic vascular disease. Vasc Med 13:127–140
81. Tso E, Flamm SD, White RD, Schvartzman PR, Mascha E, Hoffman GS (2002) Takayasu arteritis: utility and limitations of magnetic resonance imaging in diagnosis and treatment. Arthritis Rheum 46:1634–1642
82. Villa I, Agudo Bilbao M, Martinez-Taboada VM (2011) Advances in the diagnosis of large vessel vasculitis: identification of biomarkers and imaging studies. Reumatol Clin 7(Suppl 3):S22–S27
83. Weyand CM, Goronzy JJ (2003) Medium- and large-vessel vasculitis. N Engl J Med 349:160–169
84. Willinsky RA, Taylor SM, Terbrugge K, Farb RI, Tomlinson G, Montanera W (2003) Neurologic complications of cerebral angiography: prospective analysis of 2,899 procedures and review of the literature. Radiology 227:522–528
85. Yamada I, Nakagawa T, Himeno Y, Numano F, Shibuya H (1998) Takayasu arteritis: evaluation of the thoracic aorta with CT angiography. Radiology 209:103–109
86. Yoshida S, Akiba H, Tamakawa M, Yama N, Takeda M, Hareyama M, Nakata T, Shimamoto K (2001) The spectrum of findings in supra-aortic Takayasu's arteritis as seen on spiral CT angiography and digital subtraction angiography. Cardiovasc Intervent Radiol 24:117–121

Chapter 34
Therapeutic Use of Biologic Agents in Systemic Vasculitides

John Anthonypillai and Julian L. Ambrus Jr.

Abstract Treatment of aggressive small, medium and large vessel vasculitides has relied on various chemotherapeutic agents, such as cyclophosphamide and methotrexate, that are generally effective but carry various risks for side effects. In an effort to develop safer forms of therapies for these disorders, various biological agents blocking particular lymphoid subpopulations and cytokines have been evaluated. This manuscript reviews the results of these investigations. As of now, no biological agents are yet the gold standard for treatment of any form of vasculitis, although they likely have roles in the treatment of particular patients.

34.1 Introduction

Vasculitides are diseases in which immunological injury to blood vessels leads to different degrees and types of organ dysfunction. Vasculitides are divided based on the size of the blood vessels involved and nature of the disease. In the past, glucocorticoids, cyclophosphamide, azathioprine, and methotrexate have been used for induction and maintenance therapy for medium and large vessel vasculitides. Due to side effects from these therapies, biologic agents have been developed and studied as potential alternatives. Biologic agents have been studied in the treatment of Takayasu's Arteritis (TAK), cryoglobulinemic vasculitis, Behçet's disease, and ANCA associated vasculitis (AAV). Current targets for biologics are B cells, T cells, and cytokines.

J. Anthonypillai • J.L. Ambrus Jr. (✉)
Division of Allergy, Immunology and Rheumatology, SUNY at Buffalo School of Medicine, 100 High Street, Buffalo, NY 14203, USA
e-mail: johnanth@buffalo.edu; jambrus@buffalo.edu

F. Dammacco et al. (eds.), *Systemic Vasculitides: Current Status and Perspectives*, DOI 10.1007/978-3-319-40136-2_34

34.2 B Cell Targets

B cells are involved in the pathophysiology of a variety of small and medium vessel vasculitis [1, 2]. B cell activity has been associated with disease activity in polyangitis with granulomatosis (GPA). B cells are also found in the inflammatory infiltrates in ANCA-associated vasculitides (AAV) [2–4]. Regulatory B cells which play a role in immunosuppression were found to be decreased in active AAV [5]. Memory B cells are central to the pathophysiology of both GPA and microscopic polyangiitis [4]. Cyclophosphamide which is used to treat polyangitis with granulomatosis (GPA/ Wegener's granulomatosis) works by eliminating memory B cells [6, 7].

In Takayasu's Arteritis (TAK), B cells were found in the aorta and anti-endothelial cell antibodies were found in patient sera [8–10]. A recent study showed that patients with active TAK have higher amounts of circulating plasmablasts and memory B cells compared to patients with inactive TAK or without TAK [8].

In regards to vasculitides, Rituximab and Belimumab are the main therapies being studied that target B cells. Rituximab binds CD20 and stimulates FcγRIIB, while Belimumab blocks the BAFF/BAFFR interaction.

34.2.1 *Rituximab*

Rituximab is an anti-CD20 monoclonal antibody. CD20 is a surface antigen expressed on both mature and immature B cells. However, it is not expressed on lymphoid progenitors in the bone marrow that allows for B cells to repopulate following Rituximab therapy [11–14].

Several studies were done to evaluate induction of remission in new and relapsing patients with AAV. These studies are: Rituximab in ANCA-associated Vasculitis (RAVE), Rituximab versus cyclophosphamide in ANCA-associated Renal Vasculitis (RITUXVAS), and Maintenance of Remission using Rituximab in Systemic ANCA-associated Vasculitis (MAINRITSAN).

The RAVE study was a multi-centered, randomized, double blind trial that compared Rituximab and glucocorticoid therapy vs cyclophosphamide and glucocorticoid therapy for remission induction for patients with AAV. Rituximab was dosed at 375 mg/m^2 of body surface weekly infusions for 4 weeks compared with oral cyclophosphamide dosed at 2 mg per Kg of body weight per day. This study concluded that Rituximab was not inferior to daily cyclophosphamide for induction of remission in severe AAV and may be superior for relapses. There was also no significant difference in adverse reactions between the two regimens [15].

The RAVE trial follow- up showed that a course of Rituximab was not inferior compared to cyclophosphamide for induction and remission maintenance over the course of 18 months. Relapses were observed after 6–12 months after a single Rituximab course and was correlated with B cell numbers returning to their baseline. A fixed interval treatment at every 6 months of 1 g of Rituximab for a total of

2 years was associated with a reduction in relapse and longer periods of remission. After the 2-year period ends, cessation of treatment may be considered [16–18]. Other studies have also supported this finding that relapse risk decreases with repeat dosing of Rituximab based on time-based treatment, B cells returning to normal, or an increase in serum ANCA titers [16, 19, 20].

The RITUXVAS trial was an open-label, two-group, parallel design, randomized trial. Rituximab was dosed at 375 mg/m^2 of body surface area per week for 4 weeks. The Rituximab group also received standard glucocorticoid therapy along with cyclophosphamide pulses. This was compared to the control groups that was given standard glucocorticoid therapy with IV cyclophosphamide for 3–6 months followed by azathioprine. Both therapies showed similar results in terms of maintenance of clinical remission. It was concluded that the Rituximab regimen was not associated with a reduction of adverse events and was not superior to IV cyclophosphamide for severe AAV [21]. A 2 year follow-up study for this trial was completed which showed that a rituximab-based induction regimen without maintenance therapy and a cyclophosphamide induction regimen with azathioprine maintenance therapy had similar outcomes in adverse effects and complications of AAV. In the Rituximab groups, relapse was correlated with increases in B cells and ANCA titters. This study showed Rituximab based regimens offer an alternative to standard treatments [22].

Maintenance of Remission using Rituximab in Systemic ANCA-associated Vasculitis (MAINRITSAN) was funded by the French Ministry of Health. It compared Rituximab to azathioprine for maintenance therapy. This was a randomized controlled prospective trial that compared a maintenance therapy of 500 mg of rituximab twice during the first month and then every 6 months till 18 months to daily azathioprine for 22 months. It was concluded that rituximab reduced the frequency of AAV relapses compared to azathioprine without significant differences in severe adverse events in both treatment groups. Thus, concluding that rituximab is superior for maintenance therapy, which also has been shown in other studies [23, 24]. RITAZAREM is an ongoing trial aiming to evaluate rituximab as maintenance therapy (ClinicalTrials.gov Identifier NCT01697267).

In addition to maintenance therapy, Rituximab may be used as an induction therapy for newly diagnosed severe GPA or MPA as recommended by the American College of Rheumatology where cyclophosphamide is indicated but not preferable in certain patients such as young women at risk of ovarian failure. Rituximab may also be used to treat GPA or MPA that is refractory to other immunosuppressant medications [25–27]. It has also been shown to be effective for treating AAV in patients with severe renal disease [28].

It is important to note for GPA, daily cyclophosphamide therapy given for 1 year post remission is the only treatment known to eliminate memory B cells involved in GPA pathology. Rituximab kills mature B cells but not memory cells. Future safer alternatives for GPA therapy would need to target memory B cells [18]. Furthermore, some of the trials in GPA compared rituximab to monthly IV cyclophosphamide, which was shown previously to be inferior to daily cyclophosphamide, given in doses to keep the total WBC between 3 and 4000/mm^3, in inducing and maintaining remission in GPA [29]. Daily cyclophosphamide is still the gold standard for the management of GPA and other AAV.

In regards to Hepatitis C virus (HCV)-associated cryoglobulinemic vasculitis (CV), which is a systemic vasculitis triggered by HCV, several studies have evaluated the use of rituximab to induce remission in CV. Rituximab is shown to be well tolerated and effective in treating patients with CV. It was shown to work on patients who failed antiviral treatment and can be used as mono-therapy for severe CV [30, 31]. In addition, it appears useful for long term management of CV when given during relapses [32]. HCV-associated mixed cryoglobulinemia (MC) is another systemic vasculitis that is also thought to be triggered by HCV. Treatment is usually with antivirals and Peg-interferon (IFN)/ribavirin combination, which has been shown to be effective treatment for this condition. However, Peg-interferon (IFN)/ribavirin with rituximab was shown to be superior to Peg-interferon (IFN)/ribavirin alone [33]. Several cases, however, have been reported that patients being treated for HCV associated MC developed severe systemic reactions such as serum sickness and severe flare of MC vasculitis within days of a rituximab infusion. This is thought to be due to rituximab forming a complex with IgM kappa mixed cryoglobulin. Thus, it is important to administer rituximab with caution [34].

Several case reports have shown Takayasu's arteritis (TAK) to be responsive to rituximab when prior therapy with multiple immunosuppressive agents have failed [8, 35–37]. A properly designed study needing more patients, longer follow-up, and treatment response details is needed to evaluate rituximab as a treatment for TAK. At the same time, TAK represents many different diseases that are variable in different parts of the world and with different ethnic groups [38–40]. Furthermore, since vascular damage requiring surgical repair does not improve with resolution of inflammation in TAK, it would be hard to evaluate the response to therapy [18, 41].

For Churg-Strauss, also known as eosinophilic granulomatosis with polyangiitis (EGPA), rituximab is not currently recommended as a treatment for either induction or maintenance therapy due to lack of clinical trial data. Several cases show positive responses to rituximab in refractory EGPA [25, 42, 43]. There is also some evidence that rituximab may be effective in treating Behçet's disease refractory to traditional treatments [44–46] and urticarial vasculitis [47].

The B cell depletion induced by Rituximab is generally well tolerated. Prolonged hypogammaglobulinemia may result in some patients who have B cell depletion longer than 6 months [18]. Of concern regarding the use of rituximab in vasculitis is the fact that rituximab does not eliminate memory B cells, which are the driving force in AAV as well as various other forms of vasculitis [48, 49]. Since rituximab is an immunosuppressive agent, there is an increased risk of opportunistic infections. Progressive multifocal leukoencephalopathy has been reported in patients treated with rituximab [50]. There are also risks of infusion-related adverse events such as rashes, myalgia, fevers whose risks can be reduced with administration of paracetamol, anti-histamines, and corticosteroid pre-medications. Also to consider allergic and anaphylactic reactions to the medication [25].

34.2.2 Belimumab

Belimumab is an anti-B cell activating factor (BAFF) monoclonal antibody. BAFF is a survival signal for B cells and is elevated in GPA [51]. BAFF is present on plasma cells and on other B cells at various stages of development. In vivo blockade of BAFF resulted in loss on naïve and transitional B cells in clinical trials [52]. Belimumab is approved as a treatment for Systemic Lupus Erythematous (SLE) and studies have been done evaluating it as a treatment for Sjögren's syndrome [53, 54]. The safety profile of Belimumab is very good. A study done accessing long-term safety profile showed that the most common adverse effects were arthralgia, upper respiratory tract infection, headache, fatigue, and nausea and serious infusion reactions were rare [55, 56]. There is an ongoing study, Belimumab in Remission of Vasculitis (BREVAS) (ClinicalTrials.gov Identifier NCT01663623) that aims to study belimumab in patients with AAV.

34.3 T Cell Targets

T cells can be found in inflammatory infiltrates of TAK and other vasculitides [39, 57]. In addition, impairment of T cells such as regulatory T cells have been reported in AAV, and T cell tubulitis was shown to be an independent predictor of renal outcome in sub-analysis of the RITUXVAS trial, which may be due to renal CVM infection correlating with T cell tubulitis [58, 59]. T cells are targeted by depleting T cells or by inhibition of co-stimulation [18].

34.3.1 Abatacept

Abatacept is a fusion protein of CTLA-4 and the Fc region of IgG. It works by inhibiting the B7-CD28 interaction by binding B7 on antigen presenting cells, thus blocking co-stimulation and preventing T cell activation. It is approved of for use in rheumatoid arthritis (RA) and there have been studies evaluating its efficacy in RA [60–63]. In regards to GPA, there is an open-label pilot study that evaluated Abatacept in patients with relapsing but non severe GPA. The study showed that Abatacept was well tolerated and led to disease remission in a high percentage of patients. Thus, Abatacept warrants further study as a treatment option for patients who have non-severe relapsing GPA [64].

34.3.2 Alemtuzumab

Alemtuzumab is a monoclonal antibody that binds CD52, which is present on T and B-lymphocytes and macrophages. This results in reduction of lymphocytes and macrophages. It has been used to treat chronic lymphocytic leukemia (CLL), T cell lymphoma, and multiple sclerosis [65–69]. A 5-year follow-up study of patients with relapsing/refractory AAV treated with Alemtuzumab showed that it induced remission in most patients, but relapse and adverse events were common during the follow up period [70].

34.3.3 Daclizumab

Daclizumab is an anti-CD25 monoclonal antibody. It binds to the alpha subunit of the IL-2 receptor and inhibits T cell activation by IL-2. IL-2 is a survival and maturation cytokine for T and B cells. Daclizumab has been used for preventing transplant rejection in solid organ transplantation, treating graft versus host disease, and multiple sclerosis [71–74]. A study was also done to evaluate Daclizumab in Behçet's disease that concluded that Daclizumab was no better than placebo [75]. A study titled *Daclizumab to Treat Wegener's Granulomatosis* was started in 2002. Results have not yet been reported (ClinicalTrials.gov Identifier NCT00040248).

34.4 Anti-cytokine Therapies

TNF-α, IL-6, and IL-1β are the primary cytokines targeted in vasculitides, due to the role they play in the disease process. TNF-α is a critical cytokine for granuloma formation and was first studied in GPA due to its nature as a granulomatous disease [76]. TNF-α has a number of actions which include stimulating macrophages to secrete IL-12 and IL-18. In TAK, giant cells and granulomas are found in the vascular lesions, thus allowing a role for TNF inhibitors [77]. Increased levels of TNF-α are seen in patients with Behçet's disease that have active disease [78, 79]. IL-6 is an inflammatory cytokine that drives the TH-17 pathway and induces B cell maturation to plasma cells [80]. IL-1β is another inflammatory cytokine which plays a role in cyropyrin mediated periodic syndromes and may be associated with urticaria and urticarial vasculitis [81].

34.4.1 TNF Inhibitors: Infliximab, Etanercept, Adalimumab

In TAK, Infliximab has been studied in 100 patients during various stages of the disease [82–86]. A study by Quartuccio et al. [85] evaluated long term efficacy and quality of life improvement for TAK patients treated with Infliximab. Long term follow up for the 15 patients studied was 71 +/− 44 months. This study showed long term clinical improvement along with a reduction of steroid doses needed to treat the patients. Similar results have been replicated in other studies [87]. However, the stage of disease where Infliximab should started has not been well studied. The case series that showed the lowest relapse rate in the current literature was also the study where Infliximab was started after a short period of the disease (average of 16 months) [86], thus suggesting earlier intervention may lead to better outcomes. A review of 84 patients by Comarmond et al. showed that 90 % of patients responded to TNF inhibitors; 37 % of the sample had a complete remission, while 53.5 % had a partial remission. Steroids were able to be discontinued in 40 % of the patients [82]. TNF inhibitors, such as infliximab and adalimumab, may be an alternative TAK treatment for disease that is refractory to traditional treatment [82, 88–93]. Interestingly, in patients without TAK but other immunological disease such as rheumatoid arthritis and spondylo-arthropathy that are being treated with TNF inhibitors, there have been cases developing TAK after starting anti-TNF therapy [83, 94].

TNF-α has also been found to be associated to pathogenesis in AAV, especially in GPA [95, 96]. The evidence however shows mixed results. There are a few studies that show the TNF inhibitor infliximab and adalimumab can induce remission of AAV and allow a reduction in steroid doses needed for treatment [97–99]. However, with Etanercept, a randomized, placebo controlled study involving 180 patients, carried out by the Etanercerpt Trial Group, showed that the addition of Etanercept to standard treatment was no better than placebo in maintaining remission in patients with GPA [100]. Adalimumab was also shown to be no better than standard therapy in maintaining remission in a randomized controlled trial evaluating its use in giant cell arteritis (GCA) [101, 102]. More studies are needed to better evaluate the role of TNF inhibitors in AAV.

There also have been some studies evaluating TNF inhibitors in Behçet's disease. A multicenter study involving 124 patients with refractory uveitis due to Behçet's disease evaluated Infliximab and Adalimumab. Overall, the response rate (including partial and complete) was 90 % and no significant difference was found between using the anti-TNF agents as mono-therapy or in addition with other immunosuppressive agents. Anti-TNF agents reduced the dose of steroids required to control the disease. Efficacy of the TNF inhibitors was similar between Infliximab and Adalimumab [103]. The 1-year follow up showed that 2/3 of the TNF-inhibitor-treated patients remained in remission [104]. An analysis of published data of 369 patients with Behçet's disease treated with TNF-inhibitors concluded that there is enough published experience to indicate that TNF-inhibitors are a therapeutic advancement for patients refractory to standard Behçet's treatment regimens [105].

In general, TNF-inhibitors are not curative treatments and, when stopped after remission, relapses are common. Given the positive response to anti-TNF agents, prospective, randomized controlled studies on anti-TNF agents are warranted.

34.4.2 Anti-IL6-Tocilizumab

Tocilizumab is a recombinant, humanized monoclonal antibody against the IL-6 receptor. It works by IL-6 attachment to the receptor, thus preventing further signaling down the pathway [106]. In patients with TAK, IL-6 is found in aortic tissue samples and sera. It also correlates with disease activity. IL-6 is also involved in the normal response to infection and it has been hypothesized that TAK may be due to an unusual genetically programed response to certain infections [107, 108].

Quite a few case reports and case series have documented the efficacy of Tocilizumab in TAK. Unizony et al. summarized cases of 9 patients who responded to Tocilizumab with one patient relapsing on the medication [109]. Abisror et al. reviewed a total of 44 cases with 75 % responding to Tocilizumab and allowed early steroid tapering [110]. Several other case series support this finding of Tocilizumab as an effective agent in refractory TAK, with reduction in steroid treatment course [111–113]. However, Goel et al. reviewed a total of 10 cases where 8 patients had relapse of disease after a Tocilizumab treatment was stopped. These patients had a total of 6 infusions as part of their treatment course [114]. A case report has also documented Tocilizumab efficacy in a pediatric patient [115].

Compared to TAK, there has not been much documented evidence of Tocilizumab efficacy in AAV. In vitro studies have shown that in AAV, there is ANCA induced production of IL-6. A 12-month treatment course with Tocilizumab normalized IL-6 levels and resulted in disease remission in a patient with microscopic polyangitis (MPA) [116, 117]. Two cases have been reported where Tocilizumab was effective in treating AAV resistant to standard treatments [118, 119]. Further data is needed, before a recommendation on Tocilizumab use in AAV can be made.

There have also been a number of case reports regarding Tocilizumab use in Behçet's disease (BD). Quite a few reports have documented Tocilizumab being effective in inducing remission in BD refractory to standard treatments. These cases have also shown Tocilizumab allows steroid reduction in the patients' treatment course [120–122]. However, most cases saw reduction/elimination of patients' neurological symptoms. The effect on muco-cutaneous symptoms was questionable. Cases have been documented in which Tocilizumab has not been effective in treating muco-cutaneous symptoms of BD [123].

For Giant cell arteritis (GCA), there have been a number of case reports documenting the efficacy of Tocilizumab in refractory GCA. These cases also report steroid sparing effects on Tocilizumab therapy in GCA [124–128]. Currently, there is a proposed multicenter, randomized, double-blind, and placebo-controlled study (GiACTA) designed to test the ability of Tocilizumab (TCZ) to maintain disease remission in GCA patients [129]. Ongoing studies include Tocilizumab for Patients with Giant Cell Arteritis (ClinicalTrials.gov Identifier: NCT01450137) and Study of RoActemra/Actemra (Tocilizumab) in Patients with Giant Cell Arteritis (ClinicalTrials.gov Identifier: NCT01791153). Both studies are currently recruiting participants.

34.4.3 Anti-IL-1β-Canakinumab

Canakinumab is a monoclonal anti-IL-1β antibody used as a treatment for cyropyrin mediated periodic syndrome for which IL-1β plays a role its pathogenesis. This syndrome may be associated with urticaria and urticarial vasculitis [130]. For urticarial vasculitis, an open label study involving ten patients who received a single dose of Canakinumab showed improvement in their urticarial vasculitis-associated symptoms [131]. There are also series of cases reporting that Canakinumab is effective in treating Behçet's disease refractory to standard treatment [132–134]. Further studies are needed to evaluate Canakinumab's role in these conditions.

34.4.4 Anti-IL-5-Mepolizumab

IL-5 is a cytokine that plays a role in eosinophilic development and migration. Mepolizumab has been shown to be effective in treating eosinophilic asthma [135, 136]. Mepolizumab has been used as treatment for eosinophilic granulomatosis with polyangitis (EGPA/Churg Strauss syndrome). One study involving seven patients showed that Mepolizumab was effective in inducing remission and reducing steroid dose. However, once Mepolizumab was discontinued, the patients relapsed and needed additional steroid treatments to control their disease [137]. This finding was also supported by another study involving nine patients, who received nine infusions of Mepolizumab and then switched to methotrexate. The patients were observed for a total of 22 months. The outcome they found was that, once patients were switched to methotrexate, relapses of disease occurred [138]. A case study describing a sustained response to Mepolizumab was able to prevent relapses with continuous monthly doses of Mepolizumab [139]. Thus, these reports suggest Mepolizumab may only be effective with continuous treatment. There is also a possibility that IL-5 is a downstream mediator of pathogenesis in EGPA and blocking it may not prevent relapses [18].

34.5 Conclusion

Given that multiple studies involving biologics in vasculitides have been done, none of the medications discussed here have been associated with complete maintenance of remission after initial therapy. The goal of using biologics as opposed to standard therapy is to reduce adverse effects associated with the medications of standard therapy. However, biologics may also come with their own risk of other adverse effects that is their major limiting factor for their use. Further long-term studies are needed to better understand the safety profile and effectiveness of biologics in the treatment of vasculitides. At the current time, no biological agent is yet considered the gold standard therapy for any type of vasculitis.

References

1. Eriksson P, Sandell C, Backteman K, Ernerudh J (2010) B cell abnormalities in Wegener's granulomatosis and microscopic polyangiitis: role of CD25+−expressing B cells. J Rheumatol 37(10):2086–2095
2. Voswinkel J, Mueller A, Kraemer JA, Lamprecht P, Herlyn K, Holl-Ulrich K et al (2006) B lymphocyte maturation in Wegener's granulomatosis: a comparative analysis of VH genes from endonasal lesions. Ann Rheum Dis 65(7):859–864
3. Dumoitier N, Terrier B, London J, Lofek S, Mouthon L (2015) Implication of B lymphocytes in the pathogenesis of ANCA-associated vasculitides. Autoimmun Rev 14(11):996–1004
4. Voswinkel J, Assmann G, Held G, Pitann S, Gross WL, Holl-Ulrich K et al (2008) Single cell analysis of B lymphocytes from Wegener' granulomatosis: B cell receptors display affinity maturation within the granulomatous lesions. Clin Exp Immunol 154(3):339–345
5. Wilde B, Thewissen M, Damoiseaux J, Knippenberg S, Hilhorst M, van Paassen P et al (2013) Regulatory B cells in ANCA-associated vasculitis. Ann Rheum Dis 72(8):1416–1419
6. Cupps TR, Edgar LC, Fauci AS (1982) Suppression of human B lymphocyte function by cyclophosphamide. J Immunol 128(6):2453–2457
7. Cupps TR, Fauci AS (1981) The vasculitides. Major Probl Intern Med 21:1–211
8. Hoyer BF, Mumtaz IM, Loddenkemper K, Bruns A, Sengler C, Hermann KG et al (2012) Takayasu arteritis is characterised by disturbances of B cell homeostasis and responds to B cell depletion therapy with rituximab. Ann Rheum Dis 71(1):75–79
9. Inder SJ, Bobryshev YV, Cherian SM, Wang AY, Lord RS, Masuda K et al (2000) Immunophenotypic analysis of the aortic wall in Takayasu's arteritis: involvement of lymphocytes, dendritic cells and granulocytes in immuno-inflammatory reactions. Cardiovasc Surg 8(2):141–148
10. Park MC, Park YB, Jung SY, Lee KH, Lee SK (2006) Anti-endothelial cell antibodies and antiphospholipid antibodies in Takayasu's arteritis: correlations of their titers and isotype distributions with disease activity. Clin Exp Rheumatol 24(2 Suppl 41):S10–S16
11. Abdulahad WH, Meijer JM, Kroese FG, Meiners PM, Vissink A, Spijkervet FK et al (2011) B cell reconstitution and T helper cell balance after rituximab treatment of active primary Sjogren's syndrome: a double-blind, placebo-controlled study. Arthritis Rheum 63(4):1116–1123
12. Anolik JH, Friedberg JW, Zheng B, Barnard J, Owen T, Cushing E et al (2007) B cell reconstitution after rituximab treatment of lymphoma recapitulates B cell ontogeny. Clin Immunol 122(2):139–145
13. Iwata S, Saito K, Tokunaga M, Yamaoka K, Nawata M, Yukawa S et al (2011) Phenotypic changes of lymphocytes in patients with systemic lupus erythematosus who are in longterm remission after B cell depletion therapy with rituximab. J Rheumatol 38(4):633–641
14. Leandro MJ, Cambridge G, Ehrenstein MR, Edwards JC (2006) Reconstitution of peripheral blood B cells after depletion with rituximab in patients with rheumatoid arthritis. Arthritis Rheum 54(2):613–620
15. Stone JH, Merkel PA, Spiera R, Seo P, Langford CA, Hoffman GS et al (2010) Rituximab versus cyclophosphamide for ANCA-associated vasculitis. N Engl J Med 363(3):221–232
16. Smith RM, Jones RB, Guerry MJ, Laurino S, Catapano F, Chaudhry A et al (2012) Rituximab for remission maintenance in relapsing antineutrophil cytoplasmic antibody-associated vasculitis. Arthritis Rheum 64(11):3760–3769
17. Specks U, Merkel PA, Seo P, Spiera R, Langford CA, Hoffman GS et al (2013) Efficacy of remission-induction regimens for ANCA-associated vasculitis. N Engl J Med 369(5):417–427
18. Vishwanath S, Relan M, Shen L, Ambrus JL Jr (2014) Update on the use of biologics in vasculitides. Curr Pharm Biotechnol 15(6):558–562
19. Cartin-Ceba R, Golbin JM, Keogh KA, Peikert T, Sanchez-Menendez M, Ytterberg SR et al (2012) Rituximab for remission induction and maintenance in refractory granulomatosis with

polyangiitis (Wegener's): ten-year experience at a single center. Arthritis Rheum 64(11):3770–3778
20. Rhee EP, Laliberte KA, Niles JL (2010) Rituximab as maintenance therapy for anti-neutrophil cytoplasmic antibody-associated vasculitis. Clin J Am Soc Nephrol 5(8):1394–1400
21. Jones RB, Tervaert JW, Hauser T, Luqmani R, Morgan MD, Peh CA et al (2010) Rituximab versus cyclophosphamide in ANCA-associated renal vasculitis. N Engl J Med 363(3):211–220
22. Jones RB, Furuta S, Tervaert JW, Hauser T, Luqmani R, Morgan MD et al (2015) Rituximab versus cyclophosphamide in ANCA-associated renal vasculitis: 2-year results of a randomised trial. Ann Rheum Dis 74(6):1178–1182
23. Guillevin L, Pagnoux C, Karras A, Khouatra C, Aumaitre O, Cohen P et al (2014) Rituximab versus azathioprine for maintenance in ANCA-associated vasculitis. N Engl J Med 371(19):1771–1780
24. Charles P, Neel A, Tieulie N, Hot A, Pugnet G, Decaux O et al (2014) Rituximab for induction and maintenance treatment of ANCA-associated vasculitides: a multicentre retrospective study on 80 patients. Rheumatology (Oxford) 53(3):532–539
25. Lutalo PM, D'Cruz DP (2015) Biological drugs in ANCA-associated vasculitis. Int Immunopharmacol 27(2):209–212
26. Tamura N, Matsudaira R, Hirashima M, Ikeda M, Tajima M, Nawata M et al (2007) Two cases of refractory Wegeners granulomatosis successfully treated with rituximab. Intern Med 46(7):409–414
27. Taylor SR, Salama AD, Joshi L, Pusey CD, Lightman SL (2009) Rituximab is effective in the treatment of refractory ophthalmic Wegener's granulomatosis. Arthritis Rheum 60(5):1540–1547
28. Shah S, Hruskova Z, Segelmark M, Morgan MD, Hogan J, Lee SK et al (2015) Treatment of severe renal disease in ANCA positive and negative small vessel vasculitis with rituximab. Am J Nephrol 41(4–5):296–301
29. Hoffman GS, Leavitt RY, Fleisher TA, Minor JR, Fauci AS (1990) Treatment of Wegener's granulomatosis with intermittent high-dose intravenous cyclophosphamide. Am J Med 89(4):403–410
30. De Vita S, Quartuccio L, Isola M, Mazzaro C, Scaini P, Lenzi M et al (2012) A randomized controlled trial of rituximab for the treatment of severe cryoglobulinemic vasculitis. Arthritis Rheum 64(3):843–853
31. Sneller MC, Hu Z, Langford CA (2012) A randomized controlled trial of rituximab following failure of antiviral therapy for hepatitis C virus-associated cryoglobulinemic vasculitis. Arthritis Rheum 64(3):835–842
32. Quartuccio L, Zuliani F, Corazza L, Scaini P, Zani R, Lenzi M et al (2015) Retreatment regimen of rituximab monotherapy given at the relapse of severe HCV-related cryoglobulinemic vasculitis: long-term follow up data of a randomized controlled multicentre study. J Autoimmun 63:88–93
33. Saadoun D, Resche Rigon M, Sene D, Terrier B, Karras A, Perard L et al (2010) Rituximab plus Peg-interferon-alpha/ribavirin compared with Peg-interferon-alpha/ribavirin in hepatitis C-related mixed cryoglobulinemia. Blood 116(3):326–334; quiz 504–5
34. Sene D, Ghillani-Dalbin P, Amoura Z, Musset L, Cacoub P (2009) Rituximab may form a complex with IgMkappa mixed cryoglobulin and induce severe systemic reactions in patients with hepatitis C virus-induced vasculitis. Arthritis Rheum 60(12):3848–3855
35. Ernst D, Greer M, Stoll M, Meyer-Olson D, Schmidt RE, Witte T (2012) Remission achieved in refractory advanced Takayasu arteritis using rituximab. Case Rep Rheumatol 2012:406963
36. Galarza C, Valencia D, Tobon GJ, Zurita L, Mantilla RD, Pineda-Tamayo R et al (2008) Should rituximab be considered as the first-choice treatment for severe autoimmune rheumatic diseases? Clin Rev Allergy Immunol 34(1):124–128
37. Caltran E, Di Colo G, Ghigliotti G, Capecchi R, Catarsi E, Puxeddu I et al (2014) Two Takayasu arteritis patients successfully treated with rituximab. Clin Rheumatol 33(8):1183–1184

38. Johnston SL, Lock RJ, Gompels MM (2002) Takayasu arteritis: a review. J Clin Pathol 55(7):481–486
39. Wen D, Du X, Ma CS (2012) Takayasu arteritis: diagnosis, treatment and prognosis. Int Rev Immunol 31(6):462–473
40. Dabague J, Reyes PA (1996) Takayasu arteritis in Mexico: a 38-year clinical perspective through literature review. Int J Cardiol 54:S103–S109
41. Clifford A, Hoffman GS (2014) Recent advances in the medical management of Takayasu arteritis: an update on use of biologic therapies. Curr Opin Rheumatol 26(1):7–15
42. Kaushik VV, Reddy HV, Bucknall RC (2006) Successful use of rituximab in a patient with recalcitrant Churg-Strauss syndrome. Ann Rheum Dis 65(8):1116–1117
43. Koukoulaki M, Smith KG, Jayne DR (2006) Rituximab in Churg-Strauss syndrome. Ann Rheum Dis 65(4):557–559
44. Davatchi F, Shams H, Rezaipoor M, Sadeghi-Abdollahi B, Shahram F, Nadji A et al (2010) Rituximab in intractable ocular lesions of Behcet's disease; randomized single-blind control study (pilot study). Int J Rheum Dis 13(3):246–252
45. Sadreddini S, Noshad H, Molaeefard M, Noshad R (2008) Treatment of retinal vasculitis in Behcet's disease with rituximab. Mod Rheumatol 18(3):306–308
46. Zhao BH, Oswald AE (2014) Improved clinical control of a challenging case of Behcet's disease with rituximab therapy. Clin Rheumatol 33(1):149–150
47. Saigal K, Valencia IC, Cohen J, Kerdel FA (2003) Hypocomplementemic urticarial vasculitis with angioedema, a rare presentation of systemic lupus erythematosus: rapid response to rituximab. J Am Acad Dermatol 49(5):283–285
48. Anolik JH, Barnard J, Cappione A, PughBernard AE, Felgar RE, Looney RJ et al (2004) Rituximab improves peripheral B cell abnormalities in human systemic lupus erythematosus. Arthritis Rheum 50(11):3580–3590
49. Sharma V, Dadhania D, Thomas D, Enriquez N, Menon A, Goldstein M, et al (2007) Failure to delete CD19+CD27+ memory B cells by Rituximab in sensitized renal allograft recipients. Hum Immunol 68:S110-S.
50. Toussirot É, Bereau M (2014) The risk of progressive multifocal leukoencephalopathy under biological agents used in the treatment of chronic inflammatory diseases. Inflamm Allergy-Drug Targets 13(2):121–127
51. Krumbholz M, Specks U, Wick M, Kalled SL, Jenne D, Meinl E (2005) BAFF is elevated in serum of patients with Wegener's granulomatosis. J Autoimmun 25(4):298–302
52. Jacobi AM, Huang W, Wang T, Freimuth W, Sanz I, Furie R et al (2010) Effect of long-term belimumab treatment on B cells in systemic lupus erythematosus: extension of a phase II, double-blind, placebo-controlled, dose-ranging study. Arthritis Rheum 62(1):201–210
53. Hahn BH (2013) Belimumab for systemic lupus erythematosus. N Engl J Med 368(16):1528–1535
54. Pontarini E, Fabris M, Quartuccio L, Cappeletti M, Calcaterra F, Roberto A et al (2015) Treatment with belimumab restores B cell subsets and their expression of B cell activating factor receptor in patients with primary Sjogren's syndrome. Rheumatology (Oxford) 54(8):1429–1434
55. Merrill JT, Ginzler EM, Wallace DJ, McKay JD, Lisse JR, Aranow C et al (2012) Long-term safety profile of belimumab plus standard therapy in patients with systemic lupus erythematosus. Arthritis Rheum 64(10):3364–3373
56. Wallace D, Navarra S, Petri M, Gallacher A, Thomas M, Furie R et al (2013) Safety profile of belimumab: pooled data from placebo-controlled phase 2 and 3 studies in patients with systemic lupus erythematosus. Lupus 22(2):144–154
57. Arnaud L, Haroche J, Mathian A, Gorochov G, Amoura Z (2011) Pathogenesis of Takayasu's arteritis: a 2011 update. Autoimmun Rev 11(1):61–67
58. Abdulahad WH, Lamprecht P, Kallenberg CG (2011) T-helper cells as new players in ANCA-associated vasculitides. Arthritis Res Ther 13(4):236

59. Meyer MF, Hellmich B, Kotterba S, Schatz H (2000) Cytomegalovirus infection in systemic necrotizing vasculitis: causative agent or opportunistic infection? Rheumatol Int 20(1):35–38
60. Kremer JM, Russell AS, Emery P, Abud-Mendoza C, Szechinski J, Westhovens R et al (2011) Long-term safety, efficacy and inhibition of radiographic progression with abatacept treatment in patients with rheumatoid arthritis and an inadequate response to methotrexate: 3-year results from the AIM trial. Ann Rheum Dis 70(10):1826–1830
61. Lovell DJ, Ruperto N, Mouy R, Paz E, Rubio-Perez N, Silva CA et al (2015) Long-term safety, efficacy, and quality of life with intravenous abatacept in juvenile idiopathic arthritis: up to 7 years of treatment. Arthritis Rheum 67(10):2759–2770
62. Massarotti EM (2008) Clinical and patient-reported outcomes in clinical trials of abatacept in the treatment of rheumatoid arthritis. Clin Ther 30(3):429–442
63. Smolen JS, Wollenhaupt J, Gomez-Reino JJ, Grassi W, Gaillez C, Poncet C et al (2015) Attainment and characteristics of clinical remission according to the new ACR-EULAR criteria in abatacept-treated patients with early rheumatoid arthritis: new analyses from the Abatacept study to Gauge Remission and joint damage progression in methotrexate (MTX)-naive patients with Early Erosive rheumatoid arthritis (AGREE). Arthritis Res Ther 17:157
64. Langford CA, Monach PA, Specks U, Seo P, Cuthbertson D, McAlear CA et al (2014) An open-label trial of abatacept (CTLA4-IG) in non-severe relapsing granulomatosis with polyangiitis (Wegener's). Ann Rheum Dis 73(7):1376–1379
65. Babij R, Perumal JS (2015) Comparative efficacy of alemtuzumab and established treatment in the management of multiple sclerosis. Neuropsychiatr Dis Treat 11:1221–1229
66. Gross RH, Krieger S (2015) Alemtuzumab in multiple sclerosis: an update. Neurodegener Dis Manag 5(3):225–232
67. Keating MJ, Flinn I, Jain V, Binet J-L, Hillmen P, Byrd J et al (2002) Therapeutic role of alemtuzumab (Campath-1H) in patients who have failed fludarabine: results of a large international study. Blood 99(10):3554–3561
68. Ravandi F, O'Brien S (2006) Alemtuzumab in CLL and other lymphoid neoplasms. Cancer Invest 24(7):718–725
69. Sciume M, Vincenti D, Reda G, Orofino N, Cassin R, Giannarelli D et al (2015) Low-dose alemtuzumab in refractory/relapsed chronic lymphocytic leukemia: genetic profile and long-term outcome from a single center experience. Am J Hematol 90(11):970–974
70. Walsh M, Chaudhry A, Jayne D (2008) Long-term follow-up of relapsing/refractory anti-neutrophil cytoplasm antibody associated vasculitis treated with the lymphocyte depleting antibody alemtuzumab (CAMPATH-1H). Ann Rheum Dis 67(9):1322–1327
71. Bielekova B, Howard T, Packer AN et al (2009) EFfect of anti-cd25 antibody daclizumab in the inhibition of inflammation and stabilization of disease progression in multiple sclerosis. Arch Neurol 66(4):483–489
72. Braun WE (2003) Renal transplantation: basic concepts and evolution of therapy. J Clin Apher 18(3):141–152
73. Lin YC, Winokur P, Blake A, Wu T, Romm E, Bielekova B (2015) Daclizumab reverses intrathecal immune cell abnormalities in multiple sclerosis. Ann Clin Transl Neurol 2(5):445–455
74. Wynn D, Kaufman M, Montalban X, Vollmer T, Simon J, Elkins J et al (2010) Daclizumab in active relapsing multiple sclerosis (CHOICE study): a phase 2, randomised, double-blind, placebo-controlled, add-on trial with interferon beta. The Lancet Neurology 9(4):381–390
75. Buggage RR, Levy-Clarke G, Sen HN, Ursea R, Srivastava SK, Suhler EB et al (2007) A double-masked, randomized study to investigate the safety and efficacy of daclizumab to treat the ocular complications related to Behcet's disease. Ocul Immunol Inflamm 15(2):63–70
76. Antoniu SA. Infliximab for the therapy of chronic sarcoidosis, Baughman RP, Drent M, Kavuru M et al (2006) Infliximab therapy in patients with chronic sarcoidosis and pulmonary involvement. Am J Respir Crit Care Med 74(7):795–802. Expert Opin Investig Drugs. 2007;16(5):753–756

77. Killinger Z, Cierny D, Jackuliak P, Zelinkova Z, Rovensky J, Payer J (2012) Long-term efficacy of infliximab for treating Takayasu arteritis. Isr Med Assoc J 14(12):771–773
78. O'Neill ID (2008) Off-label use of biologicals in the management of inflammatory oral mucosal disease. J Oral Pathol Med 37(10):575–581
79. Pato E, Munoz-Fernandez S, Francisco F, Abad MA, Maese J, Ortiz A et al (2011) Systematic review on the effectiveness of immunosuppressants and biological therapies in the treatment of autoimmune posterior uveitis. Semin Arthritis Rheum 40(4):314–323
80. Kovalchuk AL, Kim JS, Park SS, Coleman AE, Ward JM, Morse HC 3rd et al (2002) IL-6 transgenic mouse model for extraosseous plasmacytoma. Proc Natl Acad Sci U S A 99(3):1509–1514
81. Agostini L, Martinon F, Burns K, McDermott MF, Hawkins PN, Tschopp J (2004) NALP3 forms an IL-1beta-processing inflammasome with increased activity in Muckle-Wells autoinflammatory disorder. Immunity 20(3):319–325
82. Comarmond C, Plaisier E, Dahan K, Mirault T, Emmerich J, Amoura Z et al (2012) Anti TNF-α in refractory Takayasu's arteritis: cases series and review of the literature. Autoimmun Rev 11(9):678–684
83. Mariani N, So A, Aubry-Rozier B (2013) Two cases of Takayasu's arteritis occurring under anti-TNF therapy. Joint Bone Spine 80(2):211–213
84. Mekinian A, Néel A, Sibilia J, Cohen P, Connault J, Lambert M et al (2012) Efficacy and tolerance of infliximab in refractory Takayasu arteritis: French multicentre study. Rheumatology 51(5):882–886
85. Quartuccio L, Schiavon F, Zuliani F, Carraro V, Catarsi E, Tavoni AG et al (2012) Long-term efficacy and improvement of health-related quality of life in patients with Takayasu's arteritis treated with infliximab. Clin Exp Rheumatol 30(6):922–928
86. Schmidt J, Kermani TA, Bacani AK, Crowson CS, Matteson EL, Warrington KJ (2012) Tumor necrosis factor inhibitors in patients with Takayasu arteritis: experience from a referral center with long-term followup. Arthritis Care Res (Hoboken) 64(7):1079–1083
87. Molloy ES, Langford CA, Clark TM, Gota CE, Hoffman GS (2008) Anti-tumour necrosis factor therapy in patients with refractory Takayasu arteritis: long-term follow-up. Ann Rheum Dis 67(11):1567–1569
88. Filocamo G, Buoncompagni A, Viola S, Loy A, Malattia C, Ravelli A et al (2008) Treatment of Takayasu's arteritis with tumor necrosis factor antagonists. J Pediatr 153(3):432–434
89. Jolly M, Curran JJ (2005) Infliximab-responsive uveitis and vasculitis in a patient with Takayasu arteritis. J Clini Rheumatol 11(4):213–215
90. Maffei S, Di Renzo M, Santoro S, Puccetti L, Pasqui A (2009) Refractory Takayasu arteritis successfully treated with infliximab. Eur Rev Med Pharmacol Sci 13(1):63–65
91. Tanaka F, Kawakami A, Iwanaga N, Tamai M, Izumi Y, Aratake K et al (2006) Infliximab is effective for Takayasu arteritis refractory to glucocorticoid and methotrexate. Intern Med 45(5):313–316
92. Tato F, Rieger J, Hoffmann U (2005) Refractory Takayasu's arteritis successfully treated with the human, monoclonal anti-tumor necrosis factor antibody adalimumab. Int Angiol 24(3):304
93. Youngstein T, Peters JE, Hamdulay SS, Mewar D, Price-Forbes A, Lloyd M et al (2014) Serial analysis of clinical and imaging indices reveals prolonged efficacy of TNF-alpha and IL-6 receptor targeted therapies in refractory Takayasu arteritis. Clin Exp Rheumatol 32(3 Suppl 82):S11–S18
94. Souabni L, Ben Abdelghani K, Jradi S, Zakraoui L (2014) Takayasu's arteritis occurring under TNF-alpha blockers: a new paradoxical effect? BMJ Case Rep 2014
95. Ludviksson BR, Sneller MC, Chua KS, Talar-Williams C, Langford CA, Ehrhardt RO et al (1998) Active Wegener's granulomatosis is associated with HLA-DR+ CD4+ T cells exhibiting an unbalanced Th1-type T cell cytokine pattern: reversal with IL-10. J Immunol 160(7):3602–3609
96. Noronha IL, Krüger C, Andrassy K, Ritz E, Waldherr R (1993) In situ production of TNF-alpha, IL-1 bold beta and IL-2R in ANCA-positive glomerulonephritis. Kidney Int 43:682–692

97. Bartolucci P, Ramanoelina J, Cohen P, Mahr A, Godmer P, Le Hello C et al (2002) Efficacy of the anti-TNF-α antibody infliximab against refractory systemic vasculitides: an open pilot study on 10 patients. Rheumatology 41(10):1126–1132
98. Booth A, Harper L, Hammad T, Bacon P, Griffith M, Levy J et al (2004) Prospective study of TNFα blockade with infliximab in anti-neutrophil cytoplasmic antibody-associated systemic vasculitis. J Am Soc Nephrol 15(3):717–721
99. Laurino S, Chaudhry A, Booth A, Conte G, Jayne D (2010) Prospective study of TNFalpha blockade with adalimumab in ANCA-associated systemic vasculitis with renal involvement. Nephrol Dial Transplant 25(10):3307–3314
100. Wegener's GETW (2005) Etanercept plus standard therapy for Wegener's granulomatosis. N Engl J Med 352(4):351
101. Osman M, Pagnoux C, Dryden DM, Storie D, Yacyshyn E (2014) The role of biological agents in the management of large vessel vasculitis (LVV): a systematic review and meta-analysis. PLoS One 9(12), e115026
102. Seror R, Baron G, Hachulla E, Debandt M, Larroche C, Puéchal X et al (2014) Adalimumab for steroid sparing in patients with giant-cell arteritis: results of a multicentre randomised controlled trial. Ann Rheum Dis 73(12):2074–2081
103. Vallet H, Riviere S, Sanna A, Deroux A, Moulis G, Addimanda O et al (2015) Efficacy of anti-TNF alpha in severe and/or refractory Behcet's disease: multicenter study of 124 patients. J Autoimmun 62:67–74
104. Calvo-Rio V, Blanco R, Beltran E, Sanchez-Burson J, Mesquida M, Adan A et al (2014) Anti-TNF-alpha therapy in patients with refractory uveitis due to Behcet's disease: a 1-year follow-up study of 124 patients. Rheumatology (Oxford) 53(12):2223–2231
105. Arida A, Fragiadaki K, Giavri E, Sfikakis PP (2011) Anti-TNF agents for Behcet's disease: analysis of published data on 369 patients. Semin Arthritis Rheum 41(1):61–70
106. Alten R, Maleitzke T (2013) Tocilizumab: a novel humanized anti-interleukin 6 (IL-6) receptor antibody for the treatment of patients with non-RA systemic, inflammatory rheumatic diseases. Ann Med 45(4):357–363
107. Park MC, Lee SW, Park YB, Lee SK (2006) Serum cytokine profiles and their correlations with disease activity in Takayasu's arteritis. Rheumatology (Oxford) 45(5):545–548
108. Seko Y, Sato O, Takagi A, Tada Y, Matsuo H, Yagita H et al (1996) Restricted usage of T-cell receptor Valpha-Vbeta genes in infiltrating cells in aortic tissue of patients with Takayasu's arteritis. Circulation 93(10):1788–1790
109. Unizony S, Stone JH, Stone JR (2013) New treatment strategies in large-vessel vasculitis. Curr Opin Rheumatol 25(1):3–9
110. Abisror N, Mekinian A, Lavigne C, Vandenhende M-A, Soussan M, Fain O (2013) Tocilizumab in refractory Takayasu arteritis: a case series and updated literature review. Autoimmun Rev 12(12):1143–1149
111. Salvarani C, Magnani L, Catanoso MG, Pipitone N, Versari A, Dardani L et al (2012) Rescue treatment with tocilizumab for Takayasu arteritis resistant to TNF-alpha blockers. Clin Exp Rheumatol 30(1 Suppl 70):S90–S93
112. Loricera J, Blanco R, Castaneda S, Humbria A, Ortego-Centeno N, Narvaez J et al (2014) Tocilizumab in refractory aortitis: study on 16 patients and literature review. Clin Exp Rheumatol 32(3 Suppl 82):S79–S89
113. Tombetti E, Franchini S, Papa M, Sabbadini MG, Baldissera E (2013) Treatment of refractory Takayasu arteritis with tocilizumab: 7 Italian patients from a single referral center. J Rheumatol 40(12):2047–2051
114. Goel R, Danda D, Kumar S, Joseph G (2013) Rapid control of disease activity by tocilizumab in 10 'difficult-to-treat' cases of Takayasu arteritis. Int J Rheum Dis 16(6):754–761
115. Bravo Mancheno B, Perin F, Guez Vazquez Del Rey Mdel M, Garcia Sanchez A, Alcazar Romero PP (2012) Successful tocilizumab treatment in a child with refractory Takayasu arteritis. Pediatrics 130(6):e1720–e1724

116. Berti A, Cavalli G, Campochiaro C, Guglielmi B, Baldissera E, Cappio S et al (2015) Interleukin-6 in ANCA-associated vasculitis: rationale for successful treatment with tocilizumab. Semin Arthritis Rheum 45(1):48–54
117. Muller Kobold AC, van Wijk RT, Franssen CF, Molema G, Kallenberg CG, Tervaert JW (1999) In vitro up-regulation of E-selectin and induction of interleukin-6 in endothelial cells by autoantibodies in Wegener's granulomatosis and microscopic polyangiitis. Clin Exp Rheumatol 17(4):433–440
118. Sumida K, Ubara Y, Suwabe T, Hayami N, Hiramatsu R, Hasegawa E et al (2011) Complete remission of myeloperoxidase-anti-neutrophil cytoplasmic antibody-associated crescentic glomerulonephritis complicated with rheumatoid arthritis using a humanized anti-interleukin 6 receptor antibody. Rheumatology (Oxford) 50(10):1928–1930
119. Takenaka K, Ohba T, Suhara K, Sato Y, Nagasaka K (2014) Successful treatment of refractory aortitis in antineutrophil cytoplasmic antibody-associated vasculitis using tocilizumab. Clin Rheumatol 33(2):287–289
120. Addimanda O, Pipitone N, Pazzola G, Salvarani C (2015) Tocilizumab for severe refractory neuro-Behcet: three cases IL-6 blockade in neuro-Behcet. Semin Arthritis Rheum 44(4):472–475
121. Hirano T, Ohguro N, Hohki S, Hagihara K, Shima Y, Narazaki M et al (2012) A case of Behçet's disease treated with a humanized anti-interleukin-6 receptor antibody, tocilizumab. Modern Rheumatol 22(2):298–302
122. Urbaniak P, Hasler P, Kretzschmar S (2011) Refractory neuro-Behcet treated by tocilizumab: a case report. Clin Exp Rheumatol 30(3 Suppl 72):S73–S75
123. Diamantopoulos AP, Hatemi G (2013) Lack of efficacy of tocilizumab in mucocutaneous Behcet's syndrome: report of two cases. Presse Med 42(4P2):769
124. Işık M, Kılıç L, Doğan İ, Calgüneri M (2013) Tocilizumab for giant cell arteritis: an amazing result. Rheumatol Int 33(11):2961–2962
125. Kieffer P, Hinschberger O, Ciobanu E, Jaeger-Bizet F, Drabo A, Mostoufizadeh T et al (2014) [Clinical and biological efficacy of tocilizumab in giant cell arteritis: report of three patients and literature review]. Rev Med Interne 35(1):56–59
126. Lurati A, Bertani L, Re KA, Marrazza M, Bompane D, Scarpellini M (2012) Successful treatment of a patient with giant cell vasculitis (horton arteritis) with tocilizumab a humanized anti-interleukin-6 receptor antibody. Case reports in rheumatology 2012
127. Sciascia S, Rossi D, Roccatello D (2011) Interleukin 6 blockade as steroid-sparing treatment for 2 patients with giant cell arteritis. J Rheumatol 38(9):2080–2081
128. Vinit J, Bielefeld P, Muller G, Besancenot J-F (2012) Efficacy of tocilizumab in refractory giant cell arteritis. Joint Bone Spine 79(3):317–318
129. Unizony SH, Dasgupta B, Fisheleva E, Rowell L, Schett G, Spiera R et al (2013) Design of the tocilizumab in giant cell arteritis trial. Int J Rheum 2013
130. Kuemmerle-Deschner JB, Haug I (2013) Canakinumab in patients with cryopyrin-associated periodic syndrome: an update for clinicians. Ther Adv Musculoskelet Dis. 5(6):315–329. doi :10.1177/1759720X13502629
131. Krause K, Mahamed A, Weller K, Metz M, Zuberbier T, Maurer M (2013) Efficacy and safety of canakinumab in urticarial vasculitis: an open-label study. J Allergy Clin Immunol 132(3):751–4.e5
132. Pagnini I, Bondi T, Simonini G, Giani T, Marino A, Cimaz R (2015) Successful treatment with canakinumab of a paediatric patient with resistant Behçet's disease. Rheumatology 54(7):1327–1328
133. Ugurlu S, Ucar D, Seyahi E, Hatemi G, Yurdakul S (2012) Canakinumab in a patient with juvenile Behcet's syndrome with refractory eye disease. Ann Rheum Dis 71(9):1589–1591
134. Vitale A, Rigante D, Caso F, Brizi MG, Galeazzi M, Costa L et al (2014) Inhibition of interleukin-1 by canakinumab as a successful mono-drug strategy for the treatment of refractory Behçet's disease: a case series. Dermatology 228(3):211–214

135. Haldar P, Brightling CE, Hargadon B, Gupta S, Monteiro W, Sousa A et al (2009) Mepolizumab and exacerbations of refractory eosinophilic asthma. N Engl J Med 360(10):973–984
136. Ortega HG, Liu MC, Pavord ID, Brusselle GG, FitzGerald JM, Chetta A et al (2014) Mepolizumab treatment in patients with severe eosinophilic asthma. N Engl J Med 371(13):1198–1207
137. Kim S, Marigowda G, Oren E, Israel E, Wechsler ME (2010) Mepolizumab as a steroid-sparing treatment option in patients with Churg-Strauss syndrome. J Allergy Clin Immunol 125(6):1336–1343
138. Herrmann K, Gross WL, Moosig F (2012) Extended follow-up after stopping mepolizumab i n relapsing/refractory Churg-Strauss syndrome. Clin Exp Rheumatol 30(1 Suppl 70):S62–S65
139. Kahn J-E, Grandpeix-Guyodo C, Marroun I, Catherinot E, Mellot F, Roufosse F et al (2010) Sustained response to mepolizumab in refractory Churg-Strauss syndrome. J Allergy Clin Immunol 125(1):267–270

Chapter 35
Treatment of ANCA-Associated Vasculitides

Loïc Guillevin

Abstract ANCA-associated vasculitis (AAV) covers a group of systemic necrotizing vasculitides characterized by inflammation of small vessels, some with granuloma, and associated with autoantibodies to neutrophil cytoplasmic proteases (proteinase-3 or myeloperoxidase). Potentially lethal if not promptly diagnosed and treated, AAV in most patients can be induced into remission with the current treatment modalities. However, the risk of relapse remains high, necessitating prolonged immunosuppressive maintenance therapy, whose optimal duration remains undetermined. Herein, we review those treatment modalities for AAV. The findings of most important and recently completed therapeutic studies, including those on rituximab for maintenance, are summarized.

Keywords ANCA-associated vasculitides • Granulomatosis with polyangiitis • Rituximab • Cyclophosphamide • Glucocorticoids • Treatment

ANCA-associated vasculitides (AAV) are characterized by inflammation of small vessels and fibrinoid necrosis of the media, and are typically associated, for most patients, with circulating autoantibodies to neutrophil cytoplasmic proteases, mainly proteinase-3 (PR3-ANCA), a cationic proteolytic enzyme physiologically present in neutrophil cytoplasmic granules, in granulomatosis with polyangiitis (Wegener's) (GPA), and primarily myeloperoxidase (MPO-ANCA) in microscopic polyangiitis (MPA). This group comprises three diseases: GPA, MPA and eosinophilic granulomatosis with polyangiitis (Churg–Strauss) (EGPA).

GPA is rare, with an estimated annual incidence of 2–12 cases/million inhabitants and prevalence of 23–160 cases/million inhabitants [1, 2]. Most patients are white (Caucasian or Hispanics). MPA incidence was estimated at 3.6/million inhabitants in England [3]. In France, prevalences per million inhabitants were 25.1 for MPA, 23.7 for GPA and 10.7 for EGPA [4].

L. Guillevin (✉)
Department of Internal Medicine, Referral Center for Rare Autoimmune and Systemic Diseases, Hôpital Cochin, APHP, Université Paris Descartes,
27, rue du faubourg Saint-Jacques, 75679 Paris, France
e-mail: loic.guillevin@cch.aphp.fr

F. Dammacco et al. (eds.), *Systemic Vasculitides: Current Status and Perspectives*, DOI 10.1007/978-3-319-40136-2_35

GPA mainly affects adults 45–60 years old, but can also occur in young children or the elderly [1, 2], while MPA usually develops in older patients [5] and EGPA in younger patients [6]. Treatment modalities achieve remission in most patients. However, the risk of relapse is high and remission-maintenance therapy, whose optimal duration remains undetermined, requires more effective treatments.

35.1 Therapeutic Strategies

Before the advent of glucocorticoids (GC), and then of other immunosuppressants, AAV prognoses were very poor. Most patients died without available treatment, especially those with systemic disease. Currently used treatments can obtain remission in more than 80 % of the patients, with the 5-year overall mortality rate at 10–15 % [6, 7]. The main causes of deaths are infections and poor disease control during the first year post-diagnosis and cardiovascular complications, infections or cancers thereafter [8].

The risk of relapse, sometimes multiple in a given patient, remains a main AAV feature. Relapses are more frequent in GPA than MPA or EGPA, with 5-year relapse rates ranging from 35 % (EGPA) to 50 % (GPA).

Treatment should be adapted to the entity, severity and risk of relapse. Based on the prognostic Five-Factor Score (FFS) [9], treatment options have been designed to find the most effective regimen with the least side effects. Only MPA and EGPA are concerned because it is now known that GPA must be treated with GC and an immunosuppressant. For MPA and EGPA patients with FFS = 0, GC alone can suffice to obtain and maintain remission. However, combining GC and an immunosuppressant is mandatory when FFS ≥ 1.

35.2 Induction Therapy

The remission-induction regimen combines GC and another immunosuppressant, cyclophosphamide or, in certain settings, rituximab. Although GC can rapidly attenuate symptoms, their prolonged use at high dose is also associated with a major risk of severe adverse events, e.g. infections, osteoporosis or diabetes. The initial GC dose is 1 mg/kg/day of prednisone-equivalent, sometimes preceded, for the most severely ill patients, by intravenous (IV) methylprednisolone pulses (7.5–15 mg/kg/day) for 1–3 consecutive days. After the first 2 weeks of treatment, rapid GC-dose tapering, by ~10 % every 2 weeks, should begin. Various GC-tapering protocols are used and the optimal administration duration is still being debated. In the United States, many centers consider it useless, and even dangerous, to exceed 6 months of GC [10]. Nonetheless, most clinicians still prescribe them for much longer, but at a low dose (5 mg/day), sometimes for over 2 years. Pertinently, the results of a meta-analysis of several AAV trials suggested that continuing low-dose GC beyond 6 or

12 months after the initial disease flare could be associated with a lower subsequent risk of relapse [11].

Combined GC and cyclophosphamide remains the conventional remission-induction regimen for the systemic forms of GPA, MPA and EGPA with FFS ≥ 1. It achieves remission in most patients by 6 months [7, 12, 13]. Cyclophosphamide, administered as IV pulses at regular intervals or daily orally intake, for 3–6 months is able to induce remission in most patients [14]. Pulse cyclophosphamide is infused every 15 days for 1 month (days 1, 15 and 30) at a dose of 0.6 g/m^2 or 15 mg/kg, then 0.7 g/m^2 or 15 mg/kg every 3 weeks [15–17]. Oral cyclophosphamide is prescribed at 2 mg/kg/day. Although the two administration routes have comparable efficacies to achieve remission, IV pulses are associated with fewer infections and less frequent neutropenia than continuous oral intake [15–17]. However, long-term follow-up (median, 4.3 years) data of patients enrolled in the prospective CYCLOPS trial, comparing continuous oral versus IV pulse cyclophosphamide for induction, suggested that the subsequent relapse rate after infusions was lower, probably because of the higher cumulative dose exposure during the 6 months of continuous oral intake [18].

In addition to the risk of cytopenias and infections, other possible cyclophosphamide-associated adverse events include hemorrhagic cystitis (which can be limited with good hydration at the time of IV pulses), infertility and late bladder cancer, lymphoma or non-melanoma skin cancers [19, 20].

Rituximab is a chimeric anti-CD20 monoclonal antibody that targets and depletes B cells. It is a valid induction alternative to cyclophosphamide for adults with severe ANCA-positive GPA or MPA [12, 21], as it was found to be non-inferior to cyclophosphamide to obtain remission at 6 months, when both were combined with GC. Rituximab tolerance was good in both trials, with similar infection rates for the conventional cyclophosphamide-based and experimental rituximab-based arms reported. Moreover, the 18-month follow-up analysis of the RAVE study [22] showed that their sustained remission rates continued to be similar, i.e., 30 % in each arm had already relapsed.

Deciding whether to prescribe rituximab or cyclophosphamide for induction should be based on several factors, including their price tags and some individual patient characteristics. For now, rituximab's main indication should probably remain limited to patients with contraindications to cyclophosphamide, with relapsing disease and/or who had already received high cumulative cyclophosphamide doses, or women of child-bearing age at risk for cyclophosphamide-induced infertility. The French Vasculitis Study Group recently published recommendations for rituximab use [23]. The response to rituximab also appears to differ according to the GPA form, with granulomatous manifestations responding more poorly and/or slowly than vasculitic ones [7].

The rituximab dose is 375mg/m^2 infusions given at 1-week intervals for 3 weeks (4×375 mg/m^2, total). Possible rituximab-related adverse events include infections (mostly upper respiratory tract infections), infusion-related reactions, "immunoallergic" interstitial pneumonitis, late neutropenia and hepatitis-B virus reactivation (in previously healthy and/or chronic carriers). Some patients who had received rituximab for other conditions developed progressive multifocal (JC virus) leukoencephalopathy.

35.3 Maintenance Therapy

Once remission is achieved, treatment is switched to a maintenance regimen, to limit the risk of relapse. Maintenance therapy after cyclophosphamide-based induction is based on azathioprine (2–3 mg/kg/day, orally) or methotrexate (0.3 mg/kg/week, orally or subcutaneously, up to a maximum of 25 mg/week) [13, 14]. Both drugs can cause some side effects (e.g., opportunistic infections, liver toxicity or myelosuppression), but much less frequently than cyclophosphamide [13]. However, despite azathioprine or methotrexate maintenance, the relapse rate remains around 16 % at 18 months and continues to rise gradually thereafter: 37 % at 25 months, 52 % at 32 months, and 51–64 % at 7 years [7], with relapse-free survival rates of 49 % at 27 months and 42 % at 5 years [24].

Leflunomide or mycophenolate mofetil, also evaluated for maintenance, were equally or less effective at preventing relapses, respectively [25].

Rituximab can also be used for maintenance. A prospective trial compared azathioprine (2 mg/kg/day until month 22) to fixed-dose rituximab infusions (500 mg every 6 months for 18 months). At month 28, 29 % of the azathioprine-treated patients had relapsed versus only 5 % of those systematically given repeated rituximab infusions (P = 0.02) [26]. Because remission induction based on cyclophosphamide or rituximab is equivalent in terms of efficacy and safety, the same maintenance strategy, with systematic rituximab infusions, can probably be given to patients whose remissions were obtained with rituximab and GC. Whether rituximab-maintenance infusions could be administered as a function of ANCA status or titer or B-cell repopulation, rather than at systematically scheduled intervals to all patients in remission, is currently being evaluated in the MAINRITSAN-2 trial (ClinicalTrials.gov Identifier: NCT0173156).

Importantly, despite all these gradual refinements of maintenance-therapy strategies, its optimal duration has not yet been definitively established. Especially for patients with the greatest risk(s) of relapse, other immunosuppressant(s) (e.g., azathioprine or repeated rituximab infusions) should be given for at least 18 months, if not much longer.

The relapse risk is higher for PR3-ANCA–positive GPA patients than those with MPO-ANCA, and for patients who previously relapsed, have lung and/or ENT disease manifestations than those with glomerulonephritis or without impaired renal function. Co-trimoxazole (trimethoprim–sulfamethoxazole) cannot replace immunosuppressive maintenance therapy, but prescribed at "high doses" (320 mg/1600 mg daily of trimethoprim/sulfamethoxazole), with or after the standard immunosuppressive regimen for GPA, can lower the relapse rate [27].

35.4 Treating Limited GPA Forms

It is possible to prescribe a "lighter" regimen for patients with localized and/or limited non-life–threatening GPA, primarily when only granulomatous ENT manifestations are present. GC can obtain some attenuation in more than 50 % of them, but sustained remission is very rarely achieved. Because these forms are often recurrent, can be locally erosive and tend eventually to evolve into more generalized disease, it seems reasonable to treat them more systematically with a combined regimen of GC and an immunosuppressant, e.g. methotrexate. When methotrexate is effective, it must be continued for several years.

35.5 Treating Relapses, Refractory Disease and Other Agents

Relapses while on maintenance therapy or when immunosuppressants are no longer being taken could be treated according to the same remission-induction strategies described above or with rituximab [12].

Treating subglottic stenosis is also complex, especially because this manifestation can recur or worsen in the absence of other signs of active disease. Systemic agents, including GC, cyclophosphamide or rituximab, can be effective for about a third of these GPA patients. For most of the remaining two-thirds, whose chronic lesions are composed of fibrotic scar tissue, only local treatments, based on dilations combined with local submucosal GC injections, provide some symptomatic relief.

Intravenous immunoglobulins have mostly been prescribed to treat refractory or relapsing GPA, with concurrent serious infections or contraindication(s) to receiving other immunosuppressants, during pregnancy for patients with active AAV. They are contraindicated when renal insufficiency is severe (creatinine clearance <30 ml/min) [28].

When severe rapidly progressive glomerulonephritis and/or alveolar hemorrhage are present, plasma exchanges are sometimes prescribed together with other remission-induction treatments. When renal impairment is severe (defined as oliguria), necessitating dialysis, and/or serum creatinine exceeds 500 μmol/l, plasma exchanges contributed to improving renal function and recovery at 12 months, but these improvements were not sustained [29].

Financial Disclosures Dr. Guillevin reports receiving fees for serving on GlaxoSmithKline and Actelion advisory boards, and lecture fees from Roche, Actelion, Pfizer, CSL Behring, LFB Pharma, and Octapharma. As of January 2014, Dr Guillevin has no conflicts of interest with pharmaceutical companies to disclose.

References

1. Mohammad AJ, Jacobsson LT, Westman KW, Sturfelt G, Segelmark M (2009) Incidence and survival rates in Wegener's granulomatosis, microscopic polyangiitis, Churg–Strauss syndrome and polyarteritis nodosa. Rheumatology (Oxford) 48(12):1560–1565
2. Mohammad AJ, Jacobsson LT, Mahr AD, Sturfelt G, Segelmark M (2007) Prevalence of Wegener's granulomatosis, microscopic polyangiitis, polyarteritis nodosa and Churg–Strauss syndrome within a defined population in southern Sweden. Rheumatology (Oxford) 46(8):1329–1337
3. Watts RA, Jolliffe VA, Carruthers DM, Lockwood M, Scott DG (1996) Effect of classification on the incidence of polyarteritis nodosa and microscopic polyangiitis. Arthritis Rheum 39(7):1208–1212
4. Mahr A, Guillevin L, Poissonnet M, Aymé S (2004) Prevalences of polyarteritis nodosa, microscopic polyangiitis, Wegener's granulomatosis, and Churg–Strauss syndrome in a French urban multiethnic population in 2000: a capture–recapture estimate. Arthritis Rheum 51(1):92–99
5. Guillevin L, Durand-Gasselin B, Cevallos R, Gayraud M, Lhote F, Callard P et al (1999) Microscopic polyangiitis: clinical and laboratory findings in eighty-five patients. Arthritis Rheum 42(3):421–430
6. Comarmond C, Pagnoux C, Khellaf M, Cordier JF, Hamidou M, Viallard JF et al (2013) Eosinophilic granulomatosis with polyangiitis (Churg–Strauss): clinical characteristics and long-term followup of the 383 patients enrolled in the French Vasculitis Study Group cohort. Arthritis Rheum 65(1):270–281
7. Holle JU, Gross WL, Latza U, Nolle B, Ambrosch P, Heller M et al (2011) Improved outcome in 445 patients with Wegener's granulomatosis in a German vasculitis center over four decades. Arthritis Rheum 63(1):257–266
8. Flossmann O, Berden A, de Groot K, Hagen C, Harper L, Heijl C et al (2011) Long-term patient survival in ANCA-associated vasculitis. Ann Rheum Dis 70(3):488–494
9. Guillevin L, Pagnoux C, Seror R, Mahr A, Mouthon L, Le Toumelin P (2011) The Five-Factor Score revisited: assessment of prognoses of systemic necrotizing vasculitides based on the French Vasculitis Study Group (FVSG) cohort. Medicine (Baltimore) 90(1):19–27
10. McGregor JG, Hogan SL, Hu Y, Jennette CE, Falk RJ, Nachman PH (2012) Glucocorticoids and relapse and infection rates in anti-neutrophil cytoplasmic antibody disease. Clin J Am Soc Nephrol 7(2):240–247
11. Walsh M, Merkel PA, Mahr A, Jayne D (2010) Effects of duration of glucocorticoid therapy on relapse rate in antineutrophil cytoplasmic antibody-associated vasculitis: a meta-analysis. Arthritis Care Res (Hoboken) 62(8):1166–1173
12. Stone JH, Merkel PA, Spiera R, Seo P, Langford CA, Hoffman GS et al (2010) Rituximab versus cyclophosphamide for ANCA-associated vasculitis. N Engl J Med 363(3):221–232
13. Pagnoux C, Mahr A, Hamidou MA, Boffa JJ, Ruivard M, Ducroix JP et al (2008) Azathioprine or methotrexate maintenance for ANCA-associated vasculitis. N Engl J Med 359(26):2790–2803
14. Jayne D, Rasmussen N, Andrassy K, Bacon P, Tervaert JW, Dadoniene J et al (2003) A randomized trial of maintenance therapy for vasculitis associated with antineutrophil cytoplasmic autoantibodies. N Engl J Med 349(1):36–44
15. Guillevin L, Cordier JF, Lhote F, Cohen P, Jarrousse B, Royer I et al (1997) A prospective, multicenter, randomized trial comparing steroids and pulse cyclophosphamide versus steroids and oral cyclophosphamide in the treatment of generalized Wegener's granulomatosis. Arthritis Rheum 40(12):2187–2198
16. Haubitz M, Schellong S, Gobel U, Schurek HJ, Schaumann D, Koch KM et al (1998) Intravenous pulse administration of cyclophosphamide versus daily oral treatment in patients with antineutrophil cytoplasmic antibody-associated vasculitis and renal involvement: a prospective, randomized study. Arthritis Rheum 41(10):1835–1844

17. de Groot K, Harper L, Jayne DR, Flores Suarez LF, Gregorini G, Gross WL et al (2009) Pulse versus daily oral cyclophosphamide for induction of remission in antineutrophil cytoplasmic antibody-associated vasculitis: a randomized trial. Ann Intern Med 150(10):670–680
18. Harper L, Morgan MD, Walsh M, Hoglund P, Westman K, Flossmann O et al (2012) Pulse versus daily oral cyclophosphamide for induction of remission in ANCA-associated vasculitis: long-term follow-up. Ann Rheum Dis 71(6):955–960
19. Faurschou M, Sorensen IJ, Mellemkjaer L, Loft AG, Thomsen BS, Tvede N et al (2008) Malignancies in Wegener's granulomatosis: incidence and relation to cyclophosphamide therapy in a cohort of 293 patients. J Rheumatol 35(1):100–105
20. Le Guenno G, Mahr A, Pagnoux C, Dhote R, Guillevin L (2011) Incidence and predictors of urotoxic adverse events in cyclophosphamide-treated patients with systemic necrotizing vasculitides. Arthritis Rheum 63(5):1435–1445
21. Jones RB, Tervaert JW, Hauser T, Luqmani R, Morgan MD, Peh CA et al (2010) Rituximab versus cyclophosphamide in ANCA-associated renal vasculitis. N Engl J Med 363(3):211–220
22. Specks U, Merkel PA, Seo P, Spiera R, Langford CA, Hoffman GS et al (2013) Efficacy of remission-induction regimens for ANCA-associated vasculitis. N Engl J Med 369(5):417–427
23. Charles P, Bienvenu B, Bonnotte B, Gobert P, Godmer P, Hachulla E et al (2013) Rituximab: recommendations of the French Vasculitis Study Group (FVSG) for induction and maintenance treatments of adult, antineutrophil cytoplasm antibody-associated necrotizing vasculitides. Presse Med 42(10):1317–1330
24. Sanders JS, Slot MC, Stegeman CA (2003) Maintenance therapy for vasculitis associated with antineutrophil cytoplasmic autoantibodies. N Engl J Med 349(21):2072–2073; author reply –3
25. Hiemstra TF, Walsh M, Mahr A, Savage CO, de Groot K, Harper L et al (2010) Mycophenolate mofetil vs azathioprine for remission maintenance in antineutrophil cytoplasmic antibody-associated vasculitis: a randomized controlled trial. JAMA 304(21):2381–2388
26. Guillevin L, Pagnoux C, Karras A, Khouatra C, Aumaitre O, Cohen P et al (2014) Rituximab versus azathioprine for maintenance in ANCA-associated vasculitis. N Engl J Med 371(19):1771–1780
27. Stegeman CA, Tervaert JW, de Jong PE, Kallenberg CG (1996) Trimethoprim-sulfamethoxazole (co-trimoxazole) for the prevention of relapses of Wegener's granulomatosis. Dutch Co-Trimoxazole Wegener Study Group. N Engl J Med 335(1):16–20
28. Martinez V, Cohen P, Pagnoux C, Vinzio S, Mahr A, Mouthon L et al (2008) Intravenous immunoglobulins for relapses of systemic vasculitides associated with antineutrophil cytoplasmic autoantibodies: results of a multicenter, prospective, open-label study of twenty-two patients. Arthritis Rheum 58(1):308–317
29. Jayne DR, Gaskin G, Rasmussen N, Abramowicz D, Ferrario F, Guillevin L et al (2007) Randomized trial of plasma exchange or high-dosage methylprednisolone as adjunctive therapy for severe renal vasculitis. J Am Soc Nephrol 18(7):2180–2188

Index

F. Dammacco et al. (eds.), *Systemic Vasculitides: Current Status and Perspectives*, DOI 10.1007/978-3-319-40136-2

The manufacturer's authorised representative in the EU is Springer
Nature Customer Service Centre GmbH, Europaplatz 3, 69115 Heidelberg,
Germany. If you have any concerns regarding our products, please
contact ProductSafety@springernature.com

Printed and bound by CPI Group (UK) Ltd, Croydon, CR0 4YY
30/09/2026
02225651-0015